Theoretical and Experimental Insights into Immunology

NATO ASI Series

Advanced Science Institutes Series

A series presenting the results of activities sponsored by the NATO Science Committee, which aims at the dissemination of advanced scientific and technological knowledge, with a view to strengthening links between scientific communities.

The Series is published by an international board of publishers in conjunction with the NATO Scientific Affairs Division

A Life Sciences B Physics	Plenum Publishing Corporation London and New York
C Mathematical and Physical Sciences D Behavioural and Social Sciences E Applied Sciences	Kluwer Academic Publishers Dordrecht, Boston and London
F Computer and Systems Sciences G Ecological Sciences H Cell Biology I Global Environmental Change	Springer-Verlag Berlin Heidelberg New York London Paris Tokyo Hong Kong Barcelona Budapest

NATO-PCO DATABASE

The electronic index to the NATO ASI Series provides full bibliographical references (with keywords and/or abstracts) to more than 30000 contributions from international scientists published in all sections of the NATO ASI Series. Access to the NATO-PCO DATABASE compiled by the NATO Publication Coordination Office is possible in two ways:

- via online FILE 128 (NATO-PCO DATABASE) hosted by ESRIN, Via Galileo Galilei, I-00044 Frascati, Italy.
- via CD-ROM „NATO-PCO DATABASE“ with user-friendly retrieval software in English, French and German (© WTV GmbH and DATAWARE Technologies Inc. 1989).

The CD-ROM can be ordered through any member of the Board of Publishers or through NATO-PCO, Overijse, Belgium.

Series H: Cell Biology, Vol. 66

Theoretical and Experimental Insights into Immunology

Edited by

Alan S. Perelson

Theoretical Division
Los Alamos National Laboratory
Los Alamos, NM 87545, USA

Gérard Weisbuch

Laboratoire de Physique Statistique
de l'Ecole Normale Supérieure
24 rue Lhomond
75231 Paris Cédex 5, France

Springer-Verlag
Berlin Heidelberg New York London Paris Tokyo
Hong Kong Barcelona Budapest
Published in cooperation with NATO Scientific Affairs Division

Proceedings of the NATO Advanced Research Workshop on Theoretical Immunology
held in Paris, September 27- October 1, 1991

ISBN-13:978-3-642-76979-5 e-ISBN-13:978-3-642-76977-1
DOI: 10.1007/978-3-642-76977-1

Library of Congress Cataloging-in-Publication Data
Theoretical and experimental insights into immunology / edited by Alan S. Perelson, Gérard Weisbuch.
(NATO ASI series. Series H, Cell biology ; vol. 66)
Includes bibliographical references and index.
ISBN-13:978-3-642-76979-5
1. Experimental immunology–Congresses. 3. Immunology–Computer simulation–Congresses. I. Perelson, Alan S., 1947- II. Weisbuch, G. III. Series. QR180.3.T48 1992 616.07'9--dc20

Softcover reprint of the hardcover 1st edition 1992

Typesetting: Camera ready by author
31/3145 - 5 4 3 2 1 0 - Printed on acid-free paper

Contents

Contributors

Zvia Agur, Department of Applied Mathematics, Weizmann Institute of Science, Rehovot, Israel

Roy M. Anderson, Wellcome Centre for Research on Parasite Infections, Department of Pure and Applied Biology, Imperial College, London SW7 2BB, England

Stig I. Andersson, Research Group of Global Analysis and Applications, Chalmers University of Technology, University of Göteborg, S-412 88 Goteborg, Sweden

F. Andris, Laboratoire de Physiologie Animale, Université Libre de Bruxelles, 67, rue des Chevaux, 1640-Rhode St. Genése, Belgium

Michael Ascher, Viral and Rickettsial Laboratory, California Department of Health Services, 2151 Berkeley Way, Berkeley, CA 94704

Henri Atlan, Department of Medical Biophysics, Hadassah University Hospital, Jerusalem, Israel

C. Bédin, INSERM U 283, Hôpital Cochin, 27, rue du Fg St-Jacques, 75674 Paris Cedex 14, France

Ulrich Behn, Sektion Physik, Universität Leipzig, Augustusplatz, 10, 0-7010 Leipzig, Germany

Gilles Benichou, Department of Microbiology and Molecular Genetics, 5304 Life Sciences Building, University of California, Los Angeles, CA 90024

Vatsala Bhardwaj, Department of Microbiology and Molecular Genetics, 5304 Life Sciences Building, University of California, Los Angeles, CA 90024

G. Bocharov, Institute of Numerical Mathematics, Academy of Sciences, Leninskij Prospect 32-A, Moscow, 117334, Russia

Franco Celada, NYU Department of Pathology, Hospital for Joint Diseases, 301 E 17th Street, New York, NY 10003

Jeannine Charreire, INSERM U 283, Hôpital Cochin, 27, rue du Fg St-Jacques, 75674 Paris Cedex 14, France

Jian-Jun Chen, San Diego Regional Cancer Center, 3099 Science Park Road, San Diego, CA 92121

Irun R. Cohen, Department of Cell Biology, Weizmann Institute of Science, Rehovot, Israel

Laurence J. N. Cooper, Institute of Pathology, Case Western Reserve University, 2085 Aldebert Road, Cleveland, OH 44106

Antonio Coutinho, Unité d'Immunobiologie, Institut Pasteur, 25 rue du Dr. Roux, F-75724 Paris Cedex 15, France

Rob J. De Boer, Bioinformatica, University of Utrecht, Padualaan 8, 3584 CH, Utrecht, The Netherlands

Byron Goldstein, Theoretical Biology and Biophysics, Group T-10, Mail Stop K-710, Los Alamos National Laboratory, Los Alamos, NM 87545

Edward S. Golub, R.W. Johnson Pharmaceutical Research Institute, The Research Institute of Scripps Clinic, 10666 N. Torrey Pines Road, La Jolla, Ca 92037

Alf Grandien, Department of Immunology, University of Upsala, 75123 Uppsala, Sweden

Neil S. Greenspan, Institute of Pathology, Case Western Reserve University, 2085 Aldebert Road, Cleveland, OH 44106

Zvi Grossman Department of Physiology and Pharmacology, Sackler School of Medicine, Tel Aviv University, Tel Aviv 69978, Israel

. Leo van Hemmen, Physik-Department, TU München, James-Frank-Straße, W-8046 Garching bei München, Germany

Geoffrey Hoffmann, Departments of Microbiology and Physics, The University of British Columbia, Vancouver, B.C., Canada, V6T 1Z3

Pauline Hogeweg, Bioinformatica, University of Utrecht, Padualaan 8, 3584 CH, Utrecht, The Netherlands

James M. Hyman Theoretical Division, Los Alamos National Laboratory, Los Alamos, NM 87545

Rita Jores, Institut Pasteur, 25 rue du Docteur Roux, 75634 Paris Cedex 15, France

Jayant Kalagnanam, Department of Engineering and Public Policy, Carnegie Mellon University, Schenley Park, Pittsburgh, PA 15213

Marcelline Kaufman, Université Libre de Bruxelles, Service de Chimie-Physique II, Campus Plaine C.P. 231, Boulevard du Triomphe, 1050 Bruxelles, Belgium

Srinivas-Venkatesh Kaveri, San Diego Regional Cancer Center, 3099 Science Park Road, San Diego, CA 92121

Tracy A. Kion, Department of Microbiology, The University of British Columbia, Vancouver, B.C., Canada, V6T 1Z3

Sidney N. Klaus, M.D., Department of Dermatology and Venereology, Hadassah Hospital, Jerusalem IL-91120, Israel

Heinz Kohler, San Diego Regional Cancer Center, 3099 Science Park Road, San Diego, CA 92121

Karl E. Kürten, Institute für Theoretische Physik, Johannes-Kepler-Universität Linz, Austria

O. Leo, Laboratoire de Physiologie Animale, Université Libre de Bruxelles, 67, rue des Chevaux, 1640-Rhode St. Genése, Belgium

Frederique Lisacek, Centre de Genetique Moleculaire, CNRS, 91190 Gif sur Yvette, France

Samuel Litwin, Institute for Cancer Research, Fox Chase Cancer Center, 7701 Burholme Avenue, Philadelphia, PA 19111

Gadi Mazor, School of Mathematical Sciences, Raymond and Beverly Sackler Faculty of Exact Sciences, Tel Aviv University, Israel

Angela R. McLean, Zoology Department, S. Parks Road, Oxford University, Oxford OX1 3PS, England

Isaac Meilijson, School of Mathematical Sciences, Raymond and Beverly Sackler Faculty of Exact Sciences, Tel Aviv University, Israel

Benoit F. Morel, Department of Engineering and Public Policy, Carnegie Mellon University, Schenley Park, Pittsburgh, PA 15213

Penelope Morel, Pittsburgh Cancer Institute, Division of Basic Research, Basic Science Tower, Lathrop and Terrace, Pittsburgh, PA 15213

Avidan U. Neumann, Department of Applied Mathematics, Weizmann Institute of Science, Rehovot, Israel

Jerome K. Percus, Courant Institute of Mathematical Sciences and Department of Physics, New York University, 251 Mercer Street, New York, NY 10012

Ora E. Percus, Courant Institute of Mathematical Sciences, New York University, 251 Mercer Street, New York, NY 10012

Alan S. Perelson, Theoretical Biology and Biophysics, Group T-10, Mail Stop K-710, Los Alamos National Laboratory, Los Alamos, NM 87545

A. Romanyukha Institute of Numerical Mathematics, Academy of Sciences, Leninskij Prospect 32-A, Moscow, 117334, Russia

Stephen P. Schoenberger, Department of Microbiology and Molecular Genetics, 5304 Life Sciences Building, University of California, Los Angeles, CA 90024

A. Nicola Schweitzer, West Wing, Department of Biology, Imperial College, London SW7 2BB, England

Lee A. Segel, Department of Applied Mathematics, Weizmann Institute of Science, Rehovot, Israel

Philip E. Seiden, IBM Research Center, P.O. Box 218, Yorktown Heights, NY 10598

Eli E. Sercarz, Department of Microbiology and Molecular Genetics, 5304 Life Sciences Building, University of California, Los Angeles, CA 90024

Haynes Sheppard, Viral and Rickettsial Laboratory, California Department of Health Services, 2151 Berkeley Way, Berkeley, CA 94704

John Stewart, Department of Immunobiology, Institut Pasteur, 25 rue du Docteur Roux, 75634 Paris Cedex 15, France

Bernhard Sulzer, Physik-Department, TU München, James-Frank Straße, W-8046 Garching bei München, Germany

Jonathan Swinton, Wellcome Centre for Research on Parasite Infections, Department of Pure and Applied Biology, Imperial College, London SW7 2BB, England

H. Tang, INSERM U 283, Hôpital Cochin, 27, rue du Fg St-Jacques, 75674 Paris Cedex 14, France

B. Texier, INSERM U 283, Hôpital Cochin, 27, rue du Fg St-Jacques, 75674 Paris Cedex 14, France

Gérard Weisbuch, Laboratoire de Physique Statistique, Ecole Normale Supérieure, 24 rue Lhomond, F-75231 Paris Cedex 05, France

Margot Zöller, Unité d'Immunobiologie, Institut Pasteur, 25 rue du Dr. Roux, F-75724 Paris Cedex 15, France

Preface

From September 26 through October 1, 1991, sixty theoretical and experimental immunologists met at the Gulbenkian Foundation (Centre Culturel Portugais) in Paris as part of a NATO Advanced Research Workshop on Theoretical Immunology. This workshop was held in order to discuss the current state of modeling and theoretical work in immunology, the impact of theory on experiment, and future directions for theoretical research.

The participants learned that theory in immunology encompasses different approaches: abstract mathematical models written in terms of equations, computer models embodied as complex computer simulation codes, or conceptual models posed loosely in words. Some models were constructed to answer well-defined experimental questions and made predictions that could be compared with experiment. Other models were abstract and not easily testable, except perhaps with the invention of new techniques that would allow one to monitor the state of many cell subpopulations and measure the concentrations of many lymphokines. As organizers we were particularly gratified to see that many experimental immunologists were interested in theoretical immunology, thought that theory could help in the interpretation and design of future experiments, and were anxious to help in any way that they could in shaping the future of this vigorous young field.

We thank the Gulbenkian Foundation for their generous support and the use of their beautiful facilities. The meeting was also supported by NATO Advanced Research Workshop Program (ARW 910339) and the Centre National de la Recherche Scientifique. In particular we would like to thank Maria de Lourdes Belchior, Georges Gulbenkian, and Nicole de Barbuat from the Gulbenkian Foundation, Anne Desmond for secretarial assistance, Carol O'Rourke, Emmanuelle Weisbuch, and Benoit Malherbe for their assistance during the workshop, and Elissa Perelson for assistance in the preparation of this volume. The editors would also like to acknowledge support from NATO collaborative research grant CRG 900908, NIH grants RR06555 and AI28433 to ASP and Inserm grant 879002 to GW.

Organizing Committee

A. Coutinho, Institut Pasteur, Paris, France

R. De Boer, University of Utrecht, Netherlands

A. Perelson, Los Alamos National Laboratory, Los Alamos, NM, USA

J. Urbain, Université Libre de Bruxelles, Belgique

G. Weisbuch, Ecole Normale Supérieure, Paris, France

Introduction

Immunology is largely a science of observation and experimentation, with major advances being driven by the advent of new experimental techniques. While reductionist approaches have lead to great increases in our knowledge of the genes, molecules and cells of the immune system, they have as yet failed to provide a grand view of the coordinated operation of the immune system as a system. The understanding of the immune system in terms of its development, the build-up and control of the immune response, and the nature of recognition and memory, is still a formidable challenge to immunologists. Some kind of theory is indispensable. Experimentalists and theorists that met in this NATO Advanced Research Workshop did not readily agree on what theoretical immunology should be. The mathematicians, physicists and computer scientists by and large viewed theory as an exercise in mathematical modeling whose aim is to demonstrate the logical consequences of a set of well defined hypotheses. The approaches of Bocharov and Romanyukha, Kauffman *et al.*, Perelson and Goldstein, and Morel *et al.*, are classical and correspond to what most people expect a priori from theory: a model with a detailed description of part of the immune system that can fit quantitative data obtained from experimental measurement, and which is capable of quantitative predictions. At the meeting there were also discussions of generic approaches, based on minimal assumptions, which were aimed at understanding what assumptions were needed to get a particular behavior of the immune system. In these models, one could not say that the immune system followed the rules that the modeler set out, but only that these rules were sufficient to generate a given behavior. The real immune system might, and probably does, have redundant mechanisms. Seiden and Celada's model as well as those of Weisbuch, McLean, Kürten, Segel *et al.*, and Schweitzer *et al.* belongs to this class of minimalist models.

Different techniques were introduced to study models. Seiden and Celada introduced a simulation model that included T cells, B cells and antigen presenting cells. Binary strings were used to represent antigen receptors and MHC molecules on the surface of the cells, as well as to represent antigens. The model was very detailed and made a number of intriguing predictions. This type of detailed simulation model could be called Artificial Immunology, in the sense that one follows the evolution of large numbers of individually defined cells, with realistic receptors, and well defined rules of cellular behavior. De Boer, Hogeweg and Perelson also used a cellular automaton model but restricted their attention to the effect of the influx of large numbers of newly created B cells on the generation of idiotypic networks. Weisbuch and Neumann, also in the context of idiotypic networks, introduced ideas from statistical mechanics such as generic behavior, localized attractors, and scaling laws. Behn *et al.*, McLean, and Kürten developed differential equation models and studied their behavior. Percus, Percus and Perelson used ideas from probability theory to address questions about self non-self discrimination and the optimal size of epitopes. Agur *et al.* used ideas from optimal control theory to predict that somatic hypermutation needs to be turned on and off in a controlled manner in order to generate efficient affinity maturation. For some authors, Grossman, Hoffmann, Ascher and Sheppard, and Gollub, theory was reduced to verbal hypotheses that did not need any mathematical test, but which could nevertheless be directly verified by experiment. Despite the approach used, most agreed that theory should contribute new ideas and new ways of looking at immunological phenomena and should suggest new experiments. The point was made that all of us are biologists interested in understanding the immune system through the use of different tools.

The workshop was a confrontation of different attitudes and philosophies on some specific problems. Having representatives from both the experimental and computational communities allowed us to cover subjects from both points of view.

This book, which represents both the invited talks as well as the contributed papers, is divided in six sections that we briefly describe here.

General Perspectives

The idea and purposes of theoretical modeling in immunology are discussed in the introductory paper by Celada. The following paper, by Gollub, discusses the role of the immune system. Golub and suggests that the immune system may have functions other than protection against pathogenic organisms. The existence of such alternative functions could help explain the evolutionary origins of the immune system. This theme is taken up by Stewart and Coutinho, who discuss the evolutionary origin of the immune system. Seiden and Celada, developed a simulation model with which they study the effect of the number of MHC genes on the efficiency of the immune system. The idea is that if there are too few MHC genes, not very many antigens can be presented, while if there are too many MHC genes too many T cells will be eliminated by negative selection in the thymus. Grossman presents a new view of learning in the immune system, in which he suggests that reinforcement type learning can occur at the level of single cells and that such learning may account for the pleiotropic effects of various lymphokines.

T Cell Models

Although the majority of models in immunology deal with the humoral response, we know that a large part of the regulation of the immune system rests with T cells. Six papers relate specifically to T cells and their interactions. Kaufman *et al.* model biochemical mechanisms that may lead to anergy, i.e. unresponsiveness of T cells. Perelson and Goldstein examine the IL-2 receptor on T cells and ask whether the high affinity IL-2 receptor is a preformed heterodimer composed of p55 and p77 chains in the absence of IL-2, or whether the high affinity receptor only exists when IL-2 forms a ternary complex with the p55 and p75 chains as proposed in the affinity conversion model of Honjo. Using a theory of the binding of ligands to multichain receptors they deduce that the existing experimental data is consistent with the preformed heterodimer model but is not consistent with the affinity conversion model. A model of T cell memory in which memory cells are not necessarily long-lived is given by McLean. The model also describes T helper cell activation and proliferation and can account for the fact that T cell memory can be established in vivo but not in vitro. Kürten presents a dynamical model of the cytotoxic T cell response to a growing antigen, where the response is regulated by helper T cells. This model is only a caricature of reality but points out some of the dynamical possibilities. There exist two types of T helper cells: Th1 and Th2. The dynamics of the immune response depends of which type predominantly proliferates. Morel *et al.* and Schweitzer *et al.* discuss this question.

Networks and Attractors

One difficulty in the understanding of the immune system is the large number of its components and their interactions. For a modeler, the immune system is a complex dynamical system, whose behavior is a result of the cooperativity among its components. Idiotypic networks are a prototypic example of a system of this type, and were studied by Weisbuch and Neumann, Behn *et al.*, and De Boer *et al.*. Different models and different choices of parameters leads to different predictions of dynamical behavior. For example, some models suggest that the internal dynamics of an idiotypic network should after a long time settle down to a steady state, where clone sizes and antibody concentrations remain constant in time. Other models suggest that the long time behavior could be oscillatory, either in the form of limit cycle oscillations, or more widely fluctuating chaotic behavior. Excited states of a network can be localized, where only a limited number of clones are at levels substantially different from their virgin state, or percolation can result where one clone, say AB_1, excites AB_2, which excites Ab_3, etc., and so the entire network becomes excited. Models are good at identifying possible qualitative behaviors, and parameter regimes where such behaviors can be expected. The available data is not definitive and the models allow for many possibilities. Thus network dynamics is an area that could benefit from more collaboration between experimentalists and theorists.

From Molecular to Cellular Immunology

What fractions of a protein would give an epitope recognizable by the immune system? Lisacek addressed these question by systematic experimental studies and tried to generalize the data by traditional artificial intelligence techniques. Greenspan and Cooper discussed the nature of epitopes, and raised the issue of whether an epitope is simply a set of amino acids that contacts an antibody or if it is the constellation of features that determine whether an antibody will bind to an antigen in a given region. They point out that mutations outside the binding site can affect the affinity of an antibody for an antigen, and that in the case of IgG3 antibodies the density of epitopes on a the surface of a bacterium can affect the ability of the antibody to bind. Litwin and Jores approached another practical question in immunochemistry, measures of amino acid diversity in antibody combining sites. They showed that measures based on Shannon information may have advantages over the more standard Kabat-Wu plots. Describing the interactions between receptors, ligands and membranes is complex. Andersson provides some preliminary insights into how abstract mathematical techniques, such as differential geometry, may play a role.

Theories and Applications to Clinical Immunology

Many different techniques were used to model immune response to pathogens. Bocharov and Romanyukha used standard systems of differential equations to describe the dynamics of the response to viral infections in the case of hepatitis and influenza. The role of affinity maturation in a response to a parasitic infection was discussed by Agur *et al.* Agur and coauthors also examined the question of when hypermutation should be turned on and turned off if it were to operate in an "optimal" manner. Various diseases can be uniquely characterized by the type of rash they produce. Measles and chicken pox are common examples. No theory yet exists to explain the spatial patterns of rashes. Segel *et al.* examine in a very general way the features needed in a model to generate the spatial patterns characteristic of rashes. AIDS, of course, is a major concern to the entire experimental and theoretical communities. Hoffmann and Kion presented a novel hypothesis about the origin of AIDS in which destabilization of network interactions involving MHC and MHC-specific clones play a major role. They linked this hypothesis to experimental data on both AIDS and systemic lupus erythematosis. Sheppard and Ascher present a hypothesis of T cell activation and clonal dynamics that unifies thymic selection, AIDS pathogenesis, the effects of superantigens, and the response to alloantigens. They suppose that the interaction of gp120 on the surface of HIV with CD4, on helper T cells, gives an abnormally high affinity costimulatory signal to the T cells that leads to excessive programmed cell death.

Some contributions reflected collaborations between experimentalists and theorists. Atlan and Cohen discussed a model for adjuvant arthritis that involves a small networks of clones. Tang *et al.* take up a similar theme but with regard to experimental autoimmune thyroiditis, and present intriguing data suggesting that network interactions maybe important in the etiology of the disease.

Repertoire Selection

Repertoire selection in relation with autoimmunity was discussed in the last four, mostly experimental, papers. The crypticity of certain determinants explain why they are not recognized under normal circumstances by T cells. Benichou *et al.* discuss the general notion of cryptic, dominant and sub-dominant epitopes and the potential role of cryptic determinants in autoimmune disease and cancer therapy. Early presentation of immunoglobulin in the thymus biases the T cell repertoire. Transgenic experiments with the transgene expressing specific immunoglobulins are described by Chen *et al.*, Zöller, and Grandien *et al.* Chen *et al.* examine the possible role of immunoglobulin self peptides in the generation of the T cell repertoire. They show that the repertoire of normal mice contains T cells specific for a self peptide derived from the polymorphoric region of a germline encoded immunoglobulin. This suggests that self immunoglobulin may provide a diverse set of peptides that can be involved in thymic education. Zöller also suggests that tolerance towards non-MHC self antigens may be sustained by naturally activated autoreactive B cells. Grandien *et al.* look at the phenomenon of idiotypic mimicry, and present a model based on the bitstring representation of idiotypes to explain the elevated number of B cells expressing the same idiotype as the transgene.

B cell repertoire selection was also discussed at the meeting by Urbain. Reasoning from a set of experiments done in Brussels with the arsonate system and other data obtained by Kearney with the T15 system, he showed that idiotypes expressed during the secondary response depend upon B cell populations expressed during the early life of the mouse. Urbain suggests that as the immune system develops, a first B cell repertoire is selected, which is essentially anti-self and composed of anergic cells. This set of B cells can function to shape a second, anti-idiotypic repertoire, which is in essence anti-anti-self. This second repertoire, which Urbain calls the adapton, is adaptive and responsible memory responses. A third set of B cells, which is not selected positively or negatively, is postulated to be responsible for primary responses. Urbain's contribution is not part of this volume since it is being published separately in International Reviews in Immunology.

A. Perelson
G. Weisbuch
Santa Fe, March, 1992

I. General Perspectives

COMPUTER MODELLING OF THE IMMUNE SYSTEM: WHO ARE THE "FRUITORI"?

Franco Celada
NYU Department of Pathology
Hospital for Joint Diseases
301 E. 17th Street
New York, New York 10003, USA

INTRODUCTION

Fruitore is an Italian word straight from the Latin (the verb *fruor* means to utilize with advantage, to metaphorically harvest the fruits. I use this uncommon foreign substantive because it happens to be the most concise and precise, as compared with the English and the French (exploitateur? utiliseur? jouisseur?). Having explained the word, my assignment is to convince the reader that every experimental immunologist is a potential fruitore of the models (and will become an actual one in the near future). This was unforeseeable in the past, and I would not have written this sentence one year ago. The following factors, of which this NATO-Gulbenkian workshop is the best proof, have played into my hands. **a)** Theoretical immunology has been experiencing a substantial boom during the past several years, thanks to many new groups the world over joining what had been a restricted club of friends. **b)** The focus of many models has become more precisely coincident with that of experiments performed and discussed in conventional (classical) laboratories. **c)** Only in part a consequence of **a)** and **b)**, the immunologists are changing their traditional apartheid-like attitude towards the theorists.

We had a vivid impression of the newly developed interest of the experimentalists at a meeting that preceded the present onc by 8 months, at the Keystone Symposium on Self Reactivity and Its Regulation, January 27, 1991. The organizers, Eli Sercarz and Constantin Bona, had asked Alan Perelson and myself to conduct a discussion on modelling of the immunological self. We expected this to be one of several parallel sessions, where mainly the theoreticians would take part. At the last moment the structure of the meeting was changed and our workshop was given an entire morning -- prime time -- with only the competition of the ski slopes. About 250 immunologists took part. We had that growing special feeling of having made it, and after three hours we found ourselves surrounded by

NATO ASI Series, Vol. H 66
Theoretical and Experimental Insights into Immunology
Edited by A. S. Perelson and G. Weisbuch

special feeling of having made it, and after three hours we found ourselves surrounded by a number of fathers of modern immunology: Nossal, Mitchison, Golub, Hood, congratulating, objecting, asking, encouraging. Unfortunately the meeting was not recorded in a publication. But I have the text of my introductory talk to the session, that was improvised when it appeared necessary to introduce the group of theoreticians to the experimentalists. Since what I said has -- I think -- some validity now I shall quote some sections of that piece.

Where are the Modellists

Efforts to model have experienced a multicentric burst following Jerne's theoretical proposals of 1974: the Network exerted an extraordinary appeal on the theoreticians, and still does. After a number of attempts and what seemed to be a pause, the modelling activity took a steady, systematic path during the past 4 to 6 years. The works are in the hands of a handful of groups, whose surface characteristics are the following:

- they are well distributed around the world (France, New Mexico, Holland, British Columbia, New York, California...)
- they are small, manned by 2-3 researchers, of whom usually one is an immunologist and one a physicist or mathematician.
- they are interconnected and prone to inter-group recombinations, resulting in papers authored by a number of permutations of a small set of names.
- they meet often, and in attractive places, thanks to a number of motivated sponsors, like the Santa Fe Institute (SFI), the Institute for Scientific Interchange of Torino (ISI), the Gulbenkian Foundation in Paris.

Minimalism

In order to reach one layer deeper in the characterization of the immunological modellists, I have to choose an adjective describing their prevailing psychological attitude. This adjective is *minimalist*. It is at odds with nature's ways of proceeding. Nature is an inveterate tinkerer,[1] unable to streamline if that requires getting rid of entities or anachronisms that "work",with a passion for reaching the simplest goals by complex means. The modellist feels that simplicity is beautiful and beauty has the flavor of truth. When constructing an "immune system", each researcher seems to be in a contest with his peers to make it with one less ingredient than the others, and still obtain a product able to display all characters he deems essential. Individual judgment is involved in deciding what is required for a system to be an immune system: is it the capacity to respond to a zillion foreign insults, and/or the learning of memory, and/or the treatment of self, and/or the

network of connected internal specificities? Cezanne had no doubts which strokes of his brush were essential to depict an object unequivocally, and would use the absolute minimum. But then Picasso had different views about what was essential, and had they both painted the same guitar we would have certainly seen two quite different pictures.

Philosophers

This digression convinces me that the *individual hypothesis*, the *individual vision* of what is essential in the system, is even more important and better characterizes the theoreticians than their pursuit of minimalism.

If so, we are obliged to consider them akin to philosophers, in particular to the pre-Socratic naturalists. Dispersed in the Greek colonies about Asia Minor and the Italian coasts between the seventh and fifth centuries B.C., they looked at a world that had just awakened from the dreams of myth, and tried to explain it to reason. Each one of them sought to answer the question, "What is the principle of all?" Each came up with a different answer. Water (Taletes). Earth (Anaxagoras). Air (Anaximenes). Fire, the evermoving, the perpetual flux (Heraclites of Ephesus): it is impossible to swim twice in the same river: the second time the river is different water and even you are a different person! No, says Parmenides in Elea, the Being is static, all movement is illusion! Thus Zeno, his pupil, demonstrates to the people of the agora that speedy Achilles cannot reach the turtle. Even today it is impossible to deny the seeds of truth in these heated arguments, and we gasp in admiration at the most abstract concept that this philosophy -- by way of Democritos of Abdera -- produced: the atoms, indivisible particles, quantitatively distinct: when a lateral *wind (klinamen) disturbs their continuous vertical descent, they interact with each other and create the variety of all things. I hope you will agree with me -- after the session -- that even if the theoretical immunologists may lack the grandeur (and the naivete) of the old Naturalists my comparison is not completely without foundation.

The Letter P

Another similarity follows: as the philosophers spoke of their *idea* of the world, the modellists do not pretend to tackle the real, flesh-and-blood immune system. Instead they show us their simulation of it, containing all the points essential (according to each of them), not the natural product. Thus when they display curves and talk of cells, antibodies, antigens, idiotypes and anti-idiotypes, they will mean *their* curves, *their* cells, *their* antibodies. To avoid confusion, I have a suggestion: imagine a little *p* in front of each of these words, a *p* that means "personal" or "private". This *p* will not disturb your following their

presentation: it will be silent, like the one you may have seen in front of the name of a Wodehouse character, Mr. Psmith (Fig. 1). And as long as you know it is there it will put your experimentalist instinct at rest, prevent you from physically attacking any theoretician, and allow you to follow and comprehend the message of his talk.

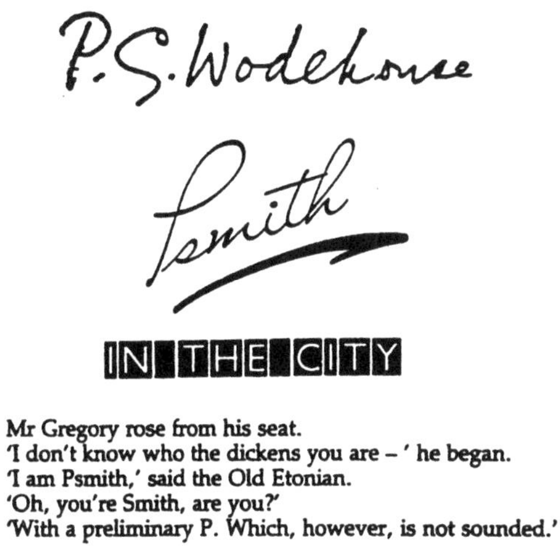

Figure 1. From <u>Psmith in the City</u>, by P.G. Wodehouse: The P is silent.

<u>Planet Immunology</u>

What is the purpose of modelling the immune system? Many purposes of course, from didactic/educational, to purely speculative. Becoming more personal, I shall try to put in words what the purpose of our own group (comprising one physicist and one immunologist) seems to be.

Planet Immunology (Fig. 2) is composed of a hard core and a large soft mantle in a semi-fluid state. (The crust is too young and unstable to be dwelled upon.) The core consists of hard proven facts and dependable figures on which consensus is almost general. The mantle contains diverse material assembled without evident order and in constant, irregular movement:

- real facts, but of unknown relevance.
- real facts certainly relevant but whose importance has been inflated out of proportion by

researcher's original sin: the tendency to make one's observation the center of the universe.

- facts that would be real if the experimental and natural temperature would coincide (they don't), or if the dose of antigen used would be attainable in any organism.

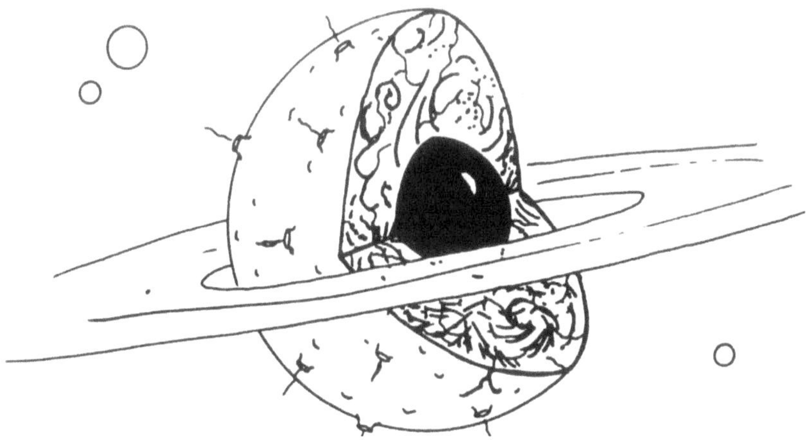

Figure 2. Planet Immunology (sectioned). For the contents of the mantel, see text.

- exciting hypotheses that would revolutionize even the core if the relevant experiment could be done; unfortunately, **a)** it is too expensive, or **b)** the technology is not yet available.

As can be seen the mantle is very interesting but difficult to tackle. We came to believe -- and are more and more convinced --that a simplified working model of the immune system allows preliminary experiments to be run *in machina*, which are for several reasons impossible *in vivo* or even *in vitro*. What would happen if self were as large as nonself? What if tolerance consists of anergic cells that can wake up? What if the strength of network interactions is of the same order as antigen-antibody? Or 1/10 of that? Or 1/100 of that? Is there an optimal number of MHC per individual? These experiments can be run rapidly and their results may guide the experimenters by showing which directions are, perhaps, interesting, and which are certainly hopeless.

Cellular Automaton

The theoretical immunologists, we saw, have different points of view. They also use quite different methods and combinations of methods. This also means that their conclusions will not be superimposable but possibly complementary. In this volume we have examples of this methodological diversity.

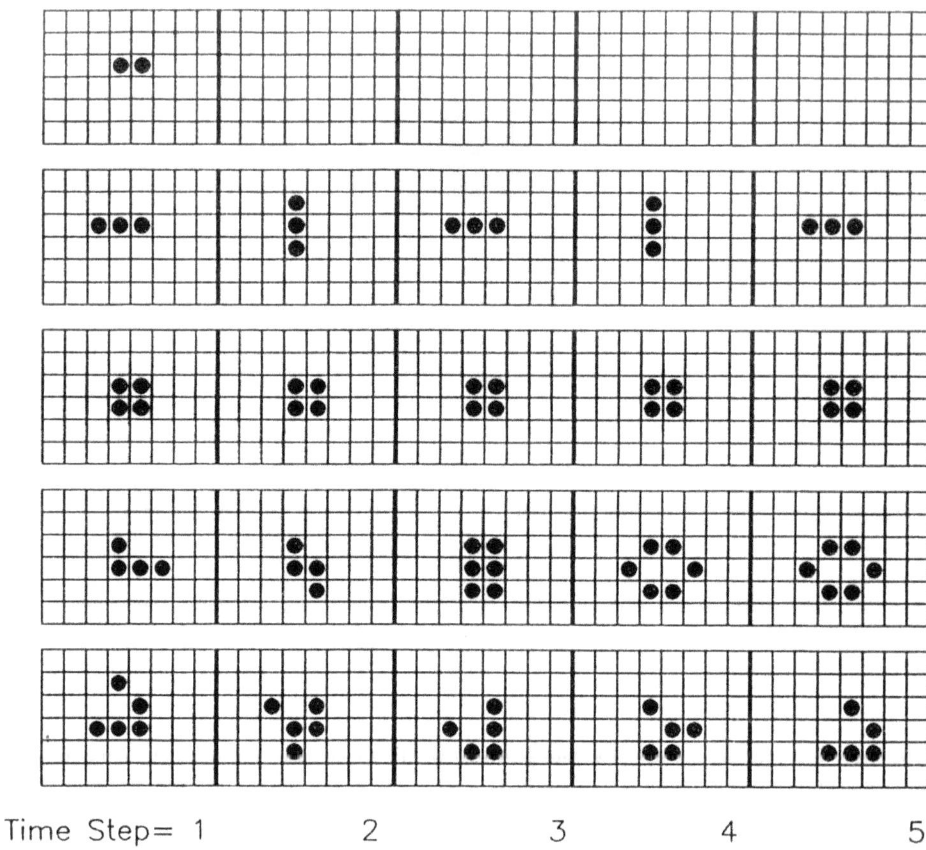

Figure 3. Conway's "Game of Life," The prototype cellular automaton. Black checkers are sprinkled on a checkerboard and obey two simple rules: 1) a black remains black if it adjacent to two or three blacks. 2) A white becomes black if it is adjacent to three blacks. See Gardner (6-7).

Seiden and myself (8-9) use a form of computer modelling based on "cellular automata," a mathematical device proposed by Von Neumann half a century ago. The first to use it in immunological modelling were Kaufmann and Varela. A typical cellular automaton is represented by space in the form of a checkerboard on which a number of checkers are distributed and can be followed for a number of timesteps. There are simple rules that govern the fate of the checkers, their persistence, disappearance, or the generation of new pieces on the board. Therefore at each timestep the number and the topology of the checkers may change. In Figure 3 these events are monitored during the course of 4 timesteps, for the case of 2, 3, 4 and 5 "checkers" initially present (the rules are listed in the figure legend). Each combination seems to acquire a life of its own. It disappears, blinks forever, stays solid, grows, or walks diagonally on the checkerboard, quite independently of us. To me, this is fascinating. And the system is immensely powerful since we can use elements more complex than a binary black-or-white. We can create a model of a B cell, associate to it a receptor with a given specificity and an MHC with a given specificity. We can tell the cell to stay or to move around. The rules will still be reasonably simple: in case an antigen happens to show up in the same square the cell will examine it, and in case it turns out to be complementary to its receptor the cell will go through a certain routine, and so on and so on. What will happen then? There is no guessing: the experiment will develop before our eyes. Fascinating.

Experimenting

Once the model is constructed, it is important to build a strategy of experimentation capable of yielding useful results or unambiguous indications. The strategy we have worked out starts by trying to reproduce a phenomenon using the minimum number of ingredients. Later, if several factors are known or suspected to be involved in the original phenomenon, they will be introduced into the model one at a time in order to evaluate their impact and relative power, and finally in combinations. This procedure does not differ from the ideal design of *in vivo* or *in vitro* experiments where one tries to isolate one factor and study the behavior of the system when this factor is varied. Immunologists have, over the years, devised a great number of partially artificial environments for those purposes, such as lethally irradiated animals as carriers of adoptive responses, semipermeable cell culture chambers to be introduced in the peritoneal cavity, monoclonal (instead of polyclonal) antibodies. Transgenic mice are a recent powerful development in this direction: by a genetic manipulation all lymphoidal cells are induced to express, e.g., the same T- or B-receptor, or to secrete the same lymphokine or the same antibody, thus eliminating the diversity characteristic of these cell clones for the purpose of conducting "simplified" experiments.

Dose and Response

A good example of this minimalist approach inside the model is the following. The antibody response to a standard dose of antigen can be reproduced in our minimum model containing A, B and T cells and obeying a set of well-tuned rules. It is easy to explore what happens if the amount of antigen introduced is increased above the optimum, and compare the model results with the outcome of many historical experiments concerned with high dose tolerance. Figure 4 shows one of these experiments, this one performed by me 20 years ago. The adoptive secondary response of primed spleen cells was elicited by a single injection of antigen (BSA) in doses varying over a wide range. The previous observation that a high density spleen cell transfer resulted in a response of higher affinity than low density transfers allowed me to test the antigen effect in both conditions. The findings of this experiment were two: **a)** the antibody production progressively decreases with high dose challenge, and **b)** the decrease is significantly more pronounced in the high-affinity response than in the low-affinity response.

In 1971 there were no clear explanations for the observed phenomenon, which was considered an example of high dose tolerance. Today the number of hypothetical mechanisms is somewhat larger, including saturation of antigen receptors, competition for antigens between memory cells and antibodies, enhanced network interactions, anergy of B cells, elimination of B and or T cells.

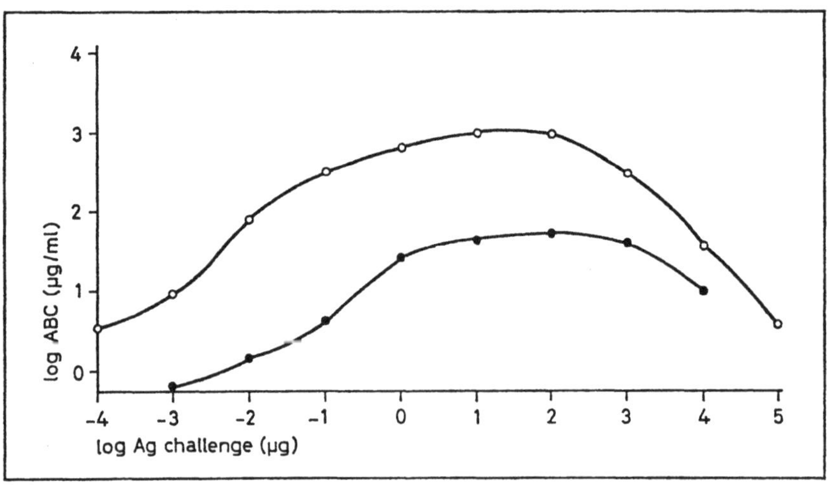

Figure 4. Profile of log peak titer against dose of HSA (log ug/mouse) used as challenge for two sizes of cell transfer from the same primed donor pool. Open circles, $3x10^7$ cells; black circles, 3×10^6. (10)

The use of the minimal model, where idiotypic interactions and anergy are not operating, to mimic high dose tolerance *de facto* isolates the first two of the mechanisms listed above and allows us to test whether receptor saturation and B cell antibody competition alone can reproduce the phenomenon.

The results of the model experiment are displayed in three-dimensional graphs in Figure 6 of the article by P. Seiden and F. Celada in this book. The similarity with the data obtained *in vivo* is remarkable. i) With increasing challenge dose the ratio of antibody secreted to antigen dose reaches a peak and then decreases. While at moderately high challenge this could be accounted for by saturation of B and T receptor sites, the decline of absolute antibody production at extreme challenge dose becomes absolute, and therefore must be caused by an additional mechanism, *i.e.*, competition. ii) The high-affinity (8-bit complementarity) response is more severely affected than the low-affinity (7-bit complementarity) response.

We consider this result very interesting, as an indication that a limited number of factors is sufficient to reproduce a regulation of the response that, we are sure, involves *in vivo* a number of additional factors. This certainly confirms the redundancy of the immune system. However, it should be noted that a complete analysis of what causes the differential decline of high and low affinity responses in the model has not yet been completed.

Affinity Maturation

Another example of model experiments that mimic *in vivo* behaviors of the immune system using only a fraction of the mechanisms operating *in vivo* concerns the affinity maturation of the antibody response. It is obvious that clonal selection is based on affinity, and that high affinity responses are favored in limiting antigen concentration. However, it is also an established fact that during, or at the end of, the primary response a number of point mutations affect the hypervariable segments of the immunoglobulin genes active in memory cells. This certainly may affect the "maturation" of the response, but is it indispensable? How much weight should we assign to it?

Our first experiment on this problem -- shown in Fig. 5 of the Seiden and Celada article -- is a good example of maturation in absence of hypermutation. After the first antigen injections the titers of the 8-bit matching (that is, the high-affinity binding) and 7-bit (low-affinity) antibodies are superimposable, while after the challenge, during the secondary response, the 8-bit matching antibody clearly dominates. This model result matches the *in vivo* observations of anti protein responses. But why is the high affinity B cell not favored during the primary response? They would be if stimulation depended solely upon the B cell-antigen encounter; but in the natural immune system and in the model the interaction is triadic. Thus, the most plausible explanation is that the double encounter in the same square of **a)** one of the 300 antigen molecules introduced and one of the virgin cells endowed with one of the few high-affinity receptors (the total B cell input is 1000, divided into 256 clonotypes in the resting state of the model; one representative of each clonotype will be found in ~4 of the 225 squares of the model) and **b)** with a T cell specific for the presented peptide, is very low. Thus, the first successful encounter will take place only after a specific T population has been built up as a consequence of many presentations operated by nonspecific macrophages, a situation that dilutes or nullifies the advantage based on B cell affinity. The situation changes drastically at the time of the secondary challenge, when all B cell clonotypes have reasonably expanded and are represented by memory cells. At this time the probability of A cell binding the antigen is lower since more squares contain specific B cells that outpower A cells, and the competition is played out mainly among B clones: obviously the 8-bit matching cells emerge as easy winners.

This simple model experiment indicates that antigen presentation and cell cooperation between APC, B cells and T cells along the lines of the present consensus produce a credible maturation of the antibody response. Consequently, it suggests that hypermutation is not essential to the phenomenon of affinity maturation and may even support the view that the net effect of mutation (which can certainly produce a better binding antibody, but also cause a drastic decrease of affinity or even disrupt the binding site) may not be conducive to optimization of binding. Therefore, our next experiment in this line will be to introduce hypermutation and observe its effects: on affinity maturation and/or fluctuation, but also on self-regulation of the response, which could well be the evolutionary payload of the phenomenon.

CONCLUSION

The two examples of in machina experimentation have been chosen because I am very familiar (actually, to an almost embarrassing degree) with both the experimental work and the computer simulation in two cases. They illustrate one way of using a model as a complement to the bench work and to the drawing board. Basically, a well-known phenomenon was reproduced quite accurately in the model, utilizing only a fraction of the factors that exist in vivo and are thought to be connected with the etiological mechanism. The first explanation that comes to be mind is that Nature is inherently pleiotropic and tends to co-develop more than one mechanism to obtain a single goal, or forgets to obliterate a mechanism even when a more efficient one is developed (2). Another explanation is that some of the factors suspected are not involved at all, or their involvement is marginal and not sufficient to justify their presence in evolution unless they fulfill other functions that are still unknown (this would be the case of hypermutation). It follows that more experiments are needed: the model has served as a stimulus for the imagination.

There are certainly other ways to use models (any of those described in this book). E.g., one may fail to obtain the equivalent of an **in vivo** observed phenomenon even introducing -- separately and then together -- all factors suspected to be involved. Thus, the emerging suggestion is to look for additional elements: to make new abductions and to seek experimental proofs. The model experiment may even offer some guidelines for the next step, it may suggest the directions more likely to yield results or the magnitude of the missing causative factor. In our case (see Seiden & Celada, this volume) **in machina** experimentation failed to behave in a consensus-like manner when idiotype-anti-idiotype interactions were allowed: there was a rapid growth of network-connected clones that clogged the system, a clear indication that a regulatory device was missing. What is this device that Jerne kept up his sleeve? There are several candidates, from T-peptide crypticity, to suppression, to B cell anergy. Before launching a search we asked the model to provide some indication of the size of regulatory force involved. This was done by introducing a completely artificial, parameter; we decreased the interaction strength between idiotype and anti-idiotype until the final picture became "consensus-like." The size of the regulatory force needed can be gathered, by analogy, from the fact that it took a 500 fold decrease of the binding strength to control the clonal expansion.

In all cases I have discussed, I believe the simulation has been used properly, as a tool to try out ideas and make simplified experiments, to prepare the way of **in vivo** or **in vitro** tests. Improper ways would be to try to substitute the model for reality, but also, to belittle it by considering it a curiosity, or a little flower on the lapel.

Who are then, the **fruitori**? If we follow these guidelines and cautions, all immunologists, and very soon.

REFERENCES

1. Jerne, N.K. (1974) Towards a network theory of the immune system. Ann. Immunol. (Inst. Pasteur) 125C, 373).
2. Celada, F. (1989) Bricolage in Cellular Immunity. Int. Rev. Immunol. 4, 107.
3. Janeway, C. (1989) Approaching the asymptote: Evolution and revolution in immunology. Cold Spring Harbor Symp. Quant. Biol. Volume LIV, 1.
4. Kaufmann, M., Urbain, J., Thomas, R. (1985) J. Theoretical Biol. 114:527.
5. Varela, F.J. (1988) Structural coupling and the origin of meaning in a simple cellular automaton. In The Semiotics of Cellular Communication in the Immune System, E.E. Sercarz, F. Celada, A.N. Mitchison, T.Tada (eds). Springer-Verlag, Berlin. pp.151-161.
6. Gardner, M. (1970) Sci. Am. 223 (Oct.), 120-123
7. Gardner, M. (1971) Sci. Am. 224 (Feb.), 112-117
8. Celada, F., Seiden, P.E. (1992) A Computer Model of Cellular Interactions in the Immune System. Immun. Today (Feb.),
9. Seiden, P.E., Celada, F. (1992) (Submitted), J. Theor. Biol.
10. Celada, F. (1971) The Cellular Basis of Immunologic Memory. Prog. Allergy, vol. 14, pp. 223-267, Karger, Basel (1971).

IS THE FUNCTION OF THE IMMUNE SYSTEM ONLY TO PROTECT?

Edward S. Golub
R.W. Johnson Pharmaceutical Research Institute
The Research Institute of Scripps Clinic
La Jolla, California,
USA

Abstract

Any theoretical analysis of the immune system must make assumptions about its function. The almost universal assumption among immunologists is that the sole function of the system is to protect but over the past quarter century there have been occasional claims that the system also has generative functions. This paper puts the human immune response into an historical context to show that the protection that the system provides is primarily against opportunistic pathogens. Since humans organized themselves into societies until the end of the nineteenth century they have been subjected to continuous decimation by infectious agents, both as endemic disease and epidemics of catastrophic proportions. The changes in society brought about as a result of industrialization and the introduction of modern sanitation measures have been the primary reasons for the control of infectious diseases in the industrialized nations. The tragic lesson of the AIDS epidemic is that in the presence of adequate public health and antibiotics, the immune system is necessary for the defense against infections by opportunistic pathogens. It is of course possible that this is the sole function of the immune system, and if it is, theorists must be skeptical of data and conclusions from experiments designed with the large doses of highly antigenic material because these may not be examining the response as it functions in nature. Similarly, for reasons discussed in the paper, theorists and experimentalists should look to other functions of the immune system. These *generative* functions mayserve in growth and differentiation of the organism.

NATO ASI Series, Vol. H 66
Theoretical and Experimental Insights into Immunology
Edited by A. S. Perelson and G. Weisbuch

Introduction

Any theoretical analysis of *how* the immune system functions must be based on some idea of *why* it functions and the approaches that are taken to the understanding of the 'how' are necessarily influenced by the answer to 'why'. Immunology, as a scientific discipline is less than one hundred years old and because it arose in the context of protection from infectious diseases, it has been assumed from the very beginning that the sole function of the immune system is to protect from disease. Interestingly, the term "immune system" only appears in the literature ca twenty-five years ago (Moulin,1989), indicating that the elevation of the status of the protective function to a separate system in the body is a recent event.

At the beginning of the twentieth century, the time when immunology was born, the industrial world was entering a new era; life expectancy was rapidly rising and infant mortality was declining leading to a demographic change of momentous proportions. The decline in deaths due to infectious diseases was the most obvious factor that was associated with these conditions, which were the prelude to the standards of health that we have come to take for granted in the twentieth century. An analysis of historical events shows very clearly that it was a combination of improved hygiene and immunization that led to the improvements in the level of health for the inhabitants of the industrialized nations at the start of this century However, almost from the beginning it was the immunization part of the combination that was perceived, in the minds of both the public and of scientists, as the dominant one. This early perception has been a crucial determinant in what immunologists think the function of the immune system because there has been a subtle change in the perceived role of the immune system during this century. From its beginnings as a means of conferring protection by vaccination, it has become perceived as responsible for protection on its own, that is, without artificial immunization. The evolutionary pressures on a system that has evolved to protect by itself are quite different than those on a system that can be manipulated to protect. These considerations are significant to experimental immunologists attempting to determine how the system works and they should be of no less significance to theorists who want to model the system.

This paper will examine some of the assumptions inherent in the view that the sole purpose of the immune system is to protect from disease and suggest that the immune system may have other important functions.

The Presence of Epidemics

In eighteenth century there was a movement by sanitationists to insititute procedures and practices of what we now call public health (Hamlin,1990; Latour,1988; Smith,1979). By the middle of the nineteenth century, when the microbial etioloty of infectious diseases was discovered, there was a struggle between the sanitationists and the bacteriologist for dominance in the methods that would be used in the control of infectious diseases (Dubos,1959; Latour,1988; McKeon,1988). The bacteriologists emerged victorious in this battle, with Louis Pasteur as the person who exemplifies the scientific (Dubos,1950). As a result of his work and personality, bacteriology and then its daughter, immunology, were credited with a greater role in protection from disease (Latour,1988).

But just how strong is the evidence that the immune system is such an effective means of fighting infectious disease that it deserves to overshadow the importance of hygiene in both the public and scientific mind? In the education of every immunologist, homage is paid to Pasteur and the roots of the science in the discovery of vaccines to induce specific immunity. This acknowledgement of past debts serves to re-enforce the protective function of the immune response in the minds of young scientists even though it has been several generations since most immunologists in industrialized nations have had first-hand contact with rampant infectious disease. It is interesting that in our post-literate scientific world perhaps the only classical literature that protoimmunologists read is a selection from Thucydides in which they learn about those that recovered from the plague of Athens.

> " Yet still the ones who felt most pity for the sick and the dying were those who had had the plague themselves and had recovered from it. They know what it was like and at the same time felt themselves to be safe, *for no one caught the disease twice, or, if he did, the second attack was never fatal.* Such people were congratulated on all sides, and they themselves were so elated at the time of their recovery that they fondly

imagined that they could never die of any other disease in the future". (emphasis added). (Thucydides, *History of the Pelloponesian Wars*)

It is significant that immunologists always focus on the part of this quotation showing that those that recovered never got the disease again. Overlooked is the fact that recovery was such a rare event that the people were congratulated on all sides and were so elated that they thought that they would never die of any other disease. All this elation should tell us is that for the lucky ones the immune system worked to prevent them from getting the disease a *second* time; the immune system apparently was not very effective in saving people who contracted the disease the *first* time.

Pasteur showed that the process of vaccination used by Jenner against small pox could be generalized to other diseases so that those that are *vaccinated* against a disease are equivalent to those that have recovered from it (Silverstein, 1989). But the historical evidence of epidemics causing death on a large scale in human history is overwhelming, a fact that tells us that the *unvaccinated* immune system was not terribly protective. For example, Lawrence Stone describing the early family; "The most striking feature which distinguishes the Early Modern family from that of today does not concern either marriage or birth; it was *the constant presence of death.* Death was at the center of life, as the cemetery was at the center of the village." (Stone, 1977, emphasis added).

The causes of death were many, but infectious disease was certainly the leader. Philip Ziegler, discussing the death rates during the plague of 1348 in England states that as "a rough and ready rule-of-thumb, therefore, the statement that a third of the population died of the Black Death should not be too misleading" (Ziegler, 1969). W.H. McNeill in *Plagues and Peoples* gives a magisterial analysis of how this constant presence of epidemic has affected human history (McNeill, 1977). There is general agreement that crowding and the unsanitary conditions that accompany it are the significant factor. In fact, until the growth of public health, conditions in the large cities of Europe and the United States were deplorable (see for example, Hobsbawm, 1968; Himmelfarb, 1991).

The immune system evolved long before humans discovered the joys of urban living and the immune system may have served a protective role in sparsely populated human societies as it may in animal populations in the wild. But the historical evidence is overwhelming that before the introduction of public health measures, the immune system, while undoubtedly important, was not a very efficient means of protection from epidemic disease.

The Origins of Public Health and the Rise of the Hygiene Movement

The deplorable living conditions which were responsible for so much of the disease began to change in the early 1800s in England and Western Europe. This change was due to the rise of the hygiene movement. The reasons for this rise are complex, but they correlate with the rise of industrialization (Hobsbawm,1968). It seems certain that the beginnings of industrialization, the growing level of rural pauperism and the migrations to the cities of the north all played a role in the hygiene movement in England. By the early part of the 19th century, for example, there was a growing concern over the purity of the water and its effect on disease in metropolitan areas such as London (Hamlin,1990). Arguably the changes in the Poor Laws in 1834 were in part motivated to allow the migration of the rural poor to urban areas as a source of labor. Industrialization thus made labor a valued commodity which the institution of sanitation measures protected (Rosen,1958).

But there was little *scientific* basis for the desired reforms in public health. Hamlin describes the conflict between the chemists who measured the hardness of the water and correlated it with disease, and the microscopists who saw the teeming masses of organisms (few of which, we now know, were pathogens) as causative agents. Since the causes of communicable diseases such as cholera that were associated with impure water and tuberculosis that were associated with bad housing were not known, there was be no clear direction that the well intentioned hygiene movement could take.

Public Health versus Immunization

Simultaneous with the perceptible improvement in living conditions brought about by public health measures, the new science of bacteriology began to have an effect. That the great diseases were infectious had been known for a very long time; now that the agents of the infection were being identified, reasonable strategies to prevent them were at hand. One of the first of these reasonable strategies came from the new discipline of immunology. Pasteur's dramatic proof at Poilly le Fort in 1881 showed that vaccination against anthrax could work and his even more dramatic treatment of young Joseph Meister with his new rabies vaccine solidified in the public mind that vaccination held the promise of being able to prevent all infectious diseases. The success of anti-diphtheria antiserum therapy erased from the mind of the public and the scientists alike the near catastophic failure of Koch's attempt to vaccinate against tuberculosis and Almoth Wright's controversial claim to have developed a vaccine against typhoid. The stunning success outweighed the failures in the early years of immunology and the merging of science and medicine in the public's mind served to establish the path that immunology would travel for the entire century.

As the field of immunology grew from a practical subject, devoted to devising and testing vaccines (with more failures than successes) the importance of *immunization* to prevent disease became transformed into the importance of the *immune response* to fight disease. After a century of pursuing this line of thought, the time may be right to re-examine it because of what we have learned about the immune response from the AIDS epidemic.

Lessons from AIDS

Paradoxically, AIDS, which results in immunosuppression, is the dreadful experiment of nature that allows us to examine the relative importance of the immune system in human populations in which public health and antibiotics are readily available. It is well known that in this syndrome the helper T cell component is reduced and the resultant immunosuppression leads to an array of infections resulting in death. This certainly argues for the importance of the immune system. However, when one examines the nature of the infections to which AIDS patients have increased susceptibility, one notes that they are not the rampant infections that we have been discussing above. Rather, these patients succumb to infections from organisms of low pathogenicity that are always present in the population. Interestingly, intentionally immunosuppressed people undergoing transplantation suffer from the same spectrum of opportunistic infections and lymphomas (Penn,1988).

It thus appears that the immune system is not very efficient and effective against infectious agents of high pathogenicity, but it is very effective against opportunistic infections by ubiquitous organisms of low pathogenicity. Even in the presence of public health measures and antibiotics, the loss of the immune system results in these opportunistic agents causing disease and death. With proper immunization, of course, the system can be driven to make antibodies of high affinity to a wide range of antigens (see Golub and Green, 1991). In this way the system can carry out the wider range of protection that vacination affords. However, in the competition between an organism of high pathogenicity and the immune system, the molecular events required to generate high affinity antibody occur at too slow a rate to be a major source of protection.

Generative Functions of the immune system

It is of course possible that the primary function of the unvaccinated immune system is to protect individuals against opportunistic infections. But there is reason to speculate that it has other important functions. Even at the very beginning of the ascendancy of immunology, Elie Metchnikoff suggested that while anti-organ antibodies at high concentrations are damaging, at low concentration they can be stimulating and Alex Carrel in 1922 postulated a *trophic* function for lymphocytes (cited in Prehn and Lappe', 1971). In 1963 Burwell postulated a role for lymphoid tissue in *morphostasis*, "...the morphological expression in the mammal of ill-understood homeostatic mechanisms which lead to a balance of differentiated tissue.." (Burwell, 1963). In the 1970s Richmond Prehn developed the rather original idea in tumor immunology that not only is the immune response active in destroying tumors, but that under certain conditions the immune response was responsible for the enhancement of tumor growth. He argued that the coin of immunity had two sides; one destroyed cells and inhibited growth and the other stimulated cell division and growth Prehn and Lappe', 1971).

Reading this old literature can be an adventure in historiography unless one also reads contemporary papers on the subject published in these journals. Read in the context of the 'hard science' of the times one sees that these ideas of a non-defense function were built on the very most up to date ideas of the time. As a continuation of this tradition, the topic has been recently revisited by two groups; Green and Wegmann (1986) in Edmonton and Varela and Coutinho (1991) in Paris.

Building on their thinking, I suggest that we now know enough about the cellular and molecular basis of the immune response to foreign antigen to begin to ask if there are in fact other non-defense functions of the immune system, and what they might be. I will call these other functions *generative* to distinguish them from the *protective* functions discussed above.

Green and Wegmann have attempted to put the generative function into the system as we are beginning to understand it; "...tissue generation is signalled by an alteration in the levels of major histocompatibility antigens. T lymphocytes, activated by altered MHC expression, release lymphokines which act directly... to promote growth of the target tissue." (Green and Wegmann,1986)).

There are many examples of what might be the generative properties of the immune system. For example, in bone marrow transplantation it is well known that removal of all of the donor T cells reduces the incidence of graft-versus-host disease (GVHD). Unfortunately, the removal of T cells from the graft also reduces the likelihood of engraftment of the transferred cells. These facts seem to point to the anomalous idea that a modest immune response is necessary for the cells to establish themselves. Obviously, the engraftment of foreign bone marrow cannot be an evolutionarily conserved function of the immune system, but it may be analogous to one that is.

The role of the immune response in fertilization, implantation, placentation and fetal survival has been commented upon for many years (see Prehn and Lappe',1971 for review). There is growing evidence that women have great difficulty conceiving if their HLA profiles are too similar to that of the father of the child and a recent therapy to has had success in overcoming this condition by immunizing the mother to the HLA of the father. One can envision the need for the local accumulation of growth factors at the site of implantation and the inflammatory response caused by implantation triggering the release of these growth factors (lymphokines) from accumulating lymphocytes. The slight degree of histoincompatiblity could be necessary to ensure that enough lymphocytes are present to secrete enough of the proper factors. A similar role for the delivery of growth factors in wound healing can be envisioned.

Central to the idea that the immune system has a generative property is the fact that T lymphocytes are known to either synthesize or induce a very wide range of growth factors. The problem that earlier arguments for a generative role faced was how these factors could be delivered to sites in which they are needed. Placentation does not pose this problem but generalized wound healing does. A potential means of overcoming this problem can be found if yet another sacred assumption of immunology is re-examined; the ability of the immune system to distinguish self from non-self.

The Role of Autoreactivity

Varela and Coutinho have been developing the idea of *second generation immune networks* in which they argue that the original idea of immune networks that Jerne proposed must be altered to see the *emergent properties* of the immune system. The core of the idea is that "external stimuli modify only the local environment of individual components [of the immune system] (which behave according to conventional stimulus-response patterns), but the system's operation and properties will always emerge from global co-operativity. This passage from local rules to a new emergent property of the entire system, which cannot be reduced to the quality or performance of single components, is the very heart of network mechanisms. To miss such emergent, global properties (as was the case in first generation studies) is to doom a network view to trivial information"(Varela and Coutinho,1991) [See Golub,1991 for an argument against first generation network ideas based on the size of the repertoire and evolutionary considerations.]

Based on these ideas, Varela and Coutinho argue that the V regions of B cells (immunoglobulin) and T cell receptor connectivity results in a degree of intrinsic activity of the lymphocytes that is independent of environmental antigens. The network, in this new view, consists not only of connectivity between V regions (as in the original theory) but also includes autoreactivities to other somatic components. Thus, perhaps the most sacred of immunolgocial mantras, *horror autotoxicus*, is called into question. This turns out to be a good time to question this earliest of assumptions, that the body does not react against itself, because of the growing evidence for autoreactivty.

The literature on autoreactive antibodies and autoreactive T cells have been recently reviewed (Avrameas,1991; Zauderer,1989). One of the most striking features of the accumulated data is that there is a very large amount of *autoreactivity* but very little autoimmune *disease*. Taken along with the recent attention focussed on the polyspecific and autoreactive immunoglobulilns secreted by CD5+ B cells, there seems to be little doubt that the immune system spends a significant part of its effort in self responses. Assuming that negative selection, tolerance, and clonal anergy. account for removal of much of the harmful autoreactivity, that which remains merits examination. The possibility remains that it reflects an important part of the connectivity and serves a general homeostatic function. The specificities of natural autoantibodies found in the serum may point to an immediately accessible role of autoreactivity in the generative role of the immune system.

The use of autoreactivity in generative functions

Much of the autoreactive natural antibody found in the serum reacts with internal components of the cell (Avrameas,1991). Because these components are only available to the immune system when the cells are damaged, one can visualize an accumulation of self-antigen-specific cells at sites of injured tissue during inflammation, infection and even tissue senescence. Thus one can see a role of the immune system in delivering the vast array of growth factors to needed sites where they are released. Because of the two sides of the coin notion of the immune response, too intense an insult leads to too large an accumulation of cells and tissue destruction.

Enough aspects of the idea of a generative role for the immune system are testable so that this need not be an excercise in mere spculation. If there is indeed a core of truth in the idea, it alters not only the way we design and interpret experiments, but also how we think of using the immune system for therapeutic purposes.

Summary

From its beginnings, immunology has worked on the assumption that the sole function of the immune system is protection from disease. A cursory analysis the historical data however shows that until the early twentieth century there was little evidence of protection. Human populations, until then, could expect a very high rate of infant mortality, the continuous decimation of the population from epidemic diseases, and a mean life expectancy of forty years. It was the institution of sanitation measures and the rising standard of living associated with industrialization that brought about the changes in these conditions that we associate with modern life. By the twentieth century passive serum therapy and immunization contributed to a lowering of the infant mortality rate and antibiotics reduced the mortality rates from infections. The AIDS epidemic shows us that in the presence of proper public health measures and antibiotics, the immune system functions to protect from infectious agents of low pathogenicity; the so called opportunistic infections. Given the complexities of the immune system, this paper explores the idea that we should be looking for some other function that this very complicated system carries out. This is especially true for those who are interested in constructing theoretical models of the immune system. The idea that the immune system also has a *generative* function is proposed.

In an attempt to entice investigators into thinking about experimental tests of the hypothesis, one possible example of how the generative role could be carried out is given. T lymphocytes can be thought of as mobile delivery systems for a wide array of growth factors that are involved in growth, differentiation and wound healing. The specificity elements of the immune system may have been selected in evolution as the means of retaining cells producing these growth factors at sites where they are needed. Since much of the activity of the immune system is directed to anti-self components that are mostly intracellular, it is proposed that when there is tissue damage cells reactive to the exposed internal components accumulate at the site and are involved in healing. In this way appropriate concentrations and mixtures of growth factors can be delivered to a localized area.

References

Avrameas, S. Natural autoantibodies: from 'horror autotoxicus' to 'gnothi seauton'. Immunol. Today *12*,154.1991

Burwell, R.G. The role of lymphoid tissue in morphostasis. Lancet,*2*,69.1963

Dubos, R. *Louis Pasteur, Free Lance of Science*. Da Capo Press, New York,1950.

Dubos R. *The Mirage of Health. Utopias, Progress and Biological Change* New York, Harper.1959).
Hobsbawm, E.J. *Industry and Empire*. Penquin Books, 1968
Golub, E.S. Scan. J. Immunol.*33*,498. 1991.
Golub, E.S. and D.R. Green, *Immunology: A Synthesis* Second Ed. Sinauer Assoc., Sunderland MA. 1991.
Green,D.R. and T.G. Wegmann. Beyond the immune system: The immunotrophic role of cells in organ generation and regeneration. Progress in Immunol. IV. pp1100-1112. 1986.
Hamlin, C. *A Science of Impurity. Water Analysis in Nineteenth Century Britain.* Univ. California Press, Berkeley. 1990.
Himmelfarb, G. *Poverty and Compassion. The Moral Imagination of the Late Victorians*. Knopf. 1991.
Latour, B. The Pasteurization of France. Harvard Univ. Press, Cambridge. 1988.
McKeown, T., *The Origins of Human Disease*. Blackwell, Oxford. 1988.
McNeill,W. H. Plagues and Peoples, New York, Doubleday, 1977.
Moulin, A-M. Immunology old and new: the beginning and the end. In. Mazumdar, P.H. ed. *Essays on the History of Immunology*. Wall and Thompson, Toronto. 1989. p291.
Penn, I.Tumors of the immunocompromised host. Ann. Rev. Med. *39*,63.1988.
Prehn, R.T. and M.A. Lappe'. An immunostimulation theory of tumor development. Transplant Rev *7*,26-54. 1971.
Rosen, George. *A History of Public Health.* MD Publication, New York. 1958.
Silverstein, A. *A History of Immunology*, Academic Press. 1989.
Smith, F.B. *The People's Health. 1830-1910*. Weidenfeld and Nicolson, London, 1979.
Stone, L.. *The Family, Sex and Marriage in England 1500-1800*. New York, Harper & Row. 1977. pg66.
Varela, F.J. and A. Coutinho. Second generation immune networks. Immunol Today *12*,159.1991
Zauderer, M. Origin and significance of autoreactive T cells. Adv. Immunol *45*;417. 1989.
Ziegler, P. *The Black Death*, New York, Harper & Row, 1969.

THE IMMUNE SYSTEM IN AN EVOLUTIONARY PERSPECTIVE.

J. Stewart
Department of Immunobiology,
Pasteur Institute
25 rue du Dr Roux
75634 Paris Cedex 15
France

I. INTRODUCTION.

I. 1. The significance of the evolutionary dimension.

There are two reasons why the evolution of the immune system is an important question for theoretical immunology. Firstly, it is a fascinating subject in its own right; and whether we are theoreticians or experimentalists, we can hardly claim to be true biologists if we have no knowledge of the evolutionary origin of the system we are studying. Secondly, a major requirement for effective theoretical modelling of the immune system lies in the judicious choice of radical simplifications which nevertheless grasp essential features of the system. A consideration of evolution is a tool in this direction: when the immune sytem first appeared, it presumably existed in a relatively simple "primordial" form which we might be able to take as the basis for a simplified model.

Even the most superficial examination of the empirical evidence reveals a striking fact: a fully-fledged immune system, producing circulating immunoglobulins, exists in all vertebrates but in no invertebrates. This is, indeed, something of an enigma: it is almost as though the immune system never existed in a primordial form, but sprang miraculously out of nowhere at the time of the first vertebrates. Nevertheless, we know on solid theoretical grounds that evolution proceeds only by single steps. This paper is a speculative attempt to resolve this enigma, by reconstructing the "missing links" which led to the primordial immunoglobulin-producing system.

I. 2. Structure and Function.

In evolution, what is conserved is structure, not function. Structures, be they molecules, cells, organs or bodies, do of course change through evolution; but they change more or less continuously in such a way that homologies can be reliably recognized. Functions, on the other hand, are typically not

NATO ASI Series, Vol. H 66
Theoretical and Experimental Insights into Immunology
Edited by A. S. Perelson and G. Weisbuch

conserved; they evolve discontinuously. In the opportunistic tinkering which is so characteristic of the evolutionary process, structures are frequently redeployed for qualitatively different (and on occasion completely novel) functions. This being so, it is most important not to base an evolutionary investigation on a particular pre-specified function, because apparently similar functions can be performed by evolutionarily unrelated structures; and conversely, structures that are actually homologous and evolutionarily related can come to serve apparently unrelated functions.

To take account of this important feature of evolutionary processes, the strategy of this paper will be as follows. Firstly, I will endeavour to trace the evolutionary genealogy of the essential structures of what we now call the "immune system". Secondly, with this structural history as a relatively secure and empirically-based backdrop, I will endeavour to understand the function(s) of these structures at successive stages of their evolution. This second part of my argument is of necessity speculative; the aim is to provoke experimental work in this area designed to either confirm or refute my hypotheses.

I. 3. The VRM-system.

The methodological prescription that the object of evolutionary studies should not be identified in terms of a specific function does, of course, pose one awkward question. At the present time, the consensual definition of what the immune system is refers constitutively to the function of defence - an assumption so deeply engrained that it is inscribed in the very name we give to the system. If we wish to postulate that the principal function of the immune system has not necessarily always been that of defence, how are we to identify the object we are talking about?

In this paper, I shall define the focus of our attention as the system of molecular "variable-regions" and the cells that produce them (i.e., as far as we know at the present time: immunoglobulins (Igs) produced by B-lymphocytes), and T-cell receptors (TCRs) produced by T-cells. The structural diversity of the "V-regions" of Ig s and T-lymphocyte receptors is at least as great, and probably far greater, than that of all the other molecules in the vertebrate body put together. This diversity is generated during somatic lymphocyte development, breaking the central dogma of DNA conservation. The biochemical machinery necessary to construct and express V-region genes in somatic cells has been identified, so that the object of our study is adequately defined. I shall refer to such diversified molecules by coining the term "Variable Region Molecules" (or VRM for short), and call the system which produces these molecules the "VRM-system".

II. THE STRUCTURES OF THE VRM-SYSTEM.

In this section I shall trace the genealogy of the structures of the VRM-system. Variable-region molecules are proteins produced by straightforward transcription of the (rearranged) genes; it is therefore convenient to start with a description of these genes.

II.1. VRM GENES.

The genes necessary for VRM synthesis have been studied in great detail. I shall not review the empirical data as such (i.e., structural homologies in the different classes of vertebrates), but rather give a synthetic account of the evolutionary process as it can be reconstructed on the basis of this data (Klein 1982, Du Pasquier 1989).

II.1.1. The common ancestral lineage.

We shall start with the common ancestral lineage of all contemporary VRMs, up to the point of divergence between Ig s and TCRs.

II.1.1.a. One-domain and two-domain molecules.

At the beginning there was a primordial gene, P, which probably carried information for some 110 amino acids - the present length of a single V or C domain. At an early stage, perhaps when the most advanced invertebrates began to branch off into vertebrates, the P gene gave rise to two genes by tandem duplication. The two genes, separated by a primordial intron, were transcribed as a unit into one messenger RNA. The duplication allowed the two genes to evolve separately: one became the ancestor of all V genes and the other the ancestor of all C genes.

II.1.1.b. The generation of diversity: somatic DNA rearrangement.

Each V-region in a VRM is translated from a single message which also encodes the respective C-region. V-region genes, however, do not exist in germ-lines, but are formed somatically by an extraordinary process of DNA arrangement that brings together two or three gene segments in the genome of the differentiating lymphocyte. The chemistry of this recombination process has revealed a very important feature. In all rearranging groups of gene segments known to date, and in all species analyzed (including sharks, frogs, chicken and mammals), the same "signals" (nucleotide sequences at the joining sites; length of the "spaces" between them) are conserved. Furthermore, mutants have recently been isolated in the mouse that fail to accomplish such rearrangements in all gene complexes, and thus produce no VRMs at all; this shows that the same enzymes and chemical processes are uniformly used for all VRM genes. It follows that in all likelihood, this extraordinary process of somatic DNA rearrangement was invented only once in evolution, and was thereafter recurrently used for all V-regions.

The first evolutionary step (after the initial tandem duplication) therefore seems to have been the eclosion of the DNA recombination mechanism, perhaps after the genomic integration of an episome or viral sequence, inserted into the V-like gene and separating it in two segments (V- and J-) that need to be juxtaposed by somatic recombination before expression. This allowed for combinatorial possibilities, as soon as the V-gene and the J-gene segments duplicated and diverged. Furthermore, it led to clonality in VRM expression, since the recombination signals only allow for one (V-J) joining at each gene complex.

II.1.1.c. Duplication and heterodimer VRMs.

This primitive complex of V-J-C genes has subsequently duplicated to provide possibilities for combinatorial association of two different VRM chains in the same molecule. Another major step took place immediately: this event is the duplication of the "episome" in the V-gene of one of the V-J-C complexes to generate V- and D-gene segments, thereby increasing significantly the combinatorial assortment of gene segments into the same mature (rearranged) V-region gene. In all mammalian VRMs known today (Igs, a-b TCR and g-d TCR) the heterodimer is always composed of one V-region made up of V-D-J and another composed of V-J only. This indicates that the events of this section preceded the divergence between Igs and TCRs.

II.1.1.d. Divergence between Igs and TCRs.

All contemporary TCRs are dimers of two-domain chains which resemble the V_L-V_L dimers of Ig light chains. It is therefore likely that the divergence between Igs and TCRs occurred after another set of duplications at this stage. Not only are the genes for these different VRMs now on different chromosomes, but the duplication in question here concerns both the VJC complex and the VDJC complex; this may therefore be the result of a polyploid duplication. From this point on, Igs and TCRs were distinct heterodimer VRMs; we shall trace their subsequent histories separately.

II.1.2. Ig-gene evolution.

II.1.2.a. The emergence of a heavy chain.

The first event following the separation of the Ig genes and TCR genes was the tandem duplication of the C-region of one of the complexes coding for Igs to form the C_H1, C_H2 and C_H3 regions of the heavy chain.

II.1.2.b. The emergence of a secretory form.

Primordial single domains of the Ig family (such as Thy-1) are membrane bound, and contemporary two-domain TCRs are also membrane-bound. The evolution of secretory forms of VRMs requires that the C-termini of the two chains composing the original heterodimer receptor both lose the very hydrophobic stretch of aminoacids that assumes alpha-helical conformation across the lipid bilayer and anchors the chain to the cell membrane. In Ig's, different pathways were seemingly followed for the two receptor chains: the light chain lost the membrane-encoding exon altogether and became a molecule that can never be inserted in a membrane; the heavy chain acquired an extra nucleotide sequence encoding hydrophilic aminoacids.

II.1.2.c. Class switch in Igs.

Further evolution of Igs included the duplication and divergence of C-genes (to produce the various Ig isotypes M,G,A,E) together with a mechanism to recombine the whole rearranged VDJ-gene to any of the alternative C-region genes. This process of somatic recombination is based on "signals" and enzymatic activities that are different from those operating in VDJ recombination and are thus likely to have evolved independantly.

II.1.2.d. Hypermutation in Ig genes.

In higher, homeothermic vertebrates at least, yet another mechanism has recently been disclosed which accounts for increased diversity among V-regions of Igs. This is a process of somatic hypermutation (10^{-3} bp/cell generation) of rearranged V(D)J genes, the chemistry of which remains to be investigated.

II.1.3. TCR-gene evolution.

In mammals, T-cells are differenciated into two classes: a-b and g-d cells. The mechanisms accounting for the VRM diversity are essentially the same in these two classes, and hence provide no arguments concerning their evolutionary relationship. The diversity of the g-d receptors is much less than that of the a-b receptors; this difference arises in part because g-d TCRs are assembled from a significantly lower number of germ-line gene segments than those available to a-b TCRs (or Ig's), a fact which suggests that the g-d receptor is probably the more primitive form. This conclusion is strongly confirmed by the detailed gene organization: the whole of the d locus (V, D, J and C) is imbedded in the a locus, so that the rearrangement of Va-Ja (Ca) necessary for the expression of a TCR a-chain eliminates all the d genes from that chromosome. This gene organization is natural if the a-chain genes arose by tandem duplication from the d-chain genes, but would be completely bizarre if the evolutionary relationship were the other way round.

II.2. Cells.

In contemporary mammals, there are three classes of VRM-producing cells: B-lymphocytes, a-b T-lymphocytes and g-d T-lymphocytes. The purpose of this section is to establish the genealogical relationship between these three classes.

II.2.1. B-cells and T-cells.

We have seen that on the basis of the genes coding for the VRM receptors, B-cells and T-cells have a common ancestral lineage at least during the period when gene rearrangement was invented, up to and including the point when the second episome insertion gave rise to the D-segment in the V-D-J-C complex on one of the two chromosomes carrying these genes.

On the other hand, still from the receptor viewpoint, the simple two-domain structure of the T cell receptor, the lack of complicated class switch, and the lack of secretory forms or of somatic mutations make this molecule a candidate for a conserved primitive recognition molecule (Du Pasquier 1989). This suggests that just before the time of divergence, the common ancestor to B-cells and T-cells resembled present T-cells more than present B-cells.

II.2.2. a-b and g-d T-cells.

As regards the evolutionary relationship between a-b and g-d T-cells, we have already mentioned the genetic evidence suggesting an earlier origin of g-d T-cells or, rather, their higher relatedness to the primordial T-lymphocyte from which both T cell lines evolved. Four other characteristics support this conclusion. First, g-d T-cells precede a-b T-cells in development. Second, in contrast to a-b T cells, some g-d T cells differenciate in the absence of a thymus and could thus have preceded evolutionarily the rather late development of this organ. Third, most g-d T cells do not express CD4 or CD8 accessory cell-adhesion molecules which are essential in the process of thymic selection of a-b T cells. Fourth, a-b TCR are predominantly directed at MHC genes (either Class I or Class II) that are expressed by most cells in the body, are highly polymorphic in the species, and are "recognized" by a-b VRMs in association with CD4 and CD8 molecules. It is thus likely that a-b T cells, and specific forms of B cell-T cell interaction such as T-helper activity, co-evolved with the polyporphism of those MHC loci in the species; I shall argue that this was a relatively late pattern in vertebrate evolution.

II.2.3. Lymphocyte phylogeny.

In conclusion, we may reconstruct the scenario of the evolution of VRM-producing lymphocytes as follows:

(a) The common ancestor to all three types most closely resembled contemporary g-d "T" cells (at this stage there was no thymus).

(b) Ig-producing B-cells diverged from this ancestor at an early stage, probably that of the common ancestor to all contemporary vertebrates from hagfish to mammals. The divergence between B-cells and primitive "T" cells occurred with the appearance of Igs with light and heavy chains and the emergence of a secretory form.

(c) Finally, a-b T-cells derived from the gd-like ancestral line at a later stage.

II.2.4. Lymphocyte ontogeny.

Since evolution occurs by modifications in the processes of development, there is always a relationship between phylogeny and ontogeny.

In order of evolutionary appearance, the first type of VRM-producing lymphocyte is the g-d "T" cell. These cells are typically made outside the thymus (hence the inverted commas on "T"), and subsequently migrate, notably to the surfaces of the body: digestive and urogenital tracts, lungs and skin. The precise site where g-d cells are made is not known (it is not the bone-marrow; they could actually be produced locally) : this recalls the diffuse mesodermic origin of haemopoietic cells in evolutionarily primitive animals.

The second type of VRM-producing lymphocyte to appear in evolution is the Ig-secreting B-cell. In adult mammals, B-lymphocytes are made in the bone-marrow. However, during ontogeny, the successive waves of B-lymphopoiesis occur in the coelom (para-aortic islets and omentum), liver, spleen and finally bone-marrow - a sequence which is indeed a striking recapitulation of phylogeny. Moreover, if the bone-marrow is selectively destroyed by strontium-90, B-lymphopoiesis can actually revert to spleen, and upon splenectomy to the liver, which can be interpreted as another reflection of evolutionary history.

Finally, the third type of VRM-producing lymphocyte to appear in evolution is the ab T-cell. These cells are unique in absolutely requiring two specific organs for their production: initially formed in the bone-marrow, where their commitment to colonize the thymus rather than to produce Igs occurs, they then migrate to the thymus, an obligatory step for them to complete their maturation. This striking characteristic clearly corresponds to their status as evolutionary late-comers, and is sufficient to virtually rule out any interpretation which would attribute them the status of evolutionarily primitive cells.

Mammalian lymphocyte ontogeny cannot be a direct recapitulation of phylogeny (as summarized in II.2.3.), because the definitive commitment to one of the three types (B, ab-T or gd-T) depends on which of the three mutually exclusive forms of gene rearrangement (necessary for VRM production) occurs, and this is a single irreversible event in the ontogeny of any cell-line. However, as we have seen, the ontogeny of the three types of VRM-producing lymphocytes is clearly coherent with their phylogeny; and this in turn confirms our reconstruction of that phylogeny.

III. THE FUNCTIONS OF THE VRM-SYSTEM.

III.1. Phase 1 : the endogenous emergence of a primordial VRM-system.

III.1.1. A problem to be solved.

With this background of structural history in place, we now address the enigma noted in the introduction. How can we explain, firstly, the relative rapidity with which the first VRM system appeared; and secondly, why none of the transitional stages, which must have existed, have given rise to living descendants?

In general outline, my explanation will take the following form. I will suggest that the initial constitution of the VRM system was what I shall call an endogenous process: by this, I mean that the transitional missing links arose not because of any particular functionality or positive selection, but rather because they could not be avoided and did not fatally compromise the viability of the organisms concerned. The VRM system only acquired a positive function when it reached the stage of contemporary primitive vertebrates, i.e. when free soluble VRMs (Igs) were produced. This function was thus rather long in coming; but when it did arrive, it was so important that from then on no vertebrate could do without it.

III.1.2. The starting-point: cell-lines vs organisms.

What then was this "endogenous process"? I take as a starting-point a theme developped by Buss (1987). The first eukaryotic cells appeared 900 million years (Myr) ago. There followed a relatively short period of 200 Myr which saw the rapid evolution of a variety of different metazoans, multicellular organisms made up of a number of terminally differentiated cell-types. The body plans of contemporary metazoans fall into a number of discrete categories. These fundamental "bauplans" (together with a number of other bauplans, bizarrely unlike any contemporary forms, which rather rapidly became extinct) all appeared during this short period: by 700 Myr there are metazoans clearly recognizable as members of all the phyla that are extant today. Since then, no new body plan has arisen: for over half a billion years, metazoan development has continued to follow the same basic formulae. Buss (1987) has remarked on this striking pattern, emphasizing that it calls for explanation. He propounds the view that a cardinal consideration in this respect is the potential conflict between natural selection at the level of cell lines, and selection at the level of the multicellular individual: the latter requires that the "interests" of cell-lines be subordinated to the requirements of the individual organism as a whole. He has argued convincingly that the evolutionary patterns of metazoan development can be largely understood in terms of a necessary resolution of this potential conflict.

The first multicellular organisms faced an obvious problem: how can it come about that individual cell-lines (which had so recently been independant entities each directly subject to natural selection) effectively sacrifice their own future (for example by undergoing terminal differenciation) in the interests of the whole organism? In fact, all multicellular organisms, including contemporary higher vertebrates, continue to face this problem; the only reason that it does not seem to be a major issue at present is that we take for granted the solutions to this problem that present-day organisms have inherited from their ancestors.

The paradigmatic illustration of this problematic theme is the case of cancer. In a multicellular organism, if any cell-line undergoes a cell-heritable modification such that it escapes from negative-feedback regulation, it will implacably proliferate until something intervenes to stop it - if nothing else, the death of the whole organism. Hence the vital importance of regulatory cell-cell interactions such as those giving rise to contact inhibition. The problem becomes particularly acute with the evolutionary appearance of a substantial extracellular space. We may recall here that the development of a circulatory compartment with blood and lymph is a salient feature of the first vertebrates. Cells which are able to proliferate in this extracellular space - lymphocytes being of course the prime example - are particularly liable to go astray and to form cancers because, not being inserted in a cellular matrix, inhibitory cell-to-cell contacts are at best relatively rare. This aspect is underscored by the fact that even today leukemias are a particularly virulent form of cancer.

III.1.3. A precarious passage.

Let us consider in more detail this problem of cancerous cell-lines that are able to proliferate in extracellular spaces. Since we are postulating that in the case of these early vertebrates the lymphocytes have no particular function, it might seem that the simplest solution would simply be to dispense with them altogether - why run the risk? The answer may be that it is unsound ecologically to leave a space entirely vacant; it is much better to populate it in a balanced manner.

Accepting this, how would the system work? My hypothesis is that the lymphocytes in blood and lymph would proliferate up to the point where their density is such that cell-to-cell encounters are sufficiently frequent to inhibit further proliferation. (Lymphocyte populations in contemporary mammals are noteworthy for their capacity to dynamically regulate total cell-numbers). To work on the basis of cell-cell contacts, the schema requires that such interactions have a reasonably high affinity, and also that the inhibitory effect lasts for some time. I postulate that the cell-cell affinity, and subsequent inhibition, were mediated by a cell-surface receptor molecule, similar to Thy1, which was the precursor of the subsequent VRMs. The first gene duplication, giving rise to a two-domain molecule, was probably advantageous to the organism because it allowed for higher affinity. It is relevant to note here that apart from Thy1, all the other prototypes of the products of the Ig-gene superfamily, such as the Cell Adhesion Molecules (CAMs) which play an important role in embryology, are multidomain molecules which have the particular property of displaying homophilic binding and autoaffinity (Edelman 1987). It is therefore plausible to suppose, as is required by my hypothesis, that the primitive precursor of the VRMs also possessed this property.

The hypothetical situation thus created exemplifies the thematic interplay between the interests of individual cell-lines and those of the organism as a whole. Since cell-cell interactions are inhibitory, it is in the interest of any particular cell-line to display a variant form of the cell-surface molecule with a reduced affinity, since this will allow a greater proliferation of the cell-line in question. However, any such development is constrained by the fact that the overall result should not be such as to increase the total numbers in the lymphocyte population to a point where it becomes a serious burden for the organism. Thus, the different cell-lines (as defined by the specificities of their cell-surface molecules) are effectively competing against each other. I suggest that it is the combination of these contradictory requirements which accounts for the selective advantage of the diversification of the VRMs (by the mechanism of DNA rearrangement described above). This system is subject to the constraint that all the VRMs thus produced should retain autoaffinity and also have a high affinity with at least some of the other VRMs produced by the lymphocyte population. The internal logic of such a process is such that diversification will proceed until the repertoire of VRMs is virtually "complete", i.e. there are no more molecular shapes which do not interact with at least one or other of the VRMs in the system. The advantage of such a situation for the organism as a whole, which would account for the transmission of the necessary genes in the germ-line, is that the "ecological" regulation of the total number of potentially proliferating cells in the body fluids would extend to all cells which interact with the diversified VRM receptors.

III.1.4. Cross-linking and secretion.

The next stages in the evolutionary process were: (i) the set of gene duplications allowing for the separation between Ig-producing B-cells and TCR-producing "T"-cells; (ii) the duplication of the C-regions to form a heavy-chain Ig; and (iii) the appearance of a secretory form of Ig. The events (ii) and (iii), taken together, had a consequence of the utmost importance.

The point at issue is the effect of VRM molecules in free solution on cells with membrane-bound receptors. If the soluble VRM molecules are simply two-domain heterodimers (such as TCRs), then no cross-linking of the receptors is possible. Consequently, cellular response can only be related to receptor occupancy and the dose-response curve will be essentially monotonic. However, the new VRMs (i.e. the first Igs) were tetrameric molecules with two heavy and two light chains, and consequently the characteristic "Y" shape of all contemporary Igs. Such molecules, even in solution, are capable of cross-linking cell-surface receptors. Moreover, this gives rise to a bell-shaped dose-response curve: at high doses, the receptors are largely occupied by different ligands and hence are no longer cross-linked (Perelson 1984).

The consequence of this is that the meaning of the signal to the cell can be radically altered: whereas before the logic of the system absolutely required that the signal be inhibitory, in this new situation the signal can be stimulatory. In other words, the stage is set for the innovation of programmed cell-death: lymphocytes die unless they are stimulated. This modification dramatically attenuates the potential conflict between cell-lines and the whole organism: firstly, because the mechanism of programmed cell-death obviously reduces the risk of cancerous variants; secondly, because the bell-shaped dose-response curve guarantees that the system is automatically self-limiting (if the total population starts to expand, this will push the system to the upper portion of the bell-shaped curve where cell-cell interactions, via their secreted VRM's, are inhibitory). At the same time, this new regime is perfectly viable from the point of view of the individual cell-lines: the co-operative nature of the cell-cell interactions at low dose levels means that all cell-lines connected to the system can survive. In fact, it seems intuitively plausible that this new regime will actually consolidate the pressure towards diversification until the repertoire is complete.

In a dynamic system of the sort we have envisaged, the innovation of a secreted form of the VRM molecule, which vastly multiplies the effectivenes of the influence that a cell exerts on other members of the population, is greatly advantageous to any cell that possesses it. At the same time, this innovation is also advantageous for the organism as a whole, for it means that the system reaches a dynamic equilibrium with a total lymphocyte population considerably smaller than if the existence of a secretory form had not multiplied the effectiveness of cell-cell interactions.

On this interpretation, then, the twin innovations of a tetrameric Y-shaped molecule and the emergence of a secretory form of the molecule (which in terms of molecular structure are unrelated events) were functionally related. This may explain why the two events were more or less concomitant in evolution.

III.1.5. The conclusion to Phase 1.

I consider that this is the end of the first phase of the evolution of the VRM system. We have arrived at a picture of what I shall call the "primordial VRM system" which is in accordance with the essential characteristics of the VRM system in primitive vertebrates, and also with the VRM system in the early stages of development in more advanced vertebrates. Specifically, these animals have a repertoire of circulating Igs which is virtually complete; but it is to be noted that these Igs are essentially "natural antibodies", i.e. their presence does not depend on stimulation by external antigens. Concomitantly, their "immune responses" (i.e. the increases in specific antibody levels following immunization) are relatively poor, both quantitatively and qualitatively.

At the end of this first phase in its evolution, the VRM system is consolidated for the first time in its history. In general terms, this consolidation is due to the fact that whereas at the outset the situation was precarious - the potential conflict between cell-lines and the organism was liable to erupt violently at any moment, with disastrous consequences for all concerned - by the end of this period these conflicting interests have been reconciled and indeed in many respect are already convergent. Earlier, I raised the problem of the "missing links": why is it that none of the transitional stages have given rise to living descendants, representative of those rudimentary forms of the VRM system which must have existed? I suggest that the relative precarity of the initial and intermediate stages already constitutes a substantial part of the explanation for this awkward fact.

III.2. Function of the primordial VRM system.

III.2.1. The regulation of physiological complexity.

I have hypothesized that the primordial VRM system arose by an autonomous, epigenetic process. Once it appeared, however, the VRM system acquired a function so important that from then on no vertebrate could do without it - as witnessed by the extinction of the "missing links". What was this biological function?

In order to answer this question, we need to consider the general biology of the first vertebrates. Briefly, the situation is the following. In the first vertebrates, the main features of the physiological organization that will be further developed and refined in the course of subsequent evolution are already laid out. On the one hand, there is a localization of discrete functions in specialized organs - heart, gills, kidney, eyes, ears, nose. On the other hand, and simultaneously, we see the emergence of "central systems" which co-ordinate these differentiated functions: the circulatory system, the central nervous system and the endocrine system.

There can be little doubt that this quantum jump in physiological complexity had some direct relevance to the relationship between the organism and its external environment. In general terms, we may suppose that these innovations were relevant to the active maintenance, by homeostasis, of a constant internal environment in the face of variation in the external environment. There is a fairly obvious selective advantage to such internal homeostasis: it permits an extension of what constitutes a viable habitat. Similarly, the central co-ordination of diverse sensory inputs also has a fairly obvious selective advantage, this time in terms of an improved capacity for locomotion and finely adapted behaviour, both in the search for food and in escape from predators.

It is important to realise, however, that such primary constraints from external selection processes must have been multiplied many times over by what we may term derivative, endogenous selection processes. Physiological complexity can surely contribute to an increased capacity for homeostasis; but at the same time, it inevitably increases the fragility of the organisms, so that it concomitantly creates an increased requirement for effective homeostasis. This is illustrated by the great vulnerability of contemporary Agnatha to sudden changes in temperature: unlike simpler invertebrates, normal methods of fishing represent for them a shock so brutal that thereafter they cannot survive more than a day or so in an aquarium (Fontaine, 1948).

The logic of this process means that physiological complexity is, to a very large extent, its own cause: an initial increase in complexity positively requires further increases in complexity. This runaway tendency is reinforced by the Malthusian propensity of living organisms that so impressed Darwin: however great the extension of a habitat, natural populations will generally proliferate until the habitat is saturated, so that they always return to a situation where they are in confrontation with the limits of their habitat. Similarly, innovative behaviour has an initial effect of permitting a diversification of ecological niches; but the very act of occupying these new niches itself creates new requirements - and new opportunities - for further innovation. This endogenous creation of self-renewing selection pressure means that complexity, and behaviour, are a motor for an ongoing and largely autonomous process of evolution, as Piaget (1976) among others has pointed out.

III.2.2. Molecular ecology.

The upshot of all this is that the over-riding, compelling evolutionary problem faced by the first vertebrates was arguably neither more nor less than that of handling and regulating unprecedented physiological complexity. We may recall also the perennity of the theme of co-ordinating competing populations of cell-lineages, and regulating the potential conflict between cell-lines and the organism as a whole. These considerations may suffice to render plausible the central hypothesis of this paper: I suggest that the principal function of the primordial VRM system was to contribute to the regulation of the internal molecular environment. It is worth pointing out that such integration is only partially and inadequately conceptualized as "homeostasis", which implies return to a single fixed-point attractor. What I have in mind is rather a dynamic emerging pattern, which may be locally homeostatic but also oscillatory or chaotic, and which globally corresponds to "drift" within a domain of viability. I propose to coin the term "molecular ecology" to of the VRM system occurs in the presence of molecular shapes which are continuously present (as is the case, *par excellence* , for the molecules of the somatic self of the whole organism), then these shapes will be harmoniously integrated into the characteristic structures of the emergent "CIS". Consequently, the CIS constitutes a "molecular identity" which integrates the somatic self; it thus corresponds to what is familiarly called the "immunological self", which reinforces our notion that the VRM system, and the CIS in particular, constitutes a process of self-assertion.

It is to be noted that, contrary to what is usually supposed, this "VRM-self" is not genetically determined. The relative stability of the CIS is dynamic, not fixed once and for all nor pre-determined; it can be and normally is continually modulated as a function of molecular events occurring in the body. Thus, the CIS integrates the total history of the molecular composition of the organism, including (without distinction) genetically encoded proteins and events that are the result of environmental perturbations (such as the effects of dietary regime). In other words, the CIS is an emergent, epigenetic entity, which is not and cannot be transmitted, as such, in the germ line.

III.3.3. Properties of the PIS.

The "peripheral immune system", or PIS, is in the first instance negatively defined: it is simply that part of the VRM-system which is not integrated or significantly connected to the CIS. The relationship between the CIS and the PIS is well illustrated by considering the fate of newly-formed, diversified lymphocyte clones (in mammals, the B-cells that emerge from the bone-marrow). If a clone of this sort interacts with the Igs of the CIS (and this will systematically be the case for all clones which are "self-reactive" in the sense of having affinity for the somatic self of the organism), then this clone will either be integrated into the CIS (in which case it will be homeostatically regulated), or, if that part of the CIS is already "saturated", the clone will be directly "suppressed" by over-stimulation. Alternatively, if a newly-formed clone does not interact significantly with the CIS, it will be stimulated only sporadically or not at all, and hence will die out more or less rapidly. The "PIS" is, initially, nothing more than the collection of these moribund clones: an incidental by-product of the increased diversity and specificity of Igs which led to the emergence of the CIS.

What then are the properties of this "PIS" ? Firstly, almost by definition, it is non-reactive with all the components of the molecular self-identity of the organism. Secondly, apart from this non-reactivity to the "self" as defined by the CIS (which now occupies a relatively small, restricted portion of shape-

designate processes of this sort, which (like eco-systems of animals and plants) possess overall coherence without necessarily being homeostatic.

My suggestion that the function of the primordial VRM-system was the institution of an internal molecular ecology is, of course, heterodox in the extreme: as I have already noted, the assumption that the principal function of the "immune system" is defence is so deeply engrained that it has given its name to the scientific discipline of immunology. But I have two major arguments against attributing a "defence" function to the primordial VRM system. The first is that the invertebrate animals which were the immediate ancestors of the first vertebrates must already have had an entirely adequate defence system, otherwise they could not have existed. The innovations associated with the emergence of the first vertebrates do not, as far as I can see, create any strikingly novel defence requirements. The relationship between a lamprey and potentially pathogenic micro-organisms would not seem to be significantly different from that of a crayfish or an octopus, for example.

The second argument is that even if there was a need for a more sophisticated defence system, the primordial VRM system as I have postulated it would not fit the bill. As I have noted, this system is incapable of mounting a significant "immune response". Moreover, this absence of an aggressive immune response is not an accidental feature. The primordial VRM system, as we have pictured it, does not yet have the capacity to make a "self/nonself" distinction: its repertoire is "complete" and essentially homogeneous. Hence if this system did respond to antigens with aggressive immune responses, it would immediately run into severe problems of auto-immunity.

On the other hand, the primordial VRM system is suited to the role of instituting a molecular ecology. As Jerne (1974) has remarked, the fact that the repertoire of the VRM-system is complete (as we have seen, for its own intrinsic reasons) means that the system will necessarily form a connected network. Thus, it will automatically play a "central system" role, co-ordinating and integrating the effects of all the molecular events that occur in the body.

III.3. "Central" and "Peripheral" VRM-systems.

III.3.1. The occupation of shape-space.

Finally, we come to the question of the emergence of the joint capacities for self/nonself discrimination and mounting immune responses. I shall suggest that these capacities did not first arise in evolution because of any particular functionality, but rather by an endogenous process. As in the case of the primordial VRM-system, it was only after these novel capacities had already arisen that they were deployed - in this case, for the classical function of defence against bacteria and viruses.

My hypothesis is that the key element in this endogenous process was increasing diversity in the V-region of Igs. The evidence from comparative immunology concerning the precise mechanisms involved is not entirely adequate (probably an increased capacity for combinatorial joining of the V, D and J regions, and more sophisticated junctional diversity), but the overall fact is not in doubt: the diversity of Igs increased steadily throughout vertebrate evolution.

In order to explain the consequence of increased diversity, it is necessary to introduce the concept of "shape-space" (Segel & Perelson, 1989). According to this concept, all possible molecular "shapes" can be represented as points in an abstract multi-dimensional "space" : molecules with similar shapes will be close to each other in this space, molecules with different shapes (and hence different patterns of molecular interaction) will be far apart. I call "metadynamics" the process by which newly generated lymphocyte clones can be selectively recruited into the active network, and existing clones can be amplified, maintained or eliminated. As a result of the metadynamics, the molecules produced by the VRM-system will occupy "shape-space" in a particular way.

In the case of the "primordial VRM-system" as I have portrayed it, the whole of shape-space is occupied (i.e. the repertoire is "complete") by a diffuse cloud of points with little internal structure. With the innovations in the generation of the diversity of VRM molecules, a much more fine-grained occupation of "shape-space" becomes possible. Under these conditions, recent theoretical investigations of the "metadynamics" of an immune network (Stewart & Varela, 1991) suggest that the occupation of shape-space may no longer be homogeneous. Specifically, in certain privileged regions of shape-space, the "shapes" are densely packed in coherent, dynamically quasi-stable configurations. By contrast, elsewhere in shape-space, the occupation is sparse and transient. In terms that are at this stage purely descriptive, we may label this distinction as one between a "central immune system" (CIS) and a "peripheral immune system" (PIS). III.3.2. Function of the CIS.

Let us now examine in more detail the intrinsic properties of the "central immune system" so defined, in order to gain some clues as to its potential function. An important and far from trivial result of our theoretical investigations (Stewart & Varela, 1991) is that if the metadynamical ontogeny space), the "repertoire" of the clones which make up the PIS is otherwise more or less complete. Thirdly, the clones in the PIS are by and large under-stimulated; moreover, they are not "connected", either amongst themselves, or to any other regulatory network. Putting these three properties together, we find that if an antigen with epitopes situated in this unoccupied, unconnected portion of shape-space is introduced into the system, it will predictably provoke a major, unlimited "immune response" on the part of certain clones in the PIS. We may define such an antigen, which does not interact with the CIS, as "foreign": i.e. by "foreign" we mean neither more nor less than "unrelated to 'self' as asserted by the CIS". Thus, the endogenous process which lead to the differenciation of the VRM-system into a CIS and a PIS has as a natural consequence that the VRM-system as a whole will mount immune responses to foreign substances (via the PIS) while being "tolerant" to somatic self-antigens (via the CIS).

III.3.4. Function of the PIS.

This emergent property - which as I have portrayed it is an incidental by-product of the emergence of the CIS - means that the PIS is a ready-made candidate well suited to the task of triggering defence reactions. It is interesting to note that the PIS made its first appearance, in amphibians and reptiles, at the time when the vertebrates were moving from water to land. Langman (1989) has argued convincingly that this move posed a serious challenge to vertebrate systems of defence against microbial pathogens. The point is that the dry environment of land is hostile for these pathogens, so that the bodies of vertebrates represent an "oasis" that is not just agreeable but a vital necessity; consequently, pathogens will not be easily dissuaded from attempting to colonize vertebrates. This argument is prolonged and strengthened when birds and mammals invent homeothermic regulation of their bodies.

On this view, the conjunction between the availability of the PIS on one hand, and a novel defence requirement on the other, was essentially an evolutionary accident. Accidental or not, however, the conjunction was certainly fortuitous: I suggest that from this time onwards the PIS was indeed progressively integrated into the defence systems of vertebrates. Initially, the role of PIS was probably limited to that of triggering pre-existing defence mechanisms; but as time went on, the PIS was more and more intimately integrated into ever wider aspects of the defence system, until we reach a state in mammals where non-functionality of the PIS actually cripples the whole defence system.

III.3.4. T-cells and MHC.

Finally, the deployment of the PIS as a trigger for immune effector functions had feedback consequences for the organization of the VRM-system itself. Cohn (in Langman1989) has forcibly made the extremely important point that if a part of the VRM-system, which as a whole can and does recognize "self", is used to trigger destructive immune responses, then the question of self-nonself discrimination becomes an absolutely central issue. He has also insisted that a single discriminatory mechanism (such that "nonself" is simply negatively defined as that which is not "self", or conversely that "self" is only that which is not "nonself") is not enough; "self" and "nonself" must both be positively identified, which implies that there must be a dual recognition system. I suggest that the evolutionary appearance of ab T-cells (derived from the ancestral gd-like lineage) is best understood in this context, i.e. as a consequence of the deployment of the PIS for defence functions.

The subsequent evolution of ab T-cells was geared to their co-operation with other lymphocytes, particularly the development of helper activity with respect to B-cells in the PIS. We may cite here the development of Class II MHC polymorphism and presenting function in B-cells, and their role in the switch of Ig isotypes (cf the "third law of the immune system" concerning class regulation, in Langman 1989). T_H-B cell collaboration in general takes place only in germinal centres; this is consistent with the fact that the appearance of well-organized lymph nodes is also a late evolutionary development.

Of particular interest is the essential role of ab T-cells in promoting somatic hypermutation of Ig-genes; T-independant antigens do not induce such mutation. Biologically, it would appear that somatic hypermutation only occurs in some types of immune response, and provides for the generation of variants in the VRMs that are rapidly amplified by selection during the course of such responses. Bacteria replicate rapidly enough to generate variants that could readily escape the original immune response; however, the rates of hypermutation of VRM genes are so high that they compensate for the difference in generation time between lymphocytes and bacteria. This is coherent with my view that the special mechanisms involved are a late evolutionary development, related to the deployment of the PIS as part of the system of defence against infectious micro-organisms.

IV. CONCLUSION.

In conclusion, I suggest that the complexities of the interactions between B-lymphocytes and ab T-cells which occur in the context of immune responses to external antigens, and which are the major focus of contemporary immunology, may correspond to a relatively late evolutionary development. I put forward the hypothesis that the system of variable-region molecules first arose in evolution by an endogenous, self-organizing process; by virtue of its intrinsic properties, this primordial VRM-system instituted an internal molecular ecology, a function so important that from then on no vertebrate could do without it.

This reversal in habitual perspectives leads to two suggestions for immunological research. Firstly, a natural (if not necessarily exclusive) approach to modelling the immune system would be to follow the steps in the actual evolutionary construction of the system. On my hypothesis, these were: (i) the emergence of self-organizing idiotypic network; (ii) the institution of a molecular ecology in relation with the somatic self; and only then, building on this basis, (iii) the differentiation between a CIS and a PIS leading to self/nonself discrimination and immune responses.

Secondly, the "primordial VRM-system" of primitive vertebrates such as lampreys may be a worthy object of experimental investigation. These animals certainly do not produce very exciting immune responses, which probably explains their relative neglect up until now. However, perhaps we have been asking the wrong questions of them: it may be that their VRM-systems lay bare, in elementary form, the foundations of an internal molecular ecology.

ACKNOWLEDGEMENTS.

This work would not have been possible without the enthusiastic collaboration and immense knowledge of Antonio Coutinho. I wish to thank Francisco Varela for constructive discussion, notably on the themes of unicellular vs multicellular organization and self-assertion.

REFERENCES.

Buss LW (1987) The Evolution of Individuality. Princeton University Press, Princeton.

Du Pasquier L (1989) Evolution of the immune system. In: W.E. Paul (ed) Fundamental Immunology, Raven, New York, p139

Edelman G (1987) CAMs and Igs: cell adhesion and the evolutionary origins of immunity. Immunol Rev 100:11-45

Fontaine M (1948) Formes actuelles des cyclostomes. In: Traité de Zoologie, vol. XIII. Masson, Paris, p 13

Jerne NK (1974) Towards a network theory of the immune system. Annales d'immunologie (Institut Pasteur) 125C:373-389

Klein J (1982) Immunology : the science of self-nonself discrimination. Wiley, New York

Langman RE (1989) The immune system. Evolutionary principles guide our understanding of this complex biological defense system. Academic Press, San Diego

Perelson AS (1984) Some mathematical models of receptor clustering by multivalent ligands. In: Perelson AS, Delisi C, Wiegel FW (eds) Cell Surface Dynamics, Concepts and Models. Marcel Dekker, New York, p 223

Piaget J. (1976). Le comportement, moteur de l'évolution. Gallimard, Paris.

Segel LA, Perelson AS (1989) Shapespace analysis of immune networks. In: Goldbeter A (ed) Cell to Cell Signalling: from experiments to theoretical models. Academic Press, New York

Stewart J, Varela FJ (1991) Morphogenesis in shape-space: elementary meta-dynamics in a model of the immune network. J Theor Biol (in press).

A SIMULATION OF THE HUMORAL IMMUNE SYSTEM

Philip E. Seiden
IBM Research Center
P.O. Box 218,
Yorktown Heights, NY 10598, USA

Franco Celada
Hospital for Joint Diseases
301 E. 17th St.
New York, NY 10003, USA

INTRODUCTION

The idea of theoretically describing the immune system has become popular in the past decade, and in fact the existence of the present volume is evidence for the emergence of the field. With the advent of modern high-speed computers it no longer seems impossible to model and simulate the behavior of the immune system. Previous attempts at modeling the immune system have generally followed the course of focussing on a phenomenon or class of phenomena, writing down the differential equations for these phenomena and then trying to obtain solutions (Perelson 1988, 1990). This approach has led to better understanding of the phenomena and how they might be applied to the immune system, but often the mathematical approximations required to obtain solutions have removed much of the connection with the actual biological system, or at least made the connections much more opaque.

There has also emerged another class of calculations based on the cellular automaton approach (Kaufman *et al.* 1985, Weisbuch & Atlan 1988, Neumann 1989, Chowdhury & Stauffer 1990). This type of calculation, although appearing very abstract at the outset, actually allows one to probe the connections and relations between entities in the immune system. They allow one to determine what might happen in a system having the set of connections assumed in the computation.

A major difficulty that raises its ugly head is that the immune system is *messy*. The simple clean models usually embraced by theorists do not often express all the complexity or special cases encountered in biology. By its very nature a mathematical model or computer simulation must be a considerable abstraction from the concerns of a practicing immunologist. However, we want to create a model which has enough resemblance to the immune system so that the interaction between the theorist and experimentalist can be carried out in the language of immunology rather than mathematics. This, we believe, will make the interaction between theory and experiment stronger and easier to manage.

NATO ASI Series, Vol. H 66
Theoretical and Experimental Insights into Immunology
Edited by A. S. Perelson and G. Weisbuch

Our approach springs from the cellular automaton idea and has much of its spirit. We refer to it as a *generalized* cellular automaton although its implementation is far more complex than the automata usually analyzed by mathematicians.

It is not possible to model the entire immune system. That is far beyond the capabilities of even today's supercomputers. Therefore, it is necessary to decide on what functions of the immune system in which one is interested and build a model focussed on those functions. We are primarily interested in the cell-cell interactions in the immune system, in particular, those interactions that lead to the humoral immune response. We start by building a system that corresponds to a consensus of what immunologists today believe is at the heart of the immune response. Therefore, our model will focus on T cells, B cells and professional presenting cells. Although we start from the premise of clonal selection we will try very hard to allow the possibility and implementation of other types of interactions to be introduced. We want our model to be a test bed to be able to examine various immunological ideas.

Experimental immunology is very time consuming, labor intensive and expensive. Therefore, many desirable experiments never get performed. This means that many interesting ideas do not get tested and many interesting experimental results do not get confirmed or searched for in other systems. We thought that it would be helpful to be able to build a system where experiments on the immune system may be performed in a computer in order to predict the effects of experimental modifications on the immune system, We would like our technique to be used as a third experimental technique - experiments *in machina* - experiments in the computer. It is not that we believe that our computer experiments will be able to give definitive answers to replace biological experimentation, but that a fast inexpensive large series of experiments that can be done in a computer will be able to determine the area of influence of an idea (such as a a new interaction or entity), and thus offer a criterion for the selection of the most likely meaningful tests to be done *in vivo* or *in vitro*. That is, *in machina* experiments may precede, guide, help interpret -- but obviously not substitute for -- the biological experiment.

THE MODEL

The system we choose is a *modified* or *generalized* cellular automaton. The spirit of the model is certainly along the lines of a cellular automaton since we allow the state of each object to be modified by the local interactions between the neighboring objects. A traditional cellular automaton has the five governing rules given below (Wolfram 1984).

They consist of a discrete lattice of sites.
They evolve in discrete time steps.
Each site takes on a finite set of possible values.
The value of each site evolves according to the same *deterministic* rules.
The rules for the evolution of a site depend only on *a local neighborhood of sites around it.*

Our generalizations are the following:
Deterministic is replaced by *probabilistic*.
A local neighborhood of sites around it. is replaced by *the site itself.*
And we add a sixth rule - entities can move from site to site.

The model starts out by establishing a 2-dimensional triangular grid that represents the body. It is populated with a number of distinct entities of several types. Interactions, which are restricted as to the type of the entities involved, can take place only between entities in the same compartment of the grid. As is commonly done for cellular automata, all entities in the system are examined, the rules for interaction are applied, and the results to be obtained are determined for all entities. Then all changes are made simultaneously. The entities are then allowed to move by diffusing from one region to another. This constitutes a *time step*. The process is then repeated for as many time steps as desired.

The Cast of Characters

The cellular entities that we include are T cells, B cells and nonspecific antigen presenting cells (A cells) such as macrophages. The cells are endowed with binary receptors.[1] For the results reported here all receptors contained eight bits. A schematic of the three types of cells is shown in Figure 1. The eight bit segment labelled receptor is the specific clonotypic receptor which T and B cells possess. In addition we include MHC molecules which also have the form of eight bit receptors. A B cell contains both a receptor and MHC molecules. A T cell has only a receptor. An A cell is nonspecific so it does not have a specific receptor but it will have the same MHC molecules as possessed by the B cell. In addition an A cell possesses an Fc receptor. It is not specifically denoted on the cell since it is the same for all A cells. However, it is allowed to bind to the Fc region of an antibody when the antibody is complexed to an antigen.

Antigens are represented by a number of eight bit segments representing epitopes and peptides. The epitopes are the segments to which the B and A cells can bind. A B cell will bind if its receptor matches the antigen in a complementary fashion, i.e., zero binds to one and *vice-versa*. The peptides are the segments of the endocytosed antigen that are presented on the MHC molecule. In reality they are obtained by breaking down the antigen. However, since we do no biochemistry in the model we must specify the peptides along with the epitopes when defining the antigen. Examples of antigens are shown in Figure 2.

---

[1] The idea of a binary receptor was introduced by Farmer *et al.* (1986). In their model they allowed a *sliding* interaction between entities. For example, for an N-bit antigen and an M-bit receptor, with $M<N$ they would look for the best M-bit fit available along the N-bit antigen. We insist that the antigen and receptor be in register and be of the same length. An antigen may consist of P-segments but each segment consists of the same number of bits and each segment is presented to a receptor in the same alignment.

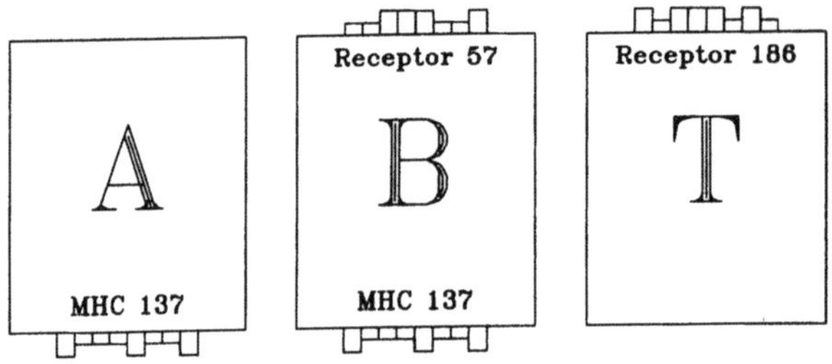

Figure 1. **A schematic of an A, B, and T cell with eight bit receptors: The integer names of the eight bit segments are the decimal values of the binary strings.**

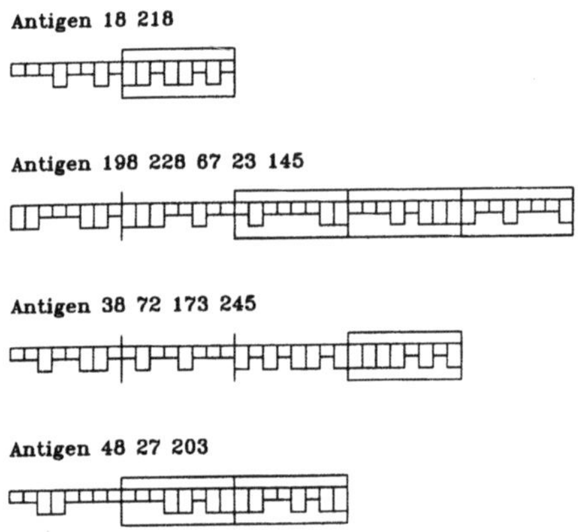

Figure 2. **Examples of antigens for eight bit segments: The epitopes are shown as exposed segments and the peptides as boxed segments. The integer names of the eight bit segments are the decimal values of the binary strings.**

Under the proper conditions, as described below, our B cells will produce antibodies. These antibodies will contain an epitope identical to the receptor of the B cell producing it and a peptide. They also contain an Fc epitope which need not be explicitly noted since it is the same for all antibodies.

The Processing of Antigen

The core of our model is concerned with the capturing and processing of antigens and how that processing affects the population and activity of the cellular components. A cartoon of the steps in antigen processing by B cells is shown in Figure 3. Antigen capture and processing are performed by B cells that bear a receptor specific for the antigen. To begin a B cell attempts to bind to the epitopes of the antigen. This binding depends on the number of matching bits. Typically we allow the binding strength to decrease with the number of mismatching bits. In the data to be shown in this paper the interaction strength is unity for an eight bit match, 0.05 for a seven bit match and zero otherwise.

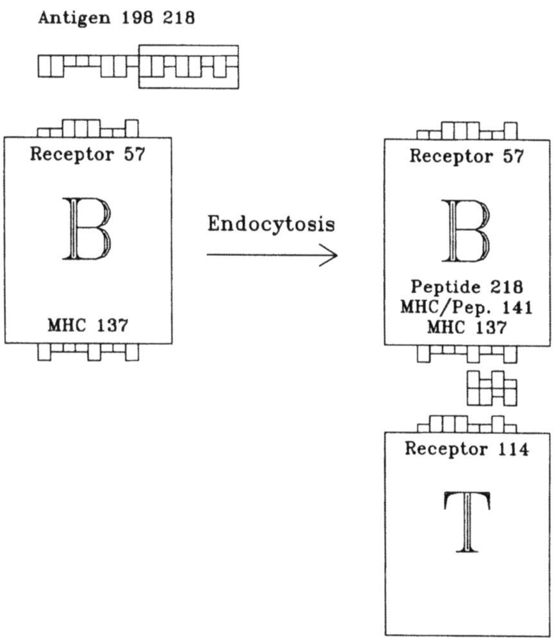

Figure 3. Steps in the recognition and processing of antigen by B cells and T cells.

The binding and processing of antigens can also be carried out nonspecifically by A cells in a two ways: directly with a low interaction strength (0.001) or in an antibody facilitated fashion (by means of an Fc receptor) with intermediate interaction strength (0.02). Once the antigen is bound by either a B or A cell the peptides are processed for presentation by the MHC.

In the immune system a T cell (class II helper cell) does not recognize free antigen but rather recognizes the presence of antigen by seeing peptides presented by the MHC molecules. A T cell binds to the MHC/peptide complex so that part of the T-cell receptor binds to the presented peptide and part to bare regions of the MHC. To simulate this we divide the MHC molecule in half. The left-hand four bits represent the bare part of the MHC to which the T-cell receptor will bind. The right-hand four bits represents the MHC groove onto which the peptide binds. If there is one MHC and one peptide the binding can take place in two ways, either the leftmost four bits or the rightmost four bits of the peptide can bind to the right-hand four bits of the MHC. The binding strength in this case also declines with mismatching bits, but not so strongly. Typically the interaction strength declines by only a factor of two for each mismatching bit. When binding takes place between the MHC and one half of the peptide the other half is presented to the T cell. Therefore, as shown in the right hand side of Figure 3, the T cell sees an eight bit segment consisting of the left-hand four bits of the MHC and the four bits of the peptide which did not bind to the MHC. The T-cell receptor is then allowed to bind to this MHC/peptide complex with the same rules as a B-cell receptor binding to an epitope.

When a successful T-cell binding takes place both the T cell and B cell are allowed to divide to establish clones of their own type (if presentation takes place by an A cell only the T cell divides). Typically we allow the T cells to divide three times resulting in eight cells. The B cells divide four times but half become B-memory cells (as for the T cells) while half become plasma cells and produce antibody.

Cellular Diversity

In the model self-nonself discrimination is obtained by preselection of one of the two lymphoid populations. The B cells are chosen randomly with complete diversity (i.e., 256 types for eight bits). The T cells are also chosen in the same way but then they are passed through the *thymus*. In our thymus they are exposed to self-epitopes and self-peptides and if they bind they are eliminated proportional to the strength with which they bind. Since the thymus is a dense organ we can allow the same T cell to be exposed to all the MHC/peptide combinations many times. If the number of times is great enough the thymus will be very efficient and most of the self-reactive T cells will be eliminated. As in the immune system our thymus is the first defense against autoimmune disease.

A typical experiment is set up as follows. First the system is populated with the desired number of A, B, and T cells. Second a schedule of *injections* of antigen is chosen. Third the simulation itself begins. In each square of the body all possible interactions are considered. If a single entity is capable of having more than one successful interaction the one that actually happens is determined stochastically. After all interactions are decided upon the birth of new cells is allowed to proceed. In addition to clonal growth a number of new virgin cells are born. Cells are also allowed to die depending on their age. After all births and deaths the entities can diffuse into the surrounding squares. This constitutes a time step. The process can be repeated for as many time steps as desired.

The above is only an abbreviated description intended to let the reader understand the nature of our approach. A fuller more detailed description can be found in Seiden and Celada (1992).

This is a good place to reemphasize the experimental nature of the model. We have mentioned a number of variables, e.g. the types and number of cells and antigens and the interaction strengths between them. These are all available for experimentation, they can be varied at any time, either singly or in combination, and the results of such variations observed. Thus for any effect observed, the source can be tracked down and the cause of any behavior can be precisely determined in the context of the model. Lastly, new entities (e.g. killer T cells) and interactions (e.g. idiotypic/anti-idiotypic interactions, see below) can be added as desired. The model is designed as an experimental system and we expect to use it in such a manner.

RESULTS

Immunization

The first result we will discuss is that of immunization. To show how this works we inject a system with the same antigen twice, once at time step zero and the second at 100. The results of the experiment are shown in Figure 4. At the first injection the response is slow as it takes time for the clones of the responding cells to develop from their small virgin populations. A small amount of antibody is produced but it takes some time for the antigen to be removed from the system. At the second injection, however, a large number of the appropriate B and T cells are present so that the response is rapid. Much more antibody is produced and the antigen is eliminated quite rapidly.

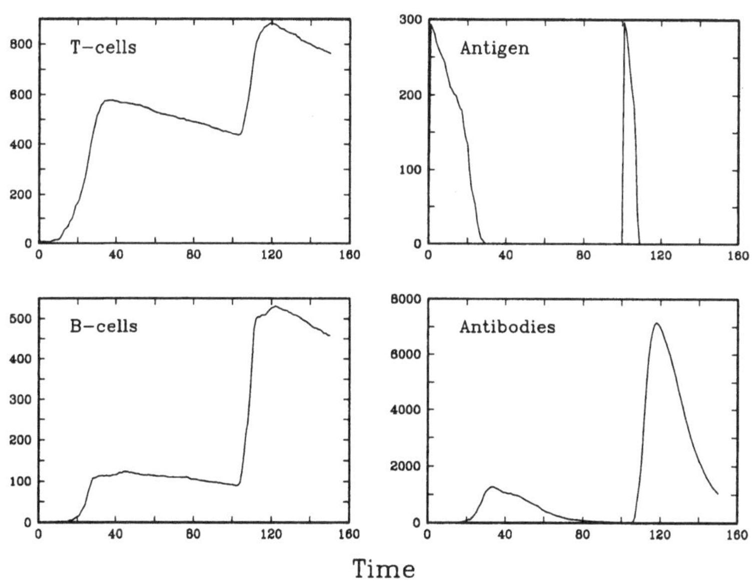

Figure 4. Time track of the cellular, antigen, and antibody population of a system injected with antigen at time steps 0 and 100.

Figure 4 shows the total antibody produced. However, there are nine different kinds of antibodies that can be produced (for a one epitope antigen), one perfect eight bit match and eight different seven bit matches. Figure 5 shows each antibody plotted separately. We call the eight bit match the high affinity antibody and the seven bit matches the low affinity antibodies. The solid line in the figure shows the high affinity antibodies and the other eight curves show the eight low affinity antibodies. An interesting point to note is that the high affinity antibodies do not dominate the primary response and in fact a number of low affinity antibodies have a larger population than the high affinity antibody. It is the secondary response that is dominated by the high affinity antibodies. This is because at the time of the challenge there is a larger pool of specific cells and in the competition for antigen it is the high affinity B cells that are favored. This happens without implementing affinity maturation via hypermutation which would accentuate the population shift even more. A similar result was found by Bell (1970) where affinity maturation occurred if the antigen dose was prolonged over long times.

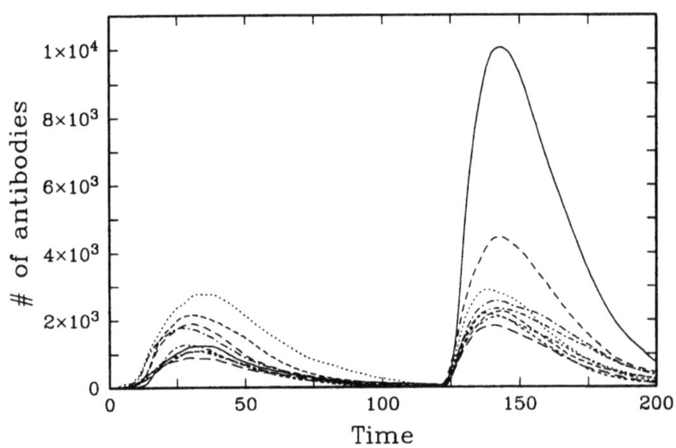

Figure 5. Two examples of antibody responses to immunization experiments: The high affinity antibodies (8-bit matches) are shown as the solid lines. The other eight lines show individually the eight low affinity antibodies (7-bit matches).

It should be noted here that these experiments (as well as those described below) are stochastic. Since many of the interaction strengths are not unity the decision as to whether an interaction takes place or not is made by choosing a random number between zero and one and comparing it to the interaction strength. This means that the results will differ for a different set of random numbers. The general character of the results will be the same but the specifics can differ. For example, for a different set of random numbers the number of high-affinity antibodies in the primary response in Figure 5 can vary to greater or lesser values so that they can be relatively more or less important for the primary response. However, the fact that the high-affinity antibody strongly dominates the secondary response is unchanged.

The regulation of the response

A relatively simple experiment that can become very informative about the functioning of the model is to expose the system to large variations in the antigen dose used for the first immunization and for the challenge. A number of experiments similar to the one above were conducted in which the first and second antigen doses were varied independently over a range from 10 to 10^5. Figure 6 shows the results represented as the peak antibody titer obtained after the secondary stimulation and expressed as the number of antibodies divided by the number of antigens injected. The graph on the left shows the high affinity antibodies and the one on the right the total antibody population (sum of the high affinity antibody and all eight low affinity antibodies).

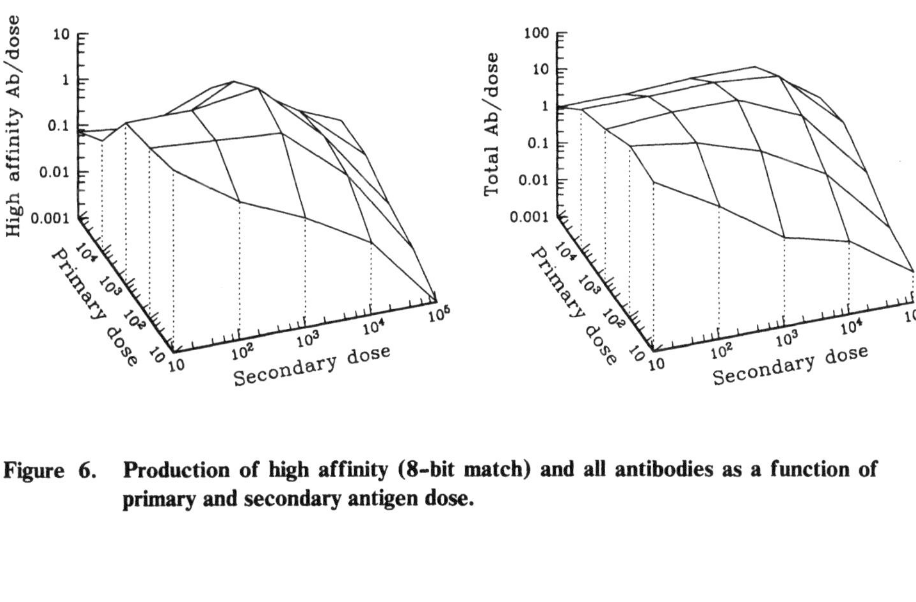

Figure 6. Production of high affinity (8-bit match) and all antibodies as a function of primary and secondary antigen dose.

Two facts emerge from the analysis of these data. First, there is an optimal response at intermediate doses which is different for high affinity and low affinity responses. The high affinity response peaks for a primary dose of 10^3-10^4 and a secondary dose of about 10^3, while the total antibody response peaks at a dose a factor of ten higher for both the primary and secondary response. Second, at higher than optimal doses the response declines, this being particularly evident for the high affinity antibodies which fall almost to zero. This is very reminiscent of the *in vivo* situation discussed in the article by F. Celada in this volume. He shows that the regulation of the response at high antigen doses and the differential sensitivity of the high and low affinity responses were observed in experiments conducted over 20 years ago.

The idiotypic network

The idiotypic network (Jerne 1974) is a very popular topic both in immunology and among theorists. Reviews of the theory of immune networks can be found in Perelson (1989) and Varela & Coutinho (1991) and a paper on the subject by G. Weisbuch is contained in this volume. The importance of the idiotypic network in shaping and regulating the immune system is not clear. Generally the work on networks and on clonal selection have been done separately. However, these two ideas are not contradictory and constitute two original contributions of immunology to the biological sciences. A serious theoretical treatment of the immune system should consider both simultaneously. The present model is particularly suitable for this task.

In the results that have been presented so far there has been no idiotypic network because we have specifically not allowed antibodies to behave as antigens. However, antibodies are just proteins and so can appear as antigens to the immune system. The constant regions should not induce any antigenic activity since they are self to which the immune system should be tolerant. The variable regions, however, are meant to be able to match foreign epitopes so they are clearly not self. We will now examine this question by turning on the B cell/antibody interaction. This will allow the B cell having a receptor that is the complement to an antibody receptor to bind to it and process the antibody as an antigen. This means that an antibody of a given idiotype can excite an anti-idiotypic B cell which will, in turn, produce anti-idiotypic antibodies. These anti-idiotypic antibodies will then reexcite the original B cells producing idiotypic antibodies which will continue the process. In fact, since a given idiotype can excite nine different B-cell types there will exist a network containing many idiotype/anti-idiotype pairs.

If the strength of the B cell/antibody interaction is the same as the B cell/antigen interaction the system is unstable and many clones grow to extremely large populations, indeed they will grow without bound unless stopped by other processes. We can examine this behavior more closely by looking at the behavior of the system as a function of the strength of the B cell/antibody interaction. This is shown in Figure 7 where we present the maximum number of antibodies produced as a function of this interaction strength. Small values of the interaction strength behave in a manner similar to the experiments of Figure 4 where the interaction strength was zero. Then at a critical value the system becomes unstable and the antibody populations experience uncontrolled growth. These data tell us that to consider antibodies simply as antigens with no special control provisions is not realistic, since the model behaves in a fashion quite different from the observed immune system. Thus a control mechanism is needed. Unfortunately, we do not know what control is operating in nature.

One possibility is B cell anergy. It is known that B cells can become anergic if subjected to too many antigenic interactions. The growth of the antibody population may be the source of the anergy, that is, it is triggered by the excessive bombardment of B cells with stimulating antibodies produced by the network interactions. We have tested whether anergy alone can bring the system under control (anergy in this case consists in the cessation of proliferation of a B cell if it experiences a certain number of antigenic *hits* during the same time step). Figure 8 shows that under these circumstances the system is brought under control. Instead of going unstable the initial large response is followed by a number of smaller oscillations in antibody production. This scenario may seem more realistic than the antibody explosions of Figure 7. However, anergy spreading onto a great number of B clones could have severe effects on the capacity of an individual to respond to external stimuli.

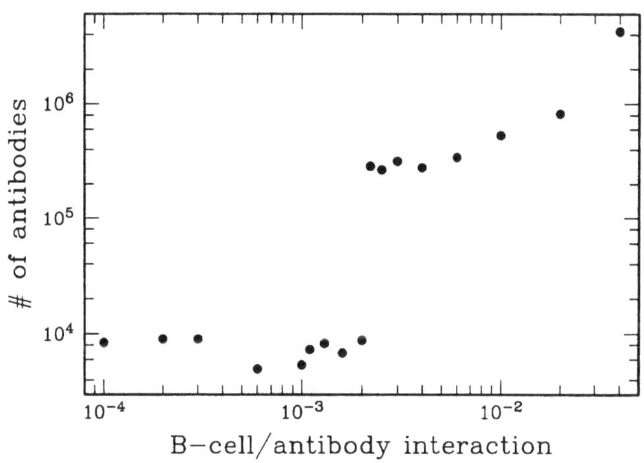

Figure 7. The maximum number of antibodies produced as a function of the B cell/antibody interaction strength: A limit is placed on the maximum population of cells per site in the body. Without this the number of antibodies would rise without limit. This limit has no effect for low values of the abcissa.

Other regulation mechanisms are possible. We mentioned above that the part of antibody associated with the receptor is certainly foreign. On the other hand, it is possible that the antibody may be poorly equipped with T epitopes, or may not be processed efficiently to yield active peptides since it occurs in the context of the rest of the antibody molecule which is primarily self. This situation may act on or safeguard against excessive anti-idiotypic interactions. This possibility will be tested in the future along with other alternatives. Clearly we have only begun to examine the problem, much more work is needed to understand the interactions between the network and clonal selection and what control mechanism or combination of mechanisms are actually working in nature. The model will allow a detailed exploration of the domain and influence of each process and mechanism as well as any synergism that may arise.

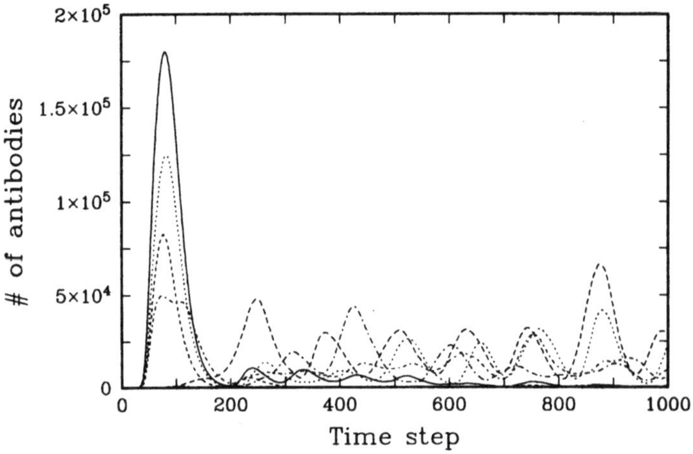

Figure 8. Oscillations in the antibody populations with time in a strongly anergic system with a large B cell/antibody interaction.

The size of self

Using a computer model it is also possible to carry out experiments that are extremely difficult or even impossible to carry out by normal *in vitro* or *in vivo* techniques. The example that we describe here is examines the possible size of self in relation to the existing diversity. We want to ask how many self peptides can we have relative to the possible number of peptides. We are not quite sure what the possible diversity of peptides is but from a simple combinatorial point of view we could have as many as $5x10^{11}$ different types for a peptide of nine amino acids. The actual effective diversity might be much different because it may not be possible to process and present all these peptides. The actual number will not be of concern here, what we will attempt to determine is the fraction of presentable peptides that can be self and still have an effective immune system. Clearly, if self encompassed the total possible diversity then we would have no clonal immune system because all the T cells would be eliminated in the thymus.

We carried out the following experiment. We picked a panel of 100 random antigens chosen from the full diversity of the system (256 for an 8-bit system). Then we subjected our system to these random antigens and calculated the *responsivity* of the system. The responsivity is the effectiveness with which the system attacks the panel of antigens. It is defined as:

$$Responsivity = \frac{1}{K}\sum_{k=1}^{K}[1 - \prod_{p}\prod_{m}\prod_{t}(1 - S_p S_t n_t)]$$

where the product p is over the panel of antigenic peptides, the product m over the MHCs, and the product t over the T cells. S_p is the binding strength of the peptide to the MHC, S_t the binding strength of the T cell to the MHC/peptide complex, and n_t the population of T cells of type t after thymus selection. The system we chose had two MHC molecules chosen at random and a variable number of self peptides also chosen at random. Since we are interested in the response of a typical system with the number of peptides chosen and not in a particular set of peptides, we ran the experiment 100 times with different choices of MHC molecules and self peptides (different types but the same number). The final responsivity is the average over all experiments (sum over k with $K = 100$) and represents the responsivity of an immune system containing that particular number of self peptides.

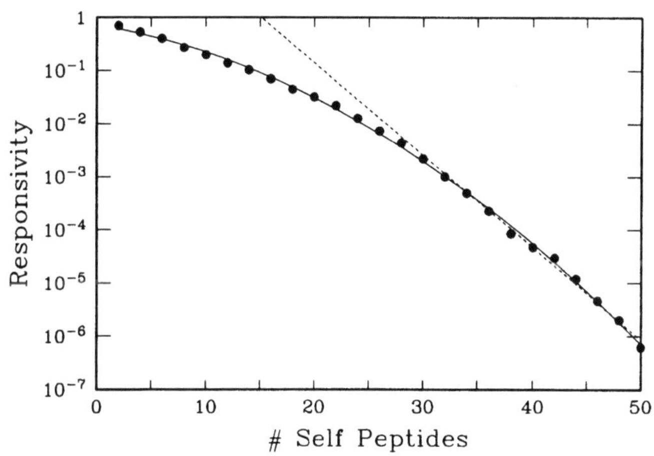

Figure 9. Response of the model immune system as a function of the number of self peptides.

We repeated the experiment for a number of self peptides running between 1-50. The results are shown in Figure 9. The responsivity drops slowly at first and then increases to approximately exponential at which point the responsivity drops so low that the immune system is ineffective. This drop is due to the reduction of the T-cell population by increased negative selection in the thymus. A rough approximation of how large a self pool is compatible with an effective immune system can be obtained by extrapolating the exponential back to unity. The intercept is at 15 peptides or 6% of the repertoire of 256. This will be an interesting number to compare with biological estimates.

CONCLUSIONS

We have described a working simulation that is capable of exploring the domain of cell-cell interactions in the immune system. We intend it to be an experimental system in which ideas about cellular interactions in the immune system can be tested quickly and cheaply. The system has performed well so far and has yielded a number of insights into the immune system and confirmed the origin of other effects previously known (Celada & Seiden 1992, Seiden & Celada 1992). Other entities and functions can easily be added as desired. We look forward to many other explorations in various areas of immune interactions.

REFERENCES

Bell GI (1970) Mathematical model of clonal selection and antibody production. Nature 228:739-744

Celada F, Seiden PE (to be published) A computer model of cellular interactions in the immune system. Immun. Today

Chowdhury D, Stauffer D (1990) Systematics of the models of immune response and autoimmune disease. J Stat Phys 59:1019-1042

Farmer JD, Packard NH, Perelson AS (1986) The immune system, adaptation, and machine learning. Physica 22D:187-204

Jerne N (1974) Towards a network theory of the immune system. Ann. Inst. Pasteur Immunol. 125C:435-441

Kaufman M, Urbain J, Thomas R (1985) Towards a logical analysis of the immune response. J Theor Biol 114:527-561

Neumann AU (1989) Control of the immune response by a threshold automata model on a lattice. Physica A 162:1-19

Perelson, AS (ed) (1988) Theoretical immunology. Addison-Wesley, Redwood City

Perelson, AS (1989) Immune network theory. Immunol. Rev. 110:5-36

Perelson, AS (1990) Theoretical immunology. In: Jen E (ed) 1989 Lectures in Complex Systems, SFI Studies in the Sciences of Complexity, vol. II. Addison-Wesley, Redwood City, p 465

Seiden PE, Celada F (submitted) A model for simulating cognate recognition and response in the immune system. J Theor Biol

Varela FJ, Coutinho A (1991) Second generation immune networks. Immunol. Today 12:159-166

Weisbuch G, Atlan H (1988) Control of the immune response. J Phys A 21:L189-192

Wolfram S (1984) Preface to Cellular Automata. Physica 10D:vii-xii

Probability of Self-Nonself Discrimination

J. K. Percus
Courant Institute of Mathematical Sciences
and Department of Physics
New York University
New York, New York 10012

Ora E. Percus
Courant Institute of Mathematical Sciences
New York University
New York, New York 10012

Alan S. Perelson
Theoretical Biology and Biophysics
Theoretical Division
Los Alamos National Laboratory
Los Alamos, NM 87545

Abstract

The binding of antibody or T-cell receptors to antigen occurs by a generalized lock and key fit of portions of the two structures. The question that we address here is: how large should the complementary regions on the two structures be? In order to estimate the size of an optimal receptor combining region, we assume that the mammalian immune system over evolutionary time has been presented with a large random set of foreign antigens that occur on common pathogens, which it must recognize, and a smaller random set of self-antigens which a mature organism must not recognize. Evolution is imagined to have coevolved the receptors in a fashion such as to maximize the probability that this task is performed. The probability of a random receptor-antigen match is estimated from this condition. For protein antigens, the genesis of the probability is traced to the complementarity of sufficiently long sequences of amino acids on the two molecules involved, and computed accordingly. The result is quite insensitive to the population sizes inserted, and results in an estimated 13-site binding region, in agreement with experimental information.

NATO ASI Series, Vol. H 66
Theoretical and Experimental Insights into Immunology
Edited by A. S. Perelson and G. Weisbuch

Introduction

In studying quantitative aspects of the mature mammalian immune system, one is struck by the large numbers — of species or categories, not merely of cells or molecules — that one has to deal with. For the mouse, it is estimated that there are (Inman, 1978; Coutinho, 1980; Perelson and Oster, 1979)

$$\begin{aligned} N &\sim 10^{16} && \text{foreign antigens to be recognized} \\ N' &\sim 10^{6} && \text{self-antigens to be ignored} \\ n &\sim 10^{7} && T\text{-cell receptor and immunoglobin variable} \\ & && \text{region sequences to carry out these tasks.} \end{aligned} \tag{1}$$

To complete the description, each receptor must recognize suitable complementary motifs or epitopes on some $10^{16}/10^{7} = 10^{9}$ antigenic species. But N, while a very large number from this functional viewpoint, is very small in comparison with the number of molecular species it is conceptually drawn from, and even small in comparison with the variety of epitopes — of the order of $10^{40} - 10^{60}$, we will later estimate — that nature might provide. The value of n, it will develop, is indeed a large number. We will return to the consequences of these observations.

Eqs. (1) represent the culmination of a long evolutionary process. We will argue that the required coevolution of the molecules represented by N, N', and n (the self-epitopes, for example, are unusual molecular segments from the viewpoint of foreign epitopes: Chalufour et al., 1987; Claverie et al., 1988) must have left an indelible mark on the overall nature of the physical processes by which the tasks of (1) are carried out. In particular, we will consider the following questions:

1. Is optimal coevolution consistent with current data?
2. Is the implied nature of the molecular combining region for recognition consistent with current data? (2)

The Nature of Optimality

We have no intention of guessing the precise mechanism of the coevolution of the antigen and receptor populations. However, the general strategy implied by the term convolution is that random populations will, by evolutionary excision, adjust their internal characteristics to maximize the rate at which a viable combined population is attained. Under these circumstances, we surmise that it is the probability

$$\Pr(N, N'; n) \tag{3}$$

of an n-receptor population performing its required tasks of recognizing N random foreign antigens and not recognizing N' self-antigens that it is to be maximized by choice of internal parameters. But since this state is aimed at by evolution, under circumstances in which the antigen populations are increasing, then the dominant receptor population should be characterized by its ability to process increasing antigen populations, i.e. by the conditional probability

$$\Pr(N, N' \,|\, N - \Delta N \;,\; N' - \Delta N'; \, n) \tag{4}$$

of successfully handling N and N' given that $N - \Delta N$ and $N' - \Delta N'$ are successfully handled.

Now it is certainly reasonable to assume that reduced antigen populations will automatically be handled successfully:

$$\Pr(N - \Delta N, \, N' - \Delta N' \,|\, N \;,\; N' \;;\; n) = 1 \;. \tag{5}$$

It follows from the standard theorem of compound probability

$$\Pr(A|B) = \Pr(A)\Pr(B|A) \,/\, \Pr(B) \tag{6}$$

that it is the ratio

$$\begin{aligned} &\Pr(N, N' \,|\, N - \Delta N, \, N' - \Delta N'; \, n) \\ &\quad = \Pr(N, N'; n) \,/\, \Pr(N - \Delta N \;,\; N' - \Delta N' \;;\; n) \end{aligned} \tag{7}$$

that is to be maximized. There is a crucial difference between (3) and (7). As we shall see, (3) is extremely small, to the extent that it is not possible to visualize a selection process that will be sensitive to its value. On the other hand, (7) need not be imperceptibly small, depending to be sure on the rate of change of N and N' during effectively stationary n. Thus, such a process can indeed be imagined, even though in practice, it still amounts to the separate optimization of (3). Our argument, it is clear, is of quite general validity, not restricted to the special circumstance under investigation.

Reduction to Elementary Events

How is $\Pr(N, N'; n)$ to be computed? For this purpose, it is important to return to the observation that N, the estimated number of foreign antigenic species encountered, while very large compared to unity, is very small compared to the full class of antigens, say of given length. Hence, we can choose this population, as well as that of the self-antigens, randomly from the full class of sequences, with the assurance that all but a negligible fraction will be distinct. Let us then define the elementary probability P_S that a random receptor recognizes a random antigen, and the corresponding complementary probability

$$P_F = 1 - P_S \tag{8}$$

that a random receptor fails to recognize a random antigen. As a consequence of the above discussion, we then have

$$\begin{aligned} \Pr(N, N'; n) &= \Pr(\text{each of } N \text{ antigens is recognized} \\ &\qquad \text{by one or more of } n \text{ receptors}) \\ &\quad \cdot \Pr(\text{none of } N' \text{ antigens are recognized} \\ &\qquad \text{by any of } n \text{ receptors}) \\ &= (1 - P_F^n)^N \, P_F^{nN'} \; . \end{aligned} \tag{9}$$

The internal operation of the recognition process is hidden in the value of the parameter P_F. If (9) is maximized with respect to P_F, we obtain at once

$$\begin{aligned} P_{max} &= \left(\frac{N}{N+N'}\right)^N \left(\frac{N'}{N+N'}\right)^{N'} \\ \text{at } P_F &= \left(1 + \frac{N}{N'}\right)^{-1/n} \sim 1 - \frac{1}{n} \ln\left(1 + \frac{N}{N'}\right) . \end{aligned} \tag{10}$$

In particular, using the data of Eq. (1), we have $P_S \sim \frac{1}{n} \ln(1 + \frac{N}{N'}) \sim 2.3 \times 10^{-6}$ [and incidentally, $P_{max} \sim 1/(10^{10}e)^{10^6}$]. This is somewhat smaller than empirical estimates. For example, Cancro et al. (1978) estimate that in $BALB/c$ mice, there are ~ 13 anti-influenza hemagglutinin specific B cells per 10^6 splenic B cells, i.e., $P_S \sim 13 \times 10^{-6}$. Since hemagglutinin is a protein found on the surface of a naturally occurring virus, it is a good choice for comparison with our theory. However, as a protein, it may have multiple epitopes, which may explain the five-fold higher measured frequency than our theory predicts.

The Antigen-Receptor Combining Region

To proceed further, we must characterize the principal recognition event. Since T-cell epitopes appear to be restricted to oligopeptides (Berzofsky and Berkower, 1989), and since peptides are a major fraction of B-cell epitopes, we will for definiteness restrict our attention to the class of protein antigens. Available evidence is certainly consistent with the straightforward process of activation by chemical affinity between significant portions of the amino acid sequences that specify exposed portions of antigen and receptor. A caricature of the measure of affinity might posit m types of amino acid combining site on the receptor, each complementary to precisely one out of m on the opposing antigen, and might demand an unbroken sequence of at least r complementing pairs (Kabat, 1976) to specify a matching antigen and trigger the required activation. Let us see the consequence of this assumption on the value of the successful recognition probability P_S.

If we denote a matching or complementary pair by x — with probability $1/m$ — a non-matching pair by y — with probability $(m-1)/m$ — and imagine a sequence of ℓ sites shared by antigen and receptor, we are then asking for the probability of at least one sequence of at least r x's out of the total of ℓ entries. This matching problem has a venerable history, going back at least to de Moivre (cf. Uspensky, 1937). We will solve it in a fashion which makes relaxation of the defining conditions relatively easy to carry out. It is convenient to switch consideration to $P_F = 1 - P_S$, and hence to the probability of having no sequence of x's longer than $r-1$ out of a sequence of ℓ x's and y's. Let us not worry about ℓ for the moment and simply write down all sequences of x's and y's satisfying this condition, a typical one being the set of $2t+1$ groupings

$$x^{a_1} y x^{a_2} y \cdots x^{a_{t-1}} y x^{a_t} \quad \text{where } t \geq 1 \,, \quad 0 \leq a_j \leq r-1 \,. \tag{11}$$

Summing over all $\{a_j\}$, we have $\frac{1-x^r}{1-x} y \cdots y \frac{1-x^r}{1-x}$, and then over t, $1/(\frac{1-x}{1-x^r} - y)$, giving us the generating function for all such sequences

$$G(x,y) = \frac{1-x^r}{1-x-y(1-x^r)} \,. \tag{12}$$

Now we assume that x has the probability weight $1/m$, y has $(m-1)/m$. If we give x and y a further weight z, we are then able to extract sequences of length ℓ by just taking the coefficient of z^ℓ. Hence, making the replacement $x \to (1/m)z$, $y \to (m-1/m)z$ in (12), we have

$$P_F = \text{coef } z^\ell \text{ in } \frac{1-\lambda z^r}{1-z+\lambda \frac{m-1}{m} z^{r+1}} \quad \text{where} \quad \lambda = m^{-r} \,. \tag{13}$$

Expanding in λ as a small parameter,

$$P_F = \text{coef } z^\ell \text{ in } \frac{1}{1-z} - \lambda \Big[\frac{m-1}{m} \frac{z^{r+1}}{(1-z)^2} + \frac{z^r}{1-z} \Big] + \cdots \,, \tag{14}$$

which we evaluate at once as

$$P_F = 1 - \lambda \Big[\frac{m-1}{m} (\ell - r) + 1 \Big] + \cdots \,. \tag{15}$$

Comparing with (10), and assuming $\ell \gg r$, dropping the negligible $((m-1)/m)r$ and 1 in (15), we see that

$$r = \ln_m n\ell - \ln_m \Big(\frac{m}{m-1} \ln(1 + \frac{N}{N'}) \Big) \,, \tag{16}$$

dominated by the total number of m-bits in n and ℓ, and in particular, very insensitive to the population sizes N and N'.

Insistence upon a single uninterrupted sequence of matches for binding ignores the observations that something more like an area composed of several linear pieces may be involved. Thus, for successful binding, a criterion requiring, say at least t sequences of at least r matches, might make more sense. Generalizing the above to such a criterion is not difficult: we "tag" any subsequence of r or more matches by a variable ζ in order to recognize it, thereby replacing the strict failure contribution $(1-x^r)/(1-x)$ in (12) by

$$\sum_{a=0}^{r-1} x^a + \zeta \sum_{a=r}^{\infty} x^a = \frac{1-x^r}{1-x} + \frac{\zeta x^r}{1-x} . \tag{17}$$

Doing so, (12) is replaced by

$$G(x,y,\zeta) = \frac{1-x^r+\zeta x^r}{1-x-y(1-x^r+\zeta x^r)} . \tag{18}$$

Failing configurations are then those in which only the powers $1, \zeta, \zeta^2, \ldots, \zeta^{t-1}$ are present. Since

$$\begin{aligned}(\text{coef } \zeta^0 + \text{ coef } \zeta^1 + \cdots + \text{ coef } \zeta^{t-1})\, G(\zeta) \\ = \text{ coef } \zeta^{t-1}\, [G(\zeta)/(1-\zeta)] ,\end{aligned} \tag{19}$$

we have, in the same fashion as (13)

$$P_F = \text{ coef } z^\ell \zeta^{t-1} \text{ in } \frac{1}{1-\zeta} \frac{1-\lambda(1-\zeta)z^r}{1-z+\lambda\frac{m-1}{m}(1-\zeta)z^{r+1}} , \tag{20}$$

from which

$$P_S = \lambda^t \Big(\frac{m-1}{m}\Big)^t \text{ coef } z^\ell \text{ in } \frac{z^{rt+t-1}/(1-z) - z^{rt+t}/m}{(1-z+\lambda\frac{m-1}{m}z^{r+1})^t} . \tag{21}$$

To leading order in λ, (21) becomes

$$P_S \sim \lambda^t \Big(\frac{m-1}{m}\Big)^t \Big[\binom{\ell+1-rt}{t} - \frac{1}{m}\binom{\ell-rt}{t}\Big] , \tag{22}$$

so that for $\ell \gg rt$,

$$ln_m P_S \sim -rt + t \ln_m \ell - t \ln_m \frac{m}{m-1} \cdots \tag{23}$$

Now, comparing with (10), we have

$$rt = \ln_m(n\ell^t) + \ln_m[(m/(m-1))^t \ln(N/N')] + \cdots . \tag{24}$$

In other words, allowing the $R = rt$ matching sites to be distributed in pieces indeed increases the required value of R, but not significantly until ℓ^t approaches the order of n.

Discussion

How reliable is the estimate (16)? A reasonable value for the size of the accessible receptor variable region is $\ell \sim 100$ amino acids, and a common assessment (Karlin et al, 1989) of the functionally distinct letters of the amino acid alphabet is $m = 4$, for acidic, basic, neutral polar, and neutral hydrophobic. If this is to be taken literally, then (16) would predict $r \sim 12.5$ for the size of the complementing region. Indeed, it is extremely consistent with observed T-cell and antibody combining site epitopes, generally small peptides of 10 to 15 contiguous amino acids (thereby drawn from a class of $4^{10} - 4^{15} \sim 10^6 - 10^9$ types of molecular fragment); see Amit et al., 1986; Claverie et al., 1988; Klein, 1990; Sheriff et al., 1987.

The closeness of the fit for r with that observed should not of course be given excessive credence, since many imponderables remain. To start with, there are the specific values of the parameters that are used. The alphabet size m is only a means of obtaining the basic matching probability which, assuming that the matching is indeed between separate amino acids, should be given by

$$\frac{1}{m} = p \rightarrow \sum w_{ij} p_i p_j \tag{25}$$

where p_i is the frequency of amino acid i and w_{ij} the contribution to the binding probability (ideally 0 or 1) from an ij pairing.

On a more conceptual level, there is the tacit assumption that the effective full antigen and receptor are both of length ℓ, in register, and requiring contiguous contact. If instead the full antigen is of length $\ell' > \ell$, there are clearly $\ell' + 2(\ell - r)$ relative positions (with neither end in register) of the linear chains which have an overlap of at least r. And if gaps in complementation are permitted (Klein, 1990) e.g. corresponding to discontinuous epitopes, the complementary configurations are again increased in number. The effect is similar to (24). On the other hand, any geometric fitting requirements would decrease the roster of complementary configurations. But one would certainly imagine that these effects are small on the logarithmic scale appropriate to (16).

We conclude that the expressed repertoire of the mammalian immune system, and the observed characteristics of the antigen-receptor combining region, are consistent with optimal coevolution of the receptor and antigen populations. Whether recognition of a set of foreign antigens and non-recognition of self-antigens were the evolutionary driving force determining the characteristics of receptor combining regions is not known, but our results raise the intriguing possibility that such coevolutionary processes may have influenced the development of the TCR and immunoglobin repertoires.

Acknowledgments

This work was performed in part under auspices of U. S. Department of Energy. It was supported by NIH Grant AI 28433 (A.S.P.).

References

Amit, A. G., Mariuzza, R. A., Phillips, S. E. V., and Poljak, R. J. (1986) Three dimensional structure of an antigen-antibody complex at 2.8 Å resolution, *Science* **233**, 747–753.

Berzofsky, J. A., and Berkower, I. J. (1989) Immunogenicity and antigen structure. In *Fundamental Immunology*, Paul, W. E., ed., Raven Press, NY, p. 185.

Cancro, M. P., Gerhard, W. and Klinman, N. R. (1978). The diversity of the influenza-specific primary B-cell repertoire in Balb/c mice, *J. Exp. Med.* **147**, 776–786.

Chalufour, A., Bougueleret, L., Claverie, J.-M., and Kourilsky, P. (1987) Rare sequence motifs are common constituents of hypervariable antibody regions, *Ann. Inst. Pasteur/Immunol.* **138**, 671–685.

Claverie, J.-M., Kourilsky, P., Langlade-Demoyen, P., Chalufour-Prochnicka, Dadaglio, G., Tekaia, F., and Bougueleret, L. (1988) T-immunogenic peptides are constituted of rare sequence patterns. Use in identification of T epitopes in the human immunodeficiency virus *gag* protein. *Eur. J. Immunol.* **18**, 1547–1553.

Coutinho, A. (1980) The self-nonself discrimination and the nature and acquisition of the antibody repertoire, *Ann. Immunol. Inst. Past.* **131D**, 235.

Inman, J. K. (1978). The antibody combining region: speculations on the hypothesis of general multispecificity. In *Theoretical Immunology*, Bell. G. I., Perelson, A. S., and Pimbley, G. H. Jr., eds., Marcel Dekker, NY, pp. 243–278.

Kabat, E. A. (1976) *Structural Concepts in Immunology and Immunochemistry*, 2nd ed., Holt, Rinehart and Winston, NY.

Karlin, S., Ost, F., and Blaisdell, B. E. (1989). Patterns in DNA and amino acid sequences and their statistical significance. In *Mathematical Methods for DNA Sequences*, M. S. Waterman, ed., CRC Press, Boca Raton, FL.

Klein, J. (1990). *Immunology*, Blackwell Scientific, Boston.

Perelson, A. S. and Oster, G. F. (1979) Theoretical studies of clonal selection: Minimal antibody repertoire size and reliability of self-nonself discrimination, *J. Theoret. Bio.* **81**, 645–670.

Sheriff, S., Silverton, E. W., Padlan, E. A., Cohen, G. H., Smith-Gill, S. J., Finzel, B. C., and Davies, D. R. (1987) Three-dimensional structure of an antibody-antigen complex, *Proc. Natl. Acad. Sci. USA* **84**, 8075–8079.

Uspensky, J. V. (1937). *Introduction to Mathematical Probability*, McGraw-Hill, NY, pp. 77–84.

CONTEXTUAL DISCRIMINATION OF ANTIGENS BY THE IMMUNE SYSTEM: TOWARDS A UNIFYING HYPOTHESIS

Zvi Grossman
Department of Physiology and Pharmacology
Sackler Faculty of Medicine
Tel Aviv University
Tel Aviv 69978, Israel†

1. INTRODUCTION

1.1 Unresolved Questions

Any comprehensive theoretical approach to the immune system should address the following questions (Grossman & Paul, 1992):

Autoreactivity. Natural antibodies and autoreactive T cells are found in healthy individuals (Shoenfeld & Eisenberg, 1989). The presence of autoreactive lymphocytes is pronounced during infections, although it does not necessarily correlate with autoimmunity (Cohen & Young, 1991). Abandonment of a strict self-nonself discrimination (Cohen, 1992) appears in conflict with the concept that lymphocytes are members of a stereotypic, antigen-oriented defense system. Furthermore, the orderly nature of autoreactivity and autoimmunity needs to be explained: the targets of both in different individuals are not random, but restricted and predictable (Cohen, 1992).

Anergy. The induction of anergy is believed to be a major self-tolerance mechanism, in addition to clonal deletion (see Science, Vol. 248, 1990, pp 1335-1379). What is the rationale for the maintenance of functionally tolerant but autoreactive cells, potentially the cause of autoimmunity?

In vivo/in vitro dichotomy. In vivo, cells are components of organized structures, at least

†and Pittsburg Cancer Institute and Graduate School of Public Health, Department of Biostatistics, University of Pittsburg, Pittsburg, Pennsylvania USA

NATO ASI Series, Vol. H 66
Theoretical and Experimental Insights into Immunology
Edited by A. S. Perelson and G. Weisbuch

part of the time. In culture they are maintained in relative isolation, and this has a conspicuous negative impact on their viability and stability. Therefore, our ability to draw inferences from in-vitro experiments and assays about the physiological mechanisms of cellular response is limited. This limitation is demonstrated by recent observations. For example, a cellular state that is characterized in vitro as non-responsive, and therefore "anergic", is in some cases correlated not with tolerance, in vivo, but with memory (Bandeira et al., 1991) and activation (Ashcer & Sheppard, 1990). Another set of examples is provided by recent studies using transgenic mice. Such experiments show a surprising resilience of animals to deficiencies in lymphokines such as IL-2 and IL-4 (Schorle et al., 1991; Kuhn et al., 1991), or to an overexpression of MHC class II antigens (Fehling et al., 1990; Bohme et al., 1989). In vitro, these lymphokines are instrumental to cell growth and differentiation, and the expression of MHC class II molecules on antigen presenting cells (APC) regulates T-cell responses in a qualitative and quantitative way. These differences need to be explained.

Multiclonal responses. Apparently unrelated clones are often involved in specific immune responses. This is attributed to "bystander" effects, to percolation of idiotypic signals or to other pathological effects. A "positive" explanation may also exist.

Suppression. The rules that determine the balance between "suppression" and "help" in any given situation are not known. The existence and nature of "suppressor cells" are controversial. Memory. "Dynamic" and "static" aspects of memory cells need to be reconciled, namely: inherently short lifetime and indications of activity on the one hand, and the paucity of cell turnover on the other (Beverley, 1990).

Cytokine pleiotropy. Many peptide factors, including lymphokines, are multi-functional, have a broad range of target cells and the activity of one peptide in a single cell depends on the other signal molecules present (Sporn & Roberts, 1988). This has to be reconciled with the requirement that messages exchanged among cells via these cytokines be well-defined and precise (Paul, 1987).

Inter-systemic signaling. The immune system communicates, in a bidirectional manner, with both the hemopoietic system and the neuroendocrine system. Such communication does not easily fit into a scheme of a stereotypic, antigen-oriented defence system.

Context discrimination. Lymphocytes can adjust their antigen responses to the context and milieu in which the antigen is encountered. An illustration is provided by the striking differences in the responses to the antigens of M. leprae in patients with tuberculoid and lepromatous leprosy (Bloom & Mehra, 1984; Bloom et al., 1992). Lymphocytes appear to recognize and classify different patterns of signals, even within the same tissue. The responding cells secrete different combinations of cytokines and perform different functions. Contextual discrimination of antigens, rather than the traditional self-nonself discrimination, emerges as the primary functional characteristic of the immune system (Cohen, 1992), but the underlying principles are at present elusive.

1.2 Why Self-Organization?

A principal theoretical issue in biology is the relative roles of strict genetic programming and self-organization in cell development and function. Given the complexity and unpredictability of the environmental contexts in which antigen is recognized, it would be advantageous for the immune system if lymphocytes could learn from the sets of stimuli to which they are exposed which response is required without the need for precise pre-programming.

The repertoire of cellular responses to infection is rich. Destruction of cells that present infection-associated antigen is just one way of operation. In fact, a sweepingly destructive immune response may be harmful rather than beneficial when the infection is wide-spread. The pathogenicity of hepatitis B, lymphocytic choriomeningitis and leprosy, for example, in the acute phase, is due to lesions caused by an immune response. Chronic infections may often be immunologically controlled, though not necessarily eliminated, and this is probably achieved through the production of factors such as interferons, tumor necrosis factor and lymphotoxin.

What is the "signal" to the immune system to use one mode of response or the other? The choice could be based, in part, on kinetic and statistical characteristics of the stimuli. Thus, a rapid initial increase in antigen concentration, or in the number of infected cells, which is typical of many real-life infections, may trigger a "stereotypical" immune response (Grossman & Paul, 1992). If the infectious agent is not eliminated and the infection becomes chronic, a variety of other cellular activities are preferentially elicited in response to quasi-stationary sets of stimuli. The same set of lymphocytes, or different sets, could mediate these different modes of response. The first possibility is more likely, as there is a growing evidence that cell function does not strictly correspond to cell lineage.

A stereotypical immune response is transient and largely aggressive. The "other" activities, to which we refer collectively as an "environment regulated" response, consist of a variety of locally and selectively applied effector functions characterized by a fine contextual discrimination of antigenic stimuli. These activities could be maintained over extended periods of time and are presumably mediated through the directed secretion of cytokines.

In the following section models are presented that describe how this functional organization of immunological responses may come about and suggest a broader range of biological implications. These models, however, fall short of explaining how cells can learn to respond, accurately and differentially, to functional messages that are conveyed by different associations of signals (Sporn & Roberts, 1988) if they have not been precisely preprogrammed to do so. It has been proposed that the classificatory challenge of context discrimination could be met by lymphocytes through such a combinatorial strategy of signaling in which the definition of functional messages is partly a result of cellular experience (Grossman, 1989). Section 3 addresses the issue of associative recognition and learning by lymphocytes.

2. MULTIPLE-STATE MODELS OF LYMPHOCYTE RESPONSE

Let us assume that the direct effect on a cell of an external stimulus is changing the metabolic state of the cell in a way that can be expressed in quantitative terms. The metabolic change will be referred to as "excitation" and it is assumed, for simplicity, that there is a scalar measure, termed "excitation level", of this excitation.*

Two classes of complementary models have been proposed: "balance-of-growth" models, and "cell-tuning" models. Both regard the lymphocyte as a multi-state system with alternative state-transition pathways, e.g.:

(a) rest → excitation → rest;

(b) rest → excitation → proliferation(=self-renewal) → rest;

(c) rest → excitation → proliferation(=amplification) → differentiation.

Only the third process is defined as "activation": the induction of proliferation and (stepwise) differentiation. Balance-of-growth models are concerned with the balance between processes (b) and (c), whereas cell-tuning models focus on (a) and (b) and their dynamic segregation from (c).

2.1 **Balance-of-Growth (BG) Models**

BG models typically depict a small number of states, for each cell type, with distinctive response characteristics. These response characteristics are such that a heterogeneous population of interacting and competing cells manifest different, biologically meaningful patterns of response to different types of challenge (Grossman & Cohen, 1980; Grossman et al., 1980; Grossman 1982, 1984, 1985, 1989; DeBoer & Hogeweg, 1987a, 1987b). In particular, the models support (i.e. demonstrate the plausibility of) the following proposition:

Proposition 1: Under certain predictable conditions, cell population-growth can be uncoupled from full activation.

---

*See the Appendix for a glossary of terms that are used in a non-standard fashion. This terminology was introduced in Grossman & Paul, 1992.

An example of a BG model is based on the following assumptions (Grossman, 1982):

a. Self-renewal and differentiation are competing cellular processes whose relative probabilities are independently regulated by extracellular signals.

b. Excitation, self-renewal and differentiation are threshold-dependent in terms of the strength of the antigenic signal, with respectively increasing thresholds.

c. Differentiation occurs in several steps and eventually impairs the cell's capacity for self-renewal.

d. Dividing cells compete in a Darwinian-like manner; suppression is generally less specific than stimulation.

e. Lymphocyte viability, growth and differentiation depend on cooperativity among several types of co-stimulated cells.

It is assumption (b) that implies that proliferation can be uncoupled from (stepwise) differentiation. Proliferation without differentiation was termed "latent proliferation". As a rule, under quasi-stationary stimulation, proliferation is essentially latent: as (by assumption c) differentiation is growth-limiting, and since (by d) growth leads to prominence, differentiation and the expression of stereotypic effector functions are dynamically restricted. A more detailed analysis suggested (Grossman, 1982, 1984; reviewed, Grossman, 1989) that latent proliferation could be the driving process in thymic selection of self-tolerant, MHC-restricted T cells; in the dynamic maintenance of memory cells; in the mostly-hidden effects of idiotypic interactions in vivo; and in the suppression of certain immune responses.

The model (in particular, assumptions a, c and e) also suggests that shifting the balance of growth from proliferation to differentiation is an effective control mechanism of an ongoing response, irrespective of other negative feedback mechanisms that can be conceived. It also predicts a non-monotonic dose-response dependency (Grossman et al., 1980) and helper-suppressor duality. Time delays in the induction of differentiation - inherent to the process of cell activation and/or associated with the accumulation and maturation of help - amplify the transient response to acute challenge, leading to an overshoot in the number of effectors and to effective elimination of the antigen (Grossman & Berke, 1989). On the other hand, infection -- or a tumor -- can "sneak through" conventional immune surveillance if its progression is slow (Grossman & Berke, 1980; Grossman, 1982, 1984). Finally, assumption (d) accounts for response focusing.

2.2 The Tunable Activation Threshold (TAT) Model

While BG models emphasize Darwinian-type selection processes and the proliferation-differentiation interplay, more recent evidence has indicated that important cellular processes may take place independently of cell activation and growth. These include the promotion of cell viability (e.g., Nunez et al., 1991; Williams et al., 1990), thymic selection of T lymphocytes during a critical phase, and memory cell maintenance. Therefore, we have refocused our modeling effort onto the "excited state": even an unactivated cell would comprise several states, characterized by the level of excitation, viability and by different environmentally regulated response characteristics. This amounts to cellular plasticity, or adaptability (Grossman & Herberman, 1986; Grossman & Paul, 1992):

Proposition 2: Lymphocytes tune and update their responsiveness to signals.

The TAT model is further guided by the following propositions (Grossman & Paul, 1992):

Proposition 3: For a stimulus to cause cell activation, the excitation level must exceed an "activation threshold".

Proposition 4: Rapidly-changing environmental perturbances tend to induce suprathreshold responses, whereas a gradually or chronically perturbed environment imposes subthreshold responses.

Proposition 5: When engaged in persistent subthreshold interactions, cells resist chance activation.

Note: the state in which a lymphocyte is resistant, or nonresponsive, to a "standard" mode of activation, even if it participates in other environment regulated responses, conforms to the operational definition of "anergy".

A "mechanism" that can accommodate these propositions is suggested by the following assumption:

Assumption 1: The activation threshold equals an "excitation index" plus some "critical value", where the excitation index is a time-dependent, weighted average of the cell's past excitation levels.

The excitation index, $\hat{E}(t)$, at time t could for example be related to the excitation, $E(t)$, by the equation:

$$(1)\ d\hat{E}(t)/dt = \alpha E(t)(E(t) - \hat{E}(t)),$$

where α is a constant. $\hat{E}(t)$ then smoothly traces the excitation $E(t)$. $E(t)$ and $\hat{E}(t)$ are positive quantities.

The excitation index could be measured by the cell through the concentration of some molecule(s) undergoing storage-degradation cycles during excitation events in a certain domain of the cell. The opposing action of tyrosine kinases and phosphatases may represent a process of this type. Let us denote the varying concentration of this molecule by $C(t)$ and suppose that its rate of production is an increasing function of the excitation $E(t)$, and that its per-cell degradation rate depends on $E(t)$ as well. We may choose the following relationship: (2) $dC(t)/dt = E(t)(\beta E(t)-\alpha C(t))$, where α and β are constants. If we change the notation so that $(\alpha/\beta)C(t) = E(t)$, eq (2) reduces to eq (1). Thus, the excitation index is proportional to $C(t)$.

It is instructive to define "perturbation" as the difference, E(t)-Ê(t), between the excitation level and the excitation index. Perturbation thus is a measure of variation. By assumption 1, the perturbation has to be larger than some (fixed) critical value to cause activation in a given cell.

The logical relationship between assumption 1 and the general "propositions" is straightforward:

(a) Rate-of-change sensitivity. Upon gradual increase in the levels of excitation, reached through a sequence of increasingly strong excitation events, a cell is not likely to be activated. This is because the perturbation associated with each excitation event is small.

(b) Activation resistance. Cells maintained at a high level of excitation have correspondingly high activation thresholds and therefore are relatively insensitive, or inert, to activation. Moreover, since there is some upper limit to the excitation level that can be reached, and since perturbations are bounded by the difference between this "saturation level" and the excitation index, cells maintaining a sufficiently high average level of excitation may be prohibited from being activated, i.e., they are completely anergic.

The other assumptions of the TAT model are concerned with the dynamics of subthreshold interactions. "Excitability" is defined as a measure of the relative overall sensitivity of a cell to stimuli at a given time, as compared to some baseline sensitivity, and also of the cell's relative capacity to excite other cells. Excitability is related to the expression of membrane-associated receptors, adhesion molecules and secretory factors.

Assumption 2: Excitation promotes viability and excitability; both effects are transient.

Generalizing proposition 1 and assumption (b) of Section 2.1, it is suggested that both proliferation and differentiation can be uncoupled from "stereotypic" activation and from each other. There is evidence for latent proliferation (e.g., Sprent et al., 1990), and also for environment instructed differentiation, apparently without proliferation -- e.g., the differentiation of thymocytes to express either CD4 or CD8 cell surface markers (Borgulya et al., 1991; Robey et al., 1991).

Assumption 3: Sufficiently strong positive perturbations, but below the activation threshold, are associated with a probability for latent proliferation.

Assumption 4: Lymphocyte differentiation during development and maturation requires (at least) subthreshold, positive perturbations of sufficient magnitude and/or persistence by appropriate stimuli.

Direct corollaries of assumptions 2-4 are: (a) in order for interacting cells to maintain viability over extended periods of time, a compound measure of the frequency of interaction and of the excitabilities of the cells must exceed a certain threshold, which we call the "cell-coupling threshold": below this threshold, both the excitability and the viability decay, between excitation events, at a faster average rate than the rate at which these quantities are promoted, on the average, during such events. (b) If certain conditions are satisfied, assumptions 2-3 can account for homeostasis in organized environments of interacting cells: that is, the local densities of the different types of cells that interact with each other in such environments will be stably maintained, so that loss of some cells will trigger a compensatory proliferation. This will occur if there exists a suitable feedback relation between cell densities and excitability. (c) Cells maintaining a high average level of excitation may be prohibited not only from being activated, but also from displaying the full range of subthreshold responses, even on maximal stimulation. Under some circumstances, such restriction may be detrimental to the cell or to its ability to sustain viability.

2.3 **Biological Implications**

Infection.

During the rapid-propagation phase of acute infection, sets of antigenic and other signals delivered to lymphocytes have the time contingencies required for the rapid induction of a high level of excitation and therefore of large, positive perturbations. The elimination of antigen would allow for down-regulation of the excitation index and for the recovery of responsiveness between challenges.

Chronic or slow infection elicits, preferentially, subthreshold cellular responses. If the strength and frequency of interactions are sufficient to exceed the cell-coupling threshold, a positive cycle of enhancement at the levels of cell adhesion and excitation is started. Therefore, the formation of local anti-infection networks of lymphocytes, target cells and accessory cells depends both on antigen recognition and on the stability of the dynamic relationship.

Under persistent stimulation, the excitation index will remain elevated and the lymphocytes will be anergic, or partly anergic, to conventional activation. This could explain the "paradoxical" repeated observation (discussed in Ascher & Sheppard, 1990), in chronic viral or parasitic infections, of in vitro assayed anergy with immune hyperactivity in vivo. Furthermore, these anergic cells act effectively as suppressor cells inhibiting the activation of non-anergized lymphocytes (see assumption (d) of section 2.1).

Autoreactivity and self-tolerance

Considering the generation of tolerance within the thymus, the outcomes of three types of intrathymic selection events are correlated with the affinity of the thymocyte's antigen receptor for the ligand expressed on the thymic APC (Sprent & Webb, 1987). The TAT model suggests a unified, simple interpretation of these alternate outcomes based on the dynamics of subthreshold interactions (Grossman & Paul, 1992).

(a) Low-affinity thymocytes whose interaction with self-MHC molecules does not surpass the cell-coupling threshold are short-lived. (b) Gradual increase in the excitability of higher-affinity thymocytes, due to increased expression the T cell's receptors and accessory molecules, results in anergy. The potency of such anergy in a given thymocyte is directly related to the affinity of the thymocyte for the self antigen. The lower affinity cells in this class are less potently anergized and therefore can later be activated by some combinations of foreign antigens and self-MHC. Developmental differentiation is facilitated by positive subthreshold perturbations. (c) High-affinity, highly excited, self-recognizing cells are eliminated because their capacity to experience positive perturbations is too severely restricted.

In the periphery, the nominal autoreactivity of some lymphocytes for self MHC molecules might be enhanced via the synergistic stimulation by infection-induced signals. This could explain, beyond a "bystander effect", the extensive participation of nonspecific lymphocytes in specific immune responses which is often observed. Such lymphocytes could participate in the control of infection, together with primarily autoreactive lymphocytes and lymphocytes directed at infection-associated antigens. The "anergy" imposed by the ongoing activity of all these cells would serve as a barrier to destructive autoimmunity.

While the expression by positively selected lymphocytes of various autoreactive receptors may be largely random, the actual involvement of autoreactive cells in sustained interaction with potential target cells in the periphery, and therefore their viability, is restricted by the requirement for an active cooperativity between the cells: only some tissue cells could provide appropriate counter-stimulation sufficient for the interaction to surpass the cell-coupling threshold. This could explain the "orderly nature" of autoreactivity and autoimmunity.

Memory

This same nominal autoreactivity could be used by memory T cells in the memory maintenance process. While earlier it was suggested (Grossman, 1982; 1984) that self-MHC molecules could drive memory cells into a controlled latent proliferation (as could anti-idiotypic stimulation; Grossman, 1984), there is little evidence that memory cells divide significantly prior to challenge. Yet, these cells appear to be "partially activated" and to require stimulation to persist (Bevereley, 1990).

The concept of "excitability" has suggested a different kind of dynamic relationship between memory cells and their environment. It implies the existence of a short-term cellular memory that can facilitate long-term memory at the population level (Grossman & Paul, 1992). Having surpassed the cell-coupling threshold during the activation phase, cells assembled under the selective influence of the antigen would continue to cooperate, even in the absence of the antigen or under weak antigenic stimulation, to provide stimuli to each other. The mechanism of tuning the activation threshold and excitability to the average excitation level acts (a) to minimize the size of the perturbations, reducing the probability for cell activation; and (b) to maintain a level of sensitivity (that is, elevated excitability and not-too-high excitation index) suitable for reactivation in response to the original antigen in an immunogenic context. The dynamic interplay between the magnitude of perturbations and the frequency of cellular encounters would act as a local homeostatic mechanism, controlling the probability for self-renewal division.

The accessory functions of non-lymphoid cells are inducible (Weaver & Unanue, 1990), as are the helper functions of T helper cells. We have suggested that they are maintained by the very same subthreshold interactions. Self-sustained "memory units" could in principle be detected and studied by in vivo/in situ methods (Van Rooijen et al., 1989). Disrupting the integrity or stability of such units (as in cell transfer experiments) is predicted to interfere with the maintenance of memory, even if the memory lymphocytes themselves are not directly targeted. The existence of phenotypically distinct "memory accessory cells" is also predicted.

3. DO LYMPHOCYTES LEARN?

"Learning" in the immune system has been implicitly defined as the selection of appropriate responses, without a precise pre-programming of the cells for such responses. So far the overall sensitivity of cells to sets of signals and its modification has been considered. A possible cellular strategy to achieve selective responses to messages conveyed by particular associations of signals, in an environment of overlapping patterns, would be based on inducible changes in the efficacy of individual signals, in analogy with neuronal learning. This section addresses itself, briefly, to (1) the physiological significance of such learning; (2) its plausibility; and (3) mechanisms.

(1) Is learning necessary? A listener present in a room with several sources of speech is able to concentrate on speech from a particular source while rejecting the other sound sources. This has been referred to as the "cocktail party" effect (Mitchell et al., 1970). In the dynamic multiple effector cell-multiple target cell communication that we envision in chronic infection (or in other kinds of quasi-stationary disturbances), it would be desirable for the effector cells to achieve a similar effect; that is, to become selectively responsive to their respective target and accessory cells, even when those do not present unique antigens. Each effector cell may learn to recognize a "feature" that distinguishes best between its own type of target cell and all the rest.

Another problem facing the system in such situations is "classification"; that is, a correct matching of target cells and effector cells, respectively (assuming an existence of cross-reactivity, or overlap, between the patterns of signals that are exchanged). If more than one matching-configuration can be formed, than selection of the physiologically correct configuration would require "feedback reinforced" learning; that is, the above described process in which effector cells learn to develop target selectivity would be influenced, or biased, by feedback signals, geared to the collective performance of the cells, from the "environment". The environmental source of feedback messages could be either local or external to the tissue.

(2) Is learning plausible? There are yet no compelling instances, at the cellular level, of this kind of learning behavior. To find such instances, one would have to adopt new experimental strategies. Nevertheless, in considering current concepts about immune response mechanisms and the elaborations required, theoretically, in order for "learning" to pertain, it is notable not only that there are no apparent contradictions, but also that such elaborations are relatively moderate and, in my view, not implausible.

It is clear that the efficacy of signals changes in the course of lymphocyte differentiation (Street et al., 1990; Borgulya et al., 1991) and activation (Weaver & Unanue, 1990; Noelle & Snow, 1990), both quantitatively and qualitatively. In some cases, there is evidence that an associative or coordinate engagement of receptors instructs and directs the occurrence and processing of such changes (Borgulya et al., 1991; Rocken et al., 1992). In order for these changes to be compatible with learning as it is believed to occur in brain cells, it is required (a) that the existence of multiple, alternative cellular states (at least two) be the rule, rather than the exception to the rule, even in the later stages of lymphocyte differentiation (Grossman & Herberman, 1986); and (b) that, in general, a commitment to any one of these states occur in several steps, rather than in one step, and be reversible, initially at least. By the latter requirement, the cell could explore its environment ("context") and avoid a chance decision. Moreover, the "environment" could simultaneously evaluate the cell's performance and influence the process via some feedback signals. Tuning of receptive pathways to signals in several small steps would allow for the achievement of a fine target selectivity, especially during multiple subthreshold excitation events.

(3) Mechanisms of learning.

(a) Rules of learning. Molecular events during lymphocyte activation appear to be consistent with the following rule (Grossman, 1992): "stimuli become linked in the cell's memory when they occur repeatedly in temporal proximity". One patterned stimulus, initiated usually by antigen, elicits a largely stereotypic response via a principal pathway. Secondary stimuli, initiated independently or facilitated by the principal stimulus, become progressively coupled to the principal pathway as the excitation process evolves. This resembles a mechanism proposed for behavioral conditioning (Alkon & Rasmussen, 1988; Alkon, 1989). The specialization of thymocytes to the expression of CD4 or CD8 cell surface markers and to responsiveness to class II or class I MHC molecules, respectively; and the differentiation of T cells into multiple phenotypes exemplified by the Th_1 and Th_2 types, which are distinguished by the lymphokines the cells produce and by their responsiveness to exogenous signals -- these examples suggest a second rule of learning: "different, repeatedly occurring sets of temporally proximal stimuli compete with each other for exclusive representation in the cell's memory". If the competing sets of signals partly overlap, as do sound patterns in a cocktail party, then achieving pattern selectivity would require "feature detection".

(b) A Hebbian-like algorithm (Hebb, 1949; Bienenstock et al., 1982; Intrator, 1992). The following exemplifies a possible mechanism by which the proposed rules of learning could be realized. Define the "strength" of a signal by its relative contribution to the excitation level during an excitation event: the excitation level is a positive definite, saturable, increasing function of the sum of strengths of all the signals. "Signal strength" is the product of the signal's intensity and the receptor's strength. "Receptor strength", in turn, is directly related to the receptor's multiplicity on the cell's membrane, its binding constant and the efficacy with which it transduces the signal. Signals that have a negative impact on the excitation level, suppressing other signals, are referred to as "negative signals". These definitions necessarily oversimplify the dynamics of signal transduction and the interactions among signal pathways. It has been proposed (Grossman, 1992) that cells adapt to perturbations by making (initially reversible) modifications of the strengths of engaged receptors in such a way that positive perturbations cause enhancement of the cell's responsiveness to the exciting signal pattern and negative perturbations cause down-regulation. More precisely, each strength is increased or decreased by an amount directly related to the size of the perturbation and to the strength of the signal (e.g., to their product, or to the product of some increasing non-linear functions of these quantities), depending on whether the perturbation is positive or negative. Positive perturbations cause enhancement of strengths of receptors for positive signals and diminution of negative signals; and vice-versa for negative perturbations.

If $S_i(t)$ and $R_i(t)$ are the strengths of the i'th signal and of the receptor for that signal, respectively, $R_i(t)$ could be modified according to a formula such as:

(3) $dR_i(t)/dt = \mathbf{f}(S_i(t)).\mathbf{g}(E(t), E(t)-\hat{E}(t))$,

where $\mathbf{f}$ is an odd function of S_i, and the function $\mathbf{g}$ changes sign at $E = \hat{E}$ and vanishes as E approaches zero.

Since, by definition, the excitation index is essentially a moving average, partitioning positive and negative perturbations there is always one or more signal pattern(s) being suppressed as long as the cell is still excitable by, and responding to, more than one pattern. Therefore, if the process converges, pattern selectivity is eventually reached.

(c) Multiple cell learning. Feedback-reinforced learning can also be formulated

(Grossman, 1992), where several lymphocytes simultaneously become selective with respect to the (generally different) patterns to which they respond. The theory does not require specific feedback signals for each cell. Rather, fluctuations in one (or a few) feedback signal(s), reflecting changes in the collective performance of the system of cells and affecting each cell in a similar way, can result in a convergence towards the desired classification (Barto et al., 1981). For instance, the signal could be a local measure of "stress" in the tissue, reinforcing an organization that becomes increasingly correlated to reduction in the stress.

(d) Learning with cell turnover. Cell proliferation and differentiation observed during inflammatory reactions may be partly manifestations of a conventional immune response, but may also represent in part the most dynamic phase of an associative search whereby the network's "architecture" is being selected. Thus, in the immune and hemopoietic systems an adaptive enhancement of connectivity, as an underlying principle of learning, could take place via dynamically regulated changes in the numbers and location of interacting cells (a mechanism not generally attributed to the adult brain) as well as by adaptive changes in receptors and mutual responsiveness (a property shared with brain cells). Quantitative models of learning in "non-neural networks" with turning-over cells have yet to be formulated.

4. CONCLUSION

Our models provide some clues to the "unresolved questions" listed at the beginning. Autoreactivity can be robustly dissociated from uncontrolled aggression and be regularly employed to subserve physiological functions; the orderly nature of autoreactivity and autoimmunity is conceptually explained by the selectivity of the physiological constraints related to the threshold of cell-coupling; the "paradoxes" of anergy are accommodated by the TAT model; "positive" explanation has been offered for multiclonal responses; "suppression" arises naturally from the competition between a "conventional" response and other, cryptic responses; the hypothesis reconciles "dynamic" and "static" aspects of memory cells; cell might learn to properly recognize associative, environmental messages.

The in-vitro/in-vivo dichotomy is a reflection of the fact that in vitro we can observe only limited aspects of cellular behavior. Redundancy alone is not sufficient to explain the resilience of in vivo processes. For example, the property of self-tuning of cellular responsiveness to the environment predicts and explains the observations in transgenic mice overexpressing MHC class II antigens.

One message to immunologists is the need to develop in-vivo/in-situ methods to study cell interactions within self-organized cellular environments. This is apparently at variance with the more optimistic recommendation of DeBoer et al. (1990), who suggested that "if experiments were performed in vitro ... we would be able to do a better job at model discrimination, probing and eventually understanding the immune response." The difference can be traced, in part, to different ways of viewing the response characteristics of individual cells: fixed and robust, or tunable and context-dependent.

A resolution of this latter issue itself could, in principle, be approached in vitro. One specific question is whether cell activation is an all-or-none phenomenon, or if subthreshold excitation makes a difference. Freshly cultured lymphocytes would have to be maintained and manipulated in a non-dividing, viable state prior to activation.

Even more ambitious experiments are required to demonstrate controlled conditioning-like phenomena at the cellular level. Suppose, for instance, that a lymphocyte is known to respond to a signal (or a set of signals), S_1, so that the readout is R_1, and also to another set S_2 with a readout R_2. Suppose also that lymphocytes of this type are repeatedly stimulated by a combination S_1+S_2 to yield R_2. Classical conditioning occurs if, subsequently, these lymphocytes respond to S_1 with R_2. It should be noted that such adaptive behavior of cells probably depends, in vivo, on subthreshold excitations. Our ability to reconstruct or monitor such conditions is extremely limited. Nevertheless, the following indications of a learning-behavior could also be explored experimentally: (a) memory cells from different clones may manifest idiosyncratic phenotypes; that is, they will display fine differences in their interaction with APC and accessory signals. (b) Properly timed "neutral" signals may affect immune responses and the memory phenotypes (Grossman et al., 1992a). (c) Different memory states may result from a manipulation of the priming process.

Pathological processes associated with chronic infection, autoimmunity and other immune deficiencies could be usefully described in terms of a destabilization of cognitive networks (Grossman et al., 1992b). The very plasticity of cellular characteristics and the network-like coupling among cells suggest that the collective profile of gene expression of interacting cells could adequately reflect the dynamic state of tissue organization. The implication for those who study these diseases is that shortcuts are not likely: a multi-dimensional, patterned process would require a multi-dimensional database to characterize it. Typically, the time-evolution of a broad range of simultaneously measured molecular and cellular parameters would have to be monitored. The role of theorists would be to develop methods for the detection of lower-dimensionality features in the empirical patterns and to find useful (i.e., predictive) correlations. This would then provide a basis to rationalize treatment protocols and to formulate conventional, cause-and-effect models.

A comment for theoretical immunologists

Mathematical models of immunological phenomena differ in their dimensionality, in the details of interactions terms, and in the use of differential equations or cellular automata to represent cell populations (Perelson, 1990; this volume). A considerable portion of these models are concerned with idiotypic networks, and others limit themselves to within the scope of clonal selection. However, most of these models do not challenge each other's basic conceptions or the "accepted knowledge" in the field by proposing alternative conceptions. In particular, self-organization is generally restricted to the stabilization of alternative clonal distributions.

If indeed "knowledge grows by a continuous critical dialogue between competing scientific research programmes, with their competing metaphysics and competing images of knowledge" (Elkana, 1981), then this kind of conformity is somewhat unfortunate. Its origin is largely a widespread belief that the body of immunologic knowledge is today "large and solid enough to offer a rather credible picture of the immune system's structure and functions" (Seiden & Celada, 1992). Consequently, theoretical immunologists appear to maintain that the major remaining challenge for theorists is constructing sufficiently detailed ("realistic") mathematical models, using the presently existing phenomenology. Since approximations are inevitable, it is suggested that "essential characteristics" can be "captured" by simplified mathematical descriptions. It is often stated that novel genuine characteristics of the phenomena being modelled "emerge" from the mathematical analysis of such models and that therefore the models can provide new insights.

The above approach has been challenged and the associated "image of knowledge" criticized (Grossman, 1989). Unresolved questions such as those listed in the Introduction have been taken as indications that currently accepted hypotheses may be "deficient or susceptible to modification" (Humphrey, 1984; see also Paul, 1987, and Coutinho, 1989). The modified set of propositions in Section 2 is based on the premise that new strategies of self-organization should be explored to resolve those questions.

Mathematical models as such have been argued to be inherently limited in their ability to resolve basic questions about immunity (Grossman, 1989); in most cases, they do not even pose such questions. Genuinely new insights have not "emerged" from the analysis of simple models; at most, such models have been proved useful as metaphors to animate preconceived ideas. Interesting insights into the nature of high-dimensional networks do appear to be gained from simulations of such networks (see this volume); however, such insights are useful only as they become relevant to the understanding of how the immune system itself functions.

The essential part of a theory is a hypothesis, stated verbally and pictorially. The logical consequences of varying presuppositions about the immune system can be followed to a considerable length by way of deductive reasoning, in the sense of Medawar (1969; see Grossman, 1989), without explicit mathematical analysis. In this way, it is possible, at first, to avoid over-specification of the models by ad-hoc assumptions. The non-mathematical analysis outlined in this article is believed to be sufficient to instigate various interpretations and suggest experimental strategies. However, explicit mathematical representations of various aspects of the models are necessary in order to convincingly demonstrate the self-consistency of conjectures that have been made (e.g., corollaries (a) and (b) to assumptions 2-4 of the TAT model) and the feasibility of proposed mechanisms (e.g., for lymphocyte learning). Such a mathematical analysis would facilitate a better understanding of the hypotheses and of their theoretical implications. To the mathematical modeler, such analysis could be quite interesting: the loosely defined relationships among cell excitation, self-renewal and activation offers a latitude for numerous variations that could result in qualitatively distinct solutions at the population level.

Appendix: Glossary of Terms

ENVIRONMENT REGULATED RESPONSE: a pattern of cellular responses characterized by a variety of locally applied effector functions and a fine discrimination of antigenic stimuli by the context of other associated signals. EXCITATION: a change in the metabolic state of a cell caused by an integrated external stimulus. EXCITATION EVENT: a process whereby a cell is transiently subject to a significantly increased contact with a source of stimulation, cellular or molecular. EXCITATION LEVEL: a quantitative measure of excitation at any given time; interchangeably used also for the peak excitation level reached on an excitation event. EXCITATION INDEX: a time-dependent, weighted-average of the cell's past excitation levels. PERTURBATION: the difference between the excitation level and the excitation index at any given time; interchangeably used also for the peak perturbation reached on an excitation event. EXCITABILITY: a measure of the relative overall sensitivity of a cell to stimuli at a given time, as compared to some baseline sensitivity; also considered to be a measure of the cell's relative capacity to excite other cells. SATURATION LEVEL: the maximum excitation level that a given cell can reach. SELF-RENEWAL DIVISION: cell division that is not associated with differentiation. AMPLIFICATION DIVISION: cell division that is a part of a differentiation process. LATENT PROLIFERATION: cell population growth largely through self-renewal division. STEPWISE DIFFERENTIATION: a marked, qualitative change in the pattern of genes that a cell is expressing. ACTIVATION: the induction of cell proliferation and stepwise differentiation. ACTIVATION THRESHOLD: the minimum excitation level required for activation. SUBTHRESHOLD INTERACTION/RESPONSE: a process where-by a cell is repeatedly excited by other cells without being activated; the consequences may include latent proliferation. ACTIVATABILITY: a measure of the cell's susceptibility to the induction of activation (inversely related to the activation threshold). ANERGY: a cellular state in which the activation threshold is elevated; interchangeably used for a complete anergy -- a (generally transient) state in which the cell cannot be activated. VIABILITY: the cell's (variable) life expectancy at any given time; also used in the usual sense.

CELL-COUPLING THRESHOLD: the minimum value of a compound measure of the (average) frequency and "strength" of stimuli, which needs to be exceeded for a cell to maintain its viability. BALANCE-OF-GROWTH: the concept that, upon supra-threshold stimulation, the relative probability of cell differentiation versus self renewal, and hence the cell population's rate of growth, is variable and dynamically regulated. MEMORY UNIT: an organized environment of interacting lymphocytes and accessory cells, where the cell-coupling thresholds have been exceeded by the interacting cells and homeostasis is maintained.

Acknowledgement

This work was supported by the Harry Palley Endowment Fund of the United Jewish Federation Foundation.

REFERENCES

Alkon DL (1989) Memory storage and neural systems. Sci Am 261:42-50.
Alkon DC, Rasmussen H (1988) A spacial-temporal model of cell activation. Science 239:998-1005.
Ascher MS, Sheppard HW (1990) AIDS as immune system activation. II. The panergic imnesia hypothesis. J AIDS 3:177-191.
Bandeira A, Mengel J, Burlen-Defranoux O, Coutinho A (1991) Proliferative T cell anergy to Mls-1a does not correlate with in vivo tolerance. Int Immunol 3:923-931.
Barto AG, Sutton RS, Brouwer PS (1981) Associative search network: a reinforced learning associative memory. Biol Cybern 40:201-211.
Bevereley PCL (1990) Is T-cell memory maintained by crossreactive stimulation? Immunol Today 11:203-205.
Bienenstock EL, Cooper LN, Munro PW (1982) Theory for the development of neuron selectivity: orientation specificity and binocular interaction in visual cortex. J Neurosci 2:32-48.
Bloom BR, Mehra V (1984) Immunological unresponsiveness in adults. Immunol Rev 8:5-28.
Bloom BR, Modlin RL, Salgame P (1992) Stigma variations: observations on suppressor T cells and leprosy. Annu Rev Immunol, to appear.
Bohme J, Haskins K, Stecha P, von Euijk W, LeMeur M. Gerlinger P, Benoist C, Mathis P (1989) Transgenic mice with I-A on islet cells are normoglycemic but immunologically intolerant. Science 244:1179-1183.
Borgulya P, Kishi H, Muller U, Kirberg J, von Boehmer H. (1991) Development of the CD4 and CD8 lineage of T cells: instruction versus selection. EMBO J 10:913-918.
Cohen IR (1992) Introduction to a cognitive paradigm of the immune system, to appear.
Cohen IR, Young DB (1991) Autoimmunity, microbial immunity and the immunological homunculus. Immunol Today 12:105-110.
Coutinho A (1989) Beyond clonal selection and network. Immunol Rev 110:63-87.
DeBoer RJ, Kerrekidis IG, Perelson AS (1990) A simple idiotypic model with complex dynamics. Chem Eng Sci 45:2375-2382.
DeBoer RJ, Hogeweg P (1987a) Self-nonself discrimination due to immunological nonlinearities: the analysis of a series of models by numerical methods. IMA J Math Appl Med Biol 4:1-32.
DeBoer RJ, Hogeweg P (1987b) Immunological discrimination between self and non-self by

precursor depletion and memory accumulation. J Theor Biol 124:343-369.
Elkana Y (1981) A programmatic attempt at an anthropology of knowledge.in:Mendelsohn E, Elkana Y (eds) Sciences and Cultures. Sociology of the Sciences. Vol. V, 1-76.
Fehling HJ, van Ewijk W, Pasquali JL, Waltzinger C, LeMeur M, Gerlinger P, Benoist C, Mathis D (1990) Functional consequences of overexpressed Ia antigens in AK alpha/AK beta transgenic mice. J Immunol 144:2865-2874.
Grossman Z, Berke G (1980) Tumor escape from immune elimination. J Theor Biol 83:267-296.
Grossman Z, Cohen IR (1980) A theoretical analysis of the phenotypic expression of immune response genes. Eur J Immunol 10:633-640.
Grossman Z, Asofsky R, DeLisi C (1980) The dynamics of antibody secreting cell production: regulation of growth and oscillations in the response to T-independent antigens. J Theor Biol 84:49-92.
Grossman Z (1982) Recognition of self, balance of growth and competition: horizontal networks regulate immune responsiveness. Eur J Immunol 12:747-756.
Grossman Z (1984) Recognition of self and regulation of specificity at the level of cell populations. Immunol Rev 79:119-138.
Grossman Z (1985) Balance of growth models of cell populations: the significance of simple mathematical considerations. Lecture notes in Biomath 57:312-319.
Grossman Z, Herberman RB (1986) Natural killer cells and their relationship to T-cells: hypothesis on the role of T-cell receptor gene rearrangement on the course of adaptive differentiation. Cancer Res 46:2651-2658.
Grossman Z (1989) The concept of idiotypic network: deficient, or premature?in:Atlan H, Cohen IR (eds) Theories of Immune Networks. Springer-Verlag, Berlin, p 38-52.
Grossman Z, Paul WE (1992) Adaptive cellular interactions in the immune system: the tunable activation threshold and the significance of subthreshold responses. Submitted.
Grossman Z (1992) Context dependent recognition and response in the immune system: programmed or learned? (to be submitted
Grossman Z, Herberman RB, Livnat S (1992a) Neural modulation of immunity: conditioning phenomena and the adaptability of lymphoid cells. Int J Neurosci, to appear.
Grossman Z, Herberman RB, Bentwich Z (1992b) From infection to AIDS: are manifestations of effective immune resistance to HIV infection misinterpreted? Submitted.
Hebb DO (1949) The Organization of Behavior. Wiley, New York.
Humphrey JH (1984) Serendipity in immunology. Annu Rev Immunol 2:1-21.
Intrator N (1992) Feature extraction using an unsupervised neural network. Neural Computation, to appear.
Kuhn R, Rajewsky K, Muller W (1992) Generation of interleukin-4 deficient mice, Science, to appear.
Medawar PB (1969) Induction and Intuition in Scientific Thought. Methuen, London.
Mitchell OM, Ross CA, Yates GH (1971) Signal processing for a cocktail party effect. J Acoust Soc Am 50:656-660.
Noelle RJ, Snow EC (1990) Cognate interactions between helper T cells and B cells. Immunol Today 11:361-368.
Paul WE (1987) Between two centuries: specificity and regulation in immunology. J Immunol 139:1-6.
Perelson AS (1990) Theoretical immunology.in:Jen E (ed) Lectures in Complex Systems, SFI Studies in the Sciences of Complexity, Vol. II. Addison-Wesley, Redwood City, pp 465-499.
Robey EA, Fowlkes BJ, Gordon JW, Kioussis D, von Boehmer H, Ramsdell F, Axel R (1991) Thymic selection in CD8 transgenic mice supports an instructive model for commitment to a CD4 or CD8 lineage. Cell 64:99-107.
Rocken M, Saurat J-H, Hauser C (1992) A common precursor for CD4+ T cells producing interleukin 2 or interleukin 4. J Immunol, to appear.
Schorle H, Holtschke T, Hunig T, Schimpl A, Horak I (1991) Development and function of T cells in mice rendered interleukin-2 deficient by gene targeting. Nature 352:621-624.

Seiden PE, Celada F (1992) A model for simulating cognate recognition and response in the immune system. J Theor Biol, submitted.
Shoenfeld Y, Isenberg D (1989) The mosaic of autoimmunity. Elsevier, Amsterdam-Oxford-New York.
Sporn MB, Roberts AB (1988) Peptide growth factors are multifunctional. Nature 332:217-219.
Sprent J, Webb SR (1987) Function and specificity of T cell subsets in the mouse. Adv Immunol 41:39-133.
Sprent J, Gao EK, Webb SR (1990) T cell reactivity to MHC molecules: immunity versus tolerance. Science 248:1357-1363.
Street NE, Schumacher JH, Fong TA, Bass H, Fiorentino DF, Leverah JA, Mosmann TR (1990) Heterogeneity of mouse helper T cells. Evidence from bulk cultures and limiting dilution cloning for precursors of Th1 and Th2 cells. J Immunol 144:1629-1639.
van Rooijen N, Claassen E, Kraal G, Dijkstra CD (1989) Cytological basis of immune functions of the spleen. Immunocytochemical characterization of lymphoid and non-lymphoid cells involved in the "in situ" immune response. Prog Histochem Cytochem 19:1-71.
Weaver CT, Unanue ER (1990) The costimulatory function of antigen presenting cells. Immunol Today 11:49-55.

II. T Cell Models

Theoretical Insight into Antigen-Induced T-Cell Unresponsiveness

M. Kaufman, F. Andris* and O. Leo*
Université Libre de Bruxelles
Service de Chimie Physique - C.P. 231
Boulevard du Triomphe
1050 Bruxelles - Belgium

1. Introduction

It is now well established that immune responses involve intricate, highly regulated interactions between functionally different sets and subsets of lymphocytes. The available evidences indicate that *helper* T lymphocytes, in particular, play an essential role in the initiation of most antibody or cell mediated responses. Activation of mature T helper cells leads to the production of regulatory factors that have an effect on the activation and function of other cells of the immune system. Numerous studies therefore are devoted to the conditions of activation and inactivation of these lymphocytes. In the present work we focus more specifically on the induction of long-lasting unresponsiveness after interaction of Interleukin-2 (Il-2) producing T helper cells with antigen. This phenomenon is important for the downregulation of a normal immune response, and might also participate in the mechanisms of self-tolerance.

In section 2 we recall a few basic features of the biochemical pathways involved in T cell signalling, and present some experimental results on the modulation of antigen-responsiveness in T cell hybridomas. In section 3 we present a theoretical model for the understanding of antigen-induced T cell hyporesponsiveness, based on a phosphorylation cascade (Kaufman et al., submitted). An important ingredient of this model is the presence of an autophosphorylating protein tyrosine kinase. A molecular mechanism both

---

* Laboratoire de Physiologie Animale, Université Libre de Bruxelles.
67, rue des Chevaux, 1640 - Rhode St Genèse, Belgium.

NATO ASI Series, Vol. H 66
Theoretical and Experimental Insights into Immunology
Edited by A. S. Perelson and G. Weisbuch

for desensitization in the presence of antigen, and persistent lowering of cell responsiveness after stimulus removal is described*. Section 4 is devoted to a short discussion.

Although at a primitive stage our approach provides a theoretical framework for viewing T cell anergy in terms of an unconpling of the receptors from the intracellular signalling events, and illustrates several concepts that have been proposed in other contexts (Delbrück, 1949; Lisman,1985; Segel et al.,1986; Knox et al., 1986; Kaufman & Thomas, 1987; Hervagault & Canu, 1987; Li & Goldbeter,1989).

2. Antigen-Induced T-Cell Unresponsiveness.

2.1 T-cell receptor signaling pathways.

In normal physiological conditions, T cell activation occurs via a dual recognition system : antigen is recognized only in association with self-histocompatibility molecules on the surface of antigen presenting cells (APC). Activation requires the simultaneous presence on the T cell surface of CD4 (or CD8) proteins recognizing the MHC encoded molecules, and of the T cell receptor (TCR) specific for the antigen. This T cell receptor is itself a multimeric complex consisting of the antigen-binding, clonotypic $\alpha\beta$ heterodimer, Ti, and the noncovalently associated CD3 invariant polypeptides. Transmission of the external signal to the interior of the cell is presently thought to occur via these CD3 chains. T cell activation can also be initiated without the MHC context, by antibody directed against the Ti or CD3 components of the T cell receptor.

The ligand-receptor interactions at the plasma membrane lead to immediate intracellular changes. In particular, the engagement of the TCR complex activates a phosphoinositide-specific phosphodiesterase (PI-PLC-γ_1) that generates two intracellular messengers : inositol 1,4,5 - triphosphate (IP_3) and diacylglycerol (DAG). The resulting increase in cytoplasmic free calcium

----

* Part of this work has been presented at the International Titisee Conference "Deterministic Chaos in Biology and Medecine" in February 1990 (Titisee, Germany), and at the First European Conference on "Mathematics Applied to Biology and Medecine" in January 1991 (l'Alpe d'Huez, France), Abstract volume p.246.

concentration and activation of protein kinase C (PKC), in turn, initiate a cascade of biochemical events that result in new gene expression, and ultimately, in functional responses such as lymphokine production, new surface-receptor expression and proliferation (Imboden & Stobo, 1985).

In addition to the phosphatidyl inositol (PI) pathway, T cell stimulation via the TCR also activates a protein-tyrosine kinase (PTK) pathway which results in phosphorylation of a number of intracellular substrates on tyrosine residues (June et al., 1990). Several lines of evidence suggest an interplay between these two signal transduction pathways. As will be discussed in more detail in section 3, tyrosine phosphorylation/dephosphorylation reactions are now believed to play an important role in initiating and regulating TCR - mediated cell activation.

2.2. T-cell inactivation.

T cell interaction with antigen may also lead to specific unresponsiveness or tolerance (Finkel et al., 1991). Recently, several in vitro and in vivo systems have been described in which interaction of ligand with the TCR results in functional inactivation of Il-2 - producing T helper lymphocytes (Wilde et al., 1984; Jenkins & Schwartz, 1987; Schwartz, 1990; Blackman et al., 1991; Dewit et al.,1992).

A detailed study of the induction of long - term unresponsiveness after activation of Murine T cell hybridomas has been reported in Andris et al. (1991). T cell hybridomas require less stringent conditions for activation than normal untransformed cells and constitute therefore a convenient experimental model to evaluate the role of TCR engagement in cell activation. Hybridomas are optimally activated independently of co-stimulatory factors from APC, are insensitive to most soluble growth factors such as Il-2, and proliferate constitutively. Their activation is thus adequately characterized by their Il-2 production.

The data on fig. 1 show that hybridomas that have been previously stimulated to Il-2 production through the TCR, respond poorly to a subsequent restimulation irrespective of the agonist used. Both APC - dependent (Ag/MHC) and APC - independent, mAb-mediated, primary activations lead to a decreased sensitivity to further stimulations. Hyporesponsiveness is induced in $CD4^+$ $CD8^-$ and in $CD4^-$ $CD8^-$ T cell hybridomas.

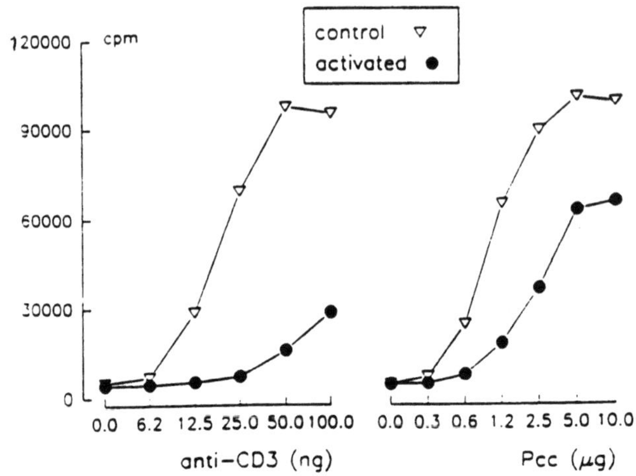

Fig.1. Il-2 production of control cells (Δ) and previously activated cells (•) by unsolubilized anti-CD3 mAbs. After 48h cells were restimulated with unsolubilized anti-CD3 mAbs (left) or Ag in presence of APC (right).

Experiments aimed at clarifying the origin of this functional inactivation have been undertaken. Some main results are summarized below (Andris et al., 1991).

(1) Hyporesponsiveness does not result from reduced surface receptor expression between control and anergic cells.

(2) Hyporesponsiveness is cell autonomous and not mediated by soluble inhibitory factors released in the culture medium.

(3) Hyporesponsiveness persists in spite of cell division and does not require continuous exposure to ligands.

(4) Hyporesponsiveness is long lasting but reversible. It does not result from the selection of genetically - stable low responding cells.

(5) Hyporesponsive cells retain their ability to produce lymphokines in response to pharmacological agents that bypass receptor triggering. Previously activated cells which respond poorly to TCR-mediated stimulation respond normally to a combination of Ca^{2+} ionophore and PKC agonists. The down-stream enzymatic machinery for Il-2 production and secretion is thus not significantly altered.

(6) Hyporesponsiveness is not a failure to respond to the second messengers IP_3 and DAG, but is concomitant with a reduced phosphatidylinositol (PIP_2) hydrolysis.

Although the exact point of uncoupling of the TCR signal from cellular responses remains to be determined, our data strongly suggest that anergy or hyporesponsiveness results from defective signal transduction at the very first steps of the activation process. Furthermore, our observations also indicate that the reduced ability to respond to antigen should result from regulatory phenomena that remain active through cell division.

3. Theoretical Insight into Antigen-Induced T-Cell Unresponsiveness.

In parallel with this experimental work, a theoretical study of antigen-mediated T-cell hyporeactivity has been undertaken which focuses on the proximal phosphorylative events that are induced upon ligand-receptor interaction. The main features of our approach arise from the observation that perturbation of the TCR-CD3 complex induces tyrosine phosphorylation of several proteins, including the zeta subunit of the receptor complex itself (June et al., 1990; Vivier et al.,1991). Recent works demonstrate that tyrosine kinase activity is needed for PLC-γ_1 activation and subsequent inositol phospholipid breakdown during receptor-mediated activation (Weiss et al., 1991). The precise role of the zeta chain relative to signal transduction, on the other hand, remains to be established. Several studies, however, indicate that it may serve a function in transmembrane signaling (Irving & Weiss, 1991). In addition, the zeta chain is constitutively phosphorylated on tyrosine residues in antigen non-responsive T cells from lpr and gld mutant mice. Tyrosine phosphorylation could thus positively and negatively regulate the signaling events proximal to TCR engagement.

The T cell receptor lacks intrinsic protein-tyrosine kinase activity, but multiple src-like kinases, localized to the inner cell membrane, have been demonstrated in T cells. Two plausible candidates for an increased tyrosine kinase activity following TCR occupancy are the tyrosine kinases p56 ^{lck} and p59 ^{fyn}, which have been shown to be physically associated with the cytoplasmic domain of, respectively, the CD4 (or CD8) molecule and the CD3 subunit (Rudd, 1990; Klausner & Samelson, 1991). Interestingly, both kinases possess, like the other members of the src-family, a key regulatory residue and a site for autophosphorylation. Dephosphorylation at the regulatory site and phosphorylation at the autophosphorylation site markedly enhance the catalytic activity of these enzymes.

Tyrosine phosphatase (PTPase) activity, on the other hand, has clearly been demonstrated for the cytoplasmic tail of another cell-surface molecule, the CD45 (Mustelin & Altman, 1990). There is now growing evidence that CD45-mediated tyrosine dephosphorylation has also an important role in the regulation of T cell activation.

Taken together these observations have led us to develop a model in terms of a phosphorylation-dephosphorylation cascade (fig. 2). A first cycle involves the autophosphorylation of a protein-tyrosine kinase. Binding of the receptors to unsolubilized ligand causes receptor aggregation, thus switching on or enhancing this autophosphorylative cycle. The most active forms of the enzyme, in turn, catalyze the phosphorylation of the T-cell receptor. The basic idea underlying our modelling is that tyrosine phosphorylation of the receptor complex or of a receptor-coupled phosphoprotein (G-like protein ? see Yokoyama & Gachelin, 1991) could negatively affect its function in signal transduction, by uncoupling to some extent the T cell receptors from the signalling apparatus (Kaufman et al., submitted).

3.1. Description of the Model.

The basic components of the model include the TCR complex associated with the p59 ^{fyn} kinase, the CD4 coreceptor linked to the p56 ^{lck} kinase and the CD45 molecule. Some important features and assumptions of the model are summarized below (see also fig. 2).

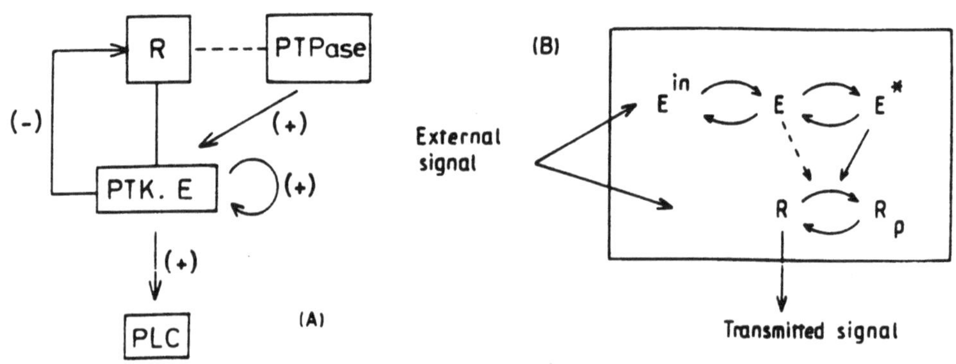

Fig.2. (a) Schematic interaction diagram. (b) Phosphorylation cascade.

(1) The external signal has a twofold characteristic : binding of the receptors to the agonist is accompanied by receptor "aggregation". Interaction of the TCR with its ligand may possibly induce a conformational change thus augmenting its signal transduction capability. This possibility is considered in section 3.4. A simple binding-diffusion mechanism that leads to an increase in receptor density is presented in section 3.2.

(2) The ligand-receptor interactions activate the fyn and p56 kinases. In the absence of precise experimental data the respective roles of these two kinases are not distinguished. They are represented by a single entity E. Specific features of the activation of this enzyme are described in section 3.3.

(3) Activation of kinase E is turned on upon engagement of the TCR *and* CD4 with their respective ligands, or upon engagement of each of these surface receptors separately in the case of $CD4^-$ cells or CD4 crosslinking. Additional costimulatory factors are not considered.

(4) A minimal role is ascribed to the CD45 PTPase that participates in the regulation of the kinase activity by dephosphorylating the regulatory site. The level of expression and isoform of CD45 remain unchanged after activation.

(5) Kinase activation leads to *reversible* phosphorylation of some cytoplasmic components of the TCR system that play a critical role for transducing the external signal. Phosphorylation occurs as well for the occupied as for the free forms of the receptors. Phosphorylated and unphosphorylated free receptors bind to the ligand with the *same* affinity.

(6) The quantitative expression of *non-phosphorylated receptor forms* is assumed to account for the level of signalling that is transmitted to the interior, and/or for the degree of coupling of the receptor system to the intracellular signalling cascade. Modified receptor forms do not (or poorly) contribute to signal transduction. This efficiency or "activity" of the receptors gives a measure of how strongly ligand-binding contributes to the generation of the physiological response i.e. Il-2 production. Some basal activity in the absence of antigen is considered (Segel et al., 1986)

3.2. Density increase upon receptor-agonist interactions.

The first step in cell activation is the interaction of the cell receptors with antigen. In this respect the case of T cells implies specific features : unlike for B cells, the T cell receptors are monovalent and T cells can only be significantly activated by antigen presented on APC or immobilized antibodies. Interaction of the T cell receptors with surface-bound ligand may contribute in several ways to favor the activation conditions. Immobilization of the ligand may focus the receptor-ligand interactions by holding the ligand near the receptors. Binding to solid-phase ligand may prevent or modify the recycling of the receptors and lengthen the time of contact with the antigen. On the other hand, multivalent presentation of the antigen becomes possible in the context of cell surfaces or aggregated antibodies. All these factors may cooperate to enhance the overall avidity of the association/dissociation process as compared to soluble agonists. Furthermore, multivalent presentation may favor receptor aggregation or oligomerization thus allowing for interactions between adjacent cytoplasmic domains which are crucial for the generation of activation signals (Watts, 1988; Graber et al., 1991).

Here, a very crude description of the spatial distribution of the receptors upon agonist-receptor interaction is presented, in terms of a "two-compartment" model. In this approach the surface of a T cell, is divided into two parts. One part, S_c, corresponds to the cell area that comes into contact with the surface-bound ligand. The other part is the free cell area $S_f = S_t - S_c \geq S_c$, with S_t the total cell area. The receptors diffuse freely between these two "compartments", in which a homogeneous distribution of the cell receptors is assumed.

In the absence of antigen, the receptors are distributed homogeneously on the whole cell surface and the concentrations in the two compartments are identical. Upon contact with the APC or artificial substitute some receptors on the contact area will bind to the ligand and become immobilized. The resulting decrease of free receptor concentration on S_c will give rise to a net influx of free receptors from S_f to S_c. If free ligand is still available these receptors may, on their turn, bind to the ligand. In this way the binding-diffusion process will cause an increase of the total receptor concentration on S_c with a concurrent density decrease on S_f.

These surface processes are described by the differential equations :

$$dA/dt = \lambda_1 (B - A) \qquad (1.a)$$

$$dB/dt = \lambda_2 (A - B) - k_+ B L + k_- C \qquad (1.b)$$

$$dC/dt = k_+ B\, L - k_- C \tag{1.c}$$

$$\text{with } L = L_T - C \tag{1.d}$$

A and B denote, respectively, the concentration (number/unit area) of free cell receptors on S_f and S_c; C represents the concentration of occupied and thus fixed receptors on Sc; L and L_T are, respectively, the concentration of free and total immobilized ligand. $\lambda_1 = \lambda/S_f$ and $\lambda_2 = \lambda/S_c$ are the mass transfer coefficients, with λ the flow rate (surface/unit time).

Neglecting in first approximation the recycling and turnover of the receptors (see M. Kaufman, in preparation) the total *number* of receptors on the cell surface is conserved, thus :

$$T = r_s A + B + C \tag{2}$$

with $r_s = S_f / S_c$. Constant T is the maximal receptor density, with all the cell receptors on the contact area.

The steady state concentrations in the absence or presence of ligand can easily be deduced from (1) and (2) :

- in absence of ligand

$$A = B = T\ /\ (1 + r_s) \text{ and } C = 0 \tag{3}$$

- in presence of ligand

$$A = B = \frac{(T - L_T) - (1 + r_s)\, k + \sqrt{D}}{2(1 + r_s)}\ , \quad C = T - (1 + r_s)\, B \tag{4}$$

with $D = (T - L_T)^2 + (1 + r_s)^2 k^2 + 2\,(1 + r_s)\, k\,(T + L_T)$, and $k = k_-\ /\ k_+$

A large excess of ligand leads to :

$$L \cong L_T\ ,\ A = B = \frac{k\,T}{(1 + r_s)\, k + L_T} \to 0\ \ ,\ C = \frac{T\ \ L_T}{(1 + r_s)\, k + L_T} \to T \tag{5}$$

Note that the same calculations hold for the interaction of CD4 receptors with MHC molecules on APC or with aggregated antibodies.

This simple approach does not consider the formation of receptor "aggregates" through the binding-diffusion processes on the contact area S_c itself. It neglects therefore the fact that receptors which remain too far apart from each other may not contribute to cell activation. Nevertheless, it provides a preliminary mechanism for an increased receptor density upon contact with the surface-bound ligand, which allows a *qualitative* study of the dynamics of T cell activation and inactivation as a function of the ligand concentration.

3.3. Bistability at the level of the autophosphorylating kinase.

Let us now focus on the kinase/phosphatase reactions depicted on fig. (2b). E^{in}, E and E* represent three "basic" states of the same kinase enzyme, with different catalytic activities. E^{in} is an inactive form, heavily phosphorylated on the regulatory site. E is dephosphorylated at this regulatory residue and has a low catalytic activity. Removal of phosphate at the regulatory site and phosphorylation at the autophosphorylation site increases the enzymatic activity several-fold and leads to the most active form, E*.

E^{in} is converted into E by the membrane-bound CD45 PTPase whose concentration is here assumed to follow the receptor concentration. E is converted into E* by intermolecular phosphorylation reactions with E itself and with E*. As shown in section 3.2, binding of the T cell receptors to unsolubilized ligand increases the receptor density at the surface of contact with antigen, thus facilitating protein-protein interactions and, in particular, the autophosphorylation of the kinase. The reverse reactions are, at this stage of the modelling, considered to occur through cytosolic enzymes located at the inner plasma membrane. In the absence of costimulatory signals it is assumed that they are not influenced by the external signal.

Assuming, by analogy with the EGF-type of kinase, that phosphorylation occurs in a sequential manner with as first substrate the peptide and second substrate ATP, and using a quasi-steady-state approximation for the enzymatic complexes the kinetic equations for the three enzymatic forms are :

$$dE^{in}/dt = -c_1 P_T E^{in} + c_2 E \tag{6.a}$$

$$dE/dt = c_1 P_T E^{in} - c_2 E - a_2 E^2 - \delta a_2 E E^* + \frac{\nu E^*}{k + E^*} \tag{6.b}$$

$$dE^*/dt = a_2 E^2 + \delta a_2 E E^* - \frac{\nu E^*}{k + E^*} \tag{6.c}$$

The kinetic constants represent combinations of the parameters characterizing the detailed enzymatic reactions. They include a constant level of ATP. P_T is the PTPase (CD45) concentration. δ is a dimensionless parameter greater than one and accounts for the higher catalytic activity of E* as compared to E. Addition of these equations gives :

$$E^{in} + E + E^* = E_T \tag{6.d}$$

where the constant E_T represents the total amount of PTK enzyme.

At the steady state equations (6) reduce to a simple cubic equation which can be written as :

$$f_a \equiv \frac{(e_T - e^*)}{(1 + \frac{c_2}{c'_1\, p_T})} \left[\delta e^* + \frac{(e_T - e^*)}{(1 + \frac{c_2}{c'_1\, p_T})} \right] = \frac{v'\, e^*}{k' + e^*} \equiv f_b \tag{7}$$

with $e^{in} = c_2 \;/\; c'_1 p_T$, $e_T = e^* + (1 + c_2 \;/\; c'_1 p_T)\, e$ and $\delta > 1$.

In this equation we have introduced the dimensionless variables :

$e^{in} = E^{in} \;/\; E_{TM}$, $e = E \;/\; E_{TM}$ and $e^* = E^* \;/\; E_{TM}$

with E_{TM} the maximal total enzyme density that can be attained.

$e_T = E_T \;/\; E_{TM}$, $p_T = P_T \;/\; P_{TM}$

$c'_1 = c_1 P_{TM}$, $k' = k \;/\; E_{TM}$ and $v' = v \;/\; (a_2 E_{TM}{}^2)$

are now scaled parameters, with $0 \le e_T \le 1$ and P_{TM} the maximal PTPase concentration.

The steady state behavior as a function of e_T can easily be understood from the graphical analysis of equation (7) shown in fig. 3. The intersections of f_a and f_b determine the steady states of system (6). For low values of e_T, the enzyme exists in its less active forms, e and e^{in}. Increasing e_T leads to a situation with two *stable* steady states. They correspond, respectively, to a high level of e / low level of e*, and vice versa. Upon further increasing e_T the most active form e* prevails at the steady state. Note that cells that overexpress the PTK will exhibit an abnormally high phosphorylative activity with predominance of the e* form. On the contrary, in cells with a low number of receptor or PTK molecules the kinase will always remain at a low catalytic activity.

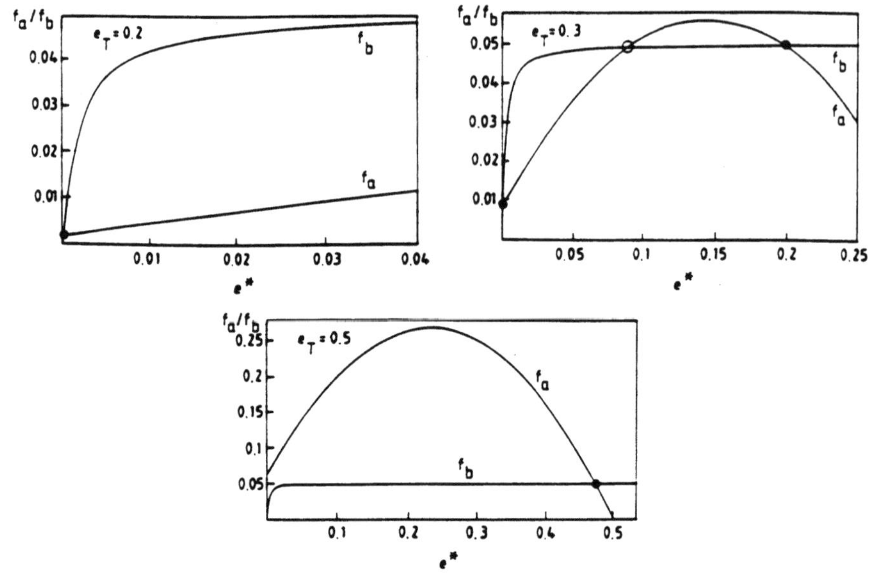

Fig.3. Graphical analysis of eq.(7) for increasing values of e_T. Stable steady states (•), unstable steady states (o).

Figure 4 illustrates the steady state behavior of e and e* as a function of the total enzyme density for the parameter values given in appendix. Although leading to more complicated equations, similar results were obtained with random or ATP/peptide sequential kinetics for the multisubstrate enzymatic reactions.

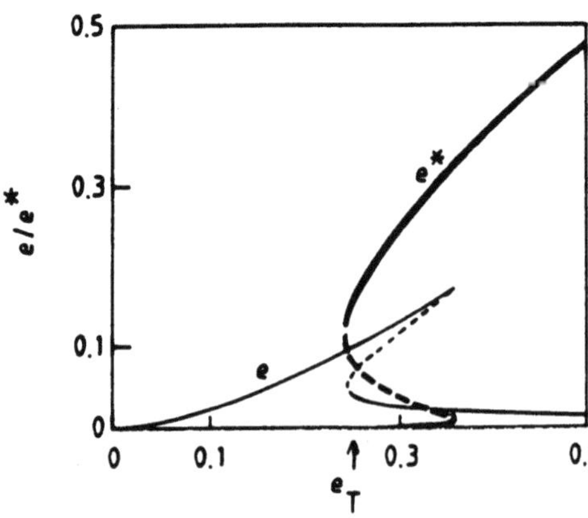

Fig.4. Stable states are indicated by solid lines, unstable states by dashed lines.

Central for bistability is the autophosphorylative activity of the kinase. Several other aspects of these kinase cycles may be modified without altering the *qualitative* picture of our results (M. Kaufman, in preparation). Coaggregation of CD45, for instance, is not an essential feature of the model and can be relaxed inasmuch the level of CD45 expression is not too low. Some extent of E- and E*-mediated conversion of E into E^{in} does not lead to significant changes in the steady state behavior. More importantly, the introduction in the model of receptor recycling and slow turnover, both accompanied by a loss of enzymatic activity, preserves the bistability feature and resulting properties of the kinase cascade.

In the following section we show that this multistationarity provides a mechanism for persistent lowering of cell responsiveness after removal of the stimulus. Antigen binding may induce an irreversible shift between two regimes of kinase activity. This transition, in turn, is reflected in the phosphorylation level of the T cell receptor complexes.

3.4. Minimal model for antigen-induced hyporesponsiveness.

The complete model integrates the binding of the ligand, the intracellular protein kinase reactions and the reversible phosphorylation of the T cell receptors. The kinetic analysis presented below applies to single cells or to a homogeneous cell population.

As before, the cell surface is divided into two compartments, S_f and S_c. The differential equations governing the time evolution of the different receptor and enzyme species in each compartment are given in appendix. They combine the various features described in the preceeding sections. In addition, both free and occupied T cell receptors can be found in a dephosphorylated or phosphorylated state.

Following Segel et al. (1986), see also Knox et al., (1986), Li & Goldbeter, (1989), we assume that the level of signalling that is transduced by the receptor system is given by a linear combination of the four receptor states :

$$S(t) = b_1 R + b_2 P + b_3 X + b_4 Y \qquad (8)$$

where the weight coefficients b_i (i = 1, 2, 3, 4) are positive constants that represent the relative contribution of each receptor state to the total "activity". R and P denote, respectively, the concentration of free unphosphorylated and phosphorylated receptors. X and Y are the corresponding concentrations of ligand-bound receptor forms. These concentrations are scaled as indicated in the appendix and S represents the scaled activity of the receptor system. A peculiar feature here is that the sum of the concentrations of the four receptor forms will vary with the strength of the external stimulus.

Consideration of unbound receptors in eq. (8) allows for some basal activity S_b prior to any exposure to antigen, that does not lead to a significant cell response. The extent of cell activation is then measured by the magnitude of the ligand-generated activity of the receptor system relative to this basal activity, S/S_b.

We consider here that the phosphorylated receptor forms do not participate in the generation of a cellular response, hence $b_2 = b_4 = 0$. Furthermore, for $b_1 = b_3$ only the ligand-induced increase in receptor density will contribute to cell activation. $b_1 > b_3$ describes, in addition, an enhanced contribution of the ligand-bound receptors.

Integration of S(t) over the time duration T of the external signal

$$S_{int} = \int_0^T [S(t) - S_b]\, dt \quad \text{for} \quad S(t) - S_b \geq 0 \tag{9}$$

is assumed to correlate with the quantity of Il- 2 that is produced during that time, and allows to determine the dose-response curves. Note that eqs. (8) and (9) apply for each compartment separately.

Some typical results are presented below. The parameter values that were used for the simulations are given in appendix, unless otherwise indicated. The strength of the stimuli are expressed in terms of the ratio of the total number of antigen-receptor (coreceptor/enzyme) molecules and ligand molecules, i.e.: N_L / N_R (N_M / N_E). Figure 5(a) displays the time course of the ligand-induced normalized activity S/S_b in compartment S_c ,for a large excess of ligands and $b_3 > b_1$. Prior to stimulation, the system is in its basal steady state which here corresponds to an initial receptor and enzyme concentration located in the bistability region (see arrow in fig. 4). At time t = 0 the stimulus is applied. As shown in figs. 5(b) and (c), the total receptor and enzyme concentrations in S_c increase rapidly. This produces a rise of the activity and the activation of the PTK which is mostly converted into its active form E*. The resulting receptor phosphorylation is then followed by a decrease of the normalized activity to a steady state value above the basal level. Desensitization remains partial, even at high ligand concentrations, since for the parameter values that have been used not all receptors become phosphorylated. At time t = 50 the ligand is removed and the system is allowed to evolve freely. The important point is that the system does not settle back to its initial steady state. Although the concentration of E* decreases substantially after stimulus removal, the switch in steady state branch and phosphorylative regime, that has occurred upon stimulation, is maintained when the initial lower receptor and enzyme density is restored. Restimulation at time t = 100, with the same ligand doses, now leads to more rapid receptor phosphorylation and a lower response. Note that low ligand concentrations do not induce a durable shift in steady state and kinase activity, in accordance with the fact that anergy is induced in T-cell hybridomas only after optimal stimulations (Andris et al., 1991).

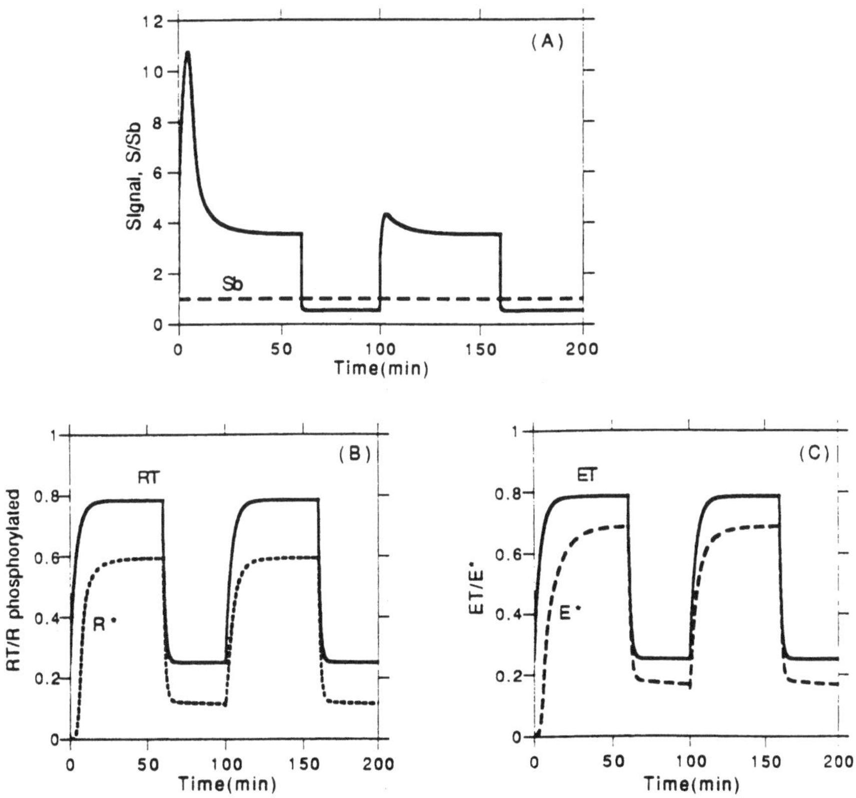

Fig.5. Time evolutions upon primary and secondary stimulation, for the ligand doses $N_L / N_R = 100$, $N_M / N_E = 100$.

When the receptors and enzymes are "aggregated" on S_c, the activity generated in compartment S_f levels off to a value below the basal level. Thus in the model the signal is transmitted only through the receptors on the contact area. This *localized* response is in good agreement with the observation that the production of cytokines in T-cells interacting with splenic B cells, for instance, is concentrated in the contact region (Kupfer et al., 1991).

Figures 6(a) and (b) give examples of dose-response curves obtained for $b_1 = 1$, $b_3 = 5$ and $b_1 = b_3 = 1$, respectively. Shown on these figures are the integrated activities (see eq. (8)) for a control system and a previously activated system that has been brought into the "less responsive" state. A shift of the dose-response curves toward higher agonist doses is observed for the previouly activated system.

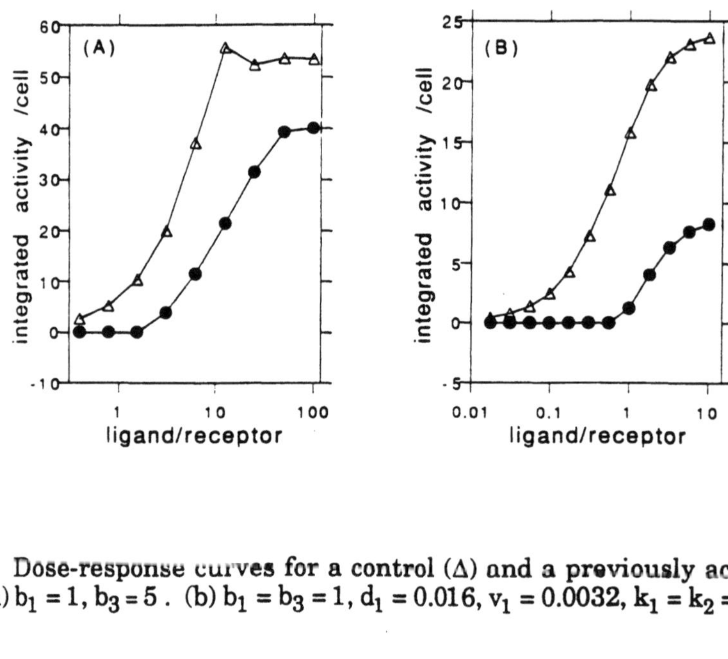

Fig.6. Dose-response curves for a control (Δ) and a previously activated system (•). (a) $b_1 = 1, b_3 = 5$. (b) $b_1 = b_3 = 1, d_1 = 0.016, v_1 = 0.0032, k_1 = k_2 = 5$.

Figure 7 illustrates a situation where a low cell responsiveness is observed at high ligand doses for the primary curve ($b_1 = b_3 = 1$). This follows from rapid and strong desensitization in the presence of an important stimulus: as indicated in fig. 7(b), during the first stimulation the activity drops to a value significantly lower than the basal activity. A strong anergy is now induced with antigen doses that lead to low primary responses.

The differences that are observed for the various dose-response curves result mainly from the level of desensitization that is established during stimulation and the rate at which this densitization occurs. This relates to the features of the receptor-ligand interactions (through parameters b_3, k_1 and k_2), to the kinetics of receptor phosphorylation and to the relative number of enzyme molecules.

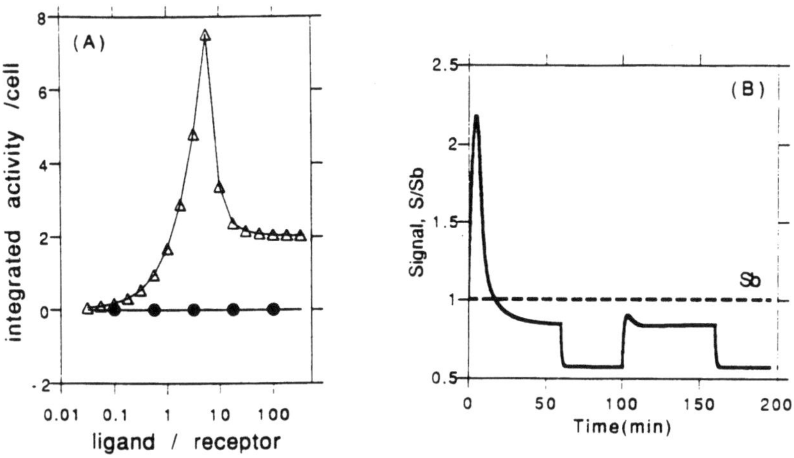

Fig.7. Dose-response curves (primary : Δ , secondary :•) and time evolution of the normalized activity, for the ligand doses $N_L / N_R = N_M / N_E = 50$. $k_1 = k_2 = 5$.

In the case of Fig. 8, the stimulus corresponds to ligand for the CD4-enzyme entity *solely*, in absence of TCR agonist ($L_T = 0$, $M_T \neq 0$, see appendix). Activation of the PTK enzyme, this time, is not accompanied by a rise in the activity S/S_b, and a low level of receptor phosphorylation is observed.

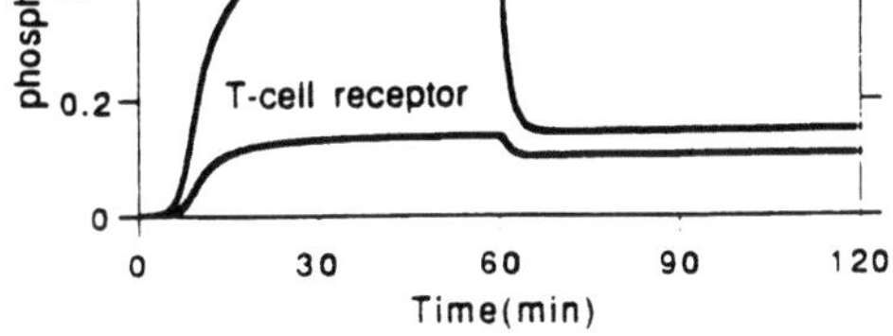

Fig.8. CD4 aggregation induces a switch in phosphorylation regime.

Nevertheless, after removal of the stimulus, a transition in phosphorylation regime has been established. A next, normal stimulation with a T-cell receptor ligand fails to evoke a significant response. Thus, in the model aggregation of CD4 by unsolubilized anti-CD4 antibodies or gp120 - antibody complexes, for example, leads to cell inactivation (Kaufman et al., submitted). This mechanism might contribute to the altered functional status of helper T-cells in HIV-infected patients (Ameisen & Gachelin, 1991)).

In the framework of this approach recovery of immunocompetence in the absence of antigen is thought to be related to fluctuations in parameters such as turnover rates or enzyme and receptor number. These fluctuations might become more important during the process of cell division and induce some cells to jump back to their basal state. Gradual recovery of immunocompetence should then arise from a heterogeneity of the cell population. Slow recovery, on the other hand, has also been observed for a system that functions *outside but close* to the transition region in a lattice model allowing for spatial fluctuations of the various receptor concentrations (Kaufman & Stauffer, submitted).

4. Concluding Remarks.

A theoretical explanation for persistent lowering of T-cell responsiveness is presented, in terms of a transition between different dynamical states of antigen-receptor tyrosine phosphorylation. The general features of the model fit those observed for experimental systems. In the framework of this analysis it is proposed that autophosphorylation of protein tyrosine kinases is responsible for epigenetic changes that can be maintained through cell division (Delbrück, 1949; Thomas & D'Ari, 1990). Many details of the phosphorylative events that accompany T-cell receptor occupancy remain to be elucidated. In particular, the respective roles of the fyn and lck kinases, possible additional regulatory features of the CD45 PTPase and the exact point of blocking of the biological response have to be clarified. Support for the role of tyrosine phosphorylation in anergy induction, underlying the model, is provided by recent experimental observations. Cell stimulation through transfected heterologous (muscarinic) receptors, that are linked to the PI pathway but do not involve the activation of a protein tyrosine kinase, does not lead to unresponsiveness, in particular upon TCR engagement (Andris et al., submitted).

Several implementations of the model are presently considered, including recycling and turnover of receptors and enzymes, and explicit activation of PLC-γ_1 by tyrosine phosphorylation. The dynamics of receptor aggregation upon agonist-receptor interactions, on the other hand, constitutes a subject of investigation in itself. This spatio-temporal problem is being studied both at the molecular level, in terms of an automata network approach, and at the macroscopic level in the framework of reaction-diffusion equations.

The model that is presented applies to the induction of T-cell refractoriness in the absence of co-stimulatory signals. Further modelling should allow to test the hypothesis that co-stimulatory factors may modify the balance of receptor phosphorylation/dephosphorylation (Kaufman et al., in preparation), and could lead to some insight into anergy versus memory induction.

Antigen-induced desensitization and unresponsiveness may serve to prevent overproduction of lymphokines that leads to physiological disturbances, and may play a role in peripheral tolerance. On the basis of our theoretical results we also propose that some helper T-cell dysfunctions in AIDS might be explained by deregulation of this normal physiological phenomenon.

Acknowledgements

This work has been supported by grants from the Belgian State, the Communeauté Francophone de Belgique and the Commission of the European Communities. M.K. acknowledge financial support by the Belgian program on interuniversity attraction poles. O.L. is a Research Associate from the National Fund for Scientific Research (Belgium).

References

Ameisen J-C, Capron A (1991) Cell dysfunction and depletion in AIDS : the programmed cell death hypothesis. Immunol Today 12 : 102-105

Andris F, Van Mechelen M, Legrand N, Dubois P-M, Kaufman M, Urbain J, Leo O (1991) Induction of long-term but reversible unresponsiveness after activation of murine T-cell hybridomas. Int Imm 3 : 609-616

Blackman M-A, Finkel T-H, Kappler J, Cambier J, Marrack P (1991) Altered antigen receptor signaling in anergic T-cells from self-tolerant T-cell receptor β-chain transgenic mice. Proc Natl Acad Sci USA 88 : 6682-6686

Delbrück M (1949) Discussion. In Unités biologiques douées de continuité génétique, Editions du CNRS, Lyon

De Wit D, Van Mechelen M, Ryelandt M, Figueirido A-C, Abramowicz D, Goldman M, Bazin M, Urbain J, Leo O (1992) The injection of deaggregated gamma globulin in adult mice induces antigen-specific unresponsiveness of T helper type 1 but not type 2 lymphocytes. J Exp Med 175 : 9-14

Finkel T-H, Kubo R-T, Cambier J-C (1991) T-cell development and transmembrane signaling : changing biological responses through an unchanging receptor. Immol Today 12 : 79-85

Graber M, Bockenstedt L-K, Weiss A (1991) Signaling via the inositol phospholipid pathway by T-cell antigen receptor is limited by receptor number. J Immunol 146 : 2935-2943

Hervagault J-F, Canu S (1987) Bistability and Irreversible Transitions in a simple substrate cycle. J Theor Biol 127 : 439-449

Imboden J-B, Stobo J-D (1985) Transmembrane signalling by the T-cell antigen receptor. Perturbation of the T3-antigen receptor complex generates inositol phosphates and releases calcium ions from intracellular stores. J Exp Med 161 : 446-456

Irving B-A, Weiss A (1991) The cytoplasmic domain of the T-cell receptor zeta chain is sufficient to couple to receptor-associated signal transduction pathways. Cell 64 : 891-901

Jenkins M-K, Schwartz R-H (1987) Antigen presentation by chemically modified splenocytes induces antigen-specific unresponsiveness in vitro and in vivo. J Exp Med 165 : 302

June C-H, Fletcher M-C, Ledbetter J-A, Samelson L-E (1990) Increase in tyrosine phosphorylation are detectable before phospholipase C activation after T-cell receptor stimulation. J Immun 144 : 1591-1599

Kaufman M, Thomas R (1987) Model analysis of the basis of multistationarity in the humoral immune response. J Theor Biol 114 : 527-561

Klausner R-D, Samelson L-E (1991) T-cell antigen receptor activation pathways: the tyrosine kinase connection. Cell 64 : 875-878

Knox B-E, Devreotes P-N, Goldbeter A, Segel L-A (1986) A molecular mechanism for sensory adaptation based on ligand-induced receptor modification. Proc Natl Acad Sci USA 83 : 2345-2349

Kupfer A, Mosmann T-R, Kupfer H (1991) Polarized expression of cytokines in cell conjugates of helper T-cells and splenic B-cells. Proc Natl Acad Sci USA 88 : 775-779

Li Y-X, Goldbeter A (1989) Frequency specificity in intercellular communication. Biophys J 55 : 125-145

Lisman J-E (1985) A mechanism for memory storage insensitive to molecular turnover : a bistable autophosphorylating kinase. Proc Natl Acad Sci USA 82 : 3055-3057

Mustelin T, Altman A (1990) Dephosphorylation and activation of the T-cell tyrosine kinase pp56lck by the leukocyte common antigen (CD45). Oncogene 5 : 809-813

Rudd C-E (1990) CD4, CD8 and the TCR-CD3 complex : a novel class of protein kinase receptor. Immunol Today 11 : 400-406

Schwartz R-H (1990) A cell culture model for T-lymphocyte clonal anergy. Science 248 : 1349-1356

Segel A, Goldbeter A, Devreotes P-N, Knox B-E (1986) A mechanism for exact sensory adaptation based on receptor modification. J Theor Biol 120: 151-179

Thomas R, D'Ari R (1990) Biological Feedback, CRC Press, Boca Raton, Florida

Vivier E, Morin P, Tian Q, Daley J, Blue M-L, Schlossman S-F, Anderson P (1991) Expression and tyrosine phosphorylation of the T-cell receptor zeta-subunit in human thymocytes. J Immunol 146 : 1142-1148
Watts T (1988) T-cell activation by preformed, long-lived Ia-peptide complexes. J Immunol 141 : 3708-3714
Weiss A, Koretzky G, Schatzmann R-C, Kadlecek T (1991) Functional activation of the T-cell antigen receptor induces tyrosine phosphorylation of phospholiyase C-γ1. Proc Natl Acad Sci USA 88 : 5484-5488
Wilde D-B, Fitch F-W (1984) Antigen-reactive cloned helper T-cells. I. Unresponsiveness to antigenic restimulation develops after stimulation of cloned helper T-cell. J Immunol 132 : 1632
Yokoyama K, Gachelin G (1991) An abnormal signal transduction pathway in CD4⁻ CD8⁻ double-negative lymph node cells of MRL lpr/lpr mice. Eur J Immunol 21 : 2987-2992

Appendix

Kinetic equations for the time evolution of the various receptor and enzymatic forms.

(1) Compartment S_f

$$dR_1 / dt = -(d_1 V_1^* R_1 - v_1 P_1) + \lambda_1 (R_2 - R_1) \tag{A1}$$

$$dP_1 / dt = (d_1 V_1^* R_1 - v_1 P_1) + \lambda_1 (P_2 - P_1) \tag{A2}$$

$$dV_1^{in} / dt = -(c_1 p_{1T} V_1^{in} - c_2 V_1) + \lambda_1 (V_2^{in} - V_1^{in}) \tag{A3}$$

$$dV_1 / dt = (c_1 p_{1T} V_1^{in} - c_2 V_1) - \phi_1 + \lambda_1 (V_2 - V_1) \tag{A4}$$

$$dV_1^* / dt = \phi_1 + \lambda_1 (V_2^* - V_1^*) \tag{A5}$$

with $\phi_1 = a_2 V_1^2 + \delta a_2 V_1 V_1^* - \dfrac{v V_1^*}{K + V_1^*}$

(2) Compartment S_c

- Unbound species :

$$dR_2 / dt = -[d_1 (V_2^* + W_2^*) R_2 - v_1 P_2] - (k_1 R_2 L - k_{-1} X_2) + \lambda_2 (R_1 - R_2) \tag{A6}$$

$$dP_2 / dt = [d_1 (V_2^* + W_2^*) R_2 - \upsilon_1 P_2] - (k_1 P_2 L - k_{-1} Y_2) + \lambda_2 (P_1 - P_2) \quad \text{(A7)}$$

$$dV_2^{in} / dt = -(c_1 p_{2T} V_2^{in} - c_2 V_2) - (k_2 V_2^{in} M - k_{-2} W_2^{in}) + \lambda_2 (V_1^{in} - V_2^{in}) \quad \text{(A8)}$$

$$dV_2 / dt = (c_1 p_{2T} V_2^{in} - c_2 V_2) - \phi_2 - (k_2 V_2 M - k_{-2} W_2) + \lambda_2 (V_1 - V_2) \quad \text{(A9)}$$

$$dV_2^* / dt = \phi_2 - (k_2 V_2^* M - k_{-2} W_2^*) + \lambda_2 (V_1^* - V_2^*) \quad \text{(A10)}$$

with $\phi_2 = a_2 V_2 (V_2 + W_2) + \delta a_2 V_2 (V_2^* + W_2^*) - \dfrac{\upsilon V_2^*}{K + (V_2^* + W_2^*)}$

- Ligand-bound species :

$$dX_2 / dt = - [d_1 (V_2^* + W_2^*) X_2 - \upsilon_1 Y_2] + (k_1 R_2 L - k_{-1} X_2) \quad \text{(A11)}$$

$$dY_2 / dt = [d_1 (V_2^* + W_2^*) X_2 - \upsilon_1 Y_2] + (k_1 P_2 L - k_{-1} Y_2) \quad \text{(A12)}$$

$$dW_2^{in} / dt = - (c_1 p_{2T} W_2^{in} - c_2 W_2) + (k_2 V_2^{in} M - k_{-2} W_2^{in}) \quad \text{(A13)}$$

$$dW_2 / dt = (c_1 p_{2T} W_2^{in} - c_2 W_2) - \Psi_2 + (k_2 V_2 M - k_{-2} W_2) \quad \text{(A14)}$$

$$dW_2^* / dt = \Psi_2 + (k_2 V_2^* M - k_{-2} W_2^*) \quad \text{(A15)}$$

Where $L = L_T / R_{TM} - (X_2 + Y_2)$

$M = M_T / E_{TM} - (W_2^{in} + W_2 + W_2^*)$

and $\Psi_2 = a_2 W_2 [V_2 + W_2] + \delta a_2 W_2 (V_2^* + W_2^*) - \dfrac{\upsilon W_2^*}{K + (V_2^* + W_2^*)}$

R and P denote, respectively, the concentrations of free non-phosphorylated and phosphorylated receptors. X and Y are the corresponding bound-receptor concentrations. V and W are, respectively, the free and bound enzymatic forms E^{in}, E and E^*. L and L_T are the concentrations of free and total unsolubilized T-cell receptor ligand. M and M_T are the corresponding concentrations for the coreceptor ligand.

These equations are complemented with the conservation relations :

$$r_s(R_1 + P_1) + R_2 + P_2 + X_2 + Y_2 = 1$$

and $$r_s(V_1^{in} + V_1 + V_1^*) + V_2^{in} + V_2 + V_2^* + W_2^{in} + W_2 + W_2^* = 1$$

expressing that the total *number* of receptor and coreceptor (or enzyme) molecules, N_R and N_E, are conserved on the total cell surface.

The concentrations of the different receptor forms and the receptor-ligand concentration are scaled to the maximal total receptor concentration $R_{TM} = N_R / S_c$. Similarly, the concentrations of the enzymatic forms and enzyme-ligand concentration are scaled by the maximal enzyme concentration $E_{TM} = N_E / S_c$. Hence, $d_1 = d_1' E_{TM}$, $k_1 = k'_1 R_{TM}$, $k_2 = k'_2 E_{TM}$, $a_2 = a'_2 E_{TM}$, $K = K' / E_{TM}$, $v = v' / E_{TM}$ and $c_1 = c'_1 P_{TM}$. The primes refer to the unscaled parameters. P_{TM} is the maximal PTPase concentration. The scaled PTPase concentrations, p_{1T} and p_{2T}, are here considered to vary proportionally with the total T-receptor concentrations in each compartment, respectively. λ_1 and λ_2 are as defined in section 3.2.

The equations for "CD4⁻" cells with a single ligand for the TCR *and* associated PTK enzyme are obtained by introducing the additional constraints :

$$M \equiv L = L_T / R_{TM} - (X_2 + Y_2), \quad k_2 \equiv k_1$$

$$V_1^{in} + V_1 + V_1^* = R_1 + P_1$$

$$V_2^{in} + V_2 + V_2^* = R_2 + P_2$$

$$W_2^{in} + W_2 + W_2^* = X_2 + Y_2.$$

3. *Parameter Values*

Unless indicated otherwise the numerical simulations were performed with the following parameter values (in the units min^{-1}) :

$d_1 = 1.6$, $v_1 = 0.32$, $c_1 = 0.16$, $c_2 = 0.06$, $a_2 = 0.5$, $v = 0.025$,
$K = 0.002$, $k_1 = k_2 = 1.0$, $k_{-1} = k_{-2} = 10.0$, $\delta = 8$, $\lambda_1 = 0.2$,
$r_s = 3$ and $\lambda_2 = r_s \lambda_1$

Additional Abbreviations : Ag, antigen; MHC, Major Histocompatibility Complex; mAb, monoclonal antibody.

INTERACTION OF INTERLEUKIN-2 WITH ITS CELL SURFACE RECEPTORS: INTERPRETATION OF EQUILIBRIUM BINDING EXPERIMENTS VIA SCATCHARD PLOTS

Alan S. Perelson and Byron Goldstein
Theoretical Biology and Biophysics Group
Theoretical Division
Los Alamos National Laboratory
Los Alamos, NM 87545 USA

ABSTRACT

The high affinity receptor for interleukin 2 (IL-2) is composed of at least two chains, a p55 chain that binds IL-2 with low affinity and a p75 chain that binds with intermediate affinity. Two molecular mechanisms have been proposed for the formation of the high affinity receptor-ligand complex. The affinity conversion model proposed by Honjo and coworkers (Saito et al., 1988) suggests that the high affinity receptor is formed via stepwise binding reactions in which IL-2 first binds to the p55 chain and the resulting complex then associates with the p75 chain to form a high affinity ternary complex. In the preformed heterodimer model the p55 and p75 chains form a non-covalently linked high affinity heterodimer in the absence of IL-2. We analyze these models theoretically and show that they can be distinguished on the basis of equilibrium binding experiments using cell lines expressing different numbers of p55 chains. We show that published IL-2 binding studies are consistent with a model in which a large concentration of preformed heterodimers is present on the cell surface and inconsistent with a model in which preformed heterodimers are absent from the cell surface.

INTRODUCTION

Interleukin 2 (IL-2) binds to cell surface receptors on activated T cells with high affinity. Two polypeptide chains have been detected on the surface of those cells that bind IL-2, a p55 chain (Tac antigen) that binds with low affinity (10^8 M^{-1}), and a p75 chain that binds with an intermediate affinity (10^9 M^{-1}). High affinity binding (10^{11} M^{-1}) has only been observed on cells that express both chains (reviewed in Smith, 1988,

NATO ASI Series, Vol. H 66
Theoretical and Experimental Insights into Immunology
Edited by A. S. Perelson and G. Weisbuch

1989). Binding studies with cells expressing only one of the chains show no high affinity binding. Scatchard plots of these binding data are straight lines, consistent with single homogeneous populations of receptors (Teshigawara et. al., 1987; Dukovich et al., 1987; Hatakeyama et al., 1989). This and a variety of other evidence (cf. Wang and Smith, 1987; Robb et al., 1987; Tsudo et al., 1987; Hatakeyama et al., 1989) indicate that high affinity binding is due to the formation of p55–p75 heterodimers. Kinetic studies reinforce this notion. Measurements of the rate of IL-2 binding and dissociation from cells expressing only p55 chains, only p75 chains, or both, established that the high affinity receptor had a rapid on rate constant, characteristic of the p55 chain, and a slow off rate constant, characteristic of the p75 chain (Wang and Smith, 1987).

Certain aspects about the formation of the high affinity receptor are unresolved. It is not known whether the heterodimer exists preformed in the absence of IL-2, or whether it requires the presence of IL-2 for its formation. Chemical crosslinking studies indicate that the heterodimer exists in the absence of IL-2 (Saragovi and Malek, 1988; Yamaguchi et al., 1989). However, Honjo and his collaborators (Kondo et al., 1986; Saito et al., 1988), based on a variety of kinetic binding studies, believe heterodimers only form in the presence of IL-2. They have proposed an "affinity conversion model" in which heterodimers form by a two step process. First, IL-2 binds to the p55 chain with low affinity, and then the IL-2-p55 complex interacts with a free p75 chain to form a heterodimer with IL-2 bound. This latter interaction causes the IL-2 to become more tightly bound, creating the observed high affinity receptor. In this paper we shall try to resolve the issue of preformed heterodimers versus affinity conversion using a mathematical model that allows us to interpret published data in a new way. A brief account of some of these results has appeared in Goldstein et al. (1992).

A MODEL FOR THE INTERACTION OF IL-2 WITH ITS RECEPTORS

Our model has as its basis the physical chemistry of the potential interactions between IL-2 and its cell surface receptors. We start by enumerating the possible chemical reactions between IL-2 and the various receptor chains. The reactions included in our model will allow us to distinguish between the preformed heterodimer model and the affinity conversion model.

It is known that IL-2 can bind from solution to free light chains (p55) and to free heavy chains (p75). If heterodimers form in the absence of IL-2, then there are three direct binding reactions that can occur:

$$C + L \underset{k_{L-}}{\overset{k_{L+}}{\rightleftarrows}} L^* \,, \tag{1a}$$

$$C + H \underset{k_{H-}}{\overset{k_{H+}}{\rightleftarrows}} H^* \,, \tag{1b}$$

$$C + D \underset{k_{D-}}{\overset{k_{D+}}{\rightleftarrows}} D^* \,. \tag{1c}$$

Iere L, H, and D are the concentrations of free p55 chain, free p75 chain, and free ›55-p75 heterodimer. The variables L^*, H^*, and D^* are the concentrations of bound complexed with IL-2) p55 chain, bound p75 chain, and bound p55-p75 heterodimer, espectively; and C is the concentration of free IL-2 in solution. The forward and everse rate constants are indicated for each reaction. The equilibrium constants for eactions (1a), (1b), and (1c) are $K_L = k_{L+}/k_{L-}$ (low affinity), $K_H = k_{H+}/k_{H-}$ intermediate affinity), and $K_D = k_{D+}/k_{D-}$ (high affinity). Since these equilibrium :onstants characterize the binding of a ligand that is free in solution, they must have ınits of 1/(solution concentration), e.g., M^{-1}. Assuming all receptors, including the ıeterodimer, are monovalent for IL-2, then the above reactions are the only direct ›inding reactions that can occur. In the affinity conversion model, reaction (1c) does ıot occur because heterodimers do not exist unless complexed with IL-2.

Vhen p55 and p75 chains are expressed on the same cell surface, they can interact to orm heterodimers by any of the following reactions:

$$L + H \underset{k_{LH-}}{\overset{k_{LH+}}{\rightleftarrows}} D \,, \tag{2a}$$

$$L^* + H \underset{k_{L^*H-}}{\overset{k_{L^*H+}}{\rightleftarrows}} D^* \,, \tag{2b}$$

$$L + H^* \underset{k_{H^*D-}}{\overset{k_{LH^*+}}{\rightleftarrows}} D^* \,. \tag{2c}$$

n the affinity conversion model, reaction (2a) does not occur. Saito et al. (1988) ırgue that on T cells that express considerably more light chain than heavy chain, the lominant reaction for forming the high affinity binding state is reaction (2b). The quilibrium constants for these reactions (2a), (2b), and (2c) are $K_{LH} = k_{LH+}/k_{LH-}$, $K_{L^*H} = k_{L^*H+}/k_{L^*H-}$, and $K_{LH^*} = k_{LH^*+}/k_{LH^*-}$. Since these equilibrium constants haracterize bimolecular interactions among receptor chains on a cell surface, they have units of 1/(surface concentration). We take all such equilibrium constants to have units of cm^2 and measure receptor concentrations in receptors/cm^2 by dividing the number of receptors per cell by the surface area of the cell. In sample calculations we take the cell surface area $A = 5 \times 10^{-6}$ cm^2, which corresponds to a smooth spherical cell of radius some what greater than 6 μm, a reasonable size for an activated lymphocyte.

There is a third type of reaction that may occur on the cell surface; namely, a bound single chain may interact with a free heterodimer and transfer a bound IL-2 to the heterodimer:

$$L^* + D \underset{k_{L^*D-}}{\overset{k_{L^*D+}}{\rightleftarrows}} L + D^* , \tag{3a}$$

$$H^* + D \underset{k_{H^*D-}}{\overset{k_{H^*D+}}{\rightleftarrows}} H + D^* . \tag{3b}$$

The equilibrium constants for these reactions are $K_{L^*D} = k_{L^*D+}/k_{L^*D-}$ and $K_{H^*D} = k_{H^*D+}/k_{H^*D-}$. These equilibrium constants are dimensionless since they refer to reactions were two receptors interact to form two other receptors. Again, in the affinity conversion model such reactions can't occur because there are no free heterodimers.

Although eight equilibrium constants have been introduced in describing reactions (1) – (3), they are not all independent. We shall show in the next section that only four equilibrium constants are required to characterize the eight reactions.

FORMULATING THE MATHEMATICAL MODEL

At equilibrium (i.e., at long times with internalization, recycling, and synthesis blocked) a number of relationships must hold among the ligand and receptor concentrations. Using the direct binding reactions, (1a), (1b), and (1c), we find that at equilibrium

$$L^* = K_L C L , \tag{4a}$$

$$H^* = K_H C H , \tag{4b}$$

$$D^* = K_D C D . \tag{4c}$$

The bimolecular reactions, (2a), (2b), and (2c), imply that at equilibrium

$$D = K_{LH} L H , \tag{5a}$$

$$D^* = K_{L^*H} L^* H , \tag{5b}$$

$$D^* = K_{LH^*} LH^* \ . \tag{5c}$$

Lastly, the transfer reactions, (3a) and (3b), yield

$$D^* = K_{L^*D} L^* D / L \ , \tag{6a}$$

$$D^* = K_{H^*D} H^* D / H \ . \tag{6b}$$

In addition to these equilibrium relationships, which follow from the law of mass action, there are two conservation laws:

$$L_T = L + L^* + D + D^* \ , \tag{7a}$$

$$H_T = H + H^* + D + D^* \ , \tag{7b}$$

where L_T and H_T are the total (both in and out of heterodimers) concentrations of light and heavy chains on the cell surface, respectively. For the affinity conversion model $D = 0$. It is also useful to introduce

$$D_T(C) = D + D^* \ , \tag{8}$$

where $D_T(C)$ is the total concentration of heterodimers, both free and bound, when the free IL-2 concentration is C. We call the concentration of heterodimers in the absence of IL-2: D_0, i.e., $D_T(0) = D(0) = D_0$. In the affinity conversion model $D_0 = 0$. In the preformed heterodimer model we can calculate D_0 from Eq. (5a). In the absence of IL-2, $L^* = H^* = D^* = 0$ and Eq. (5a) becomes

$$D_0 = K_{LH}(L_T - D_0)(H_T - D_0) \ . \tag{9}$$

This is a quadratic equation in D_0 with solution

$$D_0 = \frac{1 + K_{LH}R_T - \sqrt{(1 + K_{LH}R_T)^2 - 4K_{LH}^2 L_T H_T}}{2K_{LH}} \ , \tag{10a}$$

where

$$R_T = L_T + H_T \ . \tag{10b}$$

There are two solutions to this quadratic equation. We have chosen the physically meaningful solution with the minus sign in front of the square root. This is the solution that vanishes when one of the chain concentrations is zero.

When $L_T \gg H_T$, as is the case for many of the experimental systems studied, Eq. (9) can be approximated by

$$D_0 = K_{LH} L_T (H_T - D_0) \ ,$$

with solution

$$D_0 = \frac{K_{LH}L_T}{1+K_{LH}L_T} H_T \; . \tag{10c}$$

(This solution can also be obtained by expansion of Eq. (10a).) When $L_T \gg H_T$, which is generally the case for activated T cells, substantial heterodimer formation in the absence of IL-2 requires that $K_{LH}L_T \geq 1$. For example, when $K_{LH}L_T = 1$, Eq. (10c) shows that approximately half of the heavy chains are incorporated into heterodimers in the absence of IL-2. If there are 10^4 p55 chains on a cell with a surface area of 5×10^{-6} cm^2, so that $L_T = 2 \times 10^9$ cm^{-2}, the inequality $K_{LH}L_T \geq 1$ is satisfied if $K_{LH} \geq 5 \times 10^{-10}$ cm^2. Such values of K_{LH} are required if direct binding of IL-2 to preformed heterodimers is to be significant at low IL-2 concentrations.

SCATCHARD PLOTS

Many of the conclusions drawn from IL-2 binding studies have been based on interpretations of Scatchard plots. Recall that a Scatchard plot is a plot of B/C versus B, where B, the total concentration of surface bound IL-2, is given by

$$B = L^* + H^* + D^* \; . \tag{11}$$

Scatchard plots are commonly used to study receptor systems because, in the simple case of a homogeneous population of noninteracting receptors, the plot is a straight line with slope equal to the ligand-receptor affinity and x-intercept equal to the total number of receptors. Deviation from a straight line plot can indicate positive or negative cooperativity in binding or heterogeneity in the receptor population. In the case of interacting receptor populations, as is the case in the IL-2 system, interpretation of these plots is more complicated; but much can still be learned from the slopes and intercepts of the straight line portions of the plot, as will be discussed in the next section.

To determine the concentrations of the bound receptors, we use Eqs. (7a) and (7b), to rewrite Eqs. (4a), (4b), and (4c) as follows:

$$L^* = \frac{K_L C(L_T - D_T(C))}{1+K_L C} \; , \tag{12a}$$

$$H^* = \frac{K_H C(H_T - D_T(C))}{1+K_H C} \; , \tag{12b}$$

$$D^* = \frac{K_D C D_T(C)}{1+K_D C} \; . \tag{12c}$$

To obtain an expression for B in terms of the IL-2 concentration, C, and the total concentrations of light and heavy chains, L_T and H_T, we need an expression for $D_T(C)$ in terms of these variables. We note that from Eqs. (4c) and (5a) that $D^* = K_D C D = K_D C K_{LH} L H$. Using Eqs. (7a) and (7b) to eliminate L and H, this expression becomes $D^* = K_D C K_{LH}(L_T - L^* - D_T(C))(H_T - H^* - D_T(C))$. Finally, using Eqs. (12a), (12b), and (12c) to eliminate L^*, H^*, and D^* we obtain

$$D_T(C) = \frac{1 + K_D C}{(1 + K_L C)(1 + K_H C)} K_{LH}(L_T - D_T(C))(H_T - D_T(C)) \; .$$

This is a quadratic equation for $D_T(C)$ with solution

$$D_T(C) = \frac{1 + F(C)R_T - \sqrt{(1 + F(C)R_T)^2 - 4(F(C))^2 L_T H_T}}{2F(C)} \; , \qquad (13a)$$

where

$$F(C) = \frac{(1 + K_D C)K_{LH}}{(1 + K_L C)(1 + K_H C)} \; . \qquad (13b)$$

Note that when $C = 0$, $F(0) = K_{LH}$ and Eq. (13a) reduces to Eq. (10a). Additionally, one can show that if $K_D > K_L + K_H$, then for small C, $D_T(C)$ is an increasing function of C, i.e., the addition of small amounts of IL-2 to a system in which it is absent induces dimerization. Conversely, if $K_D < K_L + K_H$, then the initial addition of IL-2 will lead to the breakup of preformed heterodimers.

This completes the calculation of B, the equilibrium concentration of bound IL-2 as a function of the free IL-2 concentration. The total heterodimer concentration is given by Eq. (13a). The substitution of Eq. (13a) into the sum of Eqs. (12a), (12b), and (12c) gives B. For the preformed heterodimer model we can therefore predict the shape of a Scatchard plot if we know the total concentrations of light and heavy chain, L_T and H_T, and four equilibrium constants: K_L, K_H, K_D, and K_{LH}. In the Appendix we outline the analogous derivation for the affinity conversion model. In that model the complete binding curve is determined by L_T and H_T and only three equilibrium binding constants: K_L, K_H, and K_{L^*H}.

In formulating the preformed heterodimer model we considered eight distinct chemical interactions and introduced eight equilibrium constants. We now have shown that only four of these equilibrium constants are needed to fully describe the binding. This is because we have used the thermodynamic principle that the free energy of a ligand-receptor complex is independent of the way the complex was formed. Thus the concentration of bound heterodimer at equilibrium can be calculated assuming it was formed by an

IL-2 interacting with a free heterodimer, Eq. (4c), or by a bound light chain interacting with a free heavy chain, Eq. (5b), or by a bound heavy chain interacting with a free light chain, Eq. (5c), or by a transfer reaction. If cell bound heterodimers are equivalent then there are only four independent equilibrium constants. This would mean for example, that equilibrium binding experiments cannot be used to ascertain whether transfer reactions, Eq. (3a) and (3b), occur since the equilibrium constants for these reactions are not needed to determine any of the equilibrium concentrations. However, if different pathways of bound heterodimer formation lead to states with different free energies of binding, e.g., phosphorylated or unphosphorylated receptors with different affinities for IL-2, then these different heterodimer states would need to be distinguished in the model, and additional equilibrium constants would be required to describe the binding. As yet no such additional states have been found. Indeed, it has been shown that activation of protein kinase C, which is known to lead to the phosphorylation of p55 chains, does not affect the affinity of the heterodimer for IL-2 as assessed from Scatchard analysis (Robb and Rusk, 1986).

By equating different expressions for D^*, Eq. (4c), (5b), (5c), (6a), and (6b) and substituting appropriately, the following relations among the various equilibrium constants can be obtained. These are known as detailed balance constraints (e.g., Hill, 1977).

$$K_{L^*H} = K_D K_{LH} / K_L \; , \tag{14a}$$

$$K_{LH^*} = K_{L^*H} K_L / K_H \; , \tag{14b}$$

$$K_{H^*D} = K_D / K_H \; , \tag{14c}$$

$$K_{L^*D} = K_{H^*D} K_H / K_L \; . \tag{14d}$$

In the affinity conversion model only four reactions are allowed, (1a), (1b), (2b), and (2c), and therefore only Eqs. (4a), (4b), (5b), and (5c) apply to this model. Of the four equilibrium constants in these equations only three are independent. The fourth equilibrium constant can be calculated from Eq. (14b), which holds for the affinity conversion model and the preformed heterodimer model.

These constraints must be kept in mind when comparing the two models. Saito et al. (1988) have argued that reaction (2b) is the dominant pathway for forming heterodimers, and therefore that K_{L^*H} must be large. It is important to realize that if there is any interaction between p55 and p75 chains in the absence of IL-2, even if such interactions are weak and lead to few heterodimers being formed, then Eq. (14a) holds and $K_{LH} = K_{L^*H} K_L / K_D$. Thus the equilibrium constant for formation of dimer in the absence of IL-2 cannot be arbitrarily small, but is determined by $K_{L\bullet H}$ and the equilibrium constants for direct binding to the high and low affinity receptors, K_D and K_L.

INTERPRETATION OF SCATCHARD PLOTS

Experimentalists have used Scatchard plots of binding data to draw qualitative and quantitative conclusions about the interactions of IL-2 with its receptors on T cells. Figure 1 is a schematic of an equilibrium Scatchard plot, B/C versus B, where B is the concentration of surface bound IL-2 and C is the free IL-2 concentration. Indicated in the figure are the following characteristics of the plot: the initial ($C \to 0$) and final ($C \to \infty$) slopes, S_0 and S_∞, the y-intercept, Y_{int}, the initial x-intercept, i.e., the x-intercept obtained by extrapolating the initial straight line portion of the curve to the x-axis, X_0, and the true x-intercept obtained when all binding sites are filled, $X_\infty = R_T = L_T + H_T$.

Results For The Preformed Heterodimer Model

In the absence of IL-2 ($C = 0$) the concentration of free heterodimers is D_0. We also define L_0 and H_0 as the concentrations of light and heavy chains that are not in heterodimers when $C = 0$, and R_0 as the total number of receptors on the cell surface when $C = 0$. Thus,

$$L_0 = L_T - D_0 \ , \tag{15a}$$

$$H_0 = H_T - D_0 \ , \tag{15b}$$

$$R_0 = L_0 + H_0 + D_0 = L_T + H_T - D_0 \ . \tag{15c}$$

For the preformed heterodimer model we define the average affinity, $< K >$, and the average squared affinity , $< K^2 >$, to be

$$< K >= (L_0 K_L + H_0 K_H + D_0 K_D)/R_0 \ , \tag{16a}$$

$$< K^2 >= (L_0 K_L^2 + H_0 K_H^2 + D_0 K_D^2)/R_0 \ . \tag{16b}$$

We can obtain analytical expressions for the various characteristics of the Scatchard plot in the following way. The amount of ligand bound to receptors, B, is a function of

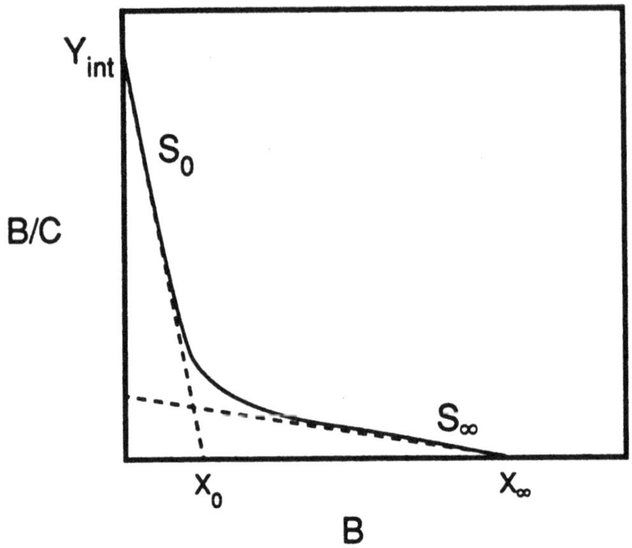

Figure 1. An equilibrium Scatchard plot. B is the concentration of bound ligand and C is the concentration of free ligand. S_0 and S_∞ are the initial ($C \to 0$) and final ($C \to \infty$) slopes, X_0 is the initial intercept obtained by extrapolating the initial straight line portion of the curve to the x-axis ($B/C = 0$), and X_∞ is the final intercept. In the absence of receptor aggregation X_∞ always equals the total number of available receptor binding sites.

the free ligand concentration, C. When C is small, we can expand B in a power series in C, i.e., $B = a_1 C - a_2 C^2 + \cdots$, so that

$$B/C = a_1 - a_2 C + \cdots \quad .$$

For convenience we introduce a minus sign in the second term of the expansion. Doing this makes no assumptions about the binding since we allow a_2 to be positive or negative. If we keep only the first two terms in the expansion, then

$$B/C \approx a_1 \left[1 - (a_2/a_1)C\right] \approx \frac{a_1}{1 + (a_2/a_1)C} \, . \tag{17}$$

The last two terms in Eq. (17) are equal when second and higher terms in the expansion are neglected. Cross-multiplying the first and last terms in Eq. (17) we have that

$$B/C \approx -(a_2/a_1)B + a_1 \ , \tag{18}$$

which is exact in the limit that $C \to 0$. From this equation, we have that the y-intercept

$$Y_{int} = a_1 \ , \tag{18a}$$

the initial x-intercept, as defined in Fig. 1,

$$X_0 = a_1^2/a_2 \ , \tag{18b}$$

and the initial slope

$$S_0 = -a_2/a_1 \ . \tag{18c}$$

To obtain a_1 and a_2 we expand B, given by Eqs. (11)–(13), in a power series in C. Using Mathematica (Wolfram Research Inc.) we find

$$a_1 = < K > R_0 \ , \tag{19a}$$

$$a_2 = < K^2 > R_0 - \frac{K_{LH} L_0 H_0}{1 + K_{LH}(L_0 + H_0)} (K_D - K_L - K_H)^2 \ . \tag{19b}$$

In a similar way we can derive an expression for the final slope of a Scatchard plot. When C is large, we can expand B in a power series in $1/C$. Since as $C \to \infty$ all sites become filled, i.e., $B \to R_T$, for large C, we have that

$$B \approx R_T - b_1(1/C) \cdots \ ,$$

where b_1 is positive since $B \leq R_T$. Keeping the first two terms in the expansion

$$B \approx R_T(1 - (b_1/R_T)(1/C)) \approx \frac{R_T}{1 + (b_1/R_T)(1/C)} \ ,$$

from which it follows that

$$B/C \approx -(R_T/b_1)B + R_T^2/b_1 \ .$$

Therefore the final slope of a Scatchard plot, S_∞ is

$$S_\infty = -R_T/b_1 \ .$$

By expanding B in $1/C$, as $C \to \infty$, we obtain b_1 and show that

$$S_\infty = -R_T \left[\frac{L_T}{K_L} + \frac{H_T}{K_H} + \frac{K_D}{K_L K_H} K_{LH} L_T H_T \right]^{-1} \ . \tag{20}$$

Note that in the absence of heterodimer formation, $D_0 = 0$ and $K_{LH} = 0$. In this case we recover the known results for a population of two noninteracting receptors, that $Y_{int} =< K > R_T$, $X_0 = R_T < K >^2 / < K^2 >$, $S_0 = - < K^2 > / < K >$, and $S_\infty = 1/ < K^{-1} >$, where the brackets stand for averages with respect to the receptor populations (Goldstein, 1975). If the two populations consist of a small population of high affinity receptors and a large population of low affinity receptors with equilibrium constants K_{high} and K_{low}, respectively, then the expressions for the slopes reduce to $S_0 \approx K_{high}$ and $S_\infty \approx K_{low}$. However, if the receptor populations interact to form heterodimers that also bind the ligand, this is no longer true as can be seen from Eqs. (18c) and (20). For example, if the additional term due to aggregation in Eq. (20) is large, the final slope will be smaller than K_L.

Consider the case where (1) $H_T/L_T \ll 1$ and (2) in the absence of IL-2 almost all the heavy chains are in heterodimers. As can be seen from Eq. (10c), conditions (1) and (2) require that $K_{LH}L_T \gg 1$. If almost all the heavy chains are in heterodimers then $H_0 \approx 0$, $D_0 \approx H_T$ and $L_0 \approx L_T - H_T$. When this is so

$$S_0 \approx -K_D \frac{1 + (K_L/K_D)^2((L_T/H_T) - 1)}{1 + (K_L/K_D)((L_T/H_T) - 1)} , \tag{21a}$$

$$X_0 \approx H_T \frac{[(1 + (K_L/K_D)((L_T/H_T) - 1)]^2}{1 + (K_L/K_D)^2((L_T/H_T) - 1)} . \tag{21b}$$

For cells expressing both p55 and p75 chains the number of p75 chains is commonly estimated by assuming that $H_T \approx X_0$, and the affinity of the heterodimer is estimated by assuming that $K_D \approx -S_0$. From our analysis we see that this interpretation is correct at equilibrium, i.e., Eqs. (21a) and (21b) reduce to $S_0 \approx -K_D$ and $X_0 \approx H_T$, only if $K_L/K_D \ll 1$ and $K_L L_T \ll K_D H_T$.

In summary, for cell lines with a preponderance of light chains, the preformed heterodimer model predicts that the initial slope of an equilibrium Scatchard plot will be approximately equal to the affinity of the heterodimer if (1) $K_{LH}L_0 \gg 1$, i.e., in the absence of IL-2 almost all the heavy chains are in heterodimers; if (2) $K_L/K_D \ll 1$, i.e., the light chain has a much lower affinity than the heterodimer; and if (3) $K_L L_T \ll K_D H_T$. We have pointed out that for a cell with 10^4 light chains on its surface condition (1) is satisfied if $K_{LH} \gg 5 \times 10^{-10}$ cm^2. Since the heterodimer is the high affinity receptor and the light chain is the low affinity receptor, condition (2) is always satisfied. Wang and Smith (1987) carried out equilibrium binding experiments on a T cell line (MT-1) expressing only light chains and found $K_L = 7.1 \times 10^7$ M^{-1}. Thus, on a T cell expressing 10^4 light chains and 2×10^3 heavy chains [normal T cell blasts studied by Wang and Smith (1987)], condition (3) is satisfied if $K_D \gg 3.6 \times 10^8$ M^{-1}. Initial slopes of dynamic Scatchard plots at 37°C and equilibrium Scatchard plots at 0°C suggest $K_D \approx 5 \times 10^{10} - 5 \times 10^{11}$ M^{-1} (Robb et al., 1984, 1987; Robb and Rusk, 1986; Wang and Smith, 1987; Tsudo et al., 1987; Hatakeyama et al., 1989). Thus, for our model T cell, condition (3) is satisfied.

Predictions of the Preformed Heterodimer Model

We illustrate the predictions of the preformed heterodimer model for a typical activated T cell with 2×10^3 heavy chains and 1×10^4 light chains on its surface.

We take $K_L = 7.1 \times 10^7$ M^{-1} and $K_H = 8.3 \times 10^8$ M^{-1}. These values were determined by Wang and Smith (1987) from equilibrium binding studies at 37°C using cells expressing only light or heavy chains. K_H was determined from membrane binding studies to ensure that there was no internalization or synthesis. These values for K_L and K_H are consistent with a number of other binding studies (e.g., Robb and Rusk, 1986; Tsudo et al., 1987; Dukovich et al., 1987; Hatakeyama et al., 1989). We take $K_D = 1 \times 10^{11}$ M^{-1} although this affinity has not been directly determined. Wang and Smith (1987) estimated that $K_D = 7.7 \times 10^{10}$ M^{-1} from binding studies at 37°C on cells expressing both light and heavy chains. In order to determine K_D, they assumed that the initial slope of their 20 min Scatchard plots, with neither internalization nor synthesis blocked, equaled K_D. Their kinetic measurements were in good agreement with this value, but since there are many possible kinetic pathways for forming a bound heterodimer, Eqs. (2a), (2b), (2c), (3a) and (3b), the forward and reverse rate constants that they determined may not equal k_{D+} and k_{D-}. This would only be true if, for the IL-2 concentrations used in their kinetic studies, the dominant reaction for forming and dissociating from a bound heterodimer was the direct reaction. Equilibrium Scatchard plots at 0°C give $K_D = 1.2 \times 10^{11}$ M^{-1}, which is consistent with estimates of K_D from dynamic Scatchard plots (Robb and Rusk, 1986). There has been no experimental estimate of K_{LH}, but, as we have discussed, if the initial slope of the Scatchard plot is to reflect direct binding to the high affinity heterodimer, a large fraction of the heavy chains must be in heterodimers. For a cell with 10^4 light chains and a surface area of 5×10^{-6} cm^2 this requires that $K_{LH} \gg 5 \times 10^{-10}$ cm^2. Shown in Fig. 2 is the model prediction of the effect of K_{LH} on the initial slope of the Scatchard plot. As K_{LH} is increased from 10^{-10} cm^2 to 10^{-8} cm^2, the initial slope varies from -2.8×10^9 M^{-1} to -9.4×10^{10} M^{-1} while the number of preformed heterodimers, D_0, increases from 2.9×10^2 per cell (2 % of the heavy chains in heterodimers) to 1.9×10^3 per cell (95 %

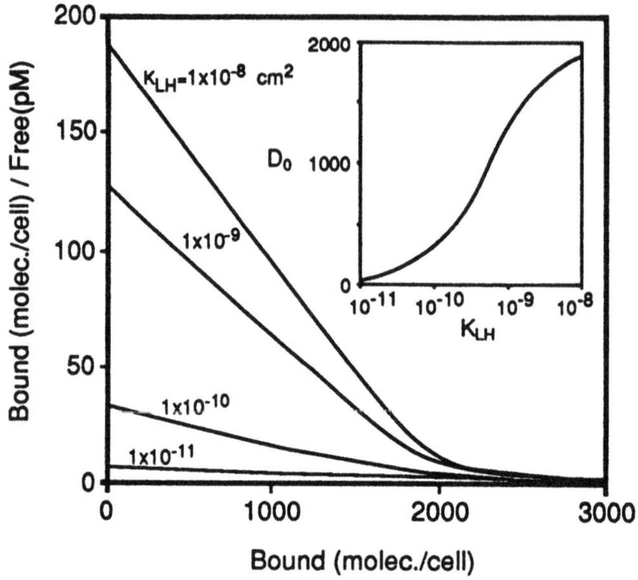

Figure 2. The effect of preformed heterodimers on the initial slope of a Scatchard plot as predicted by the preformed heterodimer model. K_{LH} is the equilibrium constant for the formation of a heterodimer from an unbound p55 chain and an unbound p75 chain, reaction (2a). The insert shows how the number of heterodimers on the cell surface in the absence of IL-2, D_0, is predicted to vary with K_{LH}. The calculations are for a model cell characterized by the following parameters: $L_T = 1 \times 10^4$ p55 chains/cell, $H_T = 2 \times 10^3$ p75 chains/cell, $K_L = 7.1 \times 10^7$ M^{-1}, $K_H = 8.3 \times 10^8$ M^{-1}, and $K_D = 1 \times 10^{11}$ M^{-1}. Receptor concentrations were converted from receptors/cell to receptors/cm^2 by taking the surface area of the cell $A = 5 \times 10^{-6}$ cm^2. The bound IL-2 concentration is the sum of the bound p55 chains, p75 chains and heterodimer concentrations, i.e., $B = L^* + H^* + D^*$. B was calculated as follows: for each value of C, the total heterodimer concentration, $D_T(C)$, was obtained from Eq. (13a). Knowing $D_T(C)$, the concentrations of bound p55 chains, p75 chains and heterodimers, L^*, H^*, and D^*, were calculated from Eqs. (12a), (12b), and (12c).

of the heavy chains in heterodimers). In the preformed heterodimer model, when the concentration of light chains is in excess, a large fraction of the heavy chains must be

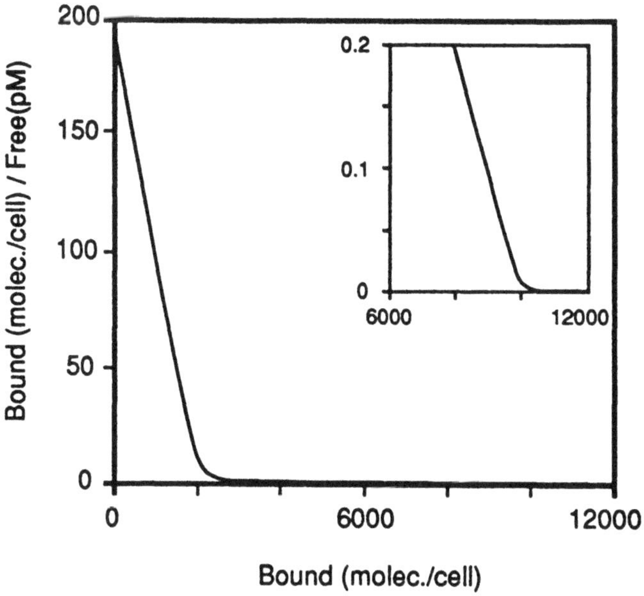

Figure 3. Scatchard plot as predicted by the preformed heterodimer model when there is a high concentration of preformed heterodimers on the cell surface. The calculations are for a model cell characterized by the following parameters: $L_T = 1 \times 10^4$ p55 chains/cell, $H_T = 2 \times 10^3$ p75 chains/cell, $K_L = 7.1 \times 10^7$ M^{-1}, $K_H = 8.3 \times 10^8$ M^{-1}, $K_D = 1 \times 10^{11}$ M^{-1}, and $K_{LH} = 1 \times 10^{-8}$ cm^2. Of the 2×10^3 p75 chains, 1.9×10^3 are in preformed heterodimers for this value of K_{LH}. Receptor concentrations were converted from receptors/cell to receptors/cm^2 by taking the surface area of the cell $A = 5 \times 10^{-6}$ cm^2. The calculations were carried out as described in Fig. 2. The insert shows the high IL-2 portion of the Scatchard plot. The slope of the curve in the insert, which appears to have an intercept at 1×10^4 molecules/cell, is approximately equal to K_L. However, this is not the final slope as defined by Eq. (20). At higher IL-2 concentrations this slope is obtained and the true intercept, $X_\infty = L_T + H_T = 1.2 \times 10^2$ molecules/cell, is reached.

in heterodimers in the absence of IL-2 if the initial slope of the Scatchard plot is to be approximately equal to the affinity of the heterodimer.

Figure 3 shows the complete Scatchard plot for $K_{LH} = 1 \times 10^{-8}$ cm^2. The initial x-intercept, X_0, is approximately 2×10^3 molecules/cell, consistent with almost all heavy chains being in preformed heterodimers. The insert shows the high IL-2 portion of the curve, which appears to have an x-intercept at approximately 1×10^4 molecules/cell. This intercept is approximately equal to the sum of the total number of receptors on

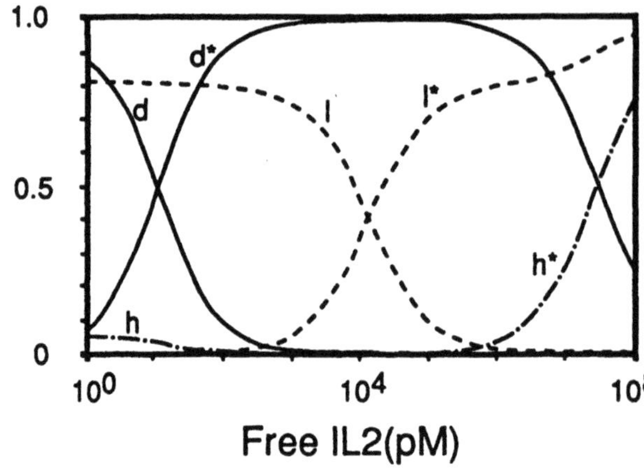

Figure 4. Distribution of free and bound IL-2 receptors as a function of the free IL-2 concentration as predicted by the preformed heterodimer model. $l = L/L_T$, $h = H/H_T$, $d = D/H_T$, $l^* = L^*/L_T$, $h^* = H^*/H_T$, and $d^* = D^*/H_T$ are the fractions of free (not bound to IL-2) and bound p55 chains, p75 chains and heterodimers, respectively. The calculations are for the same parameters as were used to calculate the Scatchard plot in Fig. 3.

the cell surface at the start of the experiment [R_0 given by Eq. (15c)]. The true x-intercept equals 1.2×10^4 molecules/cell. At sufficiently high IL-2 concentrations the model predicts that heterodimers will decrease and the total number of IL-2 molecules bound will equal the sum of the total number of light chains and heavy chains. Figure 4 shows the predicted fractions of the free and bound receptors. It can be seen that the fraction of bound heterodimers, $d^* = D_T/H_T$, does not begin to decrease until the IL-2 concentration $C > 10^6$ pM.

In general, when there is more than one population of receptors present, the initial slope of a Scatchard plot (S_0 in Fig. 1) will depend on both the affinities of the receptors and their concentrations. In Fig. 5 we plot the predicted initial slope of the Scatchard plot, as calculated from Eq. (18c), versus L_T, the total number of light chains on the cell surface. The total number of heavy chains, H_T, is held constant at 2×10^3. When the number of heterodimers is large, the slope is insensitive to the number of light chains over a wide range of light chain concentrations and is approximately equal to K_D, the affinity of the heterodimer. For example, when $K_{LH} = 1 \times 10^{-8}$ cm^2 and $K_D = 1 \times 10^{11}$ M^{-1}, the predicted slope is between -6.7×10^{10} M^{-1} and -9.7×10^{10}

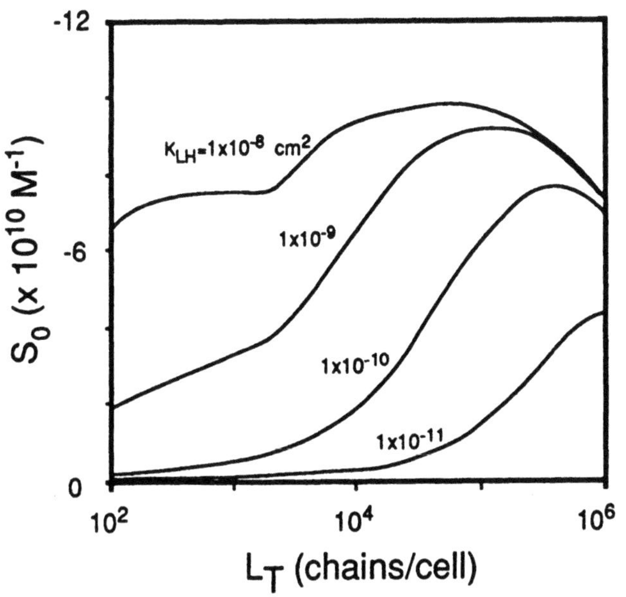

Figure 5. The dependence of the initial slope of the Scatchard plot, S_0, on the total concentration of p55 chains, L_T, as predicted by the preformed heterodimer model. The parameters used in calculating the curves are our standard ones: $H_T = 2 \times 10^3$ p75 chains/cell, $K_L = 7.1 \times 10^7$ M^{-1}, $K_H = 8.3 \times 10^8$ M^{-1}, and $K_D = 1 \times 10^{11}$ M^{-1}. Receptor concentrations were converted from receptors/cell to receptors/cm^2 by taking the surface area of the cell $A = 5 \times 10^{-6}$ cm^2. K_{LH} is the equilibrium constant for the formation of a heterodimer from an unbound p55 chain and an unbound p75 chain (see Fig. 2). S_0 was calculated from Eqs. (18c), (19a), and (19b). [First D_0 was calculated from Eq. (10a), then L_0, H_0, R_0, and $< K >$ and $< K^2 >$ were calculated from Eqs. (15a), (15b), (15c) and Eqs. (16a) and (16b), respectively.]

M^{-1} for L_T between 1×10^2 and 1×10^6 chains per cell. At lower values of K_{LH}, where the initial number of heterodimers is decreased, the slope is much more sensitive to the value of L_T.

In Fig. 6a we plot the initial x-intercept X_0 (see Fig. 1), calculated from Eq. (18b), and D_0, calculated from Eq. (10a), versus L_T for $K_{LH} = 1 \times 10^{-8}$ cm^2. As we have discussed, this is a large value of K_{LH}, which favors heterodimer formation. It can be seen that the value of X_0 is close to, but slightly lower than D_0, over a wide range of L_T values. Thus X_0 is a reasonable estimate of the number of preformed heterodimers,

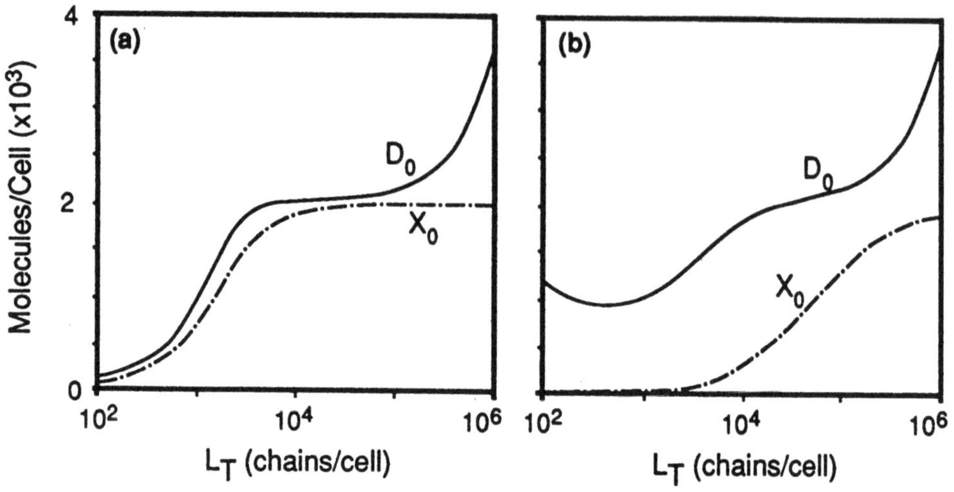

Figure 6. The dependence of the preformed heterodimer concentration, D_0, and the initial x-intercept, X_0, of the Scatchard plot on the concentration of p55 chains, L_T, as predicted by the preformed heterodimer model. D_0 was calculated from Eq. (10a) and X_0 from Eq. (18b). In both (a) and (b) the following standard parameter values were used: $H_T = 2 \times 10^3$ p75 chains/cell, $K_L = 7.1 \times 10^7$ M^{-1}, $K_H = 8.3 \times 10^8$ M^{-1}, and $K_D = 1 \times 10^{11}$ M^{-1}. Receptor concentrations were converted from receptors/cell to receptors/cm^2 by taking the surface area of the cell $A = 5 \times 10^{-6}$ cm^2. In (a) $K_{LH} = 1 \times 10^{-8}$ cm^2 while in (b) $K_{LH} = 1 \times 10^{-10}$cm^2.

for sufficiently high values of K_{LH}. However, for small values of K_{LH}, as in Fig. 6b where $K_{LH} = 1 \times 10^{-10}$ cm^2, the values of X_0 differ widely from D_0.

In summary, when $L_T > H_T$ the slope of the equilibrium Scatchard plot will be approximately equal to the affinity of the heterodimer if a large fraction of heavy chains is in dimers in the absence of IL-2. Further, the slope will be insensitive to the concentration of light chains over a wide range of concentrations. Similarly, the initial x-intercept X_0 will give a reasonable estimate of the number of preformed heterodimers.

Results for the Affinity Conversion Model

In the affinity conversion model (Shimizu et al., 1986; Kondo et al., 1986; Saito, et al., 1988), heterodimers only exist when they are bound to IL-2. There are no free heterodimers ($D = 0$ for all C) and therefore no direct binding to, or dissociation from, heterodimers. A bound heterodimer can form in two ways, by either reactions (2b) or (2c), i.e., IL-2 must first bind to one chain, e.g., p55, and then the bound chain must interact with a free chain of the other type, e.g., p75. Although there is no direct binding to free heterodimers in this model, it is useful to define an effective direct binding affinity to heterodimers, $\overline{K}_D$, and an effective initial concentration of heterodimers, $\overline{D}$, in the following way

$$\overline{K}_D = R_T K_L K_{L*H} \ , \tag{22a}$$

$$\overline{D} = H_T L_T / R_T \ . \tag{22b}$$

For the affinity conversion model, we now define the average affinity, $< K >_{AC}$, and the affinity squared average, $< K^2 >_{AC}$, to be

$$< K >_{AC} = (L_T K_L + H_T K_H + \overline{D}\,\overline{K}_D)/R_0 \ , \tag{23a}$$

$$< K^2 >_{AC} = (L_T K_L^2 + H_T K_H^2 + \overline{D}\,\overline{K}_D^{\,2})/R_0 \ . \tag{23b}$$

We also define

$$a_1 = < K >_{AC} R_T \ , \tag{24a}$$

$$a_2 = < K^2 >_{AC} R_T + 2\overline{D}\,\overline{K}_D(K_L + K_H) \ . \tag{24b}$$

(The derivations of these expressions are given in the Appendix.) Using Eqs. (24a) and (24b) we can now calculate from Eqs. (18a), (18b), and (18c), Y_{int}, X_0, and S_0 (see Fig. 1) for a Scatchard plot as predicted by the affinity conversion model. One can show that if the initial slope of the Scatchard plot is to be much larger than both K_L and K_H, then $\overline{D}\,\overline{K}_D \gg L_T K_L + H_T K_H$, which can be rewritten as

$$K_{L \cdot H} \gg 1/H_T + K_H/(K_L L_T) \ . \tag{25}$$

Under this condition

$$S_0 \approx -R_T K_L K_{L \cdot H} = -\overline{K}_D \ , \tag{26a}$$

$$X_0 \approx L_T H_T / R_T = \overline{D} \ . \tag{26b}$$

Since $R_T = L_T + H_T$, when $L_T \gg H_T$, it follows from Eq. (26b) that $X_0 \approx H_T$.

From Eq. (26a) we see that the affinity conversion model predicts that the initial slope of the Scatchard plot is proportional to the total number of cell surface receptors. We shall look at this prediction in more detail below, but first we estimate the value of $K_{L \cdot H}$, the equilibrium constant for the interaction of a bound light chain with a heavy chain to form a bound dimer. We again take $K_L = 7.1 \times 10^7$ M^{-1}, $K_H = 8.3 \times 10^8$ M^{-1},

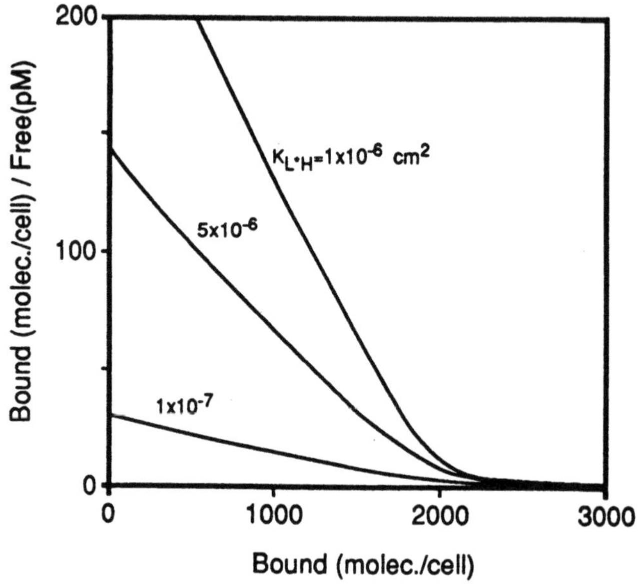

Figure 7. The effect on the initial slope of a Scatchard plot of the bimolecular equilibrium constant $K_{L\bullet H}$ as predicted from the affinity conversion model. $K_{L\bullet H}$ is the bimolecular equilibrium constant for the formation of a bound heterodimer from a p55 chain complexed with IL-2 and a free p75 chain, reaction (2b). In the calculation the standard parameter set was used: $L_T = 1 \times 10^4$ p55 chains/cell, $H_T = 2 \times 10^3$ p75 chains/cell, $K_L = 7.1 \times 10^7$ M^{-1}, and $K_H = 8.3 \times 10^8$ M^{-1}. The concentration of bound dimers, D^*, was calculated from Eq. (A5); and then the bound p55 and p75 chain concentrations, L^* and H^*, were calculated from Eqs. (A1) and (A2).

and consider a cell with 2×10^3 heavy chains and 1×10^4 light chains ($R_T = 1.2 \times 10^4$ chains/cell or 2.4×10^9 chains/cm^2 for a cell with a surface area of 5×10^{-6} cm^2). Eq. (26a) and (26b) will hold when Eq. (25) is satisfied, i.e., when $K_{L\bullet H} \gg 8 \times 10^{-9}$ cm^2. Equation (26a) predicts that in the affinity conversion model, in order for an equilibrium Scatchard plot to have an initial slope of 1×10^{11} M^{-1}, $K_{L\bullet H} = 6 \times 10^{-7}$ cm^2. Figure 7 shows the predicted effect of $K_{L\bullet H}$ on the initial slope of the Scatchard plot. As $K_{L\bullet H}$ decreases from 1×10^{-6} cm^2 to 1×10^{-7} cm^2, S_0 decreases from 1.7×10^{11} M^{-1} to 1.7×10^{10} M^{-1}.

Figure 8 shows the complete Scatchard plot for $K_{L\bullet H} = 6 \times 10^{-7}$ cm^2. The insert shows the high IL-2 portion of the curve. The curve is very similar to that obtained using the

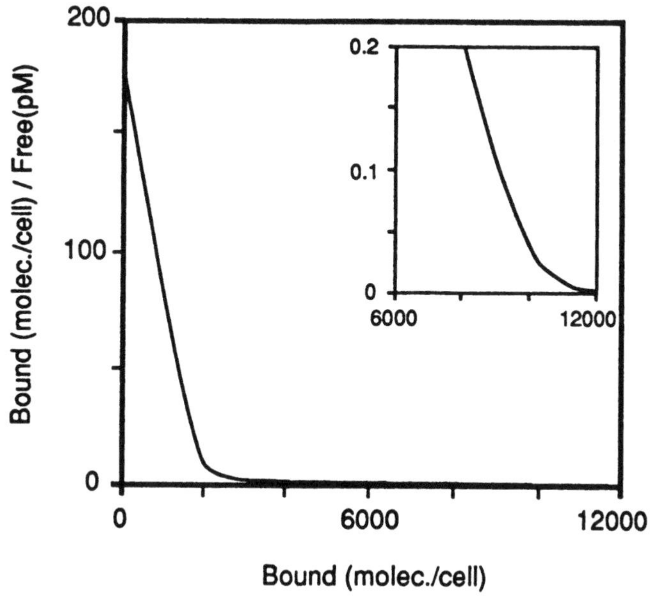

Figure 8. The Scatchard plot as predicted by the affinity conversion model. The calculations are for a model cell characterized by the following parameters: $L_T = 1 \times 10^4$ p55 chains/cell, $H_T = 2 \times 10^3$ p75 chains/cell, $K_L = 7.1 \times 10^7$ M^{-1}, $K_H = 8.3 \times 10^8$ M^{-1}, and $K_{L \bullet H} = 6 \times 10^{-7}$ cm^2. With this choice of $K_{L \bullet H}$ the initial slope of the Scatchard plot $S_0 = 1 \times 10^{11}$ M^{-1}. Receptor concentrations were converted from receptors/cell to receptors/cm^2 by taking the surface area of the cell $A = 5 \times 10^{-6}$ cm^2. The calculations were carried out as described in Fig. 7. The insert shows the high IL-2 portion of the Scatchard plot.

preformed heterodimer model (Fig. 3). Figure 9 shows the predicted concentration of the free and bound receptors. In the Appendix [Eq. (A7)] we show that the heterodimer concentration is predicted to be maximum when $C = 1/\sqrt{K_L K_H} = 4 \times 10^3$ pM (the preformed heterodimer model makes essentially the same prediction) and falls to half its maximum value when $C = 8 \times 10^6$ pM.

As Eq. (26a) indicates, the affinity conversion model predicts that the initial slope of the Scatchard plot will be strongly influenced by the concentration of chains on the cell surface. Figure 10 is a plot of the predicted initial slope, S_0, versus the light chain concentration, L_0, for the affinity conversion model. As the number of light chains increases from 10^2 to 10^6 per cell, the predicted slope increases from 1.0×10^{10} M^{-1} to

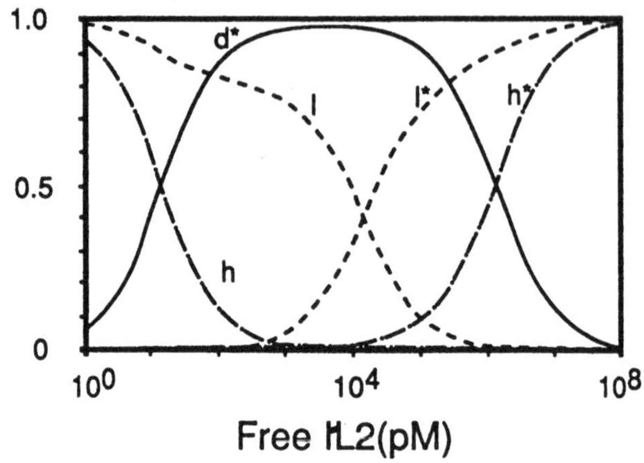

Figure 9. Distribution of free and bound IL-2 receptors as a function of the free IL-2 concentration as predicted by the affinity conversion model. $l = L/L_T$, $h = H/H_T$, $l^* = L^*/L_T$, and $h^* = H^*/H_T$ are the fractions of free (not bound to IL-2) and bound p55 chains and p75 chains, respectively. $d^* = D^*/H_T$ is the fraction of p75 chains complexed with IL-2 and p55 ($d = D/H_T = 0$ in the affinity conversion model). The calculations are for the same parameters as were used to calculate the Scatchard plot in Fig. 7.

8.5×10^{12} M^{-1}. Figure 11 shows the predicted initial x-intercept, X_0, as a function of L_T. As can be seen, when $L_T \gg H_0$, $X_0 \approx H_T$.

In summary, when Eq. (25) is satisfied, the affinity conversion model predicts that the slope of the equilibrium Scatchard plot will be given by Eq. (26a) and the intercept by Eq. (26b). Thus, S_0 is predicted to be directly proportional to the concentration of light chains on the cell surface.

Comparison of The Two Models

Equilibrium Scatchard plots at 0°C and 4°C and dynamic Scatchard plots at 37°C for IL-2 binding experiments on cells expressing both light and heavy chains, for a variety of light and heavy chain concentrations, yield initial slopes ranging from approximately -3×10^{10} M^{-1} to -3×10^{11} M^{-1}. For cells having 2×10^3 heavy chains per cell we have chosen K_{LH} and K_D in the preformed heterodimer model, and $K_{L \bullet H}$ in the affinity conversion model, so that when there are 1×10^4 light chains per cell in both models the initial slope $S_0 = 1 \times 10^{11}$ M^{-1}. Shown in Fig. 12 are the initial portions of the

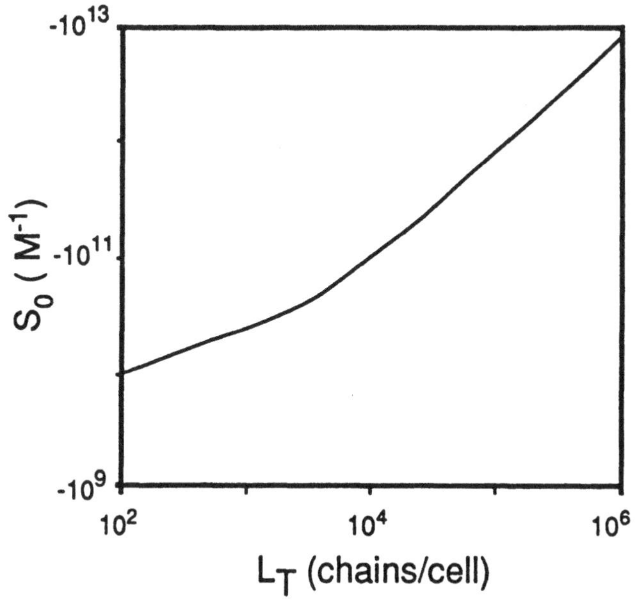

Figure 10. The dependence of the initial slope of the Scatchard plot, S_0, on the total concentration of p55 chains, L_T, as predicted by the affinity conversion model. For both models the parameters used in calculating the Scatchard plots were: $H_T = 2 \times 10^3$ p75 chains/cell, $K_L = 7.1 \times 10^7$ M^{-1}, $K_H = 8.3 \times 10^8$ M^{-1}, and $K_{L \cdot H} = 6 \times 10^{-7}$ cm^2. Receptor concentrations were converted from receptors/cell to receptors/cm^2 by taking the surface area of the cell $A = 5 \times 10^{-6}$ cm^2. S_0 was calculated from Eq. (18c), where a_1 and a_2 are given by Eqs. (24a) and (24b).

predicted Scatchard plots for $L_T = 10^3$, 10^4, and 10^5 chains per cell. One sees that S_0 is predicted to change little in the preformed heterodimer model but to decrease dramatically as L_T decreases in the affinity conversion model. While studying the regulation of p55 expression, Smith and Cantrell (1985) measured the initial slopes of Scatchard plots for two T cell populations, one with 4.3×10^4 and one with 1.0×10^4 p55 chains (determined by binding of anti-Tac antibody) and found the initial slopes to be essentially the same ($K_D = 4.8 \times 10^{10}$ M^{-1} and 4.4×10^{10} M^{-1}). The affinity conversion model would predict a difference in slopes of approximately a factor of 4 for these experiments.

Robb and his collaborators (Robb et al., 1984, 1987; Robb and Rusk, 1986) have determined the initial slope of dynamic Scatchard plots for a set of T cells exhibiting a wide

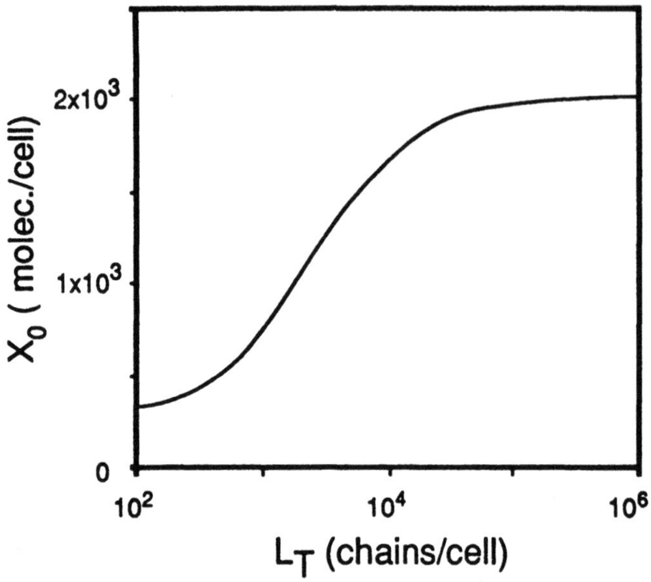

Figure 11. The dependence of the initial x-intercept, X_0, of the Scatchard plot on the concentration of p55 chains, L_T, as predicted by the affinity conversion model. The parameters are the same as in Fig. 10. X_0 was calculated from Eq. (18b), where a_1 and a_2 are given by Eqs. (24a) and (24b).

range of light chain concentrations. These data are plotted in Fig. 13 along with fits of the slopes predicted by the preformed heterodimer model and the affinity conversion model. There is no observed decrease in the experimentally determined slopes as the total number of receptors decrease.

DISCUSSION

We have considered two models for the equilibrium binding of IL-2 to its cell surface receptors. In one, the preformed heterodimer model, we allow p55-p75 heterodimers to form in the absence of IL-2, while in the other, our formulation of the affinity conversion model, heterodimers that are not complexed with IL-2 do not exist. For the two models we have derived analytic expressions for the characteristics of the predicted Scatchard plots, i.e., the initial slope, initial intercept, final slope, and final intercept. A striking difference between the two models is the predicted dependence of the initial slope of an equilibrium Scatchard plot on p55 and p75 chain concentrations. When (1) the magnitude of the initial slope is much larger than the affinities of IL-2 for isolated p55 and p75 chains (which is what is observed) and (2) the concentration of p55 chains is

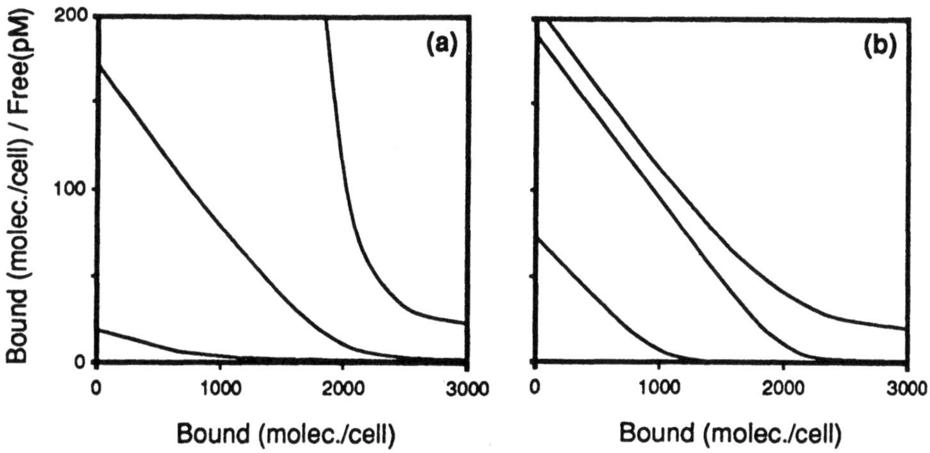

Figure 12. Scatchard plots for three p55 chain concentrations concentrations, $L_0 = 10^3$, 10^4, and 10^5 chains/cell, as predicted by (a) the affinity conversion model and (b) the preformed heterodimer model. The calculations are for a model cell characterized by the following parameters: $H_T = 2 \times 10^3$ p75 chains/cell, $K_L = 7.1 \times 10^7$ M^{-1}, and $K_H = 8.3 \times 10^8$ M^{-1}. (a) In the affinity conversion model we took $K_{L \bullet H} = 6 \times 10^{-7}$ cm^2. (b) In the preformed heterodimer model we took $K_D = 1 \times 10^{11}$ M^{-1} and $K_{LH} = 1 \times 10^{-8}$ cm^2. With these parameters both models predict an initial Scatchard plot of approximately 1×10^{11} when $L_0 = 1 \times 10^4$ p55 chains/cell.

much larger than p75 chains, the preformed heterodimer model predicts that the slope will be independent of the chain concentrations and equal to the affinity of IL-2 for the heterodimer. The preformed heterodimer model further predicts that in the absence of IL-2 almost all the p75 chains will be in heterodimers. For the same conditions the affinity conversion model predicts that the initial slope of the Scatchard plot will increase linearly with the total chain concentration. From published experiments of Robb and his collaborators (Robb et al., 1984, 1987; Robb and Rusk, 1986) and Smith and Cantrell (1985), we conclude that the affinity conversion model, where there are no preformed dimers, is inconsistent with these data. For a wide range of total chain concentrations ($10^3 - 10^5$ chains/cell) the initial slope of a Scatchard plot for the binding of IL-2 to receptors on T cells remains unchanged. This observation is consistent with the preformed heterodimer model. There is, however, one caveat. The model predictions are for equilibrium Scatchard plots. The experiments we cite were done at 37°C without blocking internalization, recycling, or protein synthesis. How well Scatchard

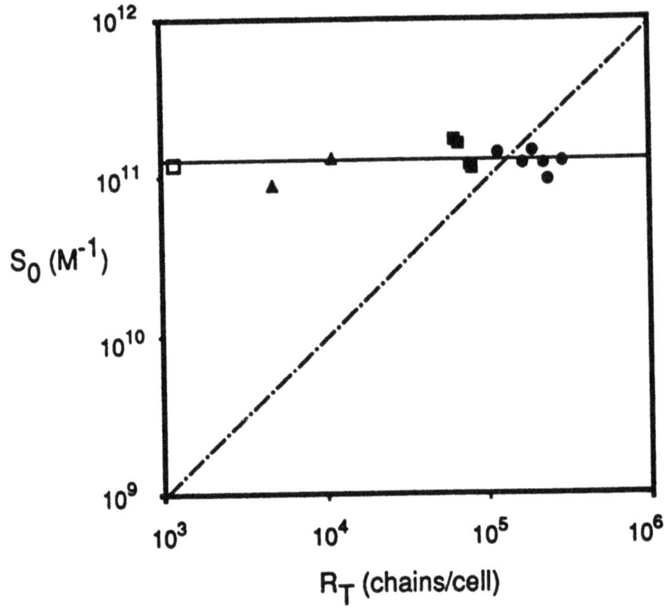

Figure 13. Comparison of model predictions with published data. S_0 is the initial slope of a Scatchard plot and $R_T = H_T + L_T$ is the total concentration of p55 and p75 chains. The data is from 37°C binding studies on T cells: HUT 102B2 (●), PHA blasts (□), PBL (■), MLC blasts (▲) (Robb et al., 1984, 1987; Robb and Rusk, 1986). To determine R_T from the data we assumed that $H_0 \approx X_0$ and that the total number of binding sites for anti-Tac antibody as determined from Scatchard plots equaled L_0. For all the data $L_0 \gg H_0$. For this case the preformed heterodimer model predicts that $S_0 \approx -K_D$. The solid line is the best fit to the data when we take $S_0 = -K_D$. (Approximating the exact slope in this way introduces an error of less than 2% for all data points except the point with the lowest R_T. This point was not used in the fit.) The affinity conversion model predicts that $S_0 \approx -R_T K_L K_{L \bullet H}$ when Eq. (25) holds. The dashed line is the best fit to the data when we take $S_0 = -R_T K_L K_{L \bullet H}$. (Approximating the exact slope in this way introduces an error of less than 5% for data points except the point with the lowest R_T. This point was not used in the fit.)

plots for such binding data approximate equilibrium Scatchard plots is unknown, although predictions of K_D from such plots are consistent with measurements at 0°C (Robb and Rusk, 1986). To accurately test our predictions, new experiments should be done with receptor internalization, recycling, and synthesis blocked. The second prediction of the preformed heterodimer model, that a large fraction of the p75 chains are in preformed heterodimers, has yet to be tested, although there are indications that preformed heterodimers exist (Saragovi and Malek, 1988; Yamaguchi et al., 1989). Since the model presented is an equilibrium model, it makes no predictions about the stability of preformed heterodimers. Indeed, since it is an equilibrium model, it assumes that all reactions are reversible, so that at equilibrium heterodimers are constantly forming and breaking apart.

A main feature of the affinity conversion model (Shimizu et al., 1986; Kondo et al., 1986; Saito, et al., 1988), which is not addressed in this paper, is the kinetic mechanism proposed for the formation of the bound heterodimer. Honjo and his collaborators argue that the dominant mechanism for such formation is, first, the binding of IL-2 to the p55 chain, reaction (1a), and then the interaction of the IL-2-p55 chain complex with a p75 chain, reaction (2b). The work presented in this paper applies only to equilibrium experiments and makes no comments about kinetic mechanisms, but since our analysis indicates that preformed heterodimers exist, both direct binding to preformed heterodimers, reaction (1c), and transfer reactions involving heterodimers, reaction (3a) and (3b), cannot be ruled out as important kinetic mechanisms.

The model we have presented applies to all receptor systems composed of two chains that bind the ligand with different affinities and that also aggregate to form a third affinity state. The first example of such a receptor system was the p55 and p75 IL-2 binding chains. It now appears that the granulocyte-macrophage colony-stimulating factor receptor system (Hayashida et al., 1990) and the interleukin-5 receptor (Takaki et al., 1991; Tavernier et al., 1991) are two other examples.

ACKNOWLEDGEMENTS

We thank Michael Riepe and David Jones for their excellent technical assistance and I. Kevrekidis and K. Smith for critical comments. This work was performed under the auspices of the United States Department of Energy. It was supported by NIH Grants AI28433 (A.P.) and GM35556 (B.G.) and the Santa Fe Institute through their Theoretical Immunology Program.

APPENDIX

The Affinity Conversion Model

From the direct binding reactions (4a) and (4b), and the conservation laws, Eqs. (7a) and (7b), it follows that for the affinity conversion model ($D = 0$):

$$L^* = \frac{K_L C}{(1 + K_L C)}(L_T - D^*) \,, \tag{A1}$$

$$H^* = \frac{K_H C}{(1 + K_H C)}(H_T - D^*) \,. \tag{A2}$$

These equations also follow from Eqs. (12a) and (12b) once it is recognized that $D_T(C) = D^*$ since $D = 0$. In the affinity conversion model, dimers only exist when bound to IL-2. In the presence of IL-2, dimers can form in two ways, by reactions (5b) and (5c). From reaction (5b) and Eq. (4c) it follows that $D^* = K_{L^*H} L^* H = CK_{L^*H} K_L LH$. Using the conservation laws, Eqs. (7a) and (7b),

$$D^* = CK_{L^*H} K_L (L_T - L^* - D^*)(H_T - H^* - D^*) \,. \tag{A3}$$

Substituting Eqs. (A1) and (A2) to eliminate L^* and H^* in Eq. (A3),

$$D^* = CK_{L^*H} K_L \frac{(L_T - D^*)(H_T - D^*)}{(1 + K_L C)(1 + K_H C)} \,. \tag{A4}$$

Solving this equation for D^* yields

$$D^*(C) = \frac{1 + F_{AC}(C) R_T - \sqrt{(1 + F_{AC}(C) R_T)^2 - 4(F_{AC}(C))^2 L_T H_T}}{2F_{AC}(C)} \,, \tag{A5}$$

where

$$F_{AC}(C) = \frac{CK_{L^*H} K_L}{(1 + K_L C)(1 + K_H C)} \,. \tag{A6}$$

It is worth noting that from the form of Eq. (A5) it follows that D^* is maximum when $F_{AC}(C)$ is maximum. If we call this concentration C_{max} (it is the value of the concentration at which $dF_{AC}(C)/dC = 0$), then

$$C_{max} = 1/\sqrt{K_L K_H} \,. \tag{A7}$$

For $K_L = 7.1 \times 10^7$ M^{-1} and $K_H = 8.3 \times 10^8$ M^{-1}, $C_{max} = 4.1 \times 10^{-9}$ M. Thus the affinity conversion model predicts that dimer formation will be maximum when the free IL-2 concentration is approximately 4×10^3 pM.

When $C \to \infty$, $F_{AC}(C) \to 0$ and

$$D^*(c) \approx L_T H_T F_{AC}(C) \approx \frac{L_T H_T K_{L^*H}}{K_H C} , \qquad (A8)$$

providing that $K_L C_{max} \gg 1$ and $K_H C_{max} \gg 1$. Calling $C_{1/2}$ the IL-2 concentration at which the $D^* = H_T/2$ (we assume $L_T > H_T$), we have from Eq. (A8) that

$$C_{1/2} \approx 2 L_T K_{L^*H}/K_H . \qquad (A9)$$

For the parameter values used to calculate the chain concentrations in Fig. 9, $C_{1/2} = 8 \times 10^6$ pM.

For the affinity conversion model we can now calculate the equilibrium concentration of any type of receptor for any free IL-2 concentration. From Eq. (A5) we can calculate D^*. Knowing D^* we can calculate L^* and H^* from Eqs. (A1) and (A2). From the conservation laws, Eqs. (7a) and (7b) we can then calculate L and H.

To obtain expressions for S_0 and X_0 (see Fig. 1) we must obtain the first two coefficients in the power series expansion in C of B/C. From Eqs. (A1), (A2), and (A4)

$$\begin{aligned} B/C =& \frac{K_L}{(1+K_L C)}(L_T - D^*) + \frac{K_H}{(1+K_H C)}(H_T - D^*) + \\ & \frac{K_{L^*H} K_L (L_T - D^*)(H_T - D^*)}{(1+K_L C)(1+K_H C)} , \end{aligned} \qquad (A10)$$

where D^* is given by Eq. (A5). When this expansion is preformed one obtains

$$\begin{aligned} B/C = L_T K_L + H_T K_H + L_T H_T K_L K_{L^*H} - [L_T K_L^2 + H_T K_H^2 + \\ L_T H_T (K_L K_{L^*H})^2 (L_T + H_T) + 2 L_T H_T K_L K_{L^*H} (K_L + K_H)] C + \cdots . \end{aligned} \qquad (A11)$$

REFERENCES

Dukovich, M., Wano, Y., Thuy, L. T. B., P. Katz, Cullen, B. R., Kehrl, J. H. and Greene, W. C. (1987). A second human interleukin-2 binding protein that may be a component of high-affinity interleukin-2 receptors. *Nature* **327**, 518–522.

Duprez, V., Cornet, V. and Dautry-Varsat, A. (1988). Down-regulation of high affinity interleukin 2 receptors in a human tumor T cell line. *J. Biol. Chem.* **263**, 12860–12865.

Goldstein, B. (1975). Theory of hapten binding to IgM: The question of repulsive interactions between binding sites. *Biophys. Chem.* **3**, 363–367.

Goldstein, B., Jones, D., Kevrekidis, I. G., and Perelson, A. S. (1992). Evidence for p55-p75 heterodimers in the absence of interleukin-2 from Scatchard plot analysis. *Int. Immunol.* **4**, 23–32..

Hatakeyama, M., Tsudo, M., Minamoto, S., Kono, T., Doi, T., Miyata, T., Miyasaka, M. and Taniguchi, T. (1989). Interleukin-2 receptor β chain gene: Generation of three receptor forms by cloned human α and β chain cDNA's. *Science* **244**, 551–556.

Hayashida, K., Kitamura, T., Gorman, D. M., Arai, K., Yokota, T., and Miyajima, A. (1990). Molecular cloning of the second subunit of the receptor for human granulocyte–macrophage colony–stimulating factor (GM–CSF): Reconstitution of a high affinity GM–CSF receptor. *Proc. Natl. Acad. Sci. USA* **87**, 9655–9659.

Hill, T. L. (1977). *Free Energy Transduction in Biology. The Steady-State Kinetic and Thermodynamic Formalism.* Academic Press, New York. p. 4.

Jones, D. A., Kevrekidis, I. G., Perelson, A. S. and Goldstein, B. G. (1989). A model of IL-2 binding and receptor-mediated internalization. Paper No. 17e. Extended Abstracts and Papers, Am. Inst. Chem. Eng. 1989 Annual Meeting, San Francisco, CA.

Kondo, S., Shimizu, A., Saito, Y., Kinoshita, M. and Honjo, T. (1986). Molecular basis for different affinity states of the interleukin 2 receptor: Affinity conversion model. *Proc. Natl. Acad. Sci. USA* **83**, 9026–9029.

Robb, R. J., Greene, W. C., and Rusk, C. M. (1984). Low and high affinity cellular receptors for interleukin 2: Implications of the level of Tac antigen. *J. Exp. Med.* **160**, 1126–1146.

Robb, R. J. and Greene, W. C. (1987). Internalization of interleukin 2 is mediated by the β chain of the high-affinity interleukin 2 receptor. *J. Exp. Med.* **165**, 1201-1206.

Robb, R. J. and Rusk, C. M. (1986). High and low affinity receptors for interleukin 2: Implications of pronase, phorbol ester, and cell membrane studies upon the basis for differential ligand affinities. *J. Immunol.* **137**, 142–149.

Robb, R. J., Rusk, C. M., Yodoi, J. and Greene, W. C. (1987). Interleukin 2 binding molecule distinct from the Tac protein: Analysis of its role in formation of high-affinity receptors. *Proc. Natl. Acad. Sci. USA* **84**, 2002–2006.

;aito, Y., Sabe, H., Suzuki, N., Kondo, S., Ogura, T., Shimizu, A. and Honjo, T. (1988). A large number of L chains (Tac) enhance the association rate of interleukin 2 to the ıigh affinity site of the interleukin 2 receptor. *J. Exp. Med.* **168**, 1563-1572.

;aragovi, H. and T. R. Malek. (1988). Direct identification of the murine IL-2 receptor)55-p75 heterodimer in the absence of IL-2. *J. Immunol* **141**, 476–482.

;himizu, A., Kondo, S., Sabe, H., Ishida, N., and Honjo, T. (1986). Structure and 'unction of the interleukin 2 receptor: Affinity conversion model. *Immunol. Rev.* **92**, .03–120.

;mith, K. A. (1988). Interleukin-2: Inception, impact, and implications. *Science* **!40**,1169–1176.

;mith, K. A. (1989). The interleukin 2 receptor. *Ann. Rev. Cell. Biol.* **5**, 397–425.

;mith, K. A. and Cantrell, D. A. (1985). Interleukin 2 regulates its own receptors. *Proc. Vatl. Acad. Sci. USA* **82**, 864–868.

[akaki, S., Mita, S., Kitamura, T., Yonehara, S., Yamaguchi, N., Tominaga, A., Miya-ima, A., and Takatsu, K. (1991). Identification of the second subunit of the murine nterleukin-5 receptor: Interleukin-3 receptor-like protein, AIC2B is a component of the ıigh affinity interleukin-5 receptor. *EMBO J.* **10**, 2833–2838.

'avernier, J., Devos, R., Cornelis, S., Tuypens, T., Van der Heyden, J., Fiers, W., and 'laetinck, G. (1991). A human high affinity interleukin-5 receptor (IL5R) is composed f an IL5-specific α chain and a β chain shared with the receptor for GM-CSF. *Cell* **66**, 175–1184.

'eshigawara, K., H.–M. Wang, K. Kato and K. A. Smith. (1987). Interleukin 2 high-ffinity receptor expression requires two distinct binding proteins. *J. Exp. Med.* **165**, 23–238.

'sudo, M., Kozak, R. W., Goldman, C. K. and Waldmann, T. A. (1987). Contribution f a p75 interleukin 2 binding peptide to a high-affinity interleukin 2 receptor complex. *'roc. Natl. Acad. Sci. USA* **84**, 4215–4218.

Tsudo, M., Karasuyama, H., Kitamura, F., Tanaka, T., Kubo, S., Yamamura, Y., Tamatani, T., Hatakeyama, M., Taniguchi, T. and Miyasaka, M. (1990). The IL-2 receptor β-chain (p70). Ligand binding ability of the cDNA-encoding membrane and secreted forms. *J. Immunol.* **145**, 599–606.

Wang, H.–M. and K. A. Smith. (1987). The interleukin 2 receptor: Functional consequences of its bimolecular structure. *J. Exp. Med.* **166**, 1055–1069.

Yamaguchi, A., T. Ide, M. Hatakeyama, T. Doi, T. Kono, T. Uchiyama, K. Kikuchi, T. Taniguchi and T. Uede. (1989). Structure of the functional interleukin-2 receptor. *Int. Immunol.* **1**, 160–168.

T MEMORY CELLS IN A MODEL OF T CELL MEMORY

Angela R. McLean
Zoology Department
Oxford University
Oxford OX1 3PS UK

INTRODUCTION

T lymphocytes can be separated into two phenotypes according to whether or not they bear a high molecular weight isoform of the leucocyte common antigen CD45 (Akbar et al 1988). This phenotypic dichotomy is matched by a functional dichotomy: it has been suggested that cells bearing the heavier isoform ($CD45R^+$) are an unprimed population whilst those with the lighter isoform ($CD45R^-$) are memory T cells (Sanders et al 1988, Clement et al 1988). Mathematical models of T helper cell activation and proliferation are able to capture many of the observed features of T helper cell population dynamics - in particular the antigen driven shift from a virgin to a memory state. However these models predict that it should be possible, in vitro, to establish continuously replicating clones of T helper cells. These predictions do not concur with the observation that in vitro cultures of T helper cells need to be re-exposed to antigen every few weeks in order to maintain proliferation. This paper presents a new mathematical model of T helper cell activation and proliferation. The model incorporates some of the recent data and theories about $CD45R^-$ (memory) T cells. It accounts for the interacting population dynamics of resting, activated and memory T helper cells, IL-2 and replicating antigen, and is able to mimic a broad range of available data on T helper cell proliferation in vivo and in vitro. The model reconciles the two observations that immune memory can be established in vivo but not in vitro. This reconciliation is achieved because the model includes the assumption that in vivo memory is maintained by constant, low level stimulation of memory cells - a stimulation that is absent in vitro.

The main part of the paper is arranged in two sections: model description and results. In the first part the model's biological assumptions and equations are described. The second part presents results of modelling the in vivo and in vitro situations. These results are

NATO ASI Series, Vol. H 66
Theoretical and Experimental Insights into Immunology
Edited by A. S. Perelson and G. Weisbuch

presented as phase plane analyses and numerical simulations. The paper ends with a discussion which compares the structure of this model with existing models to explain why this model inherits some of the properties of previous models but not others.

MODEL

The model counts five populations: resting Th cells (W), activated Th cells (X) memory Th cells (M) interleukin 2 (IL-2) (I) and antigen (A). The model is based on the antigen driven activation, but IL-2 driven proliferation of T helper cells. Recent work has identified a subset of T helper cells that bear the cell surface marker CD45R0 and appear to be of memory type. This classification is based on the observations that CD45RA (naive) cells convert to CD45R0 phenotype after activation (Akbar et al 1988), that the CD45R0 population contains most of the cells able to respond to recall antigens (Merkenschlager et al 1988), and that CD45R0 cells are functionally more potent and probably more easily activated than CD45RA cells (Jones et, 1990). With this biological background, the models assumptions and equations are set out below.

Naive Th cells (W) migrate from the thymus at constant rate Λ and (if not activated) have half-life $1/\mu$. In the presence of their specific antigen (A) they become activated at a per-capita rate proportional to the amount of antigen present αA. These assumptions give rise to the following equation for resting Th cell dynamics:

$$\dot{W} = \Lambda-\alpha AW-\mu W$$

Activated Th cells (X) can be stimulated by IL-2 (I) to divide to become 2 memory cells (M) this rate of division saturates to a constant at high levels of activated Th cells. This saturation of the division rate represents the assumption that some form of suppression acts to prevent clones of activated T helper cells from growing too large. Activated Th cells have the same half-life $1/\mu$ as all other Th cells. Memory cells can return to the activated state for two reasons. First, presentation of the specific antigen (A) by APCs to memory cells leads to their activation at a rate that is faster by factor δ than the activation of naive cells. Second, a background activation rate ε is always present. This could represent sequestered antigen or cross-reactive stimulation. These assumptions give rise to the following equation for the dynamics of activated Th cells:

$$\dot{X} = \alpha AW - \frac{\rho IX}{1+\xi X} + (\delta\alpha A+\varepsilon)M - \mu X$$

Activated Th cells divide to become two memory cells (M). These have the same half-life $1/\mu$ as other Th cell sub-sets. As stated above they are activated at per-capita rate $\delta\alpha A+\varepsilon$ representing more rapid activation by specific antigen plus some activation that is present even in the absence of specific antigen. The equation for the dynamics of memory Th cells is therefore:

$$\dot{M} = \frac{2\rho IX}{1+\xi X} - (\delta\alpha A+\varepsilon)M - \mu M$$

IL-2 is produced by activated Th cells, absorbed by activated Th cells and has half life $1/\psi$. The equation for the dynamics of IL-2 is therefore:

$$\dot{I} = \phi X - \beta IX - \psi I$$

The equation used to describe the dynamics of specific antigen (A) depends on the type of antigen that is modelled. For a replicating antigen growing at rate r in the absence of specific immunity and removed at a rate that is proportional to the number of activated Th cells specific to it, the relevant equation is:

$$\dot{A} = rA - \gamma AX$$

The model has been reduced to dimensionless form in order to perform the numerical simulations - this reduces the number of parameters and the units used in analysis become unimportant (Murray, 1989). The reduction is performed by choosing new scaling variables for W, X, M, I, A and time t. Defining $W=W'W_0$, $X=X'X_0$, $M=M'M_0$, $I=I'I_0$, $A=A'A_0$, $t=t't_0$ and choosing:

$$W_0=\frac{\mu}{\gamma} \quad X_0=\frac{\mu}{\gamma} \quad M_0=\frac{\mu}{\gamma} \quad I_0=\frac{\mu}{\rho} \quad A_0=\frac{\mu}{\alpha} \quad t_0=\frac{1}{\mu}$$

the equations for the new variables (written without primes for brevity) are:

$$\begin{aligned}
\dot{W} &= q - W - AW \\
\dot{X} &= AW - \frac{IX}{1+vX} + dAM + eM - X \\
\dot{M} &= \frac{2IX}{1+vX} - dAM - eM - M \\
\dot{I} &= w(aX - bIX - I) \\
\dot{A} &= A(c - X)
\end{aligned}$$

where the composite parameters are given by:

$$q=\frac{\gamma\Lambda}{\mu^2},\quad a=\frac{\rho\phi}{\psi},\quad b=\frac{\beta\mu}{\gamma\psi}\quad c=\frac{r}{\mu}\quad d=\delta\quad e=\frac{\varepsilon}{\mu}\quad w=\frac{\psi}{\mu}\quad v=\frac{\xi\mu}{\gamma}$$

Values for these composite parameters used in the numerical simulations are given in the figure legends.

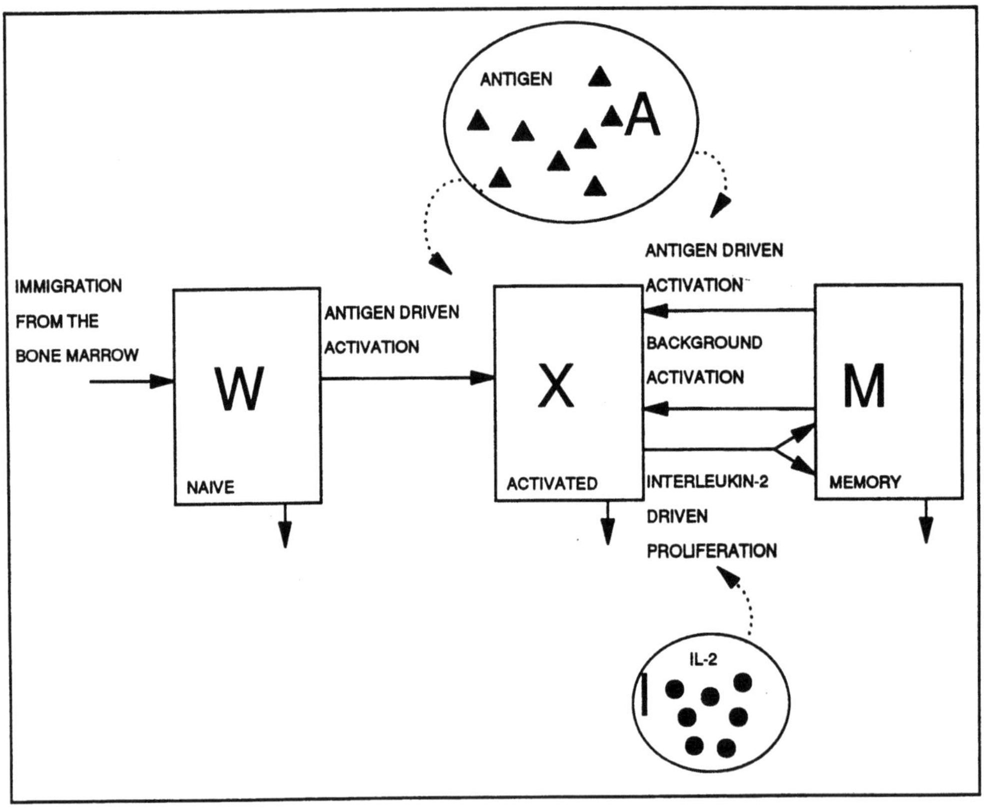

Fig.1 Model Structure. This model of the antigen driven activation and IL-2 driven proliferation of a clone of T helper cells counts five populations: naive, activated and memory T helper cells; IL-2 and antigen. Naive cells immigrate from the bone marrow, and can be activated by their specific antigen. Activated cells produce IL-2 and proliferate in response to IL-2 to produce two memory cells. Memory cells can be activated by antigen and are also subject to a background activation rate that is present even in the absence of antigen. Depending on the experimental system modelled, antigen may or may not be replicating, and it is removed at a rate proportional to the number of activated T helper cells.

This statement of the model's assumptions and its equations includes a large number of parameters. In the equations, the rate at which each interaction proceeds is governed by one or more model parameter. These parameters represent a link between the model and experimental work and their definitions and ways in which they might be estimated from experimental data are described in Table 1.

Table 1 Model Parameters

PARAMETER	BIOLOGICAL MEANING	POSSIBLE DATA SOURCES
Λ	immigration rate of T helper cells	studies of T helper cell population dynamics
μ	death rate of all T helper cells	
α	'per antigen' stimulation rate of naive and memory cells	time series of naive and memory response to non-replicating antigen
$\delta\alpha$		
ϵ	background stimulation rate of memory T helper cells	relative sizes of naive and memory clones in vivo
r	replicating antigen's growth rate	in vitro studies
γ	per activated, specific T helper cell antigen removal rate	time series studies of the removal of non-replicating antigen in vitro
ρ/ξ	'per IL-2' proliferation rate at saturating numbers of activated T helper cells	IL-2 dose response curves
$1/\xi$	T helper cell concentration at half maximum T helper proliferation rate	Il-2 dose response curves
ϕ	per activated T helper cell IL-2 production rate	culture of a single lymphocyte
β	IL-2 - activated T helper cell complex formation rate	IL-2 dose response curves
ψ	IL-2 decay rate	IL-2 half life

RESULTS

The model is used to investigate T helper population dynamics in vivo and in vitro. First a reduced version of the model with just two variables is considered so that isoclines can be inspected. Then numerical simulations are presented and, where appropriate, compared with data. In the simulation studies the full model is first studied to illustrate its behaviour akin to in vivo immune memory. Then results from a reduced version of the model (one where some of the parameter values are set to zero) are compared to some in vitro data on T cell population dynamics. In the absence of numerical estimates of the parameters in table 1, the model's behaviour is studied by characterising all possible types of behaviour. This is made easier by working with the model expressed in non-dimensional terms. A consequence of using dimensionless parameters is that 'time' in analyses and numerical simulations is expressed in terms of a model parameter (e.g. T cell lifespan) so cannot easily be related to ordinary units of time (e.g. weeks, days). For this reason the X axes in figure 3 have no units of time. The X axes in figure 4 have been labelled consistently across all simulations using units of days to allow easy comparison with the experimental data.

Equilibrium results

The two dimensional model:

$$\dot{X} = \frac{qA}{1+A} + \frac{aX^2}{(1+bX)(1+vX)}\left[\frac{dA+e-1}{dA+e+1}\right] - X$$
$$\dot{A} = A(c-X)$$

is derived by making quasi-steady state assumptions for W, M, and I so that:

$$W = \frac{q}{1+A}, \quad M = \frac{2IX}{(1+vX)(dA+e+1)}, \quad I = \frac{aX}{1+bX}$$

The model describes the interacting dynamics of activated T helper cells and antigen. In figure 2 isoclines for fixed activated T helper cell population are drawn using parameters appropriate for the in vivo and in vitro situation. The corresponding isoclines for fixed antigen population are given by $A=0$, and $X=c$. With $A=0$ the roots for the X isocline are given by $X=0$ and the two roots of the quadratic:

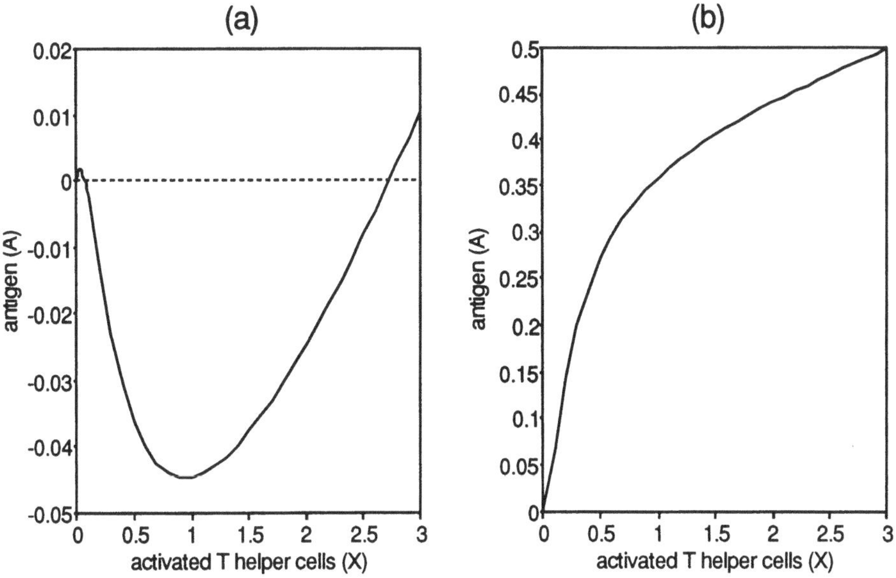

Fig.2 Isoclines for constant activated T helper cells in the reduced model with just two variables, activated T helper cells X and antigen A. (a) with in vivo parameters, (b) with in vitro parameters. Parameter values are: q=9, a=100, b=5, d=2, v=1, with e=1.5 in (a) and e=0 in (b). The parameter were chosen so that when e=1.5, the isocline would cross the axis A=0 - only under these circumstances is it ever possible to have a memory state. The isoclines for zero growth of antigen are A=0 and X=c. Thus in (a) there are four equilibria, A=0, X=0; A=0, $X=X_-$; A=0, $X=X_+$ and $A=A^*$, X=c. Here X_- and X_+ are the roots of the quadratic

$$(e+1)bvX^2+[(e+1)(b+v)-a(e-1)]X+(e+1)=0$$

and A^* is the root of

$$\frac{qA}{(1+A)}+\frac{ac^2}{(1+bc)(1+vc)}\left[\frac{dA+e-1}{dA+e+1}\right]-c=0$$

If $c<X_+$ then the point A=0, $X=X_+$ is stable and in vivo immune memory can be established as illustrated in Fig 3a. If $c>X_+$ then the point $A=A^*$, X=c is stable and the replicating antigen will persist - as illustrated in Fig 3b. These results are trivially established by a graphical analysis of the phase plane. If $c<X_-$ then there is a second stable state at $A=A^*$, X=c the stability of this state is not easy to establish graphically or analytically, but is shown numerically in Fig 3c. In (b) when e<1 there are no real roots to the quadratic, and the only equilibrium with A=0 is X=0, when c=0, (as it is in the in vitro situation) this equilibrium is stable.

$$(e+1)bvX^2+[(e+1)(b+v)-a(e-1)]X+(e+1)=0$$

Let us call these two roots X_- and X_+. When $A=0$ the system of equations will have three equilibria at: $X=0$, $X=X_-$, $X=X_+$. Graphical analysis of figure 2a can show that for a replicating antigen ($c>0$) the only one of these three that can ever be stable is $A=0$, $X=X_+$, and that for non-replicating antigens, both $A=0$, $X=0$, and $A=0$, $X=X_+$, are stable. Thus the only candidate for a memory state is $A=0$, $X=X_+$. As soon as $e<1$ the second term of the quadratic becomes always positive and there can be no real roots. Thus $e>1$ is a necessary condition for the existence of a memory state. The composite parameter e is defined by $e=\varepsilon/\mu$, where ε is the rate at which memory cells are re-stimulated by background stimulation, and μ is the death rate of memory cells. Thus $e>1$ for the existence of a memory state has an intuitively appealing biological interpretation. The background stimulation rate ε must be high enough that of the two memory cells produced when an activated cell divides, on average, at least one will be re-activated by background stimuli before the end of its lifespan. An obvious example where this will not be the case is the in vitro situation where there is no background stimulation so $e=0$. As illustrated with an example in figure 2b there can be no memory state when background stimulation is absent.

In vivo simulation results

Figure 3 summarises the possible responses of this model clone of T helper cells to a replicating antigen. The parameter values in all three parts of figure 3 are the same except for the growth rate of the replicating antigen (parameter r in table 1 and c of the dimensionless parameters). Depending on the value of the growth rate there are three possible responses.

Figure 3a shows how an antigen with a low growth rate is cleared leaving immune memory. A small amount of replicating antigen is introduced at time zero, when all cells present are naive cells. In the absence of a specific immune response the antigen initially grows exponentially. The presence of the antigen drives naive cells to become activated (hence the transient fall in the number of naive cells) and the activated cells go through

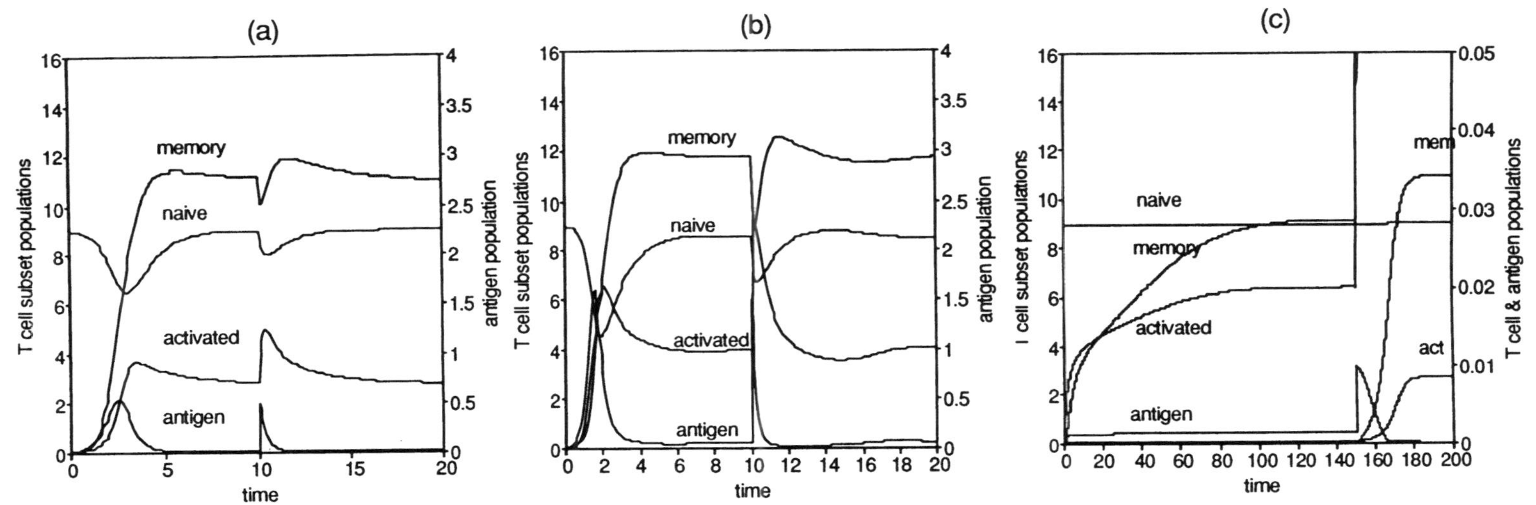

Fig.3 Model Behaviour 'in vivo'. The full model with all interactions present is used to explore the response, in vivo, to a replicating antigen. This model exhibits three possible responses depending on the growth rate of the antigen. (a) Replicating antigens with intermediate net growth rates are cleared leaving immune memory that will be maintained indefinitely. (b) Antigens with high growth rates cannot be cleared and instead establish low level, persistent infections that are never cleared. (c) There is a third group of very slowly replicating antigens that are also able to establish persistent infections if their infecting dose is very low. However these slowly replicating persistent antigens can be cleared by administering a comparatively large amount of antigen. This clearance is followed by long-lived immune memory. In (a) and (b) the left axis applies to the size of the T cell subset populations (in arbitrary units) and the right axis applies to the size of the antigen population (in arbitrary units). In (c) memory and activated T cell subset populations are of sizes as indicated on the right axis for time 0-150 units and as indicated on the left axis for time 150-200 units. (The simulations shown in (a), (b) and (c) were performed using composite parameters as described in the appendix. The parameter values used were the same for each figure for all parameters except antigen growth rate c. q=9, a=100, b=5, d=2, e=1.5, w=2, v=1, (a) c=4, (b) c=2, (c) c=0.02. Initial conditions were w=9, X=0, M=0, I=0, with A=0.01 for (a) and (b) and A=0.001 for (c).

several rounds of division. Each time an activated cell divides, two memory cells are formed, and these are re-activated immediately by antigen. Thus the presence of antigen causes a rise in the number of activated and memory cells. Activated cells orchestrate the specific immune response to the antigen so that the amount of antigen falls and is eventually completely cleared. However, in the process of expanding in response to the antigen, the activated and memory cells exceed the threshold above which they form a self-sustaining population. In this state background stimulation of memory cells balances with suppression of the proliferation of activated cells to generate a stable population of activated and memory T helper cells. This population is able to maintain immunological memory indefinitely. This memory is illustrated by the introduction of a large challenging dose of antigen at time 10 units, which is immediately cleared.

The size of the activated and memory population maintained in the absence of the replicating antigen depends only on internal interactions amongst elements of the immune system. This is why replicating antigens that grow too fast cannot be cleared and instead establish persistent infections. The response to such an antigen is illustrated in figure 3b. The course of events is the same as in fig 3a except that, instead of being cleared, the antigen population is reduced to low levels. This low level persisting antigen cannot be cleared even if the immune response is boosted with a large dose of antigen - as illustrated at time 10 units in figure 3b.

Figure 3a shows that the assumptions on which this model is built are sufficient to generate long term immune memory in the absence of persisting antigen. The 'cost' of suppression at high densities of activated cells is illustrated in figure 3b - there will be some fast replicating antigens that cannot be cleared. These two possibilities are intuitively appealing, but not very surprising. The unexpected prediction of this model is the existence of a third set of antigens with <u>very</u> low growth rates, which, if introduced at a low dose, can also establish persistent infections. One is illustrated in figure 3c. These antigens grow so slowly that they only stimulate their specific T cell clone slightly. The activated subset of the clone grows just enough to halt the further growth of the antigen, but never exceeds the threshold above which the activated and memory cells form a self sustaining population. However, a boost to the immune response with a 'large' dose of antigen (the large dose of antigen in figure 3c is the same size as the initial dose of antigen in figures 3a and b) is enough to push the activated and memory cell subsets over the threshold. In consequence the antigen is cleared leaving immune memory in the same manner as in figure 3a. This third possibility shows how the study of a mathematical representation of a set of biological assumptions can yield unexpected results.

In vitro simulation results

The behaviour of this model described so far is very similar to that of a previously published model (McLean & Kirkwood 1990). However this earlier model makes no distinction between the situation in vivo and in vitro, so cannot be compared with in vitro data. To compare the current model with in vitro data, three of the parameters are set to zero - equivalent to switching off three of the interactions. The most important of the three is the background stimulation rate for memory cells. This is set to zero representing the assumption that, in vitro, either there is no cross-reactive stimulation of memory cells or there is no presentation of sequestered antigen. The other two differences are that there is no influx of naive cells and the antigen does not grow. Under these assumptions the model no longer displays immune memory and a single exposure to antigen leads to a transient activation and proliferation. Over the course of this response all cells convert from naive to memory phenotype. Figure 4 compares in vitro experimental results with simulations.

CONCLUSIONS

This model reconciles the existence in vivo of long term immunological memory with the observation that in vitro cultures of T helper cells need to be re-exposed to antigen every few weeks if they are to maintain proliferation. This reconciliation is achieved through the background stimulation of memory cells - assumed present in vivo, but absent in vitro. Further study of the in vivo situation has revealed three possible type of model behaviour akin to immune memory, persistent infection, and slow growing persistent infection cleared after a boosting dose of antigen. This range of interactions between a replicating antigen and a clone of T helper cells is predicted by an earlier model (McLean & Kirkwood 1990). This simpler model accounts for antigen, activated T helper cells and IL-

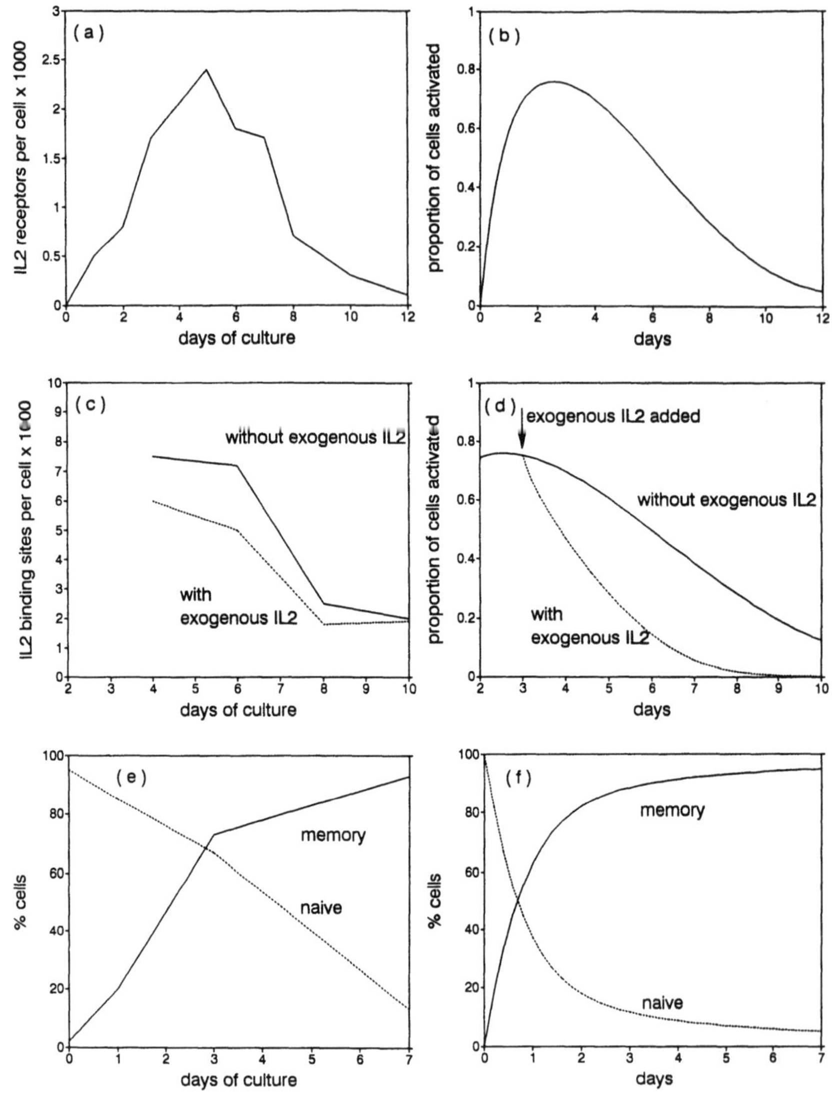

Fig.4 Model behaviour 'in vitro' compared with in vitro experiments. To compare the model's behaviour with in vitro experiments, a number of the possible interactions in the model are switched off by setting some parameters to zero. Most importantly there is no longer any background stimulation of memory cells (e=0); also there is no immigration of naive cells from the bone marrow (q=0) and the antigen is assumed not to replicate (c=0). Time series data on IL-2 binding sites per cell (a and c) are compared with simulations (b and d) by plotting the proportion of cells that are activated: in this model only activated cells respond to IL-2. Both data and simulations show a transient rise in cells able to respond to IL-2 (a and b) which can be foreshortened by the addition of exogenous IL-2 (c and d). Over the course of this transitory response, cells convert from being mostly naive to being mostly memory (f) as has been demonstrated in vitro (e). In vitro data are from (a) (Smith 1988) (c) (Cantrell & Smith 1985) (e) (Akbar et al 1988). The following parameter values were used in the simulations. q=0,a=100,b=5,c=0,d=10,e=0,w=2,v=1.

2 and makes the assumption that activated T helper cells divide to give rise to two more activated T helper cells - whether or not antigen is present. In its dimensionless form (with a quasi-steady state assumption for IL-2) the model can be written:

$$\dot{X} = A + \frac{aX^2}{(1+bX^2)} - X$$
$$\dot{A} = A(c-X)$$

where definitions of the dimensionless, composite parameters a, b and c are comparable to their equivalents in the model presented here. The two models are equivalent if the pool of naive cells (W) is fixed and the background stimulation rate of memory cells is large compared to their death rate ($e \gg 1$). It is therefore clear why the two models can produce very similar in vivo predictions, but only the newer one can also embrace the in vitro situation (where $e=0$). Because this new model can be used to explore both in vivo and in vitro situations it is a much better candidate for comparing with in vitro experimental results. It can also be used to investigate in vivo behaviour of T helper cell clones. In another paper (McLean, 1992) this model is used to address questions like: what effect would be predicted if memory cells (M) revert to naive cells (W) as has been suggested by recent experimental work (Bell & Sparshott 1990, Sprent et al 1991).

The model presented here includes a set of cells that have been called memory cells. These have some special properties, but are <u>not</u> long-lived. Their ability to allow a clone of T helper cells to maintain immunological memory comes from their low activation thresholds and subsequent susceptibility to continuous, low-level activation from sources other than the replicating antigen that originally stimulated their production. Thus this model displays memory without invoking the existence of long-lived memory cells.

The benefit of working with mathematical models like the one presented here is that they can give precise and uncompromising representations of a set of biological assumptions. Once the assumptions have been made into a mathematical model <u>every</u> consequence of those assumptions can be investigated.

REFERENCES

Akbar A.N., Terry L., Timms A., Beverley P.C.L., & Janossy G.(1988) Loss of CD45R and gain of UCHL1 reactivity is a feature of primed T cells. J. Immunology **140** 2171-2178

Bell E.B. & Sparshott S.M. (1990) Interconversion of CD45R subsets of CD4 T cells in vivo. Nature **348** 163-166

Beverly P.C. (1990) Is T-cell memory maintained by cross-reactive stimulation? Immunology Today **11** 203-205

Clement L.T, Yamashita N. & Martin A.M. (1988) The functionally distinct subpopulations of human CD4+ helper/inducer T lymphocytes defined by anti-CD45R antibodies derive sequentially from a different pathway that is regulated by activation-dependent post-thymic differentiation. J. Immunology **141** 1464-1470

Gray D. & Skarvall H. (1988) B cell memory is short-lived in the absence of antigen. Nature **336** 70-73

Jones K.R., Hickling J.K., Targett G.A.T & Playfair J.H.L. (1990) Polyclonal in vitro proliferative responses form nonimmune donors to Plasmodium falciparum malaria antigens require UCHL1+ (memory) T cells. Eur J Immunol **20** 307-315

McLean A.R. (1992) Modelling T Cell Memory (submitted to International Immunology)

McLean A.R. & Kirkwood T.B.L. (1990) A Model of Human Immunodeficiency Virus Infection in T helper Cell Clones. J. theor. Biol. **147** 177-203

Merkenschlager M, Terry L., Edwards R, & Beverly P.C.L. (1988) Limiting dilution analysis of proliferative responses in human lymphocyte populations defined by the monoclonal antibody UCHL1: implications for differential CD45 expression in T cell memory formation. Eur. J. Immunol **18** 1653-1661

Murray J.D. (1989) Mathematical Biology Springer Verlag Berlin, Heidelberg, New York.

Smith K.A. (1988) Interleukin-2: Inception, Impact, and Implications. Science **240** 1169-1176

Smith K.A. & Cantrell D.A. (1985) Interleukin-2 regulates its own receptors Proc Nat Acad Sci USA **82** 864-868

Sprent J, Schaefer M, Hurd M, Surh C.D. & Ron Y. (1991) Mature murine B and T cells transferred to SCID mice can survive indefinitely and many maintain a virgin phenotype. J Exp Med **174** 717-728

A "MINIMAL" BASIC T-CELL MODEL

Karl E. Kürten
Institut für Theoretische Physik
Johannes-Kepler-Universität Linz
A-4040 Linz, Austria
and
Institut für Neuroinformatik
Ruhr-Universität Bochum
D-4630 Bochum, FRG

ABSTRACT

We present a basic regulatory model for the cellular immune response of two T cell populations interacting with a specific antigen pool. Analytical study and corresponding stability analysis reveal that our model system accounts for the occurrence of multiple steady states in the absence as well as in the presence of the antigen population. Besides the virgin state, where no population is present, the model exhibits a state of immune memory, one state with low antigen and high helper concentration (tolerance) and one state with high antigen and low helper concentration (disease). While the latter states also allow oscillatory behaviour, injection of high antigen doses allow the system to jump from one attractor to the other. Finally, we discuss the intriguing issue of dynamic diseases.

INTRODUCTION

There is no doubt that the functioning of the immune system depends crucially on the activation and extinction of various functionally different subsets of interconnected lymphocyte populations. It is well known that T lymphocytes recognize the foreign antigen displayed on infected macrophages and, in addition, control the immune response of B lymphocytes by enabling them to develop into an antibody producing cell. On the other hand, T_H helper cells are a precondition for the action of various other T cell populations. Hence, a subsystem of T lymphocyte assemblies can be decoupled and studied separately. In fact, we do not intend to give an appropriate description for the functioning of the entire immune system but rather to study the dynamical properties of a specific subsystem. One objective of this contribution is to develop a dynamical model which is simple enough to be analyzed analytically but still complex enough to embody some important characteristics of T cell regulation. Once models dealing with only a few populations are well understood one might try to generalize their behaviour to larger systems or connect various subsystems together in order to develop more realistic models.

NATO ASI Series, Vol. H 66
Theoretical and Experimental Insights into Immunology
Edited by A. S. Perelson and G. Weisbuch

1. THE MODEL

We consider a system of specific T cell populations, where the basic variables are the time dependent concentrations of a helper population T_H, a cytotoxic killer population T_K and a specific antigen pool E. The growth or decay of each population is modelled in terms of differential equations, where the time dependent variables $h(t)$, $k(t)$ and $e(t)$ describe the concentration of the corresponding cell populations present in the organism.

It is widely believed that two distinct mechanisms play a crucial role in the development of T_H cells: An autocatalytic component based on proliferating T_H cells already present and a maturation process of resting T_H cells driven by the invading antigen population. The latter mechanism appears to be inhibited by a T_S suppressor pool, whose development requires the presence of T_H cells. We assume that the rate of maturation is proportional to the antigen concentration, whereas the rate of T_H cell proliferation by cell division is known to be limited saturating at high levels of T_H cells. The role of T_S suppressor cells, whose existence is not firmly established is implicitly taken into account by assuming that their concentration is proportional to that of the T_H population. Hence the dynamics of the T_H population can be described by the differential equation

$$h'(t) = -\alpha_h h + \beta_h F(h) + \gamma_h e(1 - F(h)) \quad . \tag{1.1}$$

The dynamics of fully differentiated T cells represented by activated cytotoxic T_K-cells, whose proliferation rate is assumed to be proportional to the product of the division rate of the T_H population and the dynamics of the antigen concentration, is modelled by the equation

$$k'(t) = -\alpha_k k + \gamma_k e F(h) \quad . \tag{1.2}$$

We further assume that the growth or natural decay of antigen is exponential, if there is no immune response present, whereas the destruction of the antigen is assumed to be proportional to the product of the antigen and a saturating T_H killer concentration reflecting the probability that a cytotoxic killer cell has to encounter the antigen before it is eventually eliminated. Accordingly, the dynamics of the antigen pool is described by

$$e'(t) = (-\alpha_e + \beta_e)e - \gamma_e e F(k) \quad . \tag{1.3}$$

The first term in Eq.(1.3) represents a growth or decay term according to the nature of the invading antigen and depends on whether or not the quantity $(\beta_e - \alpha_e)$ is positive or negative, respectively. The saturation function F is specified by a sigmoid function of the form

$$F(x) = \frac{x^n}{\theta^n + x^n} \quad , \tag{1.4}$$

though other forms would serve as well. The Hill coefficient n is chosen as 2, while the threshold parameter θ is set to 1.

2. LONG TIME BEHAVIOUR AND STABILITY

Let us first study the steady state solutions of the model system. Setting the left hand sides of Eq. (1.1)-(1.3) to zero, one obtains the following three algebraic fixed point equations:

$$\frac{\alpha_h}{\beta_h} h^* = F(h^*) + \frac{\gamma_h}{\beta_h} e^*(1 - F(h^*)) \tag{2.1}$$

$$\frac{\alpha_k}{\gamma_k} k^* = e^* F(h^*) \tag{2.2}$$

and

$$\frac{\beta_e - \alpha_e}{\gamma_e} e^* = e^* F(k^*) \tag{2.3}$$

The local stability of the resulting steady states can then be determined from the eigenvalues of the Jacobian

$$\begin{pmatrix} -\alpha_h + F'(h^*)(\beta_h - \gamma_h e^*) & 0 & \gamma_h(1 - F(h^*)) \\ \gamma_k F'(h^*) e^* & -\alpha_k & \gamma_k F(h^*) \\ 0 & -\gamma_e e^* F'(k^*) & (\beta_e - \alpha_e) - \gamma_e F(k^*) \end{pmatrix} \tag{2.4}$$

while the corresponding cubic eigenvalue problem can easily be solved with the aid of the Cardano formulas.

2.1 Steady states in the absence of antigen

We first concentrate on the situation where no antigen is present ($e^* = 0$). Equation (2.2) implies that there will be no killer cells ($k^* = 0$), while the fixed point of Eq. (2.1) for the T_H population simplifies to

$$\frac{\alpha_h}{\beta_h} h^* = F(h^*) \quad . \tag{2.5}$$

Besides the naive "virgin" state $P_1^* = (0,0,0)$, where the concentrations of all populations are zero, Eq. (2.5) admits the steady states $P_2^* = (h_2^*, 0, 0)$ and $P_3^* = (h_3^*, 0, 0)$, with $h_{2/3}^* = \frac{\beta_h}{2\alpha_h} \mp \sqrt{(\frac{\beta_h}{2\alpha_h})^2 - 1})$, provided that $\frac{\beta_h}{2\alpha_h} \geq 1$. The corresponding eigenvalues according to Eq. (2.4) are

$$\lambda_1 = -\alpha_h + \beta_h F'(h_i) \quad , \quad \lambda_2 = -\alpha_k \quad \text{and} \quad \lambda_3 = -\alpha_e + \beta_e \quad \text{for} \quad i = 1,2,3. \tag{2.6}$$

This set of steady states is only stable for $\beta_e - \alpha_e < 0$, when antigen decays exponentially in the absence of cytotoxic killer cells. While the steady state P_2^* can be shown to be always unstable independent of the choice of the net parameters α_h and β_h, the steady state P_3^* is strictly stable and thus establishes an immune memory. Accordingly, there will be an elevated level of the T_H cells after a successful immune response. Note that the autocatalytic term $\beta_h F(h)$ in Eq. (1.1) gives rise to the existence of multiple steady states [1,2]. Introducing an additional biologically justified multiplicative dependence of antigen, which prevents helper proliferation in the absence of antigen, would destroy the memory state. A more realistic model would therefore have to distinguish between *activated* and *nonactivated* helper populations [3]. However, there are experiments which indicate that T_H cells still proliferate for a few weeks after the disappearance of antigen.

2.2 Steady states in the presence of antigen

Assuming that antigen is present in finite concentrations Eq.(2.3) yields for the killer component k^*

$$k^* = F^{-1}(\frac{\beta_e - \alpha_e}{\gamma_e}) = \sqrt{\frac{\beta_e - \alpha_e}{\gamma_e - (\beta_e - \alpha_e)}} \quad , \tag{2.7}$$

whose existence requires a growth term of the antigen pool, more precisely $\frac{\beta_e - \alpha_e}{\gamma_e} > 1$. Eq.(2.2) then implies

$$e^* = \frac{\alpha_k k^*}{\gamma_k F(h^*)} \quad . \tag{2.8}$$

Note that Eq.(2.8) reveals that the long time behaviour of the concentrations of antigen and helper population are reciprocal, since low (high) antigen concentrations implies high (low) helper concentration representing a state of tolerance (disease).

Inserting Eq. (2.8) into Eq. (2.1) finally leads to an algebraic equation for the helper components h^*

$$\frac{\alpha_h}{\beta_h} h^* = F(h^*) + \varepsilon \frac{1 - F(h^*)}{F(h^*)} \quad , \tag{2.9}$$

whose left and right hand side we denote with $g_1(h)$ and $g_2(h)$, respectively. The quantity ε specified as

$$\varepsilon = \frac{\gamma_h \alpha_k}{\beta_h \gamma_k} \sqrt{\frac{\beta_e - \alpha_e}{\gamma_e - (\beta_e - \alpha_e)}} \tag{2.10}$$

absorbs seven net parameters into a unique control parameter ε, while the remaining constant α_h is incorporated in $g_1(h)$. Figure 1 shows the quantity $g_2(h) = F(h) + \varepsilon \frac{1-F(h)}{F(h)}$ for several values of the control parameter ε as well as $g_1(h) = \frac{\alpha_h}{\beta_h} h$ for $\frac{\alpha_h}{\beta_h} = 0.49$, whose intersects determine the fixed point components of h^*. Obviously, Eq.(2.9) admits at least one steady state within the whole parameter range $(\frac{\alpha_h}{\beta_h}, \varepsilon) \in [0, \infty]^2$. One can easily verify analytically by inspecting the solutions of $g_2(h^*) = g_2'(h^*)h^*$ that there is a critical value of $\varepsilon_c = 0.018$ such that there exists only one single fixed point P_4^* for $\varepsilon > \varepsilon_c$, whereas for $\varepsilon < \varepsilon_c$ the whole repertoire of three fixed points with nonzero components as shown in Fig. 1 can be found.

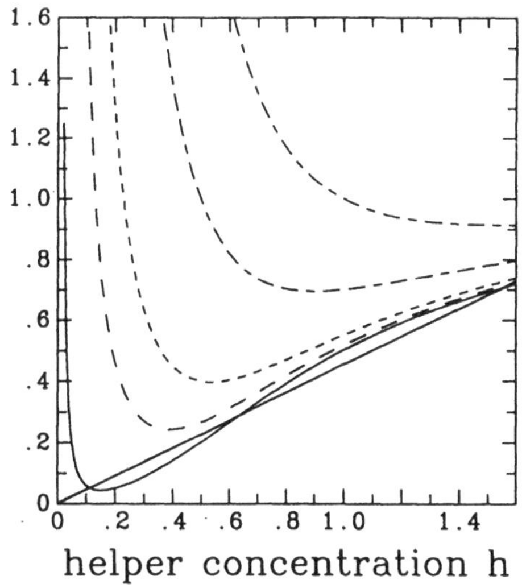

Fig. 1. $g_1(h)$ for $\frac{\alpha_h}{\beta_h} = 0.49$ and $g_2(h)$ for $\varepsilon = 0.0005, 0.018, 0.05, 0.05, 0.5$ (from below)

Hence, the dynamics of the model system described by Eq. (1.1)-(1.3) admits six fixed points that separate into two classes as demonstrated in Table 1. The first class is characterized by the absence of antigen and killer populations, whereas the latter allows steady states with nonzero components. Inspecting the eigenvalues of the Jacobian (2.4) reveals that P_4^* as well as P_6^* can give rise to complex eigenvalues. By contrast, P_5^* turns out to be always unstable, one eigenvalue always being real and positive, independent of the parameter choice.

h^*	k^*	e^*	existence
0	0	0	always
$\frac{\beta_h}{2\alpha_h} - \sqrt{(\frac{\beta_h}{2\alpha_h})^2 - 1)}$	0	0	$\frac{\beta_h}{2\alpha_h} > 1$
$\frac{\beta_h}{2\alpha_h} + \sqrt{(\frac{\beta_h}{2\alpha_h})^2 - 1)}$	0	0	$\frac{\beta_h}{2\alpha_h} > 1$
h_4^*	$\sqrt{\frac{\beta_e - \alpha_e}{\gamma_e - (\beta_e - \alpha_e)}}$	$\frac{\alpha_k k^*}{\gamma_k F(h_4^*)}$	$\frac{\beta_e - \alpha_e}{\gamma_e} > 1$
h_5^*	$\sqrt{\frac{\beta_e - \alpha_e}{\gamma_e - (\beta_e - \alpha_e)}}$	$\frac{\alpha_k k^*}{\gamma_k F(h_5^*)}$	$\frac{\beta_e - \alpha_e}{\gamma_e} > 1 \quad \wedge \quad \varepsilon < \varepsilon_c$
h_6^*	$\sqrt{\frac{\beta_e - \alpha_e}{\gamma_e - (\beta_e - \alpha_e)}}$	$\frac{\alpha_k k^*}{\gamma_k F(h_6^*)}$	$\frac{\beta_e - \alpha_e}{\gamma_e} > 1 \quad \wedge \quad \varepsilon \leq \varepsilon_c$

(2.11)

Table 1. Fixed point components h^*,k^* and e^* of the dynamics (1.1)-(1.3)

Thus, our model accounts for the existence of two steady states with nonzero components, a state of tolerance, where antigen is present in low concentrations and a state of disease, where the helper concentration remains low admitting higher antigen concentrations.

3. COMPUTER SIMULATIONS

In this report we focus on studying various dynamical behaviours for the more interesting case, where the antigen concentration is allowed to grow exponentially in the absence of an immune response ($\beta_e - \alpha_e > 0$). Moreover, we intend to concentrate on illuminating oscillatory behaviour of the model system, although the analytical study in section **2.2** showed that the corresponding parameter range is quite narrow. We provide a selection of simulations, where the system parameters have been chosen as $\frac{\alpha_h}{\beta_h} = 0.49$ and $\varepsilon = 0.005$. For our special parameter choice, P_4^* turns out to be a stable fixed point, whereas P_6^* can give rise to sustained oscillations, since one of the corresponding eigenvalues is complex with positive real part. Figure 2 depicts the primary and secondary responses obtained after two injections of antigen at time $t = 0$ and $t = 90$ (days), respectively. We observe that in contrast to the first injection, the antigen concentration does not increase after the second injection. Moreover, the antigen is cleared substantially faster, although the second antigen dose is larger than the first one.

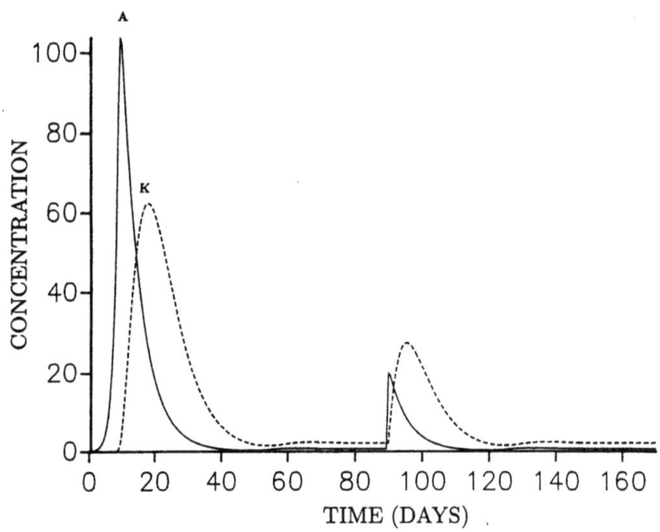

Fig. 2. Antigen (A) and killer (K) concentrations as a function of time.

Oscillatory behaviour around the steady state P_6^* is illustrated in Fig. 3, where the system undergoes a see-saw motion between low and high antigen concentrations and high and low helper concentrations, respectively. These transitions from tolerance to disease might point to the concept of periodic or dynamic diseases frequently found in physiological control systems [4]. Here, we might be confronted with a chronic disease accompanied by regular bursts of antigen proliferation. It is remarkable that the limit cycle corresponding to P_6^* is not stable with respect to antigen perturbation. Introducing a high antigen dose at time $t = 430$ enables the system to jump to the steady state corresponding to P_4^*, the stable state of tolerance. This phenomenon is quite reminiscent of therapies, where the patient is repeatedly treated with allergen.

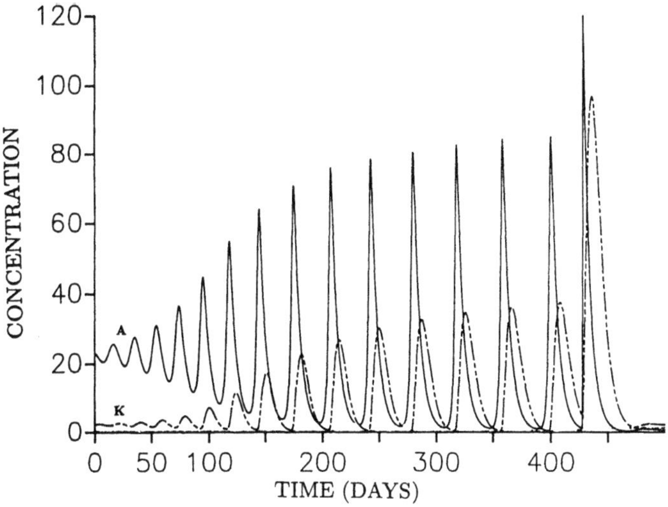

Fig. 3. Antigen (A) and killer (K) concentrations as a function of time.

It is conceivable that dynamic diseases can be evoked by a single control parameter driven out of its physiological range, while the underlying physiological control system is still intact. On the other hand, it is well known that control parameters can often be fine tuned by immunosuppressants, immunostimulation or drugs, such that the disease can be cured. Following this idea we study the change of the dynamical behaviour of our system after a slight variation of the control parameter $\frac{\alpha_h}{\beta_h}$, specifying the ratio of the proliferation and decay rate of the T_H cells. Figure 4 illustrates that the system now seems to be close to the edge of a transition from sustained oscillations to damped oscillations until at $t > 480$ the system is eventually back to the state of tolerance.

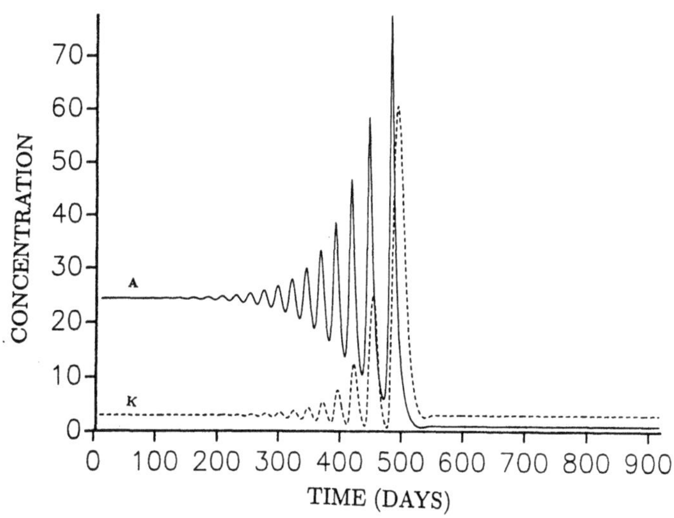

Fig. 4. Antigen (A) and killer (K) concentrations as a function of time.

However, it is questionable that an organism survives after being exposed to high antigen concentrations over a longer period. When high antigen concentrations cannot be cleared the immune system can invoke regulatory mechanisms to shut itself down. Furthermore, sustained high level responses might produce side effects that may be more dangerous to the organism than the antigen itself.

4. OUTLOOK

A wide range of experiments is still under consideration, especially the rich dynamical behaviour of our system, when certain system parameters enter domains close to their critical values and the nature of the response changes drastically. Work in progress attempts to study various suppressive mechanisms as well as at incorporating different B cell populations (see also [1]) in order to arrive at a still "minimal" but more realistic model which still admits rather complete mathematical analysis and thus a more fundamental understanding.

ACKNOWLEDGEMENTS

This work was supported by the German Science Foundation under contract number Se 251/32-1. The author gratefully acknowledges stimulating and helpful discussions with C. Romanin, A. Schinner and U. An der Heiden as well as critical comments by A. Perelson.

REFERENCES

1. M. Kauffman, J. Urbain, and R. Thomas (1985). Towards a logical analysis of the immune response, *J. Theor. Biol.* **114**, 527-561

2. A.R. Mc Lean and T.B.L. Kirkwood (1990). A model of human immunodeficiency virus infection in T helper cell clones, *J. Theor. Biol.* **147**, 177-203

3. I.G. Kevrekidis, A.D. Zecha and A.S. Perelson (1989). Modeling dynamical aspects of the immune response: T cell proliferation and the effect of IL-2. In: *Theoretical Immunology, Part I*, (Perelson, A.S., ed.) pp. 167–197, New York: Addison-Wesley

4. M.C. Mackey and U. An der Heiden (1982). Dynamical diseases and bifurcations: Understanding functional disorder in physiological systems, *Funkt. Biol. Med.* **I**, 156-162

MATHEMATICAL MODELING OF Th1-Th2 DYNAMICS$

Benoit F Morel*, Jayant Kalagnanam* and Penelope A. Morel# .

*Department of Engineering and Public Policy
Carnegie Mellon University
Schenley Park
Pittsburgh, PA 15213

Introduction:

1- The Th1-Th2 system:

One of the main functions of the immune system is the recognition and removal of foreign antigens. The first specific recognition of invading antigens is carried out by antigen-specific T lymphocytes. T cells are activated following the recognition of antigenic peptides that are bound in a cleft on molecules of the major histocompatibility complex (MHC) (Zinkernagel and Doherty, 1974; Fathman and Frelinger, 1983; Davis and Bjorkman,1988). MHC molecules are highly polymorphic cell surface glycoproteins that occur in two distinct forms: class I MHC molecules (H-2K,D,L in mice and HLA-A,B,C in humans) and class II MHC molecules (H2-I in mice and HLA-D in humans) (Hood et al,1983). Class I MHC molecules are expressed on almost all cells of the body, bind peptides derived from endogenous proteins and are recognized by T cells bearing the CD8 marker; these T cells are in general cytotoxic T cells (CTLs). Class II MHC molecules are primarily expressed on macrophages, dendritic cells, B cells and thymic epithelium, bind peptides derived from exogenous antigens and are recognized by T cells bearing the CD4 marker; these T cells are in general helper T cells (Th) (Swain, 1983). Activation of Th cells results in morphologic changes (blastogenesis) at about 12 hours, cell division by 24-48 hours and differentiation as up to 70 genes are sequentially activated for several days (Crabtree,1989). During this process a wide variety of lymphokines are secreted. These include the T cell growth factors IL-2, IL-4 and IL-7, the macrophage activator IFN-γ, and the B cell activators IL-4 , IL-5 and IL-6 (Kurt-Jones et al, 1987; Mosmann et al, 1987). These lymphokines, and many others are responsible for the activation and differentiation of cells important in the immune response. Th cells provide "help" in this fashion to CTLs, B cells , macrophages and neutrophils (Kim et al, 1985; Stout and Bottomly, 1989; Tamira et al, 1983; Vitetta et al, 1984). Recently, it has

$ This work was supprted by an NIH grant # 1RO1 AI31427-01

Address: Pittsburgh Cancer Institute, BST 9th Floor, De Soto at O'Hara Streets, Pitsburgh, PA15213.

NATO ASI Series, Vol. H 66
Theoretical and Experimental Insights into Immunology
Edited by A. S. Perelson and G. Weisbuch

emerged that, in the murine system, Th cells can be divided into two subsets on the basis of their pattern of lymphokine secretion; Th1 cells that secrete IL-2, IFN-γ and lymphotoxin and Th2 cells that secrete IL-4, IL-5, IL-6 and IL-10 (Mosmann et al,1987; Cherwinski et al, 1987). In addition Th1 cells appear to provide help for the cell-mediated responses observed in delayed type hypersensitivity (DTH) (Cher and Mosmann, 1987) whereas Th2 cells provide help for the differentiation of antigen-specific B cells (DeKruyf et al, 1989). Both types of cells secrete IL-3 , IL-7 and GM-CSF. The existence of these two types of Th cells has provided an explanation for the well-documented separation of immune responses into DTH or antibody production, which often appear to be mutually exclusive (Parish, 1972).

Recent evidence suggests that, in the response to infectious organisms, this dichotomy in the effector T helper cell response is very relevant. For example, murine parasitic infections preferentially induce Th2 responses (Finkelman et al, 1991), whereas bacterial infections stimulate Th1 responses (Scott and Kaufmann, 1991). Similar results have been shown in humans where responses to allergens are mediated by Th2 cells, and responses to the bacterially derived PPD are mediated by Th1 cells (Del Prete et al, 1991; Haanen et al,1991). In other infections, such as murine malaria, murine shistosomiasis and possibly human HIV infection, the initial response is a Th1 type response and a Th2 response is stimulated at a later stage of infection (Scott and Kaufmann, 1991). In some circumstances, the strain of mouse determines the Th response: for example in Leishmania major infection one strain of mouse develops a protective Th1 response whereas another strain develops a Th2 response that leads to exacerbation of the disease (Locksley and Scott, 1991). Similar dichotomies have been observed in the two known forms of leprosy, tuberculous and lepromatous. It appears therefore, that in some cases the nature of the antigen determines the Th response, but in others there are clearly other factors which determine whether a Th1 or a Th2 response will predominate.

In order to try and explain the fact that the Th1 and Th2 responses are mutually exclusive, many experiments have been done in isolated *in vitro* systems to characterize the interactions occurring between Th1 and Th2 cells and the lymphokines they produce. IFN-γ, produced by Th1 cells, inhibits the proliferation of Th2 cells (Gajewski and Fitch, 1988) and antagonizes the effect of IL-4 on B cells (Rabin et al, 1986). IL-10, produced by Th2 cells , inhibits the lymphokine production by Th1 cells without affecting the proliferation of these cells (Fiorentino et al, 1989). Recent evidence suggests that this effect is mediated by a downregulation of the antigen presenting ability of macrophages (de Waal Malefyt et al, 1991). IL-2 is a growth factor for both types of cells although it is only produced by Th1 cells (Mosmann et al, 1987). IL-4, produced by Th2 cells, synergizes with IL-2 on the growth of T cells (Fernandez-Botran et al, 1989). Thus while both Th1 and Th2 cells clearly produce lymphokines that antagonize the other subset, they also produce lymphokines that stimulate and activate both cells.

What is the outcome of these competing influences *in vivo*? What factors are critical in the response to a particular antigen that allows the growth and differentiation of one type of Th cell over the other? Clearly, the early activation step will be important and it may be that certain antigens will bias the immune response one way or the other (Mosmann and Coffman, 1989). In general a predominantly Th1 response will result in increased killing of intracellular parasites and tumor cells due to macrophage activation induced by IFN-γ and lymphotoxin (Murray et al, 1985; Pace et al, 1983). Preferential activation of Th2 cells should lead to high antibody levels, as well activation of several features of allergic responses (Mosmann, 1990; Sanderson et al, 1986). However there are many antigens which are equally able to activate Th1 and Th2 cells, and thus, many normal responses will involve a mixture of the two cell types. In view of the wide array of lymphokines produced by these cells and the complexity of the individual interactions, outlined above, it is not possible, using standard immunological techniques, to assess which factors are the most critical for the maintenance of the balance of Th1 and Th2 *in vivo*. In addition, it would be essential to determine which factors could be important in the development of responses that tip the balance in favor of one subset over the other. We have chosen to address this issue by attempting to develop a mathematical model encompassing Th1, Th2 cells, and the lymphokines IL-2 , IL-4, IL-10 and IFN-γ. This model will address the situation in which both types of Th cell are optimally stimulated by the same antigens and will seek to identify features of the lymphokine secretion pattern that will determine the eventual outcome of a response. We have a panel of well characterized T cell clones of both types available for use in experiments designed to test the model as it develops (Morel et al, 1987; Danska et al, 1990).

2-Mathematical modeling:

This work tries to understand the dynamics of the Th1-Th2 system. We start with the fundamental assumption that this system can be represented by the influence diagram shown in figure 1. This is the simplest, most basic outline of interactions possible and will obviously be expanded to include other cell types and lymphokines as we progress. In particular, we are not attempting to model the antigen presentation function of the APC at this time. In the experiments that we run the antigenic stimulation will not involve APCs. Therefore, in the model, we will treat the antigenic excitation as a time dependent source term. Later we plan to model this interaction more realistically. One objective of our study is to check the validity of this representation of the Th1-Th2 interactions, and to develop a methodology to permit a detailed analysis of its properties.

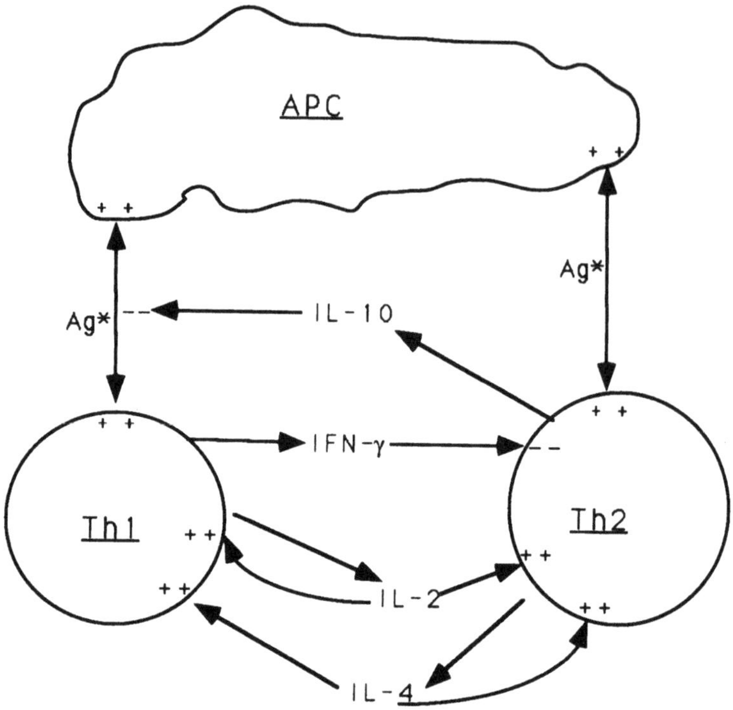

Figure 1. Diagram representing the interactions between Th1, Th2, IL-2, IL-4, IL-10 and IFN-γ. The antigen presenting cell (APC) represents both macrophages and B cells and is not considered in detail at this point as we are assuming optimal stimulation for our Th cells. Ag* refers to the antigenic excitation and is depicted by a 2-headed arrow to indicate that signals are delivered both to the APC and the Th cells and that this interaction requires cell-cell contact. ++ and -- refer to positive and negative effects delivered by the lymphokines to the Th cells. These interactions do not require cell-cell contact.

The best way to describe our model building methodology is to detail the steps we follow. We start with a crude model, and progressively refine it when required to by the experimental data. We attempt to avoid unnecessary complications, however the danger of simplifying too much is that one may miss essential features. Therefore our attitude is to let our model get progressively more and more complicated to accommodate the experimental results. Modeling of the IL-2/T-cell interaction has already been attempted in seminal works by Alan Perelson (Kevredis et al, 1988). The approach described here is more phenomenological or less theoretical as it attempts to elicit a mathematical model from experimental data. Experiments are designed to test the structure of the models and facilitate parametric estimation. The model is constructed progressively, piece by piece.

The crudest mathematical expression which could describe the influence diagram of figure 1 is shown in table 1. These equations were derived using the following assumptions :

• The state variables relevant for our analysis are the concentrations of the lymphokines IL2,4,10, IFN-γ, and the numbers (or concentrations) of Th1 and Th2 cells .

TABLE 1. Model of Th1/Th2 dynamics

1) $dTh1/dt = (-b_1(t) + b_3 IL\text{-}2 + b_5 IL\text{-}4)Th1 + b_2\, APC\, Th1$

2) $dTh2/dt = (-g_1(t) + g_3 IL\text{-}2 - g_4 IFN\text{-}\gamma + g_5 IL\text{-}4)Th2 + g_2\, APC\, Th2$

3) $dIL\text{-}2/dt = (p_1 - p_3 IL\text{-}2 - p_4 IL\text{-}10)Th1 - p_5\, IL\text{-}2\, Th2$

4) $dIL\text{-}4/dt = (m_1 + m_3 IL\text{-}2 - m_4 IL\text{-}4 - m_5\, IFN\text{-}\gamma)Th2 - m_2 IL\text{-}4\, Th1$

5) $dIL\text{-}10/dt = e_1 Th2 - e_2 IL\text{-}10\, Th1$

6) $dIFN\text{-}\gamma/dt = (n_1 + n_3 IL\text{-}2 + n_4 IL\text{-}4 - n_5\, IL\text{-}10)Th1 - n_6 IFN\text{-}\gamma\, Th2$

• The intensity of the interactions between the cells and the lymphokine molecules is proportional to the probability that they find each other, i.e. to the product of the concentrations of the two substances. Depending on the coupling considered, the parameters multiplying the product of concentrations measure either the affinity between the molecules and the cell they bind to or the probability that such a binding will affect the proliferation of the cell[1] . Most of those couplings cannot be constant, as they have to describe the saturation effect (no more receptors available) or threshold effect (a minimum number of bindings necessary to activate a cell). Another possibility is that the action of the lymphokine is indirect, and involves another mechanism (as seems to be the case with the action of IL10 on Th1; de Waal Malefyt et al, 1991). In that case, the parameter e_2 becomes an effective coupling dependent on an external variable.

There are some indications that IL2 and IL4 acts synergistically on the proliferation of Th2 and Th1 (Fernandez-Botran et al, 1989; Greenbaum et al, 1988). If this is the case a new

More precisely: the parameter p_3, appearing in the IL2 equation, measures the probability at an IL2 molecule finds a receptor on a Th1 cell to bind to, whereas the parameter b_3 is the obability that this binding activates the Th1 cell and makes it proliferate. This requires that iough IL2 molecules are already bound . Those two numbers have a different interpretation id have very different values.

term, proportional to the product of concentrations: IL2*IL4*Th2 could be added to the equation describing the proliferation of Th2. Our analysis should be able to confirm whether there is such a synergy[1].

Our goal is first to build a mathematical model which describes satisfactorily all the basic vertices and the system in vitro. Then we plan to dissect its dynamical properties.

Analysis of specific couplings, the need to complicate the model:

The first phase of the model building consists in establishing the validity of the basic features of our mathematical model. This means studying first the most basic building blocks of the system under study, namely the {IL2, IL4 ,Th1} and {IL2, IL4, Th2} systems. They look simple, but when one looks at them with a magnifying glass, one discovers that their simplicity is deceiving.

In the process of proliferating, Th1's and Th2's go through a cell cycle: they bind to IL2 and IL4 until a threshold number molecules have bound, at which point the complicated process leading to mitosis is initiated. This takes several hours, i.e. its duration is significant with respect to the time scale of the experiments. A time delay between the beginning of activation and the actual proliferation has therefore to be accounted for in the equations. Following mitosis, the cells apparently enter into a resting phase. These phases can be thought of as the G_0, G_1 and S, and G_2 and M stages for the resting, potentially activated and activated cells respectively.

The coupling structure shown in table 1 does not adequately describe this process. Instead, we propose to treat the different states of Th1 or Th2 as different state variables: namely: $Th^n i$, $Th^m i$, $Th^r i$, (i=1,2), for the potentially activated Thi, the activated Thi, undergoing mitosis and the resting Thi, respectively. To describe for example the subsystem Th1, IL2, we decided to tentatively use the following equations[2]:

$$dTh^m1(t)/dt = b_2\, IL2(t)\, Th^n1(t) - dTh^m1(t-T)/dt$$

$$dIL2(t)/dt = (p1 - p2\, IL2)\, Th^n1(t)$$

$$dTh^n1(t)/dt = 2\, dTh^m1(t-T)/dt - dTh^m1(t)/dt$$

This coupling structure is the result of several iterations and it requires some explanation.

---

[1] There may be other ways that we plan to study in details, of interpreting and modeling this apparent synergy.

[2] There are obviously equivalent equations for the subsystems: Th2, IL2, IL4; Th1, IL2, IL4; or Th2, IL4. If one adds anti-IL2 receptors and blocking thereby the autocrine effect of IL2, it is possible then to study the Th1, IL4 system in the same way.

• b_2 is the probability that the binding of IL2 to Th1 "activates" the cell to proliferate. There are 1000 to 2000 receptors for IL2 on a Th1 cell. When most of the receptors are bound, the probability of additional binding decreases as it becomes difficult for a molecule to find a free receptor. This is the saturation effect. When this saturation effect occurs, b_2 and p2 must go to zero. Those two parameters cannot be completely constant, they have go to zero when the ratio IL2/Th1 is large enough. The activation of Th1 seems to require that a minimum number of receptors be bound[1]. This is the "threshold effect". This effect means that the coupling b_2 vanishes for value of IL2/Th1 under the threshold, i.e. as long as the number of bindings is less than the threshold number. The coupling describing the activation of the T-cells by interleukins, is therefore highly non-linear, in the sense that the parameters driving the couplings depend on the relative concentrations of molecules and cells or receptors. These features are derived from available knowledge concerning the biology of IL-2 and its receptor (IL-2R). Three classes of IL-2R exist on T cells, a high affinity ($K_d = 10^{-11}$), an intermediate affinity ($K_d = 10^{-9}$M) and a low affinity receptor ($K_d = 10^{-8}$M). The high affinity receptor is formed by a complex consisting of the low (p55) and the intermediate (p75) affinity receptors. At physiological concentrations of IL-2 the high affinity IL-2R, of which there are 1000-2000/cell, is responsible for delivering the signal to proliferate. The individual chains are unable to do this at these IL-2 concentrations. Work by Cantrell and Smith (Cantrell and Smith, 1984) has shown that concentrations of IL-2 that saturate the binding to the high affinity IL-2R are necessary to stimulate the cell to cycle; the threshold effect. After IL-2 has bound its receptor both IL-2 and the IL-2R are internalized and effectively removed from the cell surface.

• Th^r1 does not appear in these equations. In fact the system could have taken the following form:

$$dTh^m1(t)/dt = b_2\, IL2(t)\, Th^n1(t) - dTh^m1(t-T1)/dt$$
$$dIL2(t)/dt = (p_1 - p_2\, IL2)\, (Th^r1(t) + Th^n1 + Th^m1)$$
$$dTh^r1(t)/dt = 2\, dTh^m1(t-T1)/dt - dTh^n1(t)/dt$$
$$dTh^n1(t)/dt = dTh^r1(t-T2)/dt - dTh^m1(t)/dt$$

Th^m1 gives rise to two Th^r1 after a time T1, and then each Th^r1 becomes a Th^n1 after time T2. So, seen from the perspective of the balance of Th^n1, at time T = T1 + T2 after the

---

The exact nature of the threshold effect has not yet been totally deciphered. We assume for the ne being that activation is triggered when a minimum number of bindings takes place. This sults in properties of the coupling b_2 which we can check on the experimental results. We did similar analysis on a simpler system we describe later.

beginning of the activation, the activation has resulted in additional Th^n1 cells. So if Th^r1 and Th^m1 have the same coupling properties to IL2, one can consistently ignore Th^r1 cells in the mathematical model, and interpret T as the sum of the duration of the mitosis and the time it takes in average for the receptors to be expressed after mitosis.

In the equation describing the sink for IL2, we neglect the contribution of Th^m1, as it seems that the threshold number is close to the saturation number of bindings as discussed above. We do not know whether IL2 binds to Th^r1. This coupling structure assumes also that only Th^n1 secretes IL2. This is supported by the fact that antigenic stimulation is required for IL-2 production, and by the fact that once Il-2 has stimulated this cell to cycle into the S phase the IL-2 production ceases (Efrat and Kaempfer, 1984).

Confrontation with experiment and parametric estimation:

The confrontation of dynamical models with experiments in immunology raises methodological problems. Experiments have to be designed whenever possible to elicit basic qualitative features of the system, to check whether the model is not fundamentally at variance with what it purports to describe.

Eventually the contact between the theoretical computations and the experimental results has to be quantitative. The parameters appearing in the models have to be estimated on the experimental results. This is clearly a delicate exercise which has to be made very carefully. We first try to ensure that we understand the couplings of IL2 and IL4 to Th1 and Th2 well. Any uncertainty in the value or structure of those couplings will affect our ability to understand the behavior of the whole system.

The precision of the experimental result affects the quality of the parameter estimations. We are planning to use the maximum likelihood approach (Bard,1974). This technique not only gives the most "probable" value for the parameters, but it also permits us to translate the experimental uncertainties into uncertainties of the values of the parameters.

The conditions of measurements introduce an element of complexity in their interpretations or in their use for parametric estimation. The proliferation of cells is measured by introducing thymidine at one point, letting the system evolve for a while and measuring the thymidine ingested by the proliferating cells during that time. The amount of radioactivity is directly proportional to the proliferation, but the proliferation does not directly measure the number of cells.

The measurement of the amount of IL2 and IL4 in the supernatant proceeds from the same logic. IL2 and IL4 activates other cells (HT2). By measuring the amount of proliferation of those cells, due to IL2 or IL4, during a certain time, it is possible to estimate the amount of IL2 or IL4 there was in the supernatant. This method has several experimental problems also, in that it is not always possible to be sure that the proliferation one is measuring is solely due to the lymphokine of interest. We are therefore switching to an ELISA technique of measurement, which will directly measure lymphokine levels in a very specific manner.

Experimental analysis:

1- Testing some qualitative features:

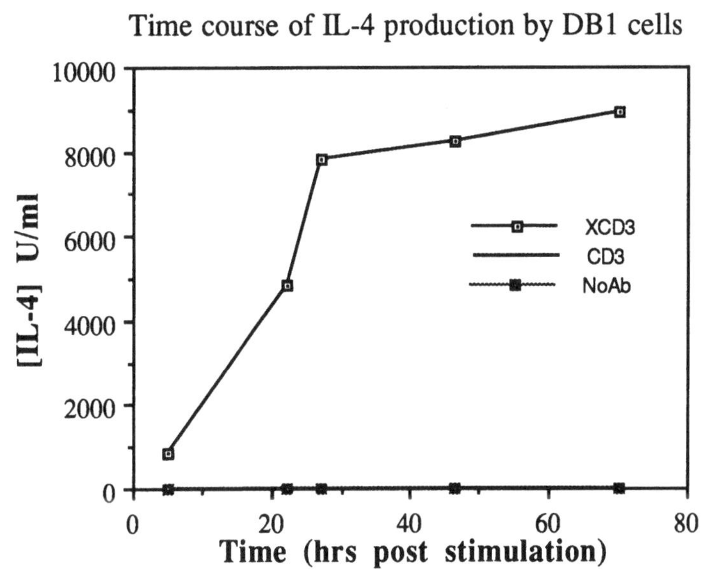

Figure 2. Time course of IL-4 production by DB1 cells. DB1 cells were washed and incubated in a 1:1 dilution of anti-CD3 monoclonal antibody (MoAb) for 30 minutes at 4°C. The cells were washed to remove excess antibody and placed in tissue culture wells (5 x 10^4 cells/well) that had been previously coated with 10μg/ml goat anti-hamster IgG. As controls cells coated with anti-CD3 (no symbols) and uncoated cells (closed squares) were incubated in wells with no second antibody. Supernatants (SNs) (6/time point) were removed 5, 22, 26, 44 and 72 hours after stimulation. IL-4 present in the SN was assessed by performing 6 1:2 dilutions of the SN and carrying out an HT-2 proliferation assay as previously described (Fernandez-Botran et al, 1986).

Figure 2 shows secretion of IL4 over time in a system where Th2 cells were initially exposed to an antigenic excitation. In these experiments DB1, an alloreactive T cell clone, was activated using cross-linked CD3 (XCD3). Supernatants were harvested at the indicated times and IL-4 was measured using the HT-2 assay. Our crude model of this situation is:

$$dTh^{m}2(t)/dt = c_3\ IL4(t)\ Th^{n}2(t)$$
$$dIL4(t)/dt = (m_1 - m_3\ IL4)\ Th^{n}2(t)$$
$$dTh^{n}2(t)/dt = 2\ dTh^{m}2\ (t-T)/dt\ -\ dTh^{m}2(t)/dt$$

The shape of the curve for IL4 shown in figure 2 could easily be explained by our crude model: The quantity of IL4 grows up until it reaches the saturation quantity where m_1-m_3 IL4 =0. Then it stabilizes. The value of the saturation concentration could be used to estimate the ratio: m_1/m_3. But this "natural" interpretation of the experimental results is probably wrong. There is evidence that the plateau in the IL4 concentration is due to the fact that IL4 is not manufactured or utilized anymore, because there are no more Th^n2 cells. The evidence supporting that other interpretation of these data is that, in separate experiments, the cells were activated by XCD3 and RNA was collected at several time points. The amount of IL-4 mRNA produced was measured using the reverse transcriptase-polymerase chain reaction. These results demonstrated that IL-4 production was detectable at 1 hour, was peaking at 4 hours and was greatly reduced by 24 hours, which was when the protein levels reached their peak (data not shown). As shown in figure 2 the protein levels remain high despite a reduction in the mRNA expression.

A possible explanation for this and other features of this system is the following: IL4 puts Th^n2 cells in the Th^m2 activated state. After a certain time the Th^m2 cells mitose to give Th^r2 cells. But IL4 by itself is not able to transform a Th^r2 into a Th^n2 cell. This requires either IL2 or an antigenic excitation. This is supported by the experimental data that either IL-2 or antigenic stimulation can induce Th2 cells to express mRNA for IL-4 (Morel P, unpublished observations).This phenomenon is not true for Th1, since IL-2 alone does not stimulate Th^r1 to express IL-2 mRNA. So, if neither antigen nor IL-2 is present, the Th2 cells progressively consist of resting cells only; cells that neither synthesize nor bind IL-2. This explanation would give a different basis to interpret the synergy between the IL2 and IL4. The corresponding equations would be :

$$dTh^{m}2(t)/dt = (c_2\ IL2(t) + c_3\ IL4(t) + c_5\ IL2(t)\ IL4(t))\ Th^{n}2(t) - dTh^{m}1(t-T1)/dt$$
$$dIL4(t)/dt = (m_1 - m_3\ IL4)\ Th^{n}2(t)$$
$$dIL2(t)/dt = -p4\ IL2(t)\ Th^{n}2(t)$$
$$dTh^{r}2(t)/dt = 2\ dTh^{m}1(t-T1)/dt - rTh^{r}2(t)\ Ag(t)$$
$$dTh^{n}2(t)/dt = rTh^{r}2(t)\ Ag(t) - dTh^{m}2(t)/dt$$

According to this scheme, the sustained proliferation of Th2 requires the presence of an antigenic excitation. IL4 per se does not have the proliferative effect on Th2 that IL2 has on Th1. This is supported by our data (figure 3)

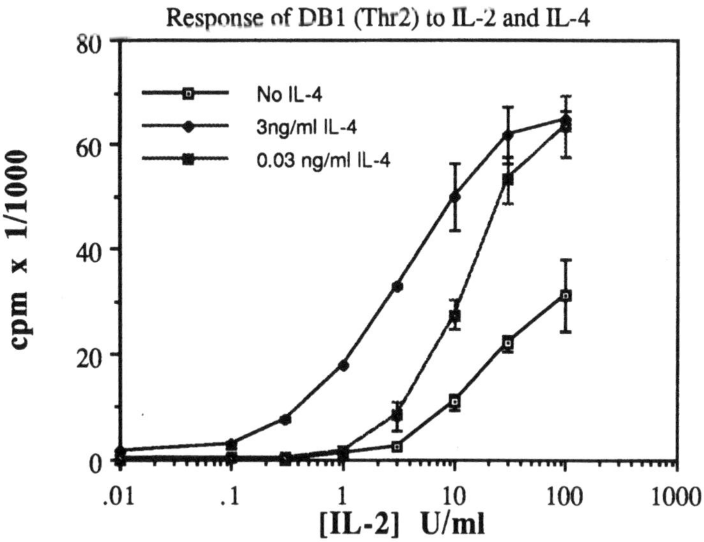

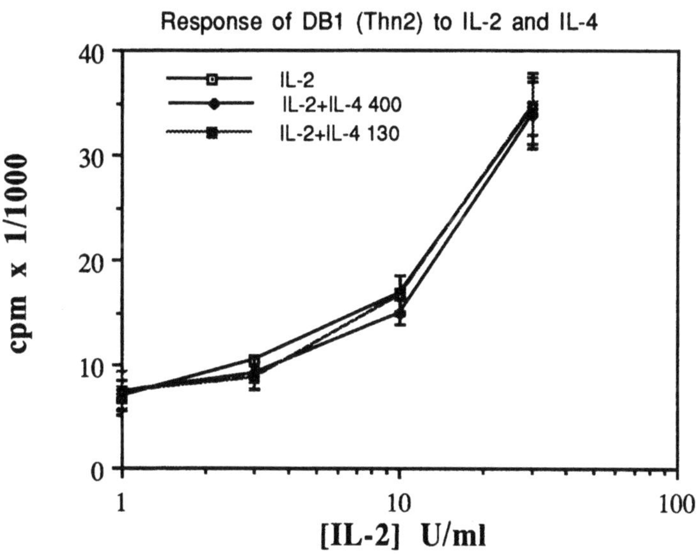

Figure 3. Response of resting ($Th^{r}2$) and activated ($Th^{n}2$) DB1 cells to IL-2 and IL-4. DB1 cells were either activated with anti-CD3 MoAb as described in the legend to figure 1 or simply washed and placed in culture (5×10^4 cells/well) in the presence of increasing concentrations of IL-2 and IL-4. After 48 hours at 37°C the cells were pulsed with 0.5μCi ^{3}H Thymidine for 16 hours. The cells were then harvested onto glass fiber strips and counted in a beta scintillation counter. The results represent the means and standard deviations of 3 replicate wells.

The synergism between the actions of IL2 and IL4 could be explained if c_5 were the main coupling of IL2 to Th2. There would be a natural explanation of the qualitative properties of our data if c_2 were equal to zero, and the proliferation of Th2 was controlled by the c_3 and c_5 terms. Obviously this hypothesis has still to pass a quantitative test which is under way.

2-Parametric estimations:

It turns out that the{IL2, HT2} and {IL4, HT2}systems build simpler versions of the subsystems we want to understand involving Th1 and Th2. HT2 cells are IL-2-dependent cells that are normally used to measure IL-2 and IL-4 levels in a supernatant. They are simpler in the sense that HT2 does not secrete IL2 or IL4. They provide a suitable laboratory to test and develop our parametric estimation techniques and study the detailed structure of the non-linearity of the couplings.

HT2 cells differ from the cells of interest for us, namely the Th1 and Th2 in many respects. Th1 and Th2 cells seem to have far more complex properties. They can be put in different states of "excitation", i.e. their response to IL2 and IL4 depends on the specifics of their states: like whether their high affinity receptors to IL2 or IL4 are expressed or not or whether they secrete IL2 or IL4 or not. The study of the interactions with IL2 and IL4 with HT2 can help us only to develop our methodology and to understand some non-linear couplings. The system which we eventually used for the parametric estimation of this situation reads:

$$dHT2^m(t)/dt = \alpha\, IL2(t)\, HT2^n(t)$$
$$dIL2(t)/dt = -\beta\, IL2(t)\, HT2^n(t)$$
$$dHT2^n(t)/dt = 2\, dHT2^m(t-T)/dt - dHT2^m(t)/dt$$

Of particular interest to us is the form of the couplings measuring the probability that IL2 HT2 binding leads to HT2 proliferation, and its interpretation in terms of saturation and threshold numbers. It turns out that the shape of the curves of proliferation of HT2 is sensitive to the details of the choice of function and value of the parameters. In this section we describe the methodology used to estimate these parameters in order to be able to reproduce the experimentally observed quantities as accurately as possible. Once we have a quantitative estimate of the parameters we can use the model for prediction.

The parameter estimation presented here is based on the two data sets presented table 2. The values of HT2 for initial values of 0 U/ml of IL2 is treated as background level of noise and subtracted from the rest of the values of HT2 proliferation. The observed quantities in columns 2 and 3 are cell counts of HT2 that are in a state of mitosis between 18 and 21 or 24 hours.

[IL2] U/ml	3 hours	6 hours
300	32780	68096
100	34127	63150
30	39312	73743
10	30489	65581
3	25341	47294
1	8741	16811
0.3	4764	6533
0.1	1809	2911

Table 2: HT2 assay of 10000 cells/well. Proliferation of HT2 for given initial values of IL2. Pulsed after 18 hours for indicated times.

Therefore the observed quantity $HT2^{observed}$ is represented by the integral

$$HT2^{calculated} = \int_{18}^{21} (dHT_2^m/dt).dt$$

In order to numerical evaluate this integral, we let the system defined by the differential equations, evolve with time on a computer, using the initial conditions for IL2 and HT2, starting with some guestimates of parameters α and β. The main aim of parametric estimation is to find estimates of α and β such that some measure of the distance between observed quantities and the calculated quantities is minimized. This problem is formulated as an optimization problem as discussed in the following subsection. The data demonstrates *two* important qualitative features. The proliferation of HT2 increases suddenly as the initial value of IL2 crosses 0.1. This is related to the *threshold effect.* The proliferation of HT2 saturates as the initial values of IL2 are about 10 U/ml. This is the *saturation effect..* Clearly, the parameter α has to reflect these two properties, and therefore cannot be a constant. α is a function of the ratio x between the concentration of IL2 and HT2 (x=HT2/IL2). One of our primary concerns is to determine the functional form of the coupling α which would allow us to model the threshold and saturation effect. After trying several forms, we finally fixed our choice on the following functional form:

$a = w/p\ (1 - e^{-ax})\ \{\mathrm{atan}(-d\ x + g) + p/2\}$

Notice that this expression for *a* introduces 4 new parameters: a, w, g and d. Figure 4 a below shows the shape of the coupling *a* as a function of IL2. Figure 4b shows the saturation effect between 0 and 1 in greater detail.

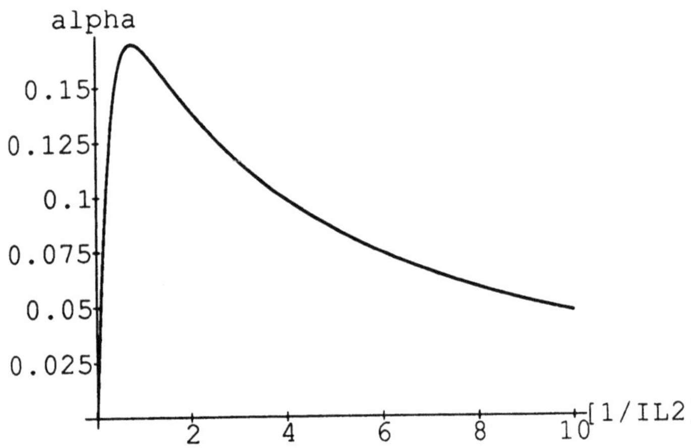

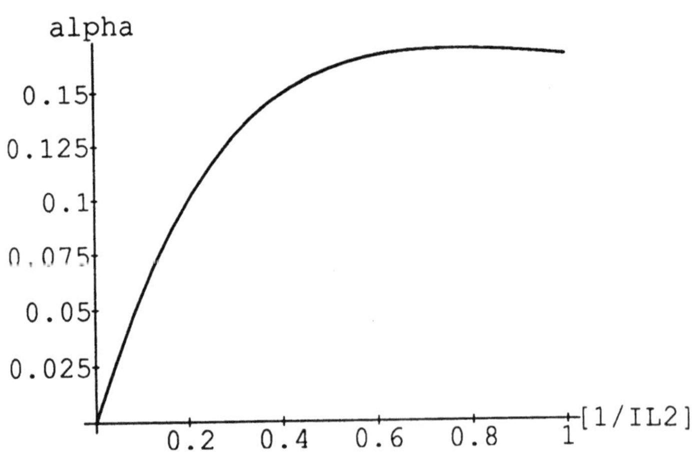

Figure 4a: The shape of the coupling *a* as a function of [1/IL2].
4b: Detail in the region 0 to 1.

In order to estimate the parameters such that the predictions of the model fit the experimental data accurately we define an objective function which measures the distance between $HT2^{predicted}$ and $HT2^{observed}$. This measure is called the objective function F which we minimize subject to the constraints that the predictions are derived by solving the delay-differential equations. This can be stated as follows:

minimize $F(HT2^{predicted}, HT2^{observed})$

subject to
$$dHT2^m(t)/dt = \alpha\, IL2(t)\, HT2^n(t)$$
$$dIL2(t)/dt = -\beta\, IL2(t)\, HT2^n(t)$$
$$dHT2^n(t)/dt = 2\, dHT2^m(t-T)/dt - dHT2^m(t)/dt$$

We have conducted the analysis by using two different forms for the objective function. The first is the *least squares function* which is as follows:

$$F = \Sigma_i \,(HT2^{predicted} - HT2^{observed})^2_i$$

The least squares objective function provides a good measures for initial analysis since it minimizes the large differences between observed and predicted values. However, the least squares function is insensitive to differences in small numbers and hence differences in the lower end are not adequately taken care of. To fine tune the parametric estimates we used the *log least squares* function shown below:

$$F = \Sigma_i \,(\log(HT2predicted) - \log(HT2observed))^2{}_i$$

The optimization was conducted by using the Sequence Quadratic Programming (SQP) algorithm (Biegler and Cuthrell, 1985). The numerical integration was done by using fourth order Runge Kutta methods.

Results

Based on the optimization procedure the estimates for the parameters are as follows. The shape of the coupling parameter α is shown in Figure 4. based on the following estimates:

$a = 0.4$, $w = 2.7$, $d = 1.8$, $g = 0.3$

begins to degrade seriously when the parameters defining a are given a value 10% off the present estimate. Similarly, b and T were estimated to be:

$b = 1x10^{-10}$, T = 5.0 hours

the quality of the fit is less sensitive to the exact value of those parameters. An uncertainty of 20% is compatible with the parametric estimation. The comparison of the experimental results and the predicted results based on the estimated parameters is shown in Figure 5.

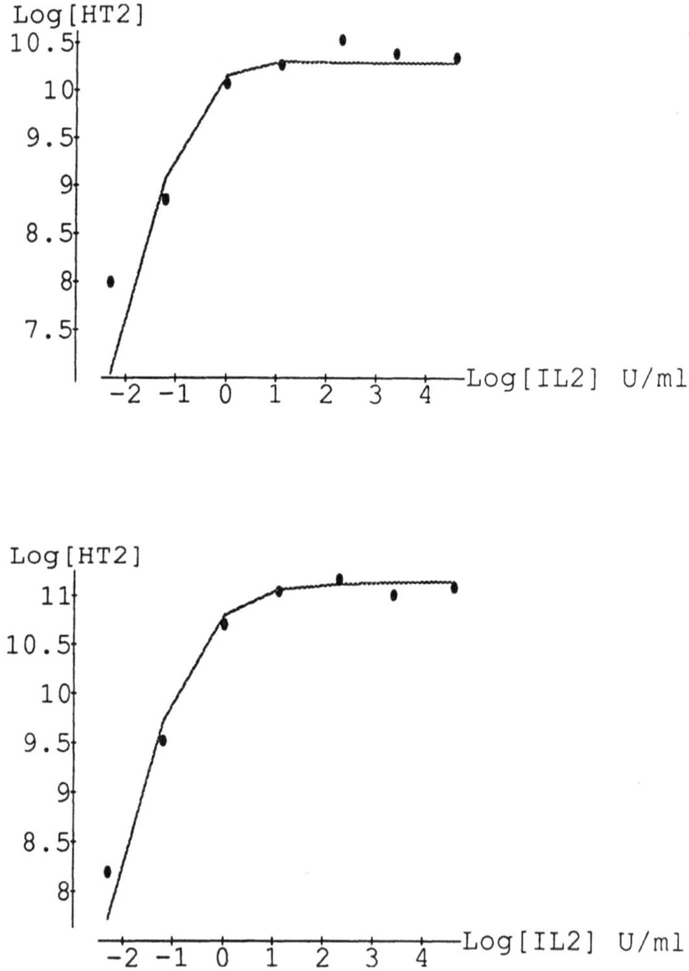

Figure 5a: Plots the proliferation of HT2 from 18 to 21 hours. 5b: The proliferation of HT2 from 18 to 24 hours. The curves represents the predictions and the o the experimental results.

Discussion

The fact that it was possible to reproduce the experimental results reasonably well, is encouraging. One may ask whether the function chosen for α, which involves four parameters needed to be that complicated. In fact this function is a rather natural choice, as it is the product of a function (1-exponential) which has the right shape to describe the saturation effect with the simplest function (an arc-tangent) which could describe the threshold effect (i.e. which would go to zero for high value of the ratio HT2/IL2). Since the proliferation of Th1 and Th2 with IL2 and IL4 is based on similar mechanisms we expect to be able to use the same kind of functional forms.

In order to estimate precisely the uncertainty on the value of the parameters, we plan to use the maximum likelihood method. It has a lot in common with our approach so far, except that to build the likelihood function it is important to know the experimental errors. So far, we have only the "statistical" errors, i.e. the errors are computed on the basis of the spread of results of several experiments made in parallel. It turns out that the resulting errors do not build a consistent set of errors. Clearly a more detailed analysis, apparently essentially new in the experimental folklore of immunology, of the systematic errors and other experimental biases, as well as the sources of variability is of essence before a meaningful maximum likelihood analysis is possible (Bard, 1974). In the Th1 and Th2 case we also intend to generate a large set of experimental results so that we would be in a position to estimate the statistics (mean, covariance etc) associated with the data matrices generated.

4- Final comments:

The goal of this study is to understand the Th1-Th2 dynamics, using mathematical modeling. Mathematical modeling has the potential to provide a unique tool to analyze the Th1-Th2 system. Not only could it help in the interpretation of its behavior, but it could permit a refined analysis of its detailed properties and deepen our understanding of how different antigenic excitations can trigger different behaviors from this system.

One difficulty may be to learn to extract the right information from this approach. Immunological reactions are the result of a very complicated and integrated regulatory systems. Experience with control systems shows that the only avenue to get real insight or predictive power of their behavior is mathematical modeling. If successful this approach could be very useful: the dynamical structures of the dynamical systems could reveal the general properties of the immunological system, the nature of the "physiological attractors", the dynamics of immunological reaction (some auto-immune diseases could be attributable to the dynamical properties of the immunological system, which would fall in a "wrong attractor"), the way to fight some infection, etc...

The early positive message of our work is that the fundamental paradigm that this approach uses to associate equations to phenomena is not at complete variance with the phenomena. But modeling processes involving subtle interactions between molecules, cells and receptors correspond necessarily to an idealization. There is an irreducible level of complexity in immunological processes that simple mathematical models cannot capture. Building more complicated models is not the solution. To be able to get the right free fall law, by studying the fall of objects from the tower of Pisa, Galileo had to subtract from the interpretation of his measurements, spurious effects he did not understand well or did not know how to model like air resistance. In a sense, we are in a somewhat similar situation. Mathematical models of immunological processes proceed from an attempt to find a simplified representation which captures the essence of the dynamics of the system.

References.

Bard Y (1974) Non-linear parameter estimation. Academic Press

Biegler, L.T., J.E Cuthrell (1985): Improved path optimization for sequential modular simulations Part II; The optimization algorithm. Comp. & Chem Eng. 9 (3) p. 257.

Cantrell DA, Smith KA (1984) The interleukin-2 T-cell system: a new cell growth model. Science 224:1312-1316

Cher DJ, Mosmann TR (1987) Two types of murine helper T cell clones. II. Delayed-type hypersensitivity is mediated by Th1 clones. J Immunol 138:3688-3694

Cherwinski HM, Schumacer JH, Brown KD, Mosmann TR (1987) Two types of mouse helper T cell clone. III. Further differences in lymphokine synthesis between Th1 and Th2 clones revealed by RNA hybridization, functionally monospecific bioassays, and monoclonal antibodies. J Exp Med 166:1229-1244

Crabtree GR (1989) Contingent genetic regulatory events in T lymphocyte activation. Science 243:355-359

Danska JS, Livingstone AM, Paragas V, Ishihara T, Fathman CG (1990) The presumptive CDR3 regions of both T cell receptor α and β chains determine T cell specificity for myoglobin peptides. J Exp Med 172:27-35

Davis MM, Bjorkman PJ (1988) T-cell antigen receptor genes and T-cell recognition. Nature 334:395-399

DeKruyf RH, Ju S-T, Hunt AJ, Mosmann TR, Umetsin DT (1989) Induction of antigen-specific antibody responses in primed and unprimed B cells. Functional heterogeneity among Th1 and Th2 T cell clones. J Immunol 142:2575-2582

Del Prete GF, De Carli M, Ricci M, Romagnani S (1991) Helper activity for immunoglobulin synthesis of T helper type 1 (Th1) and Th2 human T cell clones: the help of Th1 clones is limited by their cytolytic capacity. J Exp Med 174:809-814

deWaal Malefyt R, Haanen J, Spits H, Roncarolo M-G, te Velde A, Figdor C, Johnson K, Kastelein R, Yssel H, de Vries J (1991) Interleukin 10 (IL-10) and viral IL-10 strongly reduce antigen-specific human T cell proliferation by diminishing the antigen-presenting capacity of monocytes via downregulation of class II major histocompatibility complex expression. J Exp Med 174:915-924

Efrat S, Kaempfer K (1984) Control of biologically active interleukin 2 messenger RNA formation in induced human lymphocytes. Proc Natl Acad Sci USA 81:2601-2605

Fathman CG, Frelinger JG (1983) T lymphocyte clones. Ann Rev Immunol. 1:633-653

Fernandez-Botran R, Sanders VM, Mosmann TR, Vitetta ES (1989) Lymphokine-mediated regulation of the proliferative response of T helper 1 and T helper 2 cells. J Exp Med 168:543-556

Fernandez-Botran R, Kramner PH, Diamantstein T, Uhr JW, Vitetta ES (1986) B cell stimulatory factor-1 promotes growth of helper T cell line. J Exp Med 164:580-595.

Finkelman FD, Pearce EJ, Urban JF Jr, Sher A (1991) Regulation and biological function of helminth-induced cytokine responses.in:Ash C, Gallagher RB (eds) Immunoparasitology Today. Elsevier Trends Journals, Cambridge, p A62

Fiorentino DF, Bond MW, Mosmann TR (1989) Two types of mouse T helper cell. IV. Th2 clones secrete a factor that inhibits cytokine production by Th1 clones. J Exp Med 170:2081-2095

Gajewski TF, Fitch FW (1988) Anti-proliferative effect of IFN-γ in immune regulation. I. IFN-γ inhibits the proliferation of Th2 but not Th1 murine T lymphocyte clones. J Immunol 140:4245-4252

Greenbaum LA, Horowitz JB, Woods A, Pasqualinit, Reioch EP, Bottomly K (1988) Autocrine growth of CD4+ T cells. Differential effects of IL-1 on helper and inflammatory T cells. J Immunol 140:1555-1560

Haanen JBAG, de Waal Malefijt R, Res PCM, Kraakman E, Ottenhoff THM, de Vries RRP, Spits H (1991) Selection of a human T helper type 1-like T cell subset by mycobacteria. J Exp Med 174:583-592

Hood L, Steinmetz M, Malissen B (1983) Ann Rev Immunol 1:529-545

Kevrekidis IG, Zecha AD, Perelson AS (1988) Modeling dynamical aspects of the immune response : I. T cell proliferation and the effect of IL-2. In : Theoretical Immunology Part One. Perelson AS (ed) Addison-Wesley Publishing Company.

Kim J, Woods A. Becker-Dumm E, Bottomly K (1985) Distinct functional phenotypes of cloned Ia restricted helper T cells. J Exp Med 162:188-200

Kurt-Jones EA, Hamberg S, Ohara J, Paul WE, Abbas AK (1987) Heterogeneity of helper/inducer T lymphocytes. I. Lymphokine production and responsiveness. J Exp Med 166:1774-1786

Locksley RM, Scott P (1991) Helper T cell subsets in mouse leishmaniasis: induction, expansion and effector function. In: Ash C, Gallagher RB (eds) Immunoparasitology Today . Elsevier Trends Journals, Cambridge, p A58

Morel PA, Livingstone AM, Fathman CG (1987) Correlation of T cell receptor Vb gene family with MHC restriction. J Exp Med.166:583-588

Mosmann TR, Cherwinski H, Bond MW, Giedlin MA, Coffman RL (1987) Two types of helper T cell clones. I. Definition according to profiles of lymphokine activities and secreted proteins. J Immunol 136:2348-2357

Mosmann TR, Coffman RL (1989) Th1 and Th2 cells : Different patterns of lymphokine secretion lead to different functional properties. Ann Rev Immunol 7:145-173

Mosmann TR (1990) Helper T cells and their lymphokines. In Feldmann M, Lamb J, Owen M (eds) T cells, Wiley , New York In press

Murray HW, Spitalny GL, Nathan CF (1985) Activation of mouse peritoneal macrophages in vitro and in vivo by interferon-γ. J. Immunol. 134:1619-1625

Ohara J, Paul WE (1985) Production of a monoclonal antibody to and molecular characterization of B-stimulatory factor-1. Nature 315:333-335

Pace JL, Russell SW, Torres BA, Johnson HM, Gray PW (1983) Recombinant mouse γ-interferon induces the primary step in macrophage activation for tumor cell killing. J. Immunol. 130:2011-2017

Parish CR.(1972) The relationship between humoral and cell-mediated immunity. Transplant. Rev. 13:35-50

Rabin E, Mond J. Ohara J, Paul W (1986) Interferon-γ inhibits the action of B cell stimulatory factor (BSF-1) on resting B cells. J. Immunol. 1137:1573-1581

Sanderson CJ, O'Garra CJ, Waren DJ, Klaus GGB (1986) Eosinophil differentiation factor also has B cell growth factor activity : proposed name interleukin-4. Proc. Natl. Acad. Sci. USA 83:437-442

Scott P, Kaufmann SHE (1991) The role of T cell subsets and cytokines in the regulation of infection. Immunology Today 12:346-348

Stout RD, Bottomly K (1989) Antigen-specific activation of effector macrophges by IFN-γ-producing (Th1) T cell clones. Failure of IL-4-producing (Th2) T cell clones to activate effector function in macrophages. J. Immunol. 142:760-767

Swain SL (1983) T cell subsets and the recognition of MHC class. Immunol Rev 71:129-143

Tamira S-I, Chiba J, Kajima A, Uchida N (1983) Properties of cloned T cells that mediate delayed-type hypersensitivity against ovalbumin in mice. Cell. Immunol. 76:156-166

Vitetta ES, Brooks K, Chen YW, Isakson P, Jones S et al (1984) T cell-derived lymphokines that induce IgM and IgG secretion in activated murine B cells. Immunol Rev 78:137

Zinkernagel RM, Doherty PC (1974) Restriction of in vitro T cell-mediated cytotoxicity in lymphocytic choriomeningitis within a syngeneic or semi-allogeneic system. Nature 248:701-703

COMPLEX OUTCOMES IN MOUSE LEISHMANIASIS: A MODEL FOR THE DYNAMICS OF THE TH1 RESPONSE

A. Nicola Schweitzer*, Jonathan Swinton and Roy M. Anderson

Wellcome Centre for Research on Parasitic Infections
Department of Pure and Applied Biology
Imperial College
London SW7 2BB.

Abstract

The interaction between *Leishmania* parasites and Th1 cells is investigated using a simple mathematical model of immunological responses and parasite population growth within the host. The model generates patterns of resistance and susceptibility to infection that mirror observed trends in experimental infections of mice and humans exposed to infection in areas of endemic transmission. The heterogeneity in outcome predicted by the model can arise either through differences in the values of the parameters that characterize the genetic background of the host or as a consequence of differences in the size of the infecting inoculum of the parasite. Detailed analyses of equilibrium states and of the time course of infection within a host suggest that a limitation in the availability of precursor T cells, as a consequence of high levels of recruitment into the activated pool, may play a significant role in the progression of infection in susceptible hosts.

1 Introduction

In the last few years there has been much interest among parasite immunologists in the T helper cell dichotomy. This division of mouse and human T helper cells into two subsets is based on the identification of differing patterns of cytokine secretion and responsiveness (Cherwinski et al., 1987; Mosmann et al., 1986). Cells producing IL-2, interferon-γ and TNF have been designated Th1 cells. Those producing IL-4, IL-5 and the more recently characterized IL-10 (Fiorentino et al., 1989; Moore et al., 1990) have been designated Th2 cells. Cells with other patterns of cytokine secretion have also been observed (Street et al., 1990; Gajewski and Fitch, 1988) but are thought to result from transient stimulation, or be expressed during intermediate developmental stages.

The characterization of this dichotomy in mice is of particular importance in the context of experimental parasitic infections, where defective proliferative T cell responses or

*Correspondence: Dr A.N. Schweitzer, West Wing, Department of Biology, Imperial College, London SW7 2BB. Email: ans@bio.ic.ac.uk.

NATO ASI Series, Vol. H 66
Theoretical and Experimental Insights into Immunology
Edited by A. S. Perelson and G. Weisbuch

impaired delayed-type hypersensitivity, specific to Th1-type responses, have been associated with long term chronic or progressive infections (Ho et al., 1983; Nutman et al., 1987; Ward et al., 1988; Feldmeier et al., 1988; Castes et al., 1989). High IgE levels and marked eosinophilia, both characteristic of Th2-type responses, are prominent in humans and mice with persistent helminth infections (Sher and Colley, 1989). Examples of resistance against infection, whether natural, experimentally induced, or following drug treatment, are frequently associated with the induction of strong Th1-type responses (Scott et al., 1989; Pond et al., 1989; Locksley and Scott, 1991; Heinzel et al., 1991).

An important feature of parasitic infections of both mice and humans is the marked heterogeneity observed between hosts with respect to parasite load and resistance to infection. Such heterogeneity, where some hosts harbour few parasites and show effective immunity while others harbour high loads and appear immunologically unresponsive, has been ascribed to host genetic background and nutritional status, variability in environmental factors that determine exposure to infection or some combination of these factors (Anderson and May, 1991). However, it may be that a common mechanism underpins the influence of all these factors, perhaps linked to the differential stimulation of different T cell subsets and the regulatory interactions between these cells. Cross-regulating activities have been identified between the Th1 and Th2 subsets in mice. The most significant appear to be the capacity of interferon-γ to inhibit the proliferation of, but not cytokine secretion by, Th2-type T cells (Gajewski and Fitch, 1988; Fernandez-Botran et al., 1988), and the inhibitory effect of IL-10 and IL-4 secreted by Th2-type cells on cytokine production by Th1 cells, particularly interferon-γ (Fiorentino et al., 1989; Peleman et al., 1989). This set of interactions is summarised in Figure 1.

In this paper we examine the population dynamics of T cell proliferation, cytokine secretion, and immune response regulation as influenced by a parasitic infection that replicates within the host. Model formulation centres on the regulation of Th1-type responses and we assume that the expression of a Th2-type response represents a side effect of the absence of interferon-γ when Th1 cell abundance is low. This assumption that the Th1-type response is the key dynamical event is consistent with the failure to detect Th2-type responses even in mice susceptible to infection with *L. donovani* (Kaye et al., 1991). In such experimental infection, mice show an initial peak of Th1-type activity which subsequently disappears. Similar trends have been reported in BALB/c mice infected with *L. major* but in this system evidence of a Th1 response is much less striking, presumably due to the early rise of a Th2-type response (Heinzel et al., 1989). These observations can be interpreted as implying that the inhibitory effect of Th2 derived cytokines on Th1 function is expressed only when Th1 responses are already independently depressed. However, the precise regulatory interplay between Th1 and Th2 cells remains unclear at present.

We focus, then, on the regulation of IL-2 production by Th1-type T cells and consider the specific example of a *Leishmania* spp-mouse model for which the relationship between resistance, susceptibility and the Th1/Th2 dichotomy is perhaps best characterised. To aid in our investigation of the dynamics of the interaction between the mouse immune system and the replicating parasite, we employ simple non-linear mathematical models to describe the principal processes involved. This approach, which is not widely employed in parasite immunological studies at present, has the advantages of clearly defining the biological assumptions leading to a particular hypothesis, of identifying what needs to be

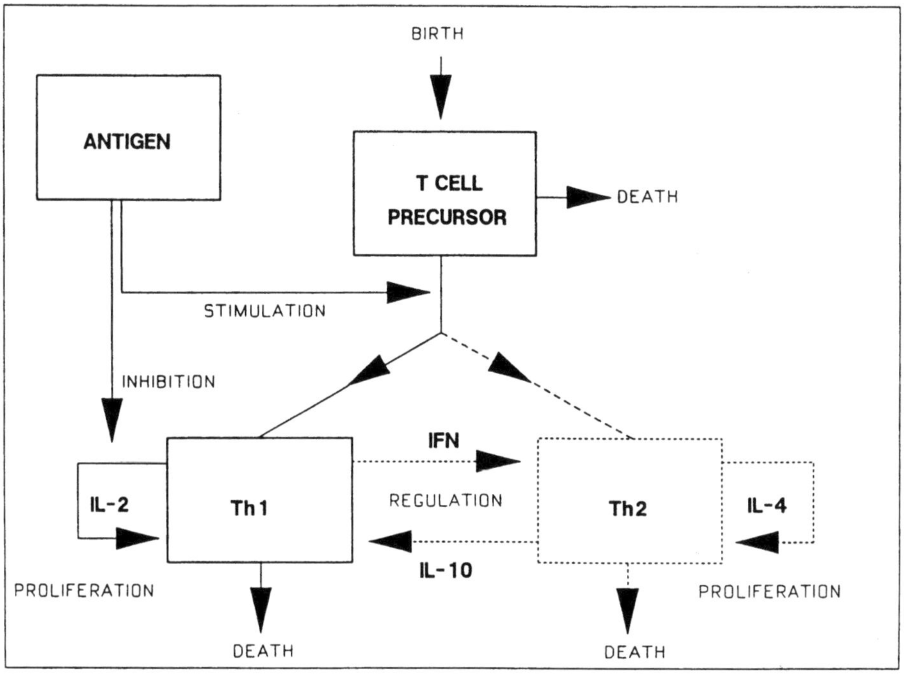

Figure 1: Flow chart summarizing current understanding of the main factors influencing the interactions between Th1 and Th2 cells, and between these cells, cytokines and foreign antigen. Model development is focused on the extent to which factors specific to Th1 cells (solid lines and boxes) influence the interaction between the parasite and the host's immunological responses. Th2 response expression is assumed to be a reflection of the production of sub-inhibitory levels of interferon γ.

measured to understand observed trends and of providing a set of analytical tools for investigating the nonlinear interactions between the major biological variables. We combine model development with an assessment of experimental observations on the *Leishmania*–mouse interaction and compare predicted patterns with observed trends.

2 Methods

We consider four variables, namely the density of precursor T-cells, $M(t)$, the density of activated parasite specific Th1 type T cells, $T(t)$, the concentration of cytokine IL-2, $C(t)$, and the abundance of the replicating parasite, $P(t)$, where t denotes time.

We make the following assumptions. Precursor cells are recruited at a constant rate R, die at a per capita rate u and are stimulated to join the activated Th1 cell pool at a net rate αMP which is proportional to the density of the parasite. These assumptions result in the following equation for the rate of change of M with respect to time:

$$\frac{dM}{dt} = R - uM - \alpha MP \tag{1}$$

Activated Th1 cells are recruited at a net rate αMP, die at a per capita rate μ, and proliferate at a per capita rate $f(C,P,T)$, assumed to be a function of the concentration of cytokine, C, the abundance of the parasite P, and Th1 cell density, T. This gives the following equation for the rate of change of T with respect to time:

$$\frac{dT}{dt} = \alpha MP + [f(C,P,T) - \mu]T \tag{2}$$

Cytokine is produced at a net rate aT by the Th1 cells and is removed at a per capita rate d. The equation for the rate of change of C with respect to time is therefore:

$$\frac{dC}{dt} = aT - dC \tag{3}$$

The parasite replicates at a per capita rate λ (where $\lambda > 0$ represents births minus natural mortality) and is removed by the immune system at a net rate βTP which is proportional to the density of Th1 cells. This gives the following equation for the rate of change of P with respect to time:

$$\frac{dP}{dt} = \lambda P - \beta TP \tag{4}$$

A key function in the model is the Th1 cell proliferation rate f. Experiments suggest a sigmoid relationship between the amount of cytokine (IL-2) and the rate of cell proliferation, where the net rate saturates to a constant value (Toribio et al., 1989; Smith, 1989). This situation is captured by the functional form $f = rC/[1 + sTC]$ where r and s are constants. However, in practice the situation is somewhat more complex than allowed for by this assumption. Experimental studies record the phenomenon of high zone tolerance, the inhibition of T cell proliferation at very high antigen concentrations (Matis et al., 1982; Matis et al., 1983; Lamb et al., 1983; Suzuki et al., 1988). We therefore make the further assumption that a density-dependent inhibiting effect acts on the rate of proliferation in correspondence with the notion of a two signal model of T cell activation originally proposed by Bretscher and Cohen (1970) and, more recently, by Mueller, Jenkins and Schwartz (1989). To mirror the two effects of a density-dependent upper limit on the level of T cell proliferation concomitant with a reduction in the rate at high antigen concentrations, we assume f has the functional form:

$$f(C,P,T) = \frac{rC}{(1+sTC)(1+pTP)} \tag{5}$$

where p is an additional constant.

To simplify the analysis of the model we assume that both precursor production and cytokine secretion occur on much faster timescales than the parasite and T cell interactions. We therefore assume that M and C attain their equilibrium values $M^* = R/(u + \alpha P)$ and $C^* = aT/d$ given by setting $dM/dt = 0$ and $dC/dt = 0$ in equations (1) and (3) respectively. The remaining two equations now become:

$$\frac{dT}{dt} = \frac{\alpha RP}{u + \alpha P} + [f(aT/d, P, T) - \mu]T \tag{6}$$

$$\frac{dP}{dt} = \lambda P - \beta TP \tag{7}$$

Simple linear rescaling of T, P, and t reduces equations (6) and (7) to the form:

$$\frac{dT}{dt} = \frac{P}{1+\eta P} + \frac{\epsilon T^2}{(1+\gamma TP)(1+\delta T^2)} - T \tag{8}$$

$$\frac{dP}{dt} = \kappa P(1-T) \tag{9}$$

where η, ϵ, γ, δ and κ are rescaled parameters, specifically $\gamma = \lambda^2 \mu pru/R\alpha\beta^2$, $\kappa = \lambda ru/R\alpha$, $\delta = a\lambda^2 s/\beta^2 d$, $\epsilon = a\lambda r/\beta d\mu$, and $\eta = \lambda\mu/R\beta$.

When $\eta = \gamma = 0$, so that the level of parasite has no suppressive effect on the amount of T cell proliferation, this system is the same as that studied by McLean and Kirkwood (1990) as representing the interaction between the human immunodeficiency virus and T helper cell clones.

3 Results

Having formulated a mathematical model, we can use it to understand types of behaviour in response to immune challenge. For a biological interpretation of the patterns generated by the model, the key observation is that that the model has four broad classes of dynamical behaviour. This behaviour depends both on parameter values — the rates that determine cell, parasite and cytokine production and removal — and also on initial conditions — the values of the various cell, cytokine and parasite populations at the beginning of the response. We now describe these broad possibilities.

We have presented elsewhere (Schweitzer et al., 1991) a detailed bifurcation analysis of this system with η set equal to zero. This simpler system corresponds to one in which recruitment into the pool of active Th1 cells is not limited by the resources available: in this case the density of precursor cells is not affected by activation. This simpler system captures many, but not all, of the features of the full system and we recapitulate the key points here. The bifurcation analysis was used to show that the simple model exhibits four major classes of behaviour (Figure 2). For some parameter values, the only possible outcome, whatever the initial conditions, is for the parasite to be maintained at an equilibrium value while the Th1 cell response remains depressed but nonzero. This situation, displayed in Figure 2a, was identified with a state of tolerance to the parasite. A second possibility, displayed in Figure 2b, is one in which the immune system clears the parasite and Th1 cells remain at an elevated level. This was identified with a state of immunity to the parasite. A third possibility (Figure 2c) is one in which both outcomes, immunity and tolerance, are possible at the same parameter value. A final possibility, which we do not discuss here for reasons of space, is for the model to admit oscillations.

There are, then, two ways in which infection may generate different outcomes in different hosts. One way is for differences in the type of parasite and the genetic background of the host. to cause differing parameter values. In these cases the particular parameter set representing the biological details of the interaction between host and parasite will predispose the host to either chronic infection or protective immunity. For example, if the magnitude of the parameter ϵ, related to the peak proliferation rate of the T-cells, is large, or conversely if γ, which measures how proliferation is downregulated in the presence of parasite, is small, then the host is likely to eliminate the parasite and to become immune.

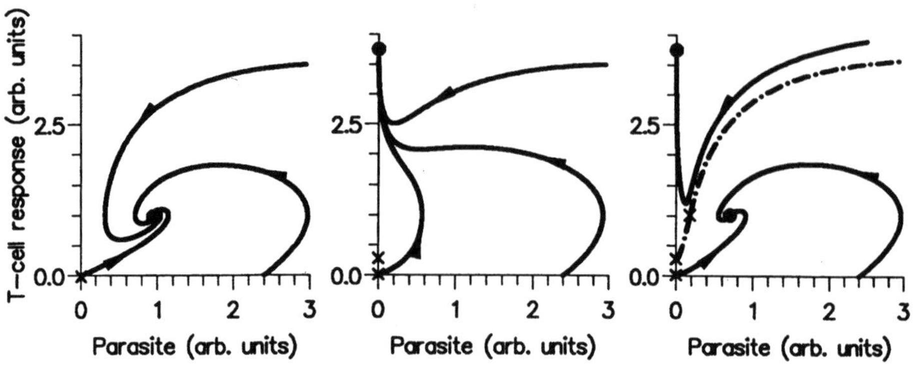

Figure 2: Classes of behaviour shown by the model without precursor limitation (equations (8) and (9) with $\eta = 0$). (a) Parasite grows to a stable, persistent level, with concomitant suppression of T cell response. $\gamma = 8$, $\delta = 1$, $\epsilon = 1$, $\eta = 0$, and $\kappa = 1$. (b) Parasite is eliminated and Th1 cell level is raised. As (a) but with $\gamma = 1$ and $\epsilon = 4$. (c) Both outcomes possible depending on the initial conditions of the system. As (a) but with $\epsilon = 4$. The closed circles indicate stable equilibria and the crosses indicate unstable equilibria. The arrowed lines indicate movement across the phase plane from the naive state. The dotted line in (c) separates those initial conditions leading to susceptibility from those leading to immunity.

Alternatively, if ϵ is small and γ is large, or if the rate of proliferation of the parasite κ is large then the system will tend to be attracted to the state where the parasite persists and the host is immunosuppressed.

A second way in which differing outcomes may be generated, of particular interest here, is the case where both immune and susceptible steady states can arise in a given host with a fixed set of parameter values for the model. In these circumstances it is possible for two different outcomes to occur for two identical hosts infected with a given parasite, but differing numbers of parasite in the initial inoculum. Then the important question is how these differing inocula affect the outcome. If we think of the outcome as determined by which of the the two stable points in the phase plane the system is attracted to, the mathematical problem is to find the line which separates those areas of the phase plane attracted to each point. This line, the *separatrix* which is mathematically the stable manifold of the unstable saddle point on $T = 1$, is shown in Figure 3 as a dash-dotted line.

3.1 Comparison of the model with a biological example: *Leishmania donovani* in mice

The model suggests that two outcomes can arise following infection with a replicating parasite, representing parasite elimination concomitant with protective immunity, and parasite persistence concomitant with continued host susceptibility. In a given host a

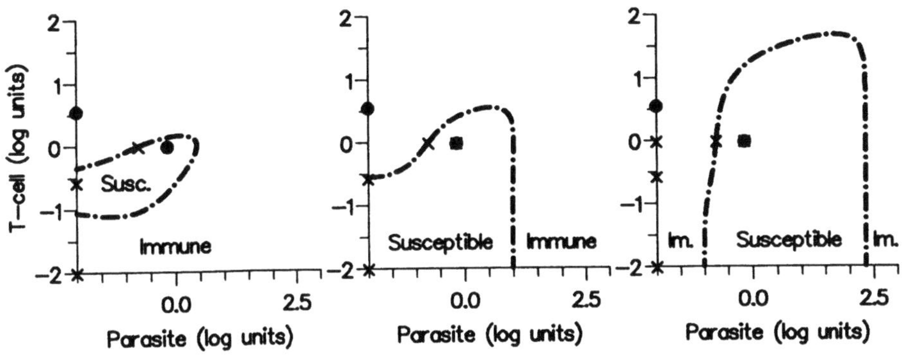

Figure 3: The relationship between initial inoculum size and final outcome when two stable points coexist. As in Figure 2, the dotted line separates those initial conditions leading to susceptibility from those leading to immunity. (a) All initial inocula lead to immunity: all solutions starting on the parasite axis are attracted to the the parasite-free fixed point. $\gamma = 8$, $\delta = 1$, $\epsilon = 4$, $\eta = 0$, $\kappa = 5$. (b) Small inocula lead to susceptibility but larger ones lead to immunity. As (a) but with $\kappa = 1$. (c) As (b), but now immunity is possible at very low inocula. As (a) but with $\kappa = 0.1$. Log scales; fixed points with with $P = 0$ represented as though on the axes.

particular outcome may be unique and independent of the size of the infecting inoculation of the parasite, or it may depend on the initial conditions of infection. The variety of outcomes generated by our simple model are a direct consequence of the very non-linear nature of the interaction between Th1-type T cells and the replicating parasite population. Infection of mice with *Leishmania donovani*, broadly speaking, generates either parasite control and immunological resistance, or persistent and chronic infection linked with immunosuppression (Bradley and Kirkley, 1977). The outcome depends on the genetic background of the mouse host, but with a fixed genetic strain different outcomes may occur depending on the size of the initial dose of the parasite (Ulczak and Blackwell, 1983). In a qualitative sense, therefore, the observed patterns are similar to those predicted by our simple model. To further explore the agreement between prediction and observation, we look in more detail at events determining the outcome of infection, in particular the effect of different initial inoculum size and the predicted temporal pattern of events following infection.

The simple model (Figure 3) predicts that large enough inocula will always generate resistance, while it is possible to produce persistence only at low or intermediate doses. In vivo observations on the *Leishmania donovani* mouse system (Ulczak and Blackwell, 1983) are in accord with the predictions of Figure 3c that small inocula induce resistance and larger inocula induce persistent infections and host susceptibility. However, there is no experimental evidence to suggest that very high inocula cause a reversion to the resistant state. There are two possible explanations for this inconsistency between observed and predicted patterns. First, published experiments may not have employed sufficiently high inocula to induce the reversion to resistance. The second, and more likely, explanation is that the model fails to take account of some important feature of the parasite-immune system interaction.

3.2 A refinement to the model: precursor limitation

We saw in the previous section that while the model is capable of predicting the co-existence of alternative equilibria, it does not realistically describe which of these two outcomes will be chosen. Since the number of precursor cells was unlimited, there was no resource constraint on the level the immune response could reach. In this section we investigate the effects of introducing such resource limitation by making η nonzero.

If η is small, which will be the case if the precursor pool, though now finite, is still large relative to depletion by activation, the qualitative picture of possible outcomes does not change significantly. Thus it is not difficult to show that the different outcomes in Figure 2 are still possible. The important difference for us, however, is the way in which the position of the separatrix between these two outcomes changes. As shown in Figure 4, it is no longer possible for very large inocula to cause clearance of the parasite. However all solutions are still attracted to one of the two stable fixed points.

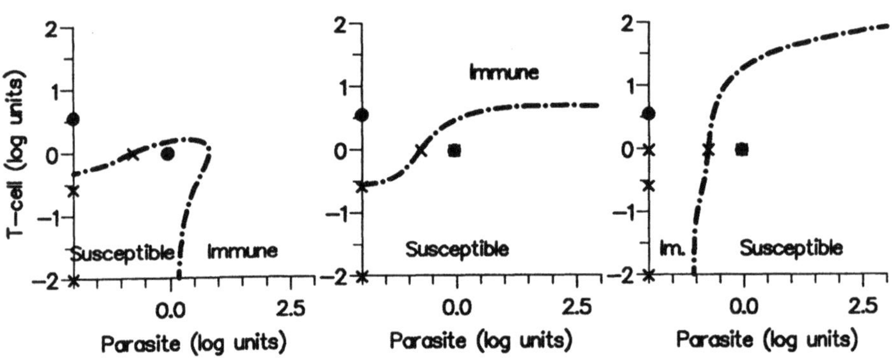

Figure 4: The effect of precursor limitation. When $\eta > 0$, it is no longer inevitable that large inocula will cause immunity. Parameter values as Figure 3 but with $\eta = 0.2$.

If η is rather larger, which would reflect a situation where depletion of the precursor pool was a significant factor, there is a further possibility. If $\eta > 1$, then it is possible for the parasite to overwhelm the host and for P to continue to grow without limit while T remains depressed.

This revised and more realistic model now generates predictions that are in accord with observed patterns both with respect to the time courses of infection and Th1-type responses, and dose-response relationships. The increased complexity of the model in allowing for the depletion of the precursor T cell pool by activation via contact with parasite antigens in fact results in a decrease in the complexity of predicted behaviour since very large inocula no longer induce resistance. Most importantly, however, the more complex model still captures observed behaviour at lower antigen doses (the induction of lowered resistance versus tolerance) and the fact that initial inoculum size may determine whether or not the host eliminates the parasite and develops protective immunity.

4 Discussion

The results of our analyses clearly demonstrate that non-linear dose-response relationships between parasite abundance and T cell proliferation rates can account for the manifestation of host resistance and susceptibility to the same parasite in a genetically heterogeneous host population or in hosts exposed to varying levels of parasite infection.

The early events observed in resistant and susceptible mice during in vivo infection with *Leishmania* spp are similar to each other, and are characterized by an initial rise in parasitaemia and the initiation of a Th1-type T cell response (Heinzel et al., 1989; Kaye et al., 1991). Some time later, however, the divergence between phenotypes becomes apparent, with the parasitaemia in resistant mice and the Th1-type response in susceptible mice, falling. The initial similarity could be a consequence of late expression of the phenotype difference during the course of infection. However, current opinion is beginning to favour the alternative that very early events determine whether resistance or susceptibility result (Locksley and Scott, 1991). The mathematical model described in this paper clearly supports the latter view.

One refinement was made to the structure of the model which enabled it to capture more realistically both the course of infection over time and the influence of the initial infecting dose of parasites. The agreement between prediction and observation created by this refinement suggests that the antigenic load confronted by the precursor T cell population during the course of infection may cause precursor cells to become limiting as a consequence of, for example, the high rate of turnover of antigen, in addition to absolute parasite abundance.

The influence of the size of the parasite inoculating dose on whether the outcome of infection is parasite elimination or the chronic persistence of infection is determined by the intersect of the separatrix on the horizontal parasite axis in the phase plane (Figure 4). If the intersect is at a high value, all but very large inoculating doses will result in parasite elimination concomitant with host immunity. Alternatively if the intersect is at a low level, or on the T cell axis, then most doses will result in chronic infection concomitant with host susceptibility. With respect to the control of infection (or the induction of immunity in susceptible individuals) it is clearly desirable to fully understand the biological process that affect the position of the intersect. The model only provides a very crude caricature of these processes and further experimental and theoretical work is required to delineate these more precisely. The influence of host genetic background in mouse–*Leishmania* systems on the rate parameters that determine Th1-type responses is an obvious area for further investigation.

Alternatively, predisposition to either susceptibility or resistance in a defined area of endemic infection in, say human hosts, may result from heterogeneity in the factors that determine the degree of initial exposure to infection. In areas of high transmission intensity, with a high value of the basic reproductive rate, R_0, there will be a correspondingly high prevalence and average intensity of infections (Anderson and May, 1991). If susceptibility is related to the level of exposure this will act as a positive-feedback mechanism which further sustains high transmission rates with the concomitant high degree of tolerance or susceptibility in the host community. An obvious area for future research is to embed models of the population dynamics of infection within individual hosts into a framework that mirrors the transmission dynamics of the parasite within a population of hosts.

Acknowledgements

We have much benefited from conversations with Paul Kaye, Jenny Blackwell, and Richard Locksley. We thank Charlotte Hetzel for a careful reading of the text. This work was supported by grants from the Medical Research Council of Great Britain and the Wellcome Trust. ANS is a Medical Research Council Training Fellow.

References

Anderson RM, May RM (1991) Infectious Diseases of Humans: Dynamics and Control. Oxford University Press.

Bradley DJ, Kirkley J (1977) Regulation of *Leishmania* populations within the host. I The variable course of *Leishmania donovani* infections in mice. Clin Exp Immunol 30:119–129.

Bretscher P, Cohn M (1970) A theory of self-non-self discrimination. Science 169:1042–1049.

Castes M, Moros Z, Martinez A, Trujillo D, Castellanos P, Rondon A, Convit I (1989) Cell-mediated immunity in localised cutaneous leishmaniasis patients before and after treatment with immunotherapy or chemotherapy. Parasit Immunol 11:211–222.

Cherwinski H, Schumacher J, Brown K, Mosmann T (1987) Two types of mouse helper T cell clone. Further differences in lymphokine synthesis between Thl and Th2 clones revealed by RNA hybridisation functionally monospecific bioassays, and monoclonal antibodies. J Exp Med 166:1229–1244.

Feldmeier H, Gasti G, Poggenese U, Daffalla A, Noguerra-Queiroz J, Capron A, Peter H (1988) Immune response in chronic schistosomiasis *haemotobium* and *mansoni*. Reversibility of alteractions after anti-parasitic treatment with praziquantel. Scand J Immunol 28:147–155.

Fernandez-Botran R, Sanders V, Mosmann T, Vitetta E (1988) Lymphokine-mediated regulation of the proliferative response of clones of T helper 1 and T helper 2 cells. J Exp Med 168:543–558.

Fiorentino D, Bond M, Mosmann T (1989) Two types of mouse T helper cell. IV. Th2 clones secrete a factor that inhibits cytokine production by Thl clones. J Exp Med 170:2081–2095.

Gajewski T, Fitch F (1988) Anti-proliferative effect of IFN-γ in immune regulation. I. IFN-γ inhibits the proliferation of Th2 but not Th1 murine helper T lymphcyte clones. J Immunol 140:4245–4252.

Heinzel FP, Sadick MD, Holaday BJ, Coffman RL, Locksley R M (1989) Reciprocal expression of interferon γ or interleukin 4 during the resolution or progression of murine leishmaniasis. J Exp Med 169:59–72.

Heinzel FP, Sadick MD, Mutha SS, Locksley RM (1991) Production of interferon γ, interleukin-2, interleukin-4, and interleukin-10 by CD4+ lymphocytes *in vivo* during healing and progressive murine leishmaniasis. Proc Natn Acad Sci USA 88(16):7011–7015.

Ho M, Koech D, Iha D, Bryceson D (1983) Immunosuppression in Kenyan visceral leishmaniasis. Clin Exp Immunol 51:107–214.

Kaye PM, Curry AJ, Blackwell JM (1991) Differential production of Th1- and Th2-derived cytokines does not determine the genetically controlled or vaccine-induced rate of cure in murine visceral leishmaniasis. J Immunol 146:2763–2770.

Lamb J, Skidmore B, Green N, Chiller J, Feldmann M (1983) Induction of tolerance in influenza virus-immune T lymphocyte clones with synthetic peptides of influenza hemagglutinin. J Exp Med 157:1434–1447.

Locksley RM, Scott P (1991) Helper T-cell subsets in mouse leishmaniasis: induction, expansion and effector function. In: Ash C, Gallagher RB (eds) Immunoparasitology Today, Elsevier Trends Journals, Cambridge, pages A58–A61.

Matis LA, Glimcher LH, Paul WE, Schwarz RH (1983) Magnitude of response of histocompatibility-restricted T-cell clones is a function of the product of the concentrations of antigen and Ia molecules. Proc Natn Acad Sci USA 80:6019–6023.

Matis LA, Jones PP, Murphy DB, Hedrick SM, Lerner EA, Janeway, Jr CA, McNicholas JM, Schwartz RH (1982) Immune response gene function correlates with the expression of an Ia antigen. II A quantitative deficiency in A_e:E_a complex expression causes a corresponding defect in antigen-presenting cell function. J Exp Med 155:508–523.

McLean AR, Kirkwood TBL (1990) A model of human immunodeficiency virus infection in T helper cell clones. J Theor Biol 147:177–203.

Moore K, Vieira P, Fiorentino D, Trounstine M, Khan T, Mosmann T (1990) Homology of cytokine synthesis inhibitory factor (IL-10) to the Epstein-Barr virus gene BCRFI. Science 248:1230–1234.

Mosmann T, Cherwinski H, Bond M, Gieldin M, Coffman R (1986) Two types of murine helper T cell clone. I. Definition according to profiles of lymphokine activities and secreted proteins. J Immunol 136:2348–2357.

Mueller DL, Jenkins MK, Schwartz RH (1989) Clonal expansion versus functional clonal inactivation: a costimulatory signalling pathway determines the outcome of T cell antigen receptor occupancy. Ann Rev Immunol 7:445–480.

Nutman T, Kumaraswami V, Ottesen E (1987) Parasite-specific anergy in human filariasis. Insights after analysis of parasite antigen-driven lymphokine production. J Clin Invest 79:1516–1523.

Peleman R, Wu J, Fargeas C, Delespesse G (1989) Recombinant interleukin 4 suppresses the production of interferon γ by human mononuclear cells. J Exp Med 170:1751–1756.

Pond L, Wassom D, Hayes C (1989) Evidence for differential induction of helper T cell subsets during *Trichinella spiralis* infection. J Immunol 143:4232–4237.

Schweitzer AN, Swinton J, Anderson RM (1991) Dynamic interaction between leishmania infection in mice and Thl-type CD4+ T-cells: complexity in outcome without a requirement for Th2-type responses. Preprint, Imperial College.

Scott P, Pearce E, Cheever A, Coffman R, Sher A (1989) Role of cytokines and CD4+ cell subsets in the regulation of parasite immunity and disease. Immunol Rev 112:161–182.

Sher A, Colley DG (1989) Immunoparasitology. In: Paul WE (ed) Fundamental Immunology, second edition, Raven Press, New York, pages 957–983.

Smith KA (1989) The interleukin 2 receptor. Ann Rev Cell Biol 5:396–425.

Street N, Schumacher J, Fong A, Bass H, Florentino D, Leverah J, Mosmann T (1990) Heterogeneity of mouse helper T cells. Evidence from bulk cultures and limiting dilution cloning for precursors of Thl and Th2 cells. J Immunol 144:1629–1639.

Suzuki G, Kawase Y, Yahara I, Kobayashi Y, Schwartz RH (1988) Antigen-induced suppression of the proliferative response of T cell clones. J Immunol 140:1359–1365.

Toribio ML, Gutiérrez-Ramos JC, Pezzi L, Marcos MAR, Martínez-A C (1989) Interleukin-2-dependent autocrine proliferation in T-cell development. Nature 342:82–85.

Ulczak OM, Blackwell JM (1983) Immunoregulation of genetically controlled acquired responses to *leishmania donovani* infection in mice: the effects of parasite dose, cyclophosphamide and sublethal irradiation. Parasit Immunol 5:449–463.

Ward D, Nutman T, Zea-Flores G, Portecarrero C, Lujan A, Ottesen E (1988) Onchocerciasis and immunity in humans: enhanced T cell responsiveness to parasite antigen in putatively immune individuals. J Infect Dis 157:536–543.

III. Networks and Attractors

GENERIC MODELLING OF THE IMMUNE NETWORK

Gérard Weisbuch and Avidan U. Neumann
Laboratoire de Physique Statistique de l'Ecole Normale Supérieure,
24 rue Lhomond, F 75231 Paris Cedex 5, France.

1. INTRODUCTION

Our experimental knowledge of the immune system allow us to describe it as:

- a cooperative complex system: the reaction of the system to any challenge involves the cooperation of a large number of different molecular and cellular species, lymphokines, immunoglobulins, macrophages, and B and T cells of different specificities.

- a dynamical system: the set of events following recognition of the antigen, aggregation and/or internalization of cell receptors, secretion of soluble molecules, cell differentiation and division, build-up of populations of committed cells, destruction of the antigen, build-up of memories, etc. that require finite amounts of time.

It is then not surprising than theorists have started to use some of the methods and concepts that have been developed for the study of complex systems.

The purpose of modeling the immune response is largely equivalent to the purposes of statistical mechanics: to deduce macroscopic properties from interactions among elementary components. The interactions among the components are extremely intricate and they are not fully understood. Yet the "macroscopic behavior" of the immune system is well characterized. The question then arises to select a simple representation of the elementary interactions that would give rise to the organized behavior observed in the immune response in terms of memories for instance. The adventure of statistical physics is full of equivalent endeavors, starting from the description of thermal properties of gases and solids based respectively on the assumption of independent particles composing a perfect gas and the coupling of harmonic oscillators, to the more recent description of neural nets. This type of approach is especially suited to theoretical immunology because of our ignorance about the detailed mechanisms responsible for the immune response. To be more specific, we are looking for generic properties among models of the immune system. We are interested in qualitative or semi-quantitative laws, such as scaling laws, which only depend on the general features of the model, and not on its details. Basically, the simple system of differential equations described in the next section plays the role of the logistic equation for chaos and turbulence: although it represent a strong simplification of the interactions present in the system it is expected to belong to the same class of universality as a "true" model of the immune system, and to exhibit the same generic properties. In the case of a dynamical system, the generic properties concern the attractors of the dynamics. Are they limits points, limit cycles or chaotic? what is their number? what are the basins of attractions? how do these properties relate to the parameters of the differential system? What about transitions among the different dynamical regimes? If our hypotheses about the universality of the model is true, the generic properties, qualitative classification of the attractors or scaling laws, should compare with their interpretation for the mammalian immune system, although the "microscopic" assumptions might look oversimplified to a molecular immunologist.

2. A MATHEMATICAL MODEL[1]

Most recent models of the Jerne network derive from the following one[1]. It is one of the most simple models that can be conceived, while still exhibiting interesting immune properties. The dynamical behavior of each element i obeys a differential equation that describes the time evolution of the population x_i of clone i

$$\frac{dx_i}{dt} = s + x_i(f(h_i) - d) \ , \tag{1}$$

where s is the source term corresponding to cells coming from the bone marrow, the function $f(h_i)$ defines the rate of proliferation, and d specifies the rate of cell population decay. For each clone 'i' we consider the total amount of anti-idiotypic stimulation as a linear combination of the populations of other interacting clones 'j'. We call h_i the field acting on clone x_i:

$$h_i = \sum_j J_{ij} x_j \ , \tag{2}$$

where J_{ij} specifies the affinity between clones x_i and x_j. The choice of a J matrix defines the topology of the network. The proliferation function $f(h_i)$ has a stimulatory and a suppressive part with a threshold for activation (i.e. θ_1) and one for suppression (i.e. θ_2). Within the stimulatory part, increasing the idiotypic field increases proliferation; within the suppressive part increasing the field decreases proliferation. Similar dose response curves are found in receptor crosslinking [2]. Fig. 1 displays the bell- shaped f(h) based [3] upon two Michaelean saturation functions with the two thresholds θ_1 and θ_2 ($\theta_2 >> \theta_1$), and a maximum rate of proliferation, p:

$$f(h) = \left(\frac{h}{\theta_1 + h}\right)\left(1 - \frac{h}{\theta_2 + h}\right) = \frac{h\theta_2}{(\theta_1 + h)(\theta_2 + h)} \ . \tag{3}$$

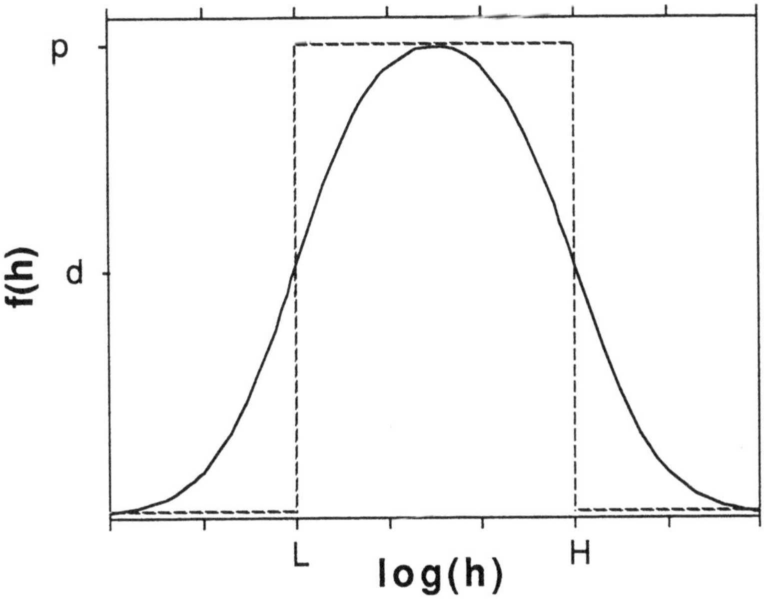

Fig. 1. f(h), the proliferation function of the field h as a solid line. The dashed line is the window automaton approximation. L and H are the field values defined in the text.

2.1. *The two populations problem*

Before handling the case of a network of 10^8 clones, let us discuss the simple case of two interacting populations, one specific for the antigen that we call idiotype x, and the other specific for the idiotype, the anti-idiotype, y. The interaction constant J_{ij} is set to 1. The system of equations (1) now becomes:

$$\frac{dx}{dt} = s + x(f(y) - d) \ , \qquad (4a)$$

$$\frac{dy}{dt} = s + y(f(x) - d) \ , \qquad (4b)$$

After presentation of the antigen, first x and then y increase. By the time the antigen is eliminated they both reach some value that we take as the initial conditions for Eq. (4). We are then interested in the attractors of system (4). The two isoclines, $\dot{x} = 0$ and $\dot{y} = 0$, intersect in five points of which only three are stable.

1. The virgin configuration is obtained when the proliferation function is very small. It has coordinates:

$$x_0 = y_0 = V = s/d \ . \qquad (5)$$

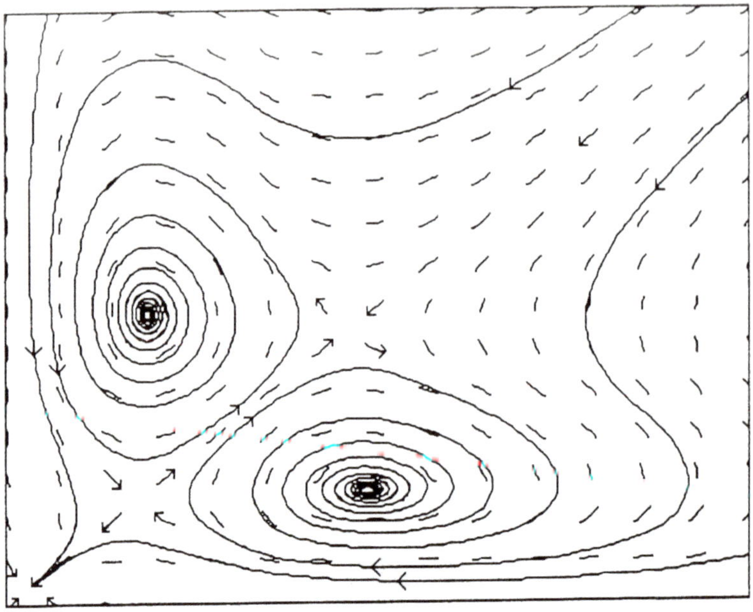

Fig. 2. Trajectories of the two-population system in the x-y plane.

In the limit of small s, the other solutions correspond to intersections of the proliferation function with the decay term (see Fig. 1).

2. In the immune configuration, the idiotype has a large population and exercises a large suppressive field on the anti-idiotype which has a small population and exercises a small excitatory field on the idiotype. Small or large field solutions are computed by approximating $f(h)$ by either $\frac{ph}{\theta_1 + h}$ or $\frac{p\theta_2}{\theta_2 + h}$ and by considering s/d as small with respect to θ_1. One obtains:

$$x_1 = H = \frac{p'\theta_2}{d} \quad , \quad y_1 = L = \frac{d\theta_1}{p'} \ , \tag{6}$$

where p' = p - d .

Attractivity towards the equilibrium is obtained by solving system (4) linearized in the vicinity of x_1, y_1. From the solution of the characteristic equation attenuated oscillations towards the attractor are then expected and indeed observed (Fig. 4). This attractor is interpreted as vaccination: when antigen is later presented, its elimination is much faster than in the virgin attractor since the relevant clone x is present at a much higher population. An interesting feature is that x population remains large even in the absence of the antigen; the memory of the antigen presentation is kept by the x-y couple which sustain each other.

3. The tolerant attractor is obtained by exchanging x_1 and y_1 (Fig. 2). It is interpreted as tolerance since when antigen is later presented, clone x is not able to react with a strong intensity because it is experiencing a suppressive field due to clone y.

4. and 5. the two symmetrical solutions with either suppressive or excitatory fields are unstable (Fig. 2).

Computer simulations indeed verify these predictions (Figs. 2 and 3). According to the conditions of antigen presentation and elimination due to the action of clone 1, different trajectories are followed by system (4); one of the three attractors of the dynamics is then reached corresponding to either return to virginity, vaccination or tolerance. The choice of the attractor as a function of the antigen presentation parameters will be further discussed within the window automaton approximation.

2.2. *The attractors of the large network*

The immune network is defined by the J_{ij}'s. Unfortunately our knowledge of the actual J_{ij}'s is very restricted. Some educated guesses can be made about the structure of the network from experiments, or simplified considerations about the basis of immune recognition. One might also be tempted to use one's favorite network structure: cellular, complete connectivity with random connection intensities, Cayley tree, etc. The Cayley tree (see Fig. 3) is in fact an intermediate approximation between the two clone model and more realistic networks. Its relevance comes from one of its features, localized attractors, which is nicely interpreted in clinical immunology. We will discuss Cayley tree nets in this section, and then discuss later the changes in dynamical behavior that are to be expected when more "realistic" architectures are used.

In order to exhibit memory an immune system model must obey certain constraints: it should be sensitive enough to change attractor under the influence of antigen presentation. It should not be too sensitive and over-react when the antigen is present at very low doses. The immune system should also discriminate among self-antigens, the molecular markers of our own cells for instance. Finally, it should be robust: memorization of previously presented antigens should not be lost when a new antigen is presented. In some sense, we are looking for independent responses to a lot of different antigens. This independence property is achieved when attractors are localized, *i.e* when the perturbation induced by antigen presentation remains localized among the clones that are close to those which actually recognize the antigen. To show that this is possible, we will then systematically search for localized attractors in a Cayley tree model.

Let us consider a Cayley tree architecture [1] with m connections per site (see Fig. 3), and J_{ij}'s which can only be 0 (no interaction) or 1 (maximum interaction). The root of the tree is selected by antigen presentation. According to their distance to the root, the clones are numbered 1 (and called Ab1 for the antigen specific clones), 2 (for the Ab2 anti-idiotypic clones, specific of some Ab1), 3 ,...., i. The different fields for clones 1, 2, 3, ..., i are given by:

$$h_1 = mx_2 \tag{7a}$$

$$h_2 = x_1 + (m-1)x_3 \tag{7b}$$

$$h_3 = x_2 + (m-1)x_4 \tag{7c}$$

....

$$h_i = x_{i-1} + (m-1)x_{i+1} \tag{7d}$$

Our approach is to classify the different attractors of the network and to interpret the transitions from one attractor to the other under the influence of antigen presentation. Let us start with the simplest configuration, corresponding to the hypothetical case where no antigen has yet been presented and all populations are s/d (all proliferation functions are 0). After presentation of the first antigen, memorization is obtained if some populations of the network reach stable populations different from s/d. In the case of a localized response there exist a patch of clones that are modified by the presentation. As long as patches corresponding to different antigens do not overlap, all these antigens can be memorized. Once the idea of localized non-interacting attractors is accepted, everything is simplified: instead of solving 10^8 equations, we only have to solve a small set of equations for those neighboring clones with large populations, supposing that those further clones that do not belong to the patch have populations s/d. The usual approach is to combine computer simulations and analytic check of the attractors by solving the field equations (see below).

2.2.1. *Vaccination.*

Let us consider the case of antigen presentation on clone Ab_1 which results in excitation of clones Ab_2, clones Ab_3 remaining close to their virgin level (see Fig. 3, left). In analogy with the two clones problem, we expect a low field solution L for Ab_1, and a large suppressor field solution H for Ab_2. The field equations allow to compute the populations:

$$h_1 = mx_2 = L = \frac{d\theta_1}{p'} \tag{8}$$

$$h_2 = x_1 + (m-1)\frac{s}{d} = H = \frac{p'\theta_2}{d}. \tag{9}$$

Of course, the solution remains localized only if the field h_3 on x_3 is much less than L

$$h_3 = \frac{L}{m} + \frac{(m-1)s}{d} < L\ , \tag{10}$$

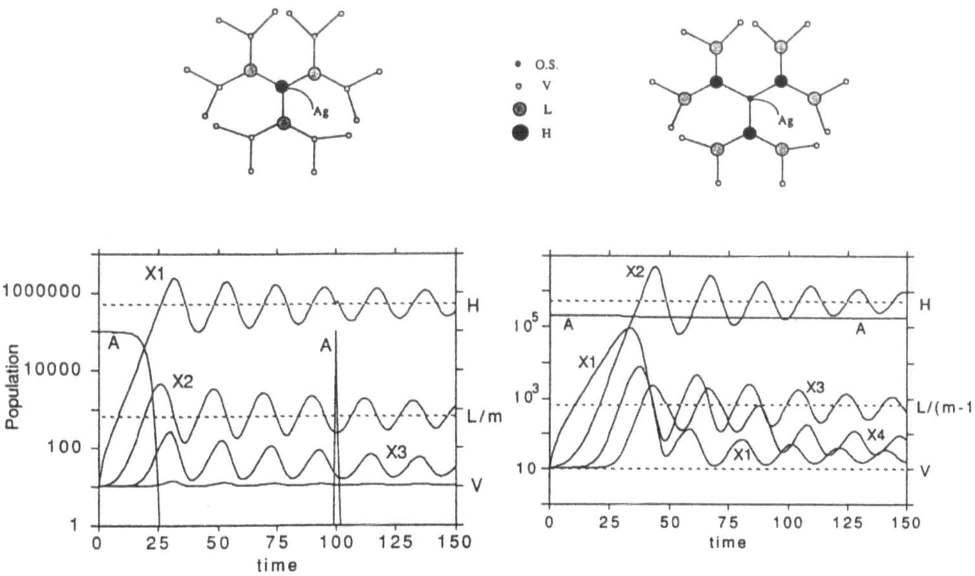

Fig. 3. Two localized attractors of the immune network corresponding to vaccination (on the left) and tolerance (on the right). On the above Cayley trees, black circles correspond to large suppressive populations (of the order of θ_2), gray circles to medium excitatory populations (of the order of θ_1) and small white circles to small virgin or oversuppressed populations (of the order of s/d). The two lower graphs are time plots of the logarithm of the populations (numerical simulation results). A is antigen concentration. Vaccination as a memory effect is illustrated on the left plot: a second antigen presentation with the same dose at time t=100 results in a faster elimination than during the first presentation. Slow decay of large antigen concentration results in tolerance.

otherwise x_3 would also proliferate. This is only possible if m is larger than 1 (a one dimensional network would then give rise to percolation; multiple connectivity is essential to localization) and if the following inequality is fulfilled:

$$\frac{p'ms}{d^2} < \theta_1 \ . \tag{11}$$

It is in fact equivalent to saying that the field due to m virgin clones should be less than the proliferation threshold.

2.2.2. Tolerance.

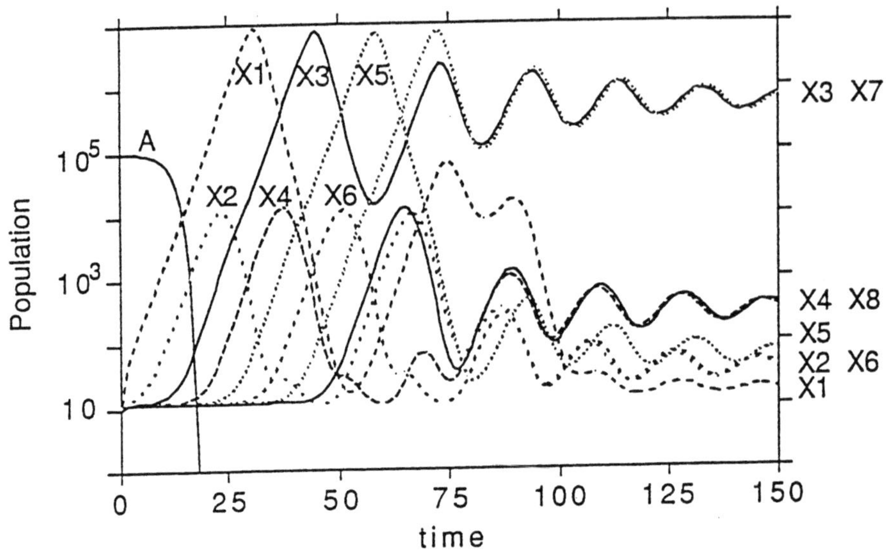

Fig. 4. Time diagram for percolation. After antigen presentation, clones 1 and 2 return to virgin state, but clone 3 and all clones with number multiple of four apart such as 7 are excited at the suppressive level. Clone 4,8,12... are sustaining the excitation of 3,7,13...

Another localized attractor corresponds to tolerance (see Fig. 3, right).

A strong suppressive field acts on Ab_1 due to Ab_2's, Ab_2's proliferate due to a low field provided by Ab_3's, but Ab_4's remain nearly virgin. The field equations once more allow us to compute the populations:

$$h_2 = x_1 + (m-1)x_3 = L = \frac{d\theta_1}{p'} \tag{12}$$

which gives x_3 by neglecting x_1 which is small.

$$h_3 = x_2 + \frac{(m-1)s}{d} = H = \frac{p'\theta_2}{d} \tag{13}$$

and then:

$$h_1 = mH \tag{14}$$

which, substituted in equation (1), shows that x_1 is of the order of s/d. h_4 is small and x_4 nearly virgin.

2.2.3. Percolation.

Some parameters allow excitations to propagate from the initial nodes across all the lattice. In such a case percolation is observed, which corresponds to a non-localized dynamics (Fig. 4). We of course do not expect this to be a healthy condition, but in some auto-immune diseases such as lupus erythematosis a large number of different auto-antibodies is detected as if there were percolation of the immune response.

2.3. *Scaling laws*

The Cayley tree model predicts localized attractors that can be interpreted in terms of vaccination or of tolerance. Because these attractors are localized they are somehow independent: starting from a fully virgin configuration, one can imagine successive antigen presentations which leave their footprints on the network by creating patches, each of these involving a set of m perturbed neighboring clones. A vaccination patch contains $1 + m$ clones, a tolerant patch $1 + m^2$ (see Fig. 3). Independence of localized attractors implies a maximum number of attractors that scales exponentially with N, the total number of clones. The following argument gives a lower bound: divide the network into $N/(1+m^2)$ spots. Each spot can be in 3 possible configurations virgin, vaccination or tolerance. This gives a number of attractors that scales as $3^{N/(1+m^2)}$. In fact, few of these attractors make sense. The relevant question is rather:

A living system must face many antigen presentations during its life. Self antigen should elicit a tolerant response; dangerous external antigens should elicit vaccination. The nature of the localized response on each individual site of the network is then determined by the fact that the presented antigen should be tolerated or fought against. In this context, how many different antigens can be randomly presented so that no overlap among different patches occurs?

A simple reasoning[4] is sufficient to get the scaling law relating n, the maximum number of antigens to N, the total number of clones. Let p be the number of clones involved in a patch, for instance $1 + m^2$. The number of configurations with n patches of p clones that do not overlap is simply:

$$\binom{N - np}{n} \tag{15}$$

(basically, each antigen blocks p clones for further positioning of next antigens). The probability P of non-overlapping distributions among random distributions is then:

$$P = \frac{\binom{N-np}{n}}{\binom{N}{n}} \tag{16}$$

which is approximated for large N by:

$$P = exp - (\frac{n^2 p}{N}) \tag{17}$$

This gaussian probability changes from one to zero when the argument of the exponential is order 1. The system can then memorize a maximum number of antigens n such that:

$$n < \sqrt{\frac{N}{p}}. \tag{18}$$

This scaling law applies to any network topology. The only assumption is the random character of the network with respect to antigens. On the other hand it can be argued that the clones expressed by the mammals have been optimized to the environment of the immune system, self molecules and most frequent parasites. This optimization during phylogenesis (*e.g* the evolution of species) can also have been carried during the epigenesis of the immune system (the build-up of the immune system during the animal lifetime) because of somatic mutations. Such an optimized immune system would have a linear scaling law:

$$n < \frac{N}{p}. \tag{19}$$

Anyhow, we expect the number of attractors to scale as a power law with a positive exponent equal to or smaller than 1, rather than a exponential scaling law.

3. ATTAINING THE ATTRACTOR: DYNAMICS WITH ANTIGEN

Now that we have characterized the attractors of the dynamics, we would like to be able to predict which attractor is obtained according to the conditions of antigen presentation. We know from immunology that, according to the age of the individual (newborn or adult), and the conditions of antigen presentation (high or low dose, presence of adjuvant, intravenous or subcutaneous injections), vaccination, tolerance or no response is obtained. In the model the influence of antigen A on clone Ab_1 is taken into account by the addition of a new term in h_1:

$$h_1 = mx_2 + JA \tag{20}$$

The antigen concentration A decreases because of interaction with Ab_1:

$$\frac{dA}{dt} = -kAx_1 \tag{21}$$

The nature of the attractors reached according to the parameters of the computer simulations are summarized in Fig. 5. The kinetic constant k and the initial antigen concentration are the parameters that define antigen presentation: theoretical analysis

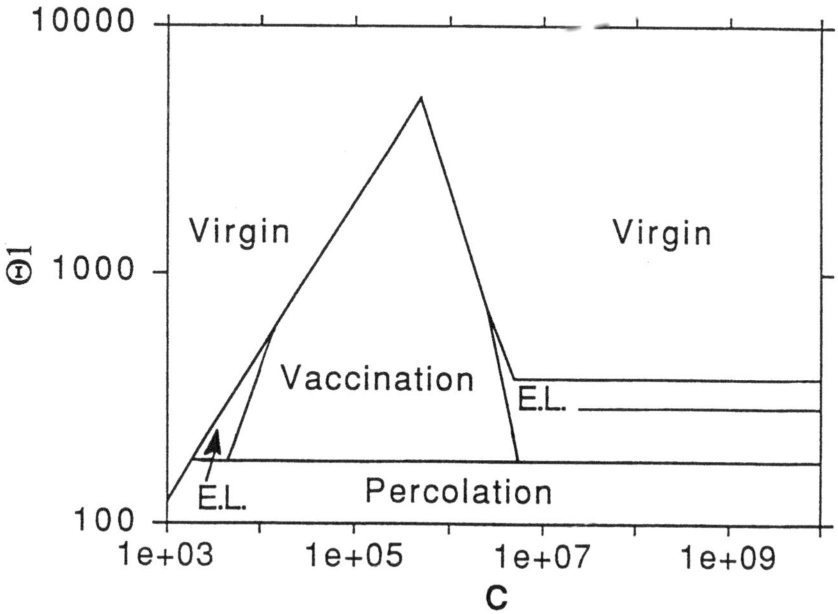

Fig. 5. Diagram of the attractor attained by the dynamics as a function of the first proliferation threshold θ_1 and C, the parameter of antigenic stimulation.

shows that they can be combined into a unique relevant parameter C defined as the Ab_1 concentration at the time of antigen elimination.

3.1. *The window automata*

By further simplifying the present model one is able to derive the scaling laws that separate the different regimes in the parameter space. The idea is to replace the proliferation function $f(h)$ by a window automaton[5] which is 0 when the field is below L and above H, and p in between. (L and H are defined by Eq. 6). Thus,

$$f(h) = \begin{cases} p, & \text{if } L < h < H; \\ 0, & \text{otherwise.} \end{cases}$$

Populations excited from the virgin level have then either exponential decay or increase (with respective exponents d and $p - d$) according to the values of the field they experience. The times between successive excitations can be estimated and from them the transitions between regimes computed. Another favorable feature of the window-automata approach is that the antigen dynamics is also simple. Starting from an initial value A_0, $L < A_0 < H$, the antigen concentration first decays very slowly until the idiotypic clone population is large enough to substantially decrease it. Since the clone population increases exponentially, further antigen decrease is then very fast. We approximate $A(t)$ by a step function, equal to A_0 up to time t_A defined by $A(t_A)=1$, and then vanishingly small,

$$\begin{aligned} t < t_A : \quad & A(t) = A_0 \\ t > t_A : \quad & A(t) = 0 \end{aligned} \tag{22}$$

The time t_A computed from Eqs. (1) and (21) by two integrations is:

$$t_A \simeq \frac{1}{p-d} \log \left[\frac{(p-d)}{k} \frac{d}{s} \log(A_0) \right] . \tag{23}$$

It can be seen that t_A is approximately the time that x_1 takes to proliferate from s/d to C cells , where

$$C = k^{-1}(p-d) \log A_0 \quad . \tag{24}$$

In most cases C is the single parameter that suffices to characterize antigenic stimulation.

3.2. *From Virginity to Excitation*

The "phase diagram" in Fig. 5 is obtained by analytical methods that we exemplify with the simplest case of the transition from a virgin attractor to an excited attractor, which could be either vaccination or some delocalized attractor. The initial configuration of the immune system is assumed to be the virgin attractor. A necessary condition for the excitation of idiotypic and anti-idiotypic clones is that the initial antigen concentration and the time to eliminate it are large enough

$$A_0 > L \quad ; \quad C > L \tag{25}$$

so that X_1 is excited by the antigen from V to a concentration C high enough to excite X_2. But excitation of X_1 above L needs to be maintained by X_2 after antigen elimination. The time for X_2 to reach L/m and sustain X_1 should then be smaller than the time for X_1 to decay back to L, thus preventing X_2 from further proliferation. The condition for escape from virginity is therefore written as (see Fig. 6):

$$t_L + t_{Lm} < t_A + t_{1d} . \tag{26}$$

On the left hand side, t_L is the time for X_1 to grow from V to L and to excite X_2, and t_{Lm} is the time for X_2 to grow from V to L/m to sustain X_1. The quantity $t_L + t_{Lm}$ is called the two-level time delay because it involves the time necessary to excite two levels (here levels 1 and 2) to a mutually sustaining fields. On the right hand side of Eq. (26), t_A is the antigen elimination time (Eq. 23), and t_{1d} is the decay time of X_1 from its maximum to L. The evaluation of t_{Lm} and t_{1d} from the parameters θ_1, θ_2, A_0

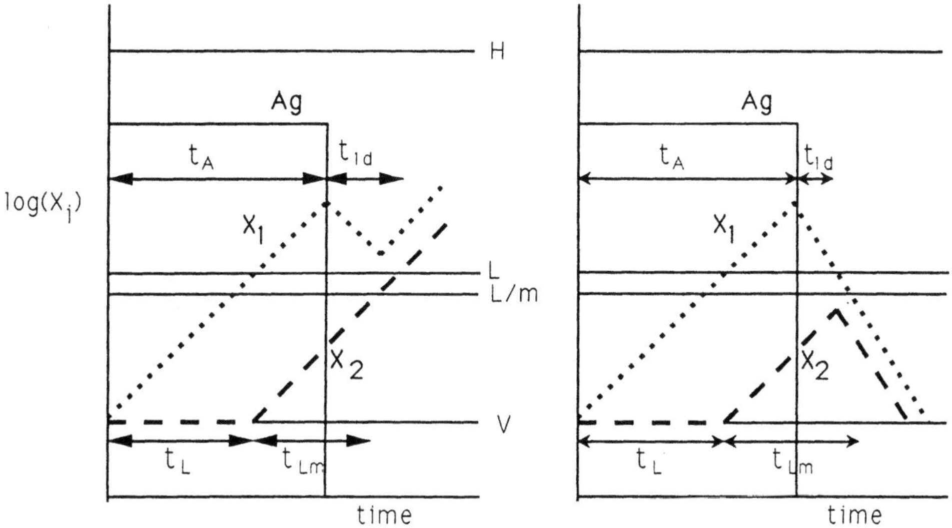

Fig. 6. Time diagrams at the transition from virginity to excitation. X_2 starts to proliferate when X_1 passes L. X_1 starts to decay when antigen is eliminated. (a) escape from virginity: excitation of X_2 to L/m (the dashed line: $t_L + t_{Lm}$) is faster than the decay of X_1 back to L (the dotted line: $t_A + t_{1d}$). (b) failure of vaccination: X_2 excitation is too slow to sustain X_1.

etc. depends upon the conditions of antigenic stimulation, but condition (26) is always valid.

In the simpler case when C (defined by Eq. 24) is smaller than H, the growth of X_2 up to L/m is monotonic. The following expressions for the time intervals are obtained:

$$t_L = \frac{1}{p-d} \log\left[\frac{L}{V}\right] \quad , \tag{27}$$

and

$$t_{Lm} = \frac{1}{p-d} \log\left[\frac{L}{mV}\right] \quad . \tag{28}$$

Proliferation and decay times (t_p and t_d respectively) between the same two concentrations are related by

$$(p-d)t_p = dt_d \ . \tag{29}$$

Then t_{1d} is computed from the time it takes for x_1 to proliferate from L to its maximum concentration reached when antigen is eliminated,

$$t_{1d} = \frac{p-d}{d} t_A - \frac{1}{d} \log\left[\frac{L}{V}\right] \quad . \tag{30}$$

One first replaces the times in inequality (14) by their expression in terms of populations A_0, V and L. These populations are then replaced by their expression (Eqs. 5 and 6) in terms of the parameters: θ_1, s, m, d and p. After reordering and exponentiation, the condition for escape from virginity appears as the following scaling law:

$$\theta_1 < \theta_1^V \equiv \frac{p-d}{d} \cdot \left(\frac{s}{d}\right)^{\alpha} \cdot (\theta_2)^{\beta} \cdot (m-1)^{\gamma} \cdot \left(\frac{(p-d)\log A_0}{k}\right)^{\delta} , \tag{31}$$

where

$$\alpha = \gamma = \frac{d}{p+d} \ ; \ \beta = 0 \ ; \ \delta = \frac{p}{p+d} \ .$$

If the threshold θ_1 is too high the idiotypic clone x_1 does not remain long enough above L to allow the relay of antigen by anti-idiotypic stimulation, and the network falls back to virginity.

The steps taken in this method are:

1. dividing the whole dynamics, into time intervals during which the logarithms of the populations vary linearly with time;

2. expressing the conditions for the different dynamical regimes as inequalities among these time sequences;

3. transposing the inequalities in terms of populations (e.g.: V,L,H,C) and then in terms of the parameters (e.g.: θ_1,θ_2,p,d etc.).

These steps allow us to derive the scaling law for the transitions from one attractor to the other and to build regime transition diagrams such as in Fig. 5. It also gives us insight into the relevant quantities that determines the dynamics:

- The antigen elimination time, and the population of the idiotypic clone at this time, C.

- $t_L + t_{Lm}$, the 2-levels time delay, which determines the possibility of excitation and of triggering.

4. MORE REALISTIC TOPOLOGIES[6]

The Cayley tree topology with equal connection strengths is a strong idealization.

4.1. *Loops*

Standard interpretation of recognition based on complementary shapes implies a network with loops.

Even loops are a likely feature of any model of recognition based on bit strings or shape space. We expect that if the idiotype recognizes by complementarity two similar anti-idiotypes, any shape complementary to one of the anti-idiotypes has a good chance to be recognized by the other one. Even loops give only quantitative changes to the phase diagram, as discussed in reference 6, and they do not change our basic conclusions about dynamics of Cayley tree networks.

Odd loops are far less likely to occur within a model based on rigid structures. Only symmetrical shapes or strings, or rather nonspecific rules for recognition can account for odd loops. On the other hand, odd loops are more probable if recognition involves soft molecular structures that can bend and have deformations to adhere to each other. In fact odd loops are present in experimental connection structures obtained by cross reactivity tests on panels of antibodies from newborn mice[7,8]. Multi-specific clones, which give rise to large connectivities and odd loops, are frequent in newborn animals, because of the favored expression of D proximal V_H genes, and seem to disappear or become infrequent in adults[8].

There are several situations where odd loops result in tolerant attractors. We then obtain a very consistent view of some possible mechanisms for the establishment of localized tolerance attractors at prenatal and neonatal stages when multi-specific clones and odd loops are present. These attractors once established remain stable, even when the multi-specific clones disappear. On the other hand, those clones that remained virgin during early development are able to react and memorize antigen presentation by reaching vaccination attractor configurations.

4.2. *The avalanche paradox*[6]

We have no a priori reason to suppose that all connection strengths are equivalent. But when one uses different connection strengths, the weaker connections seem to have more influence on the population levels than the stronger connections. Let us first consider the case of a single idiotype Ab_1 related with different connection strengths to two anti-idiotypes, Ab_2 with intensity J, and Ab_2' with intensity J_w lesser than J. After antigenic stimulation, Ab_2 gets excited as soon as Ab_1 reaches L/J the first excitation threshold. But Ab_2' gets excited when Ab_1 reaches L/J_w the corresponding excitation threshold for Ab_2' provided that the suppression threshold for Ab_2 is high enough,

$$\frac{H}{J} > \frac{L}{J_w} \tag{32}$$

(otherwise Ab_2' remains nearly virgin). Ab_2' proliferates, and so does Ab_1 until their populations get stabilized at a level such that Ab_1 represses Ab_2'. Both populations increase when J_w decreases. Furthermore, Ab_2 experiences a very strong field and is oversuppressed: its level is of order s/d rather than L. It plays no role in fixing the level of Ab_1.

Since the population of Ab'_2 is rather large it now has the possibility to be higher that the excitation threshold of a clone at level 3 with which it is connected with an intensity $J_{2'3}$ stronger than J_w. The condition for the excitation of Ab_3 is

$$\frac{L}{J_w} > \frac{L}{J_{2'3}} . \tag{33}$$

Excitation is no longer localized at levels 1 and 2 as is usual in the case of a vaccination. The equivalent process can be described from level 3 to level 5: a weak connection between levels 3 and 4 give rise to higher populations at these levels and Ab_4 is high enough to excite Ab_5. This avalanche progresses along weak/strong connections and localization of the response is then lost. The reader is referred to[6] for a full discussion of the range of parameters which allow for the avalanche process, but it can be generally assumed that a continuous distribution of J_{ij} results in avalanche. This contradicts the existence of localized attractors and all the nice properties exhibited by homogeneous Cayley trees. A number of authors [9,10] accept the idea of an immune network working in the chaotic regime which results from avalanches. But in our view, their predictions concerning the functional role of the network are unsatisfactory. For instance, they have to postulate the existence of two kinds of clones, those that are functionally connected to the net and those that are not, to account for tolerance and clonal expansion.

If we want to maintain the existence of localized attractors we have to consider improvements of the original model. Here follow a few possibilities.

We do not know much about the details of idiotypic interactions among cells. We might imagine some kind of all or none response of clones to presented antigen, even though molecular affinities are continuously distributed. This could be the result of cooperativity among the receptors at the cellular level. Cooperativity is known to occur in mechanisms such as cross-linking and capping of receptors and it does sharpen response curves as a function of concentration. If we accept the idea of an all or none response, a non-continuous affinity distribution would be obtained and a cutoff affinity would be present. Then the model does not have to be changed in order to yield localized response.

Another possible mechanism is non-specificity of suppression. This argument has been proposed by Segel and Perelson[11] as an occurrence of a general mechanism in biology (long range suppression). Suppressive factors might be secreted by T cells that react locally with little specificity. This can be translated into non specific suppression. We have simulated a modified model in which the excitation threshold θ_1 is challenged by a distribution of affinities, but the suppression threshold θ_2 is challenged by clones with equal affinities. Avalanches are not obtained since clones with strong affinities are not over-suppressed, and localized attractors are restored.

5. CONCLUSIONS

The described model is quite simple. Its main assumption is probably the proliferation function with a medium-zone response curve and low-zone and high-zone unresponsiveness. This bell-shaped response is experimentally observed. It can be derived for B cells on the assumption that cell response is due to the cross-linking of the membrane receptors by multivalent ligand. A qualitatively similar proliferation function would also be obtained with T cells under the assumption that suppression is activated at a much higher field that positive excitation.

This simple model allows one to understand how the same network can support different independent attractors that can be interpreted as virginity, vaccination and tolerance. Which attractor is reached depends on the local topology of the network and the conditions of antigen presentation. With the additional assumption that the parameter set-up is reset from a tolerant set-up during early life to a vaccination set-up during adulthood one can understand the self-nonself discrimination performed by the immune system.

"Abnormal" presentation of the antigen resulting in auto-immune diseases, whether due to previously hidden antigens, cryptic peptides, antigen mimicry, etc., can also be reformulated inside this framework. These views on auto-immune diseases are not shared by the mainstream experimentalists who usually search for TCR's or immunoglobulin's special family use, role of the MHC, particularities of the auto-antigen, and actual mechanisms for the destruction of the target. The dynamical approach shed some light on the specific role of certain MHC: since the self peptide is processed by the MHC, the local parameters, connectivity and J_{ij}'s that describe network topology and antigen presentation are under the influence of the MHC. The same reasons explain the existence of only a limited set of auto-immune diseases: although we mainly discussed a homogeneous network, it is clear the immune system is inhomogeneous. Around each node connectivity and J_{ij}'s may vary and we might expect that some tolerant attractors are more fragile than others.

ACKNOWLEDGMENTS

We thank H. Atlan, R. de Boer, B. Derrida, A. U. Neumann and A. S. Perelson for helpful discussions. GRIND[12] software was used for numerical simulations. The Laboratoire de Physique Statistique is associated with CNRS (URA 1306) and we acknowledge financial support from Inserm grant 879002 and NATO CRG 900998.

REFERENCES

1. Weisbuch, G., De Boer, R. and Perelson, A.S., Localized memories in idiotypic networks. J. Theo. Biol.146: 483-499 (1990).

2. Perelson, A.S. in: Cell Surface Dynamics: Concepts and Models. (Perelson, A.S., DeLisi, C. and Wiegel, F.W., Eds.) pp. 223-275. Marcel Dekker, New York (1984).

3. De Boer, R.J. and Hogeweg, P. , Memory but no suppression in low-dimensional symmetric idiotypic networks. Bull. Math. Biol. 51, 223-246 (1989).

4. Weisbuch G., A shape space approach to the dynamics of the immune system J. Theor. Biol. , 143(4): 507-522 (1990).

5. Neumann, A. and Weisbuch, G. Window automata analysis of population dynamics in the immune system. Bull. Math. Biol. 54: 21-44 (1992).

6. Neumann A. and Weisbuch, G. Dynamics and topology of immune networks. Bull. Math. Biol. in press.

7. Kearny, J.F and Vakil, M., Idiotype-directed interactions during ontogeny play a major role in the establishment of the of the adult B cell repertoire. Immunol. Rev. 94, 39-50 (1986).

8. Holmberg, D., G. Wennerstrom, L. Andrade and A. Coutinho, The high idiotypic connectivity of "natural" newborn antibodies is not found in the adult mitogen-reactive B cell repertoire. Eur. J. Immunol. 16:82-87 (1986)

9. Varela,F., A. Coutinho, B. Dupire and N. Vaz, Cognitive networks: immune, neural, and otherwise, pp. 359–375, in Theoretical Immunology, Part Two, A.S. Perelson, ed., Addison Wesley, Redwood City, CA (1988).

10. De Boer, R. J. and A. S. Perelson, Size and connectivity as emergent properties of a developing immune network. J. Theor. Biol. 149: 381-424 (1991).

11. Segel, L. A. and Perelson, A.S. Computation in shape space: A new approach to immune network theory. pp.321-344, in Theoretical Immunology, Part Two, A.S. Perelson, ed., Addison Wesley, Redwood City, CA (1988).

12. De Boer, R.J. GRIND: Great Integrator Differential Equations, Bioinformatics Group, University of Utrecht, The Netherlands (1983).

Growth and Recruitment in the Immune Network

Rob J. De Boer and Pauline Hogeweg
Bioinformatica
University of Utrecht
Padualaan 8
3584 CH Utrecht
The Netherlands

Alan S. Perelson
Theoretical Biology and Biophysics
Theoretical Division
Los Alamos National Laboratory
Los Alamos, NM 87545, USA

Abstract

The development of the immune repertoire during neonatal life involves a strong selection process among different clones. We investigate the hypothesis that repertoire selection is carried out during early life by the immune network. There are at least two processes in repertoire selection: clonal expansion and recruitment of clones by the bone marrow. Because both processes occur on time scales of the order of a few days, we argue that both have to be modeled. In a previous differential equation model (De Boer & Perelson, 1991), studied by numerical integration, both clonal expansion and recruitment were present but the rate of recruitment was kept low due to limitations in computational resources.

Here we present a new model based upon a two-dimensional shape space. The model is defined as an asynchronous cellular automaton (CA). In the CA model we vary (1) the rate of recruitment and (2) the specificity of the lymphocyte receptors. The networks attain an equilibrium in which the size of the repertoire remains fixed. However, the equilibrium repertoire size increases when the recruitment rate or the receptor specificity is increased. The number of functional idiotypic interactions per clone, i.e., the connectivity, is less dependent on either the receptor specificity or the recruitment rate. These observations confirm the results of our previous study. The CA model contributes to our understanding of pattern formation in immune network models because of its straightforward visualization. Using it we show that the randomness involved in lymphocyte recruitment may play a role in selecting the clones in the actual repertoire.

NATO ASI Series, Vol. H 66
Theoretical and Experimental Insights into Immunology
Edited by A. S. Perelson and G. Weisbuch

Introduction

Due to the massive production of novel lymphocyte clones in the bone marrow the composition of the immune network (Jerne, 1974) is variable. The B lymphocyte network is composed of a large number of clones, possibly as many as 10^6 to 10^8. Each clone is characterized by the variable (V) portion of its immunoglobulin receptor. B cell clones are produced in the bone marrow where reshuffling of gene segments leads to the generation of a large number of possible V regions. The diversity of receptor molecules created by gene reshuffling in the bone marrow suggests that any particular arrangement of gene segments and somatic mutations is likely to be unique. Thus, established clones are not expected to be resupplied by the bone marrow.

If stimulated properly, recirculating B cells expand in the periphery by cell division, giving rise to a clone of cells expressing the same receptor. The dynamics of recirculating B cell clones are governed by cell division and cell death. Because B cells have a life time of a few days, and because cell division typically takes less than one day, the characteristic time scale of the growth of an individual clone is on the order of days. In the adult mouse, it is estimated that the production in the bone marrow amounts to 2 – 5×10^7 B cells per day (Freitas *et al.*, 1986). This is sufficient for replacing the entire B cell population in just a few days. It thus seems that the time scale at which clones are replaced by bone marrow production and the time scale at which clones grow, are both on the order of days. In this respect, immune networks are very different from neural networks where neurons live for years but may switch from resting to firing on a time scale of milliseconds.

The importance of bone marrow recruitment as a means of generating novel clones was first stressed by Farmer *et al.* (1986; 1987) and later by Varela and co-workers (Varela *et al.*, 1988; Varela & Stewart, 1990; Stewart & Varela, 1989, 1990, 1991). In the original Varela papers recruitment was incorporated in a very rudimentary form. The bitstring models of Farmer *et al.* (1986,1987) incorporated more elaborate attempts to model recruitment but dealt with unrealistic dynamical equations. Two recent papers try to remedy the situation. In De Boer & Perelson (1991) we succeeded in combining clonal growth and recruitment in a bitstring model using a more realistic model. However, due to computational problems we were unable to study the model for realistic recruitment rates. Stewart & Varela (1991) modeled recruitment but ignored clonal growth.

The generation of novel clones by recruitment has been ignored in most mathematical models of the immune network (see De Boer, 1991; De Boer *et al.*, 1992d; and Perelson, 1989 for reviews). For reasons of simplicity, it is usually assumed that each clone has a continuous supply of cells from the bone marrow. This simplified view need not be incorrect however. First, genetically different receptors may form the same idiotype. Thus, one may view the populations of a network model as a collection of different clones having the same idiotype. Second, it is known that the B cells that appear first in ontogeny have receptors that are germ line encoded and are based upon a limited number

V_H gene families (Holmberg *et al.*, 1989). This reduces the diversity and increases the likelihood that established clones are recruited again by the bone marrow.

In this paper we take an intermediate position and assume that idiotypes (which may correspond to multiple clones) are recruited on a stochastic basis. We develop a stochastic cellular automaton model that allows us to combine clonal growth and recruitment on any time scale. By varying the recruitment rate P_R we study its impact on the model's behavior.

Bell-Shaped Activation Function

Most recent immune network models are based upon the bell-shaped interaction function proposed by De Boer (1988) or upon the two biphasic interaction functions proposed by Varela *et al.* (1988). The bell-shaped activation function $f(h)$ that we use takes the form,

$$f(h) = \frac{h}{\theta_1 + h} \frac{\theta_2}{\theta_2 + h} , \tag{1}$$

where $\theta_1 \ll \theta_2$. The first factor in $f(h)$ increases from 0 to 1, reaching its half-maximal value at $h = \theta_1$, the second factor decreases from 1 to 0, reaching its half-maximal value at $h = \theta_2$. For $\theta_2 \gg \theta_1$, the maximum is approximately one. This maximum is attained at $h = \sqrt{\theta_1\theta_2}$. The "field" h measures the effective amount of anti-idiotype a clone interacts with, and depends on both the number of anti-idiotype clones and the strength of their interaction with the idiotype (see below). Plotted as a function of $\log h$, the graph of $f(h)$ is a bell-shaped curve. An important argument for the use of a log bell-shaped function is that receptor crosslinking is involved in B cell activation. For ligands that are bivalent the cross-linking curve is bell-shaped and symmetric around its maximum (Perelson & DeLisi, 1980; Perelson, 1984).

Clonal growth is typically described in terms of differential equations. The most realistic models require several differential equations for each idiotype describing the production of antibody, B cell maturation, antibody recirculation, and so on. However, all previous models are centered around a simple differential equation describing the growth of one clone of B cells, b,

$$db/dT = b[Pf(h) - d] , \tag{2}$$

where h is the field of the clone, $f(h)$ is defined by Eq. (1), P the maximum rate of proliferation, and d is the rate of cell death. The model is non-dimensionalized by scaling the time T to the rate at which B cells turn over:

$$db/dt = b[pf(h) - 1] , \tag{3}$$

where $t = Td$ and $p = P/d$. Because $0 \leq f(h) < 1$, B cells can grow at a maximal rate $p - 1$. Thus, in order to allow for net clonal expansion p must be greater than 1. Since maximally stimulated cells divide about once every 16 h, and cells live a few days (e.g., $d = 0.5$ d^{-1}), $p = 2$ is a typical non-dimensional rate of proliferation.

Shape Space

A powerful formalism for defining the field h of each clone is the shape space theory (Perelson & Oster, 1979; Segel & Perelson, 1988, 1989, 1990; De Boer *et al.*, 1992a). Interactions amongst the clones depend on complementarities between the immunoglobulin receptors characteristic of each clone. The degree of binding of two idiotypes, usually measured by their equilibrium binding constant or affinity, depends on the generalized shapes of the two receptor molecules involved (Perelson & Oster, 1979). Thus, we let each population be characterized by a generalized shape $\mathbf{x}$. (In a multidimensional shape space $\mathbf{x}$ is a vector.) To specify the field, h, one assumes cells of shape $\mathbf{x}$ are mainly stimulated by cells of complementary or near-complementary shape $\hat{\mathbf{x}}$ centered around $\mathbf{x} = -\mathbf{x}$. The affinity or the degree of interaction between shapes $\mathbf{x}$ and $\hat{\mathbf{x}}$ usually decreases according to a Gaussian function. The Gaussian function is based upon an exponential fall off,

$$g(\mathbf{x}, \hat{\mathbf{x}}) = \exp[-\frac{|\mathbf{x} + \hat{\mathbf{x}}|^2}{\sigma^2}] , \tag{4}$$

where the standard deviation, σ, defines the rate at which the affinity falls off with the distance to the perfect match $\mathbf{x} = -\mathbf{x}$, see Segel & Perelson (1988).

An Asynchronous Cellular Automaton Model

A CA is a lattice of finite state machines. Each machine has a next state function that usually takes the local neighborhood of the machine on the lattice as input. In a previous paper (De Boer *et al.*, 1992a) we derived a discrete lattice mapping from a network model formulated in terms of partial differential equations (PDEs). This has the advantage that the conditions for stability of equilibrium states which were derived for the PDE system, provide an intuition for understanding the behavior of the lattice model. Here we follow our earlier simplifications and derive a CA model from the differential equation model defined by Eqs. (1-3).

First, we change to logarithmic variables $B = \ln b$. Second, we approximate $f(h)$ by the window automaton $w(h)$, originally proposed by Neumann and Weisbuch (1991), and defined as

$$w(h) = \begin{cases} 1 & \text{if } \exp(\theta_1) \leq h < \exp(\theta_2) \ ; \\ 0 & \text{otherwise} \ . \end{cases} \tag{5}$$

Inserting this approximation into Eq. (3), and changing to logarithmic variables, we obtain

$$dB/dt = pw(h) - 1 \ . \tag{6}$$

As a consequence dB/dt in Eq. (6) increases at a rate $p-1$ when $w(h) = 1$, and decreases at a rate -1 when $w(h) = 0$. Choosing $p = 2$, dB/dt can be only 1 or -1. Third, we discretize time and take integer variables for B to simplify Eq. (6) into the mapping

$$B(t+1) = B(t) + 2w(h) - 1 \ . \tag{7}$$

Fourth, in order to have a *finite* state machine we impose lower and upper limits on $B(t)$. A natural lower limit is $b(\mathbf{x}) = 1$, i.e., $B(\mathbf{x}) = 0$, which is thus the equivalent of a site with a single B cell. In the CA with recruitment, sites can also be empty, i.e., contain no B cells. We denote empty sites with $B(\mathbf{x}) = -1$. Empty sites never contribute to fields. As an upper limit we chose $B(t) \leq \theta_3$, which corresponds to a population maximum of $\exp[\theta_3]$ B cells. A population maximum seems reasonable because B cells are not capable of infinite proliferation and typically go through a maximum of about eight divisions (Klinman *et al.*, 1984). This gives us the next state function for the clonal growth

$$B(t+1) = \min[\theta_3, \max[-1, B(t) + 2w(h) - 1]] \,, \tag{8}$$

by which clones sizes are limited between $-1 \leq B(t) \leq \theta_3$.

In one simplification of the CA we replace the recruitment by a continuous source of cells. This is implemented by changing the lower limit for the population at a single latice site to one B cell:

$$B(t+1) = \min[\theta_3, \max[0, B(t) + 2w(h) - 1]] \,. \tag{9}$$

Finally, in combination with the next state function Eq. (8), which models clonal growth, we need a next state function for stochastic recruitment. If a recruitment event occurs, then the bone marrow adds the equivalent of one B cell, i.e., $B = 0$, to sites that are empty. Thus,

$$B(t+1) = \begin{cases} 0 & \text{if } B(t) = -1 \,; \\ B(t) & \text{if } B(t) \geq 0 \,. \end{cases} \tag{10}$$

We implement the next state functions as a 2D cellular automaton using an integer 2D shape vector $\mathbf{x}$ that is bounded between $[-N, -N] \leq \mathbf{x} \leq [N, N]$. We thus have a CA defined on a lattice of $(2N+1)^2$ sites. The self-complementary shape $[0, 0]$ is located in the center of the lattice. Because we can scale distances using the standard deviation σ, we can use $\mathbf{x}$ for both shape and for the latice site index in the automaton.

Neighborhoods. The affinity falls off with the distance to the perfect match according to the Gaussian fall-off $g(\mathbf{x}, \hat{\mathbf{x}})$ defined in Eq. (4). Thus, the contribution to the field $h(\mathbf{x})$ made by a single clone $b(\hat{\mathbf{x}}) = \exp[B(\hat{\mathbf{x}})]$ is given by

$$h(\mathbf{x}) = b(\hat{\mathbf{x}}) g(\mathbf{x}, \hat{\mathbf{x}}) \quad \Leftrightarrow \quad h(\mathbf{x}) = \exp[B(\hat{\mathbf{x}}) - G(\mathbf{x}, \hat{\mathbf{x}})] \,, \tag{11}$$

where

$$G(\mathbf{x}, \hat{\mathbf{x}}) = \frac{\mid \mathbf{x} + \hat{\mathbf{x}} \mid^2}{\sigma^2} \,. \tag{12}$$

The field $h(\mathbf{x})$ and the function $G(\mathbf{x}, \hat{\mathbf{x}})$ are again integer variables. Because $h(\mathbf{x})$ is integer, all contributions for which $\exp[B(\hat{\mathbf{x}}) - G(\mathbf{x}, \hat{\mathbf{x}})] < 1$ are ignored by integer truncation.

```
σ = 1
- - - - - - -
- - - 4 - - -
- - 2 1 2 - -
- 4 1 0 1 4 -
- - 2 1 2 - -
- - - 4 - - -
- - - - - - -
```

```
σ = 2
- - - - - - - - - - -
- - - - 4 4 4 - - - -
- - 4 3 2 2 2 3 4 - -
- - 3 2 1 1 1 2 3 - -
- 4 2 1 0 0 0 1 2 4 -
- 4 2 1 0 0 0 1 2 4 -
- 4 2 1 0 0 0 1 2 4 -
- - 3 2 1 1 1 2 3 - -
- - 4 3 2 2 2 3 4 - -
- - - - 4 4 4 - - - -
- - - - - - - - - - -
```

```
σ = 3
- - - - - - - - - - - - - - -
- - - - - 4 4 4 4 4 - - - - -
- - - 4 3 3 2 2 2 3 3 4 - - -
- - 4 3 2 2 1 1 1 2 2 3 4 - -
- - 3 2 2 1 1 1 1 1 2 2 3 - -
- 4 3 2 1 0 0 0 0 0 1 2 3 4 -
- 4 2 1 1 0 0 0 0 0 1 1 2 4 -
- 4 2 1 1 0 0 0 0 0 1 1 2 4 -
- 4 2 1 1 0 0 0 0 0 1 1 2 4 -
- 4 3 2 1 0 0 0 0 0 1 2 3 4 -
- - 3 2 2 1 1 1 1 1 2 2 3 - -
- - 4 3 2 2 1 1 1 2 2 3 4 - -
- - - 4 3 3 2 2 2 3 3 4 - - -
- - - - - 4 4 4 4 4 - - - - -
- - - - - - - - - - - - - - -
```

```
σ = 4
- - - - - - - - - - - - - - - - - - -
- - - - - - 4 4 4 4 4 4 4 - - - - - -
- - - - 4 4 3 3 3 3 3 3 3 4 4 - - - -
- - - 4 3 3 2 2 2 2 2 2 2 3 3 4 - - -
- - 4 3 3 2 2 1 1 1 1 1 2 2 3 3 4 - -
- - 4 3 2 2 1 1 1 1 1 1 1 2 2 3 4 - -
- 4 3 2 2 1 1 0 0 0 0 0 1 1 2 2 3 4 -
- 4 3 2 1 1 0 0 0 0 0 0 0 1 1 2 3 4 -
- 4 3 2 1 1 0 0 0 0 0 0 0 1 1 2 3 4 -
- 4 3 2 1 1 0 0 0 0 0 0 0 1 1 2 3 4 -
- 4 3 2 1 1 0 0 0 0 0 0 0 1 1 2 3 4 -
- 4 3 2 1 1 0 0 0 0 0 0 0 1 1 2 3 4 -
- 4 3 2 2 1 1 0 0 0 0 0 1 1 2 2 3 4 -
- - 4 3 2 2 1 1 1 1 1 1 1 2 2 3 4 - -
- - 4 3 3 2 2 1 1 1 1 1 2 2 3 3 4 - -
- - - 4 3 3 2 2 2 2 2 2 2 3 3 4 - - -
- - - - 4 4 3 3 3 3 3 3 3 4 4 - - - -
- - - - - - 4 4 4 4 4 4 4 - - - - - -
- - - - - - - - - - - - - - - - - - -
```

Figure 1. The CA's neighborhood weighting around the perfect match $\mathbf{x} = -\mathbf{x}$ as it is defined by the Gaussian function $G(\mathbf{x}, \hat{\mathbf{x}})$. The Gaussian is truncated at $\Delta = 4$: the sites indicated by a $-$ are ignored. Four neighborhoods correspond to $\sigma = 1, 2, 3, 4$ are shown. For $\sigma = 1, \ldots, 5$, the total number of sites in the neighborhood is 13, 61, 137, 241, and 385, respectively.

In immunological terms this means that the affinity of the idiotypic interaction decreases as a function of $G(\mathbf{x}, \hat{\mathbf{x}})$ in the CA. Because low affinities do not seem to lead to B cell stimulation (Fish *et al.*, 1989; Riley & Klinman, 1986; Klinman, 1972) we have implemented an affinity cut-off. In the CA formalism an affinity cut-off is naturally defined by the size of the neighborhood (see also De Boer *et al.*, 1992a). Thus, we let Δ be the value of $G(\mathbf{x}, \hat{\mathbf{x}})$ at the maximum interaction distance in shape space and ignore all larger values of $G(\mathbf{x}, \hat{\mathbf{x}})$. In Fig. 1 we show examples of our Gaussian function for $\Delta = 4$ and $\sigma = 1, 2, 3, 4$. The radius r of the neighborhood of the CA can be obtained from Fig. 1. For $\sigma = 1, 2, 3, 4, 5$, $r = 2, 4, 6, 8, 11$ respectively.

Finally, for the calculation of the field we will only consider populations for which $B(\hat{\mathbf{x}}) > 0$. We thus ignore empty lattice sites ($B = -1$) and sites that contain a single B cell ($B = 0$). Thus contributions to the field are only made by populations that have divided and are in a non-resting state. The reason for doing this is that a clone should not be stimulated to grow by a field totally comprised of virgin clones, i.e., clones that just appeared from the bone marrow. Typically, interactions between clones are not direct B cell–B cell interactions but are mediated by secreted antibodies carrying the idiotype of the B cell. Since resting clones do not secrete antibody, the exclusion of resting clones from the field seems desireable. Under this assumption, for any shape $\mathbf{x}$ the total field $h(\mathbf{x})$ correponds to

$$h(\mathbf{x}) = \sum_{\hat{\mathbf{x}} \in C(-\mathbf{x})} \exp[B(\hat{\mathbf{x}}) - G(\mathbf{x}, \hat{\mathbf{x}})] \ , \quad \forall \ B(\hat{\mathbf{x}}) > 0 \ , \tag{13}$$

where $C(-\mathbf{x})$ is the circle centered at $-\mathbf{x}$ with radius r. We assume fixed boundaries, i.e., shapes outside the shape space $[-N, -N] \leq \hat{\mathbf{x}} \leq [N, N]$ are fixed at concentration -1.

Scheduling. The sites of a CA can be updated in parallel, i.e. synchronously, or randomly, i.e. asynchronously. Most of our results are based upon asynchronous updating. We choose this method updating so that we can independently vary clonal growth and recruitment events. Thus, for each update we randomly choose one of the two next state functions Eqs. (8) or (10). The parameter P_R defines the probability with which the recruitment event is chosen. Because there are only two types of events, clonal growth is chosen with probability $1 - P_R$. The selection of the site to be updated is also random. We will only present results for which the site selection is based upon a uniform distribution in the interval $[-N, N] \leq \mathbf{x} \leq [N, N]$. Thus, the basic iteration cycle is (1) select a site to be update, (2) select a transition function, and (3) perform the update.

Parameters. We have previously studied models based upon Eq. (3) for $\theta_1 = 100$ and $\theta_2 = 10^4$. In the CA we convert θ and h values to logarthmic variables. Thus, we round these values of θ to $\theta_1 = 5$ because $\exp[5] \approx 148$ and $\theta_2 = 10$ because $\exp[10] \approx 2.2 \times 10^4$. In order to reduce the effect of the maximum population size we set $\theta_3 \gg \theta_2$. Choosing $\theta_3 = 15$, each B cell population can maximimally grow to $\exp[15] \approx 3 \times 10^6$ cells. This is probably a high estimate because B cell proliferation is typically limited to eight divisions or so (Klinman *et al.*, 1984). The affinity cut-off is set to $\Delta = \theta_3 - \theta_2 - 1$. By Eq. (11), the field contribution of a clone at maximum size, i.e., $B = \theta_3$, must be at least $h = \theta_3 - \Delta = \theta_2 + 1$. Thus, a clone at maximum size can only evoke a suppressive field. This seems reasonable because we do not want the network to sustain clones at a probably unreasonable maximum size. If equilibrium populations $B(\mathbf{x}) = \theta_3$ are allowed to have stimulatory effects on other populations, i.e., if $\Delta > \theta_3 - \theta_2$, the behavior of the CA becomes reminiscent of the model of Stewart and Varela (1991) and De Boer and Van der Laan (1992). The system forms stable lines of clones of maximum size. The distance between the lines is such that they reciprocally activate each other (not shown).

We choose $N = 50$, i.e., we work on a 101×101 lattice. This corresponds to a potential repertoire of $101^2 \approx 10^4$ clones. The results show that the CA never fills all sites of the lattice. The sites filled with B cell populations correspond to the actual repertoire. Choosing $\sigma = 3$ as the standard value for the size of the neighborhood each clone can have a maximum of 25 high affinity interactions, see Fig. 1. The total number of clones in the $\sigma = 3$ neighborhood is 137 idiotypes, which is roughly 1% of the total shape space. As a default we choose equal time-scales for proliferation and recruitment, i.e., we set $P_R = 0.5$.

Computation. The CA was implemented in the C programming language. The code is available electronically from *deboer@cc.ruu.nl.* Simulations were performed on a Silicon Graphics Personal Iris. Depending on the size of the neighborhood this took several hours. In principle the code is efficient because the various exponents required for summing the field, and for the Gaussian fall-off Eq. (12), can be stored as integer arrays that need only be computed once. The remainder of the computation boils down to stepping through the 2D shape space array, and performing the integer subtraction and addition implied by Eq. (13). Random numbers for site selection and event selection were computed with the RAN3 routine described in "Numerical Recipes" (Press, *et al.*, 1988).

Since clones expand and decrease one order of magnitude on a time scale of days, we base the time steps of our model on the clonal growth function. Time is increased one unit after $(2N+1)^2$ evaluations of the next state functions Eq. (8) or (9). This means that, on average, during each time step all clones in the CA will have updated their state. Thus, the time clock clicks independent of the recruitment. Typically the CA was studied for 500 time steps.

Previous Models

The two recent models of the recruitment of the immune system are very different from each other. In the work of De Boer and Perelson (1991) the generalized shape $\mathbf{x}$ takes the form of a bit-string, i.e., a point in a high-dimensional hypercube. Clonal growth is modeled by means of a system of differential equations based on Eq. (2). Recruitment is implemented by interrupting the numerical integration, and changing number of differential equations that model clonal growth as clones are added or deleted. The recruitment is based upon the addition of B cell populations created by an artificial bone marrow, and the deletion of populations that have grown smaller than one single cell. The main problem with this approach is that computational resources limit the number of populations that the artificial bone marrow can add to the system. Thus the system could not be studied for realistic recruitment rates.

In the work of Stewart and Varela (1991) the generalized shape $\mathbf{x}$ takes the form of a point in a 2D shape space. Clones in the shape space are either present or absent, and clonal growth is ignored. The equivalent of the bell-shaped interaction function of the clonal growth process is put directly into the recruitment process. Clones are maintained in the network if the number of interaction partners, weighted by the respective distances in shape space according to Eq. (4), falls within a window set by two parameters θ_1 and θ_2. Clones can only be added to the network if their field falls within the window. Once a clone is added to the network the fields of all established clones are checked. If a clone's field falls outside the window that clone is deleted. The check then has to be repeated because the fields have changed again. This iteration halts when all clones have a field inside the window. In the worst case all clones are deleted. Note that the algorithm depends on the order by which the clones are scanned, and hence, that it may collapse, i.e., delete all clones, in situations where better solutions are possible. However, because of the similar time scales of clonal growth and of the recruitment, the main problem with this work is that clonal growth is ignored. Since the size of a clone is important in determining the field that it generates, this model ignores a very important aspect of network dynamics.

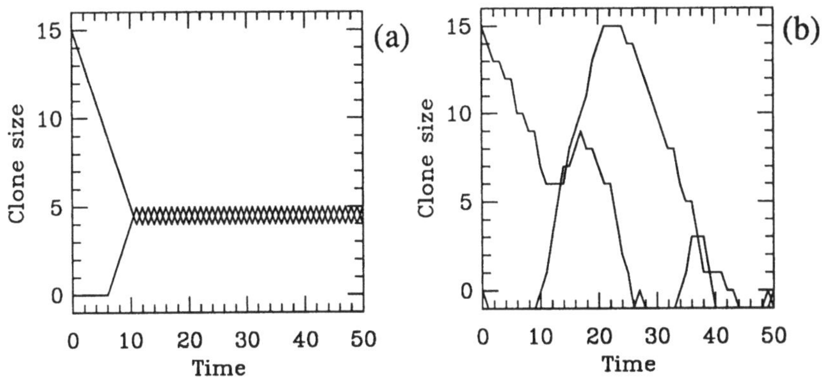

Figure 2. Time plots of a two clone system $h_1 = \exp[B_2] + \exp[A]$ and $h_2 = \exp[B_1]$. Parameters: $\theta_1 = 5, \theta_2 = 10, \theta_3 = 15$, and $P_R = 0.5$. (a) Parallel updating without recruitment, i.e., a model based upon Eq. (9). Initial conditions: $B_1(0) = 15$, $B_2(0) = 0$. (b) Asynchronous updating with stochastic recruitment, i.e., a model based upon Eqs. (8 & 10). Initial conditions: $B_1(0) = 15$, $B_2(0) = 0$.

Results

In this paper we use an asynchronous CA model of the immune network. With this model it is possible to combine clonal growth and recruitment at the correct time scales. Our principle point of interested is whether the results of the CA model differ from the results of our previous bit-string model (De Boer & Perelson, 1991). The main parameters that we varied in the bit-string model were the specificity of the receptors and the rate of production of new clones by the bone marrow. The latter parameter corresponds to the recruitment in the CA. In the bit-string model the rate of recruitment is expressed as the daily production of novel bit-strings by the artificial bone marrow. The specificity of the receptors was determined by the rules by which two bit-strings were matched. This was expressed as the "reactivity" of the receptors or $P(match)$, i.e., the fraction of other receptors each receptor is expected to bind. The reactivity was varied between binding probabilities of 0.005 and 0.205. Henceforth, we refer to our previous model as the bit-string model.

Two Interacting Clones

The basic dynamics of the next state functions Eqs. (8–10) can be studied by simulating two interacting clones instead of simulating the full CA. Thus, let $B_i(t)$, for $i = 1, 2$, be defined by Eqs. (8–10), and let the field of clone B_1 be $h_1 = \exp[B_2]$ and the field of clone B_2 be $h_2 = \exp[B_1]$. (In the orginal formulation of differential equations like Eq. (2) such two clone systems may give rise to stable equilibria, limit cycle oscillations, or chaos, depending upon parameter values, see De Boer & Hogeweg, 1989a,b; De Boer *et al.*, 1990, 1992b,c).

The simplest case is a system with parallel updating of B_1 and B_2 according to function Eq. (9). We show, in Fig. 2a, a simulation for the initial condition $B_1(0) = \theta_3$ and $B_2(0) = 0$, for the parameters $\theta_1 = 5$, $\theta_2 = 10$, $\theta_3 = 15$. Because $h_1 = 0$, B_1 decreases due to a lack of stimulation. B_2 cannot grow due to suppression, i.e., $h_2 > \theta_2$. B_2 starts to grow after $\theta_3 - \theta_2$ time steps, when B_1 has decreased to $B_1 = \theta_2$. B_1 stops decreasing after another θ_1 time steps, when B_2 has increased to $B_2 = \theta_1$. The two clones attain an attractor around θ_1 alternating between giving stimulation while decreasing due to a lack of stimulation, and not giving stimulation while increasing, see Fig. 2a. In a system with parallel updating, such a stable attractor around θ_1 is always present. The other attractors are the oscillator $B_1 = B_2 = \theta_2 \to B_1 = B_2 = \theta_2 - 1 \to B_1 = B_2 = \theta_2, \ldots,$ and the equilibrium$B_1 = B_2 = 0$. It depends on the initial conditions which of the three attractors is attained.

The same system with asynchronous updating, i.e., one based upon Eqs. (8 & 10), behaves differently. See Fig. 2b for an example with $P_R = 0.5$. The attractor around θ_1 is no longer stable because by random chance both clones may equal θ_1. They then stimulate each other reciprocally until one of them equals θ_2. Typical behavior of the asynchronous automaton is large variations in the population densities, see Fig. 2b. In this respect the behavior resembles the oscillatory/chaotic behavior of the continuous model (De Boer *et al.*, 1990, 1992c,d).

Cellular Automaton

We study the immune network as it develops during early ontogeny. In vivo this corresponds to a situation in which most self-antigens are present and B cells are absent. Network interactions could develop when autoreactive B cells are stimulated by self-antigens and subsequently start to activate anti-idiotypic B cells. Another possibility is that anti-idiotypic B cells are activated by so-called maternal antibodies which the developing system obtains from the mother. In our previous work we started the network with maternal antibodies (De Boer & Perelson, 1991). Here we start similarly by setting 1% of the shapes to the stimulatory concentration $B(\mathbf{x}) = (\theta_1 + \theta_2)/2$. These populations represent the maternal antibodies. All other sites are assumed to be empty, i.e., $B(\mathbf{x}) = -1$.

In Figs. 3 and 4 we show an example of the model's behavior. In Figs. 3a and 3b we show a snapshot of the CA after 500 time steps. The 2D distribution of the clone size $B(\mathbf{x})$ is shown in Fig. 3a and the log field size $\ln[h(\mathbf{x})]$ in Fig. 3b. The size of the population is indicated by gray-scales, where black means large and white means small. The 2D distribution of clones typically takes the form of lines and circles. The 2D

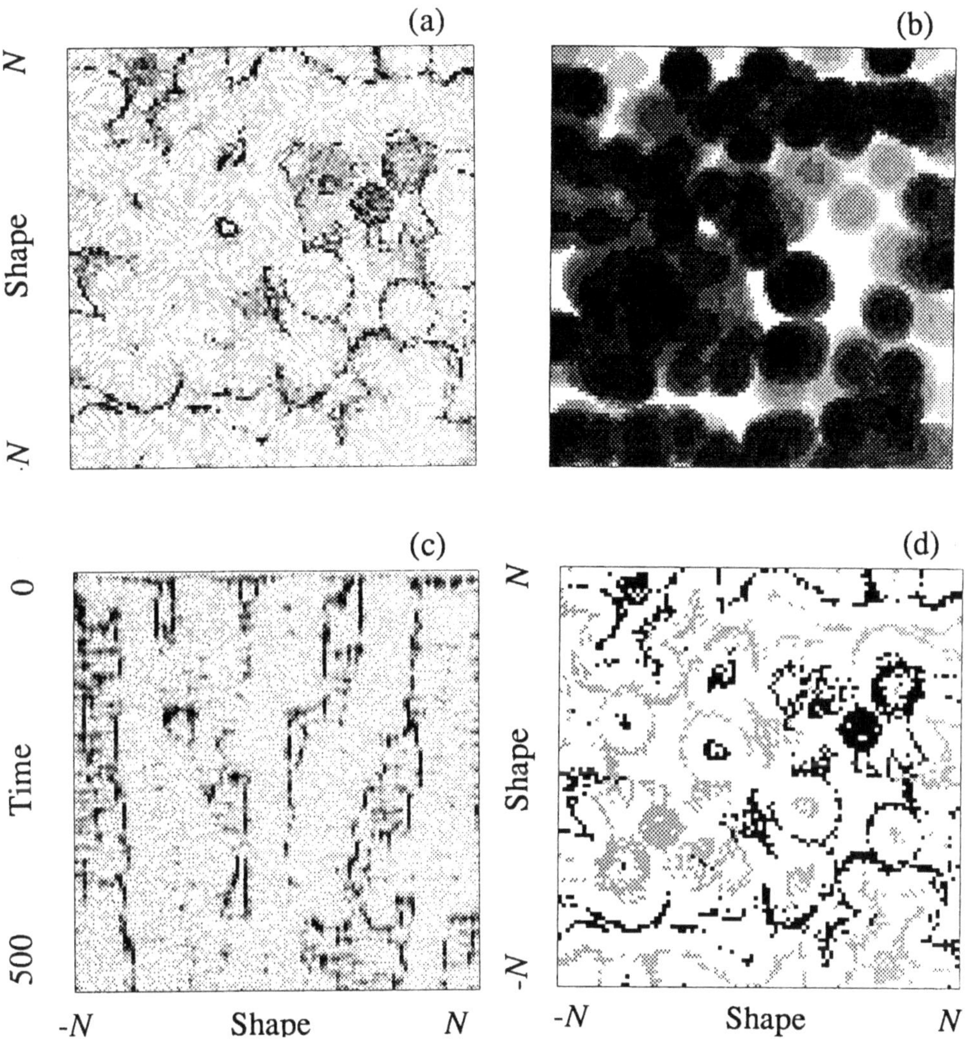

Figure 3. Shape space usage in an asynchronous CA. Parameters: $\sigma = 3, \theta_1 = 5, \theta_2 = 10, \theta_3 = 15, \Delta = 4$, and $P_R = 0.5$. (a) The 2D clone size distribution attained after 500 time steps. The gray-scales vary between white for $B(\mathbf{x}) = -1$ and black for $B(\mathbf{x}) = \theta_3$. (b) The log fields, i.e., $\ln[h(\mathbf{x})]$, corresponding to panel (a). The gray-scales vary between white for $\ln[h(\mathbf{x})] = 0$ and black for $\ln[h(\mathbf{x})] = 17$. (c) A space time plot for all clones along one diagonal of the shape space, i.e., $\{\forall\, B(x,x) \mid x = -N, \ldots, x = N\}$. (d) The 2D clone size distribution shown in panel (a). For each each point in shape space $\mathbf{x}$ the color is white if $B(\mathbf{x}) < \theta_1$ and $B(-\mathbf{x}) < \theta_1$, otherwise the color is black if $B(\mathbf{x}) > B(-\mathbf{x})$ and gray if $B(\mathbf{x}) < B(-\mathbf{x})$. Thus the black image corresponds to the real distribution of large clones. The gray pattern is the mirror image of the black pattern.

distribution of the fields typically takes the form of circular areas with the maximum field strength in the middle. This corresponds to a field landscape of hills and valleys. The comparison of the two distributions reveals that large clones are typically located at intermediate field strengths. This means that the clones live at the contour lines of intermediate height of the field landscape. The reason for this is clear: the window automaton $w(h)$ specifies that clones can only expand at intermediate field strengths. A more interesting observation is that the lines along which clones organize themselves are much longer than the size of the neighborhood of the automaton (i.e., a circle with radius $r = 6$). Thus the pattern that is attained by the self-organization has a much larger scale than the local rules of the automaton have.

In Fig. 3c we show a space-time diagram of the CA by sampling clone sizes along one diagonal of the shape space from $B(-N,-N)$ to $B(N,N)$. Thus, the self-complementary shape $B(0,0)$ is located in the middle of the space axis. Time runs along the vertical axis from 0 to 500. The figure shows that the 2D distributions fail to attain a steady state but change and move about in shape space. The vertical dark lines in the figure correspond to clones that temporarily attain a stationary state and remain large for some time. The diagonal lines correspond to travelling waves. If one views these results in the form of movies of consecutive 2D distributions, travelling waves are a striking phenomenon. Fig. 3c shows that although the 2D distributions never attains a steady state the patterns change on a time-scale that is much slower than the time-scale of recruitment and clonal growth. The travelling waves and the patterns shown in Fig. 3a are reminiscent of the self-organization of excitable media and reaction-diffussion systems (cf. Murray, 1989). This type of behavior was also seen in the bitstring model of Farmer *et al.* (1986) as discussed by Perelson (1988).

To visualize which clones interact we project the mirror image of large clones onto the opposite part of the shape space. In Fig. 3d, we color all clones $\mathbf{x}$ that are larger than θ_1 black, and project their mirror image at $-\mathbf{x}$ in gray. The white spots correpond to sites $\mathbf{x}$ for which $B(\mathbf{x}) \leq \theta_1$ and $B(-\mathbf{x}) \leq \theta_1$. Thus, the distribution of black spots corresponds to large clones, and is identical to the distribution of dark-gray spots in Fig. 3a. The distribution of the gray spots is the mirror image of the black spots reflected about $[0,0]$. The figure shows how the circles of clones enclose smaller circles of complementary clones, and how lines of clones are aligned with lines of complementary clones. The distance between the gray and the black patterns reflects the distance at which large clones evoke a stimulatory field. In Fig. 3a or 3d the self-complementary clone $[0,0]$ does not seem to behave very different than any other clone.

Sampling global properties. In the bit-string model we showed that both the number of clones in the network and the connectivity of the network attains a maximum value during the early life of the network. Following this early peak the number of clones and the connectivity approach a much lower equilibrium. These observations correspond to experimental observations that show that the network connectivity is large during early life (Holmberg *et al.*, 1989). The notion of an equilibrium size of the immune network has also been suggested by experimentalists, see the discussion in De Boer & Perelson (1991). Here we study similar global properties of the CA networks. We sampled the

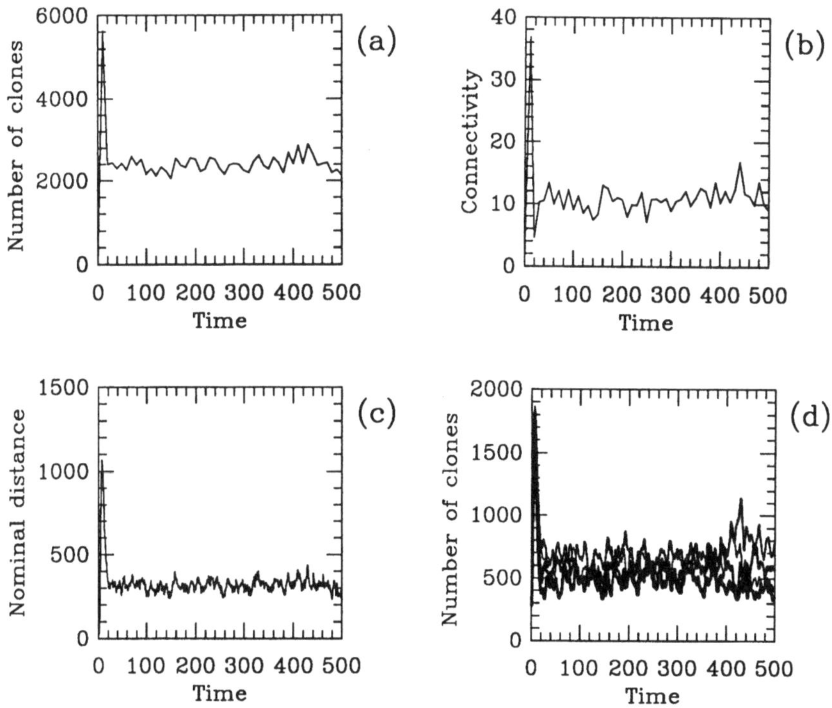

Figure 4. Global properties of an asynchronous CA. Parameters: $\sigma = 3, \theta_1 = 5, \theta_2 = 10, \theta_3 = 15, \Delta = 4$, and $P_R = 0.5$. (a) number of active clones. (b) actual connectivity. (c) nominal distance. (d) number of populations per quadrant.

number of active clones and the functional connectivity of the clones in the network. An active clone is a clone that is present and has divided, i.e., we count the number of non-zero clones. The functional connectivity is the number of active interaction partners of an active clone, i.e., we count the average number of non-zero clones in the neighborhood of each non-zero cell. Both properties are sampled every ten time steps.

For the example network of Fig. 3 we show time plots of the number of clones and the connectivity are shown in Figs. 4a and 4b. The results are very similar to the results of the bit-string model. Both the size of the network, i.e., the number of clones, and the connectivity go through an early peak, and then approach equilibrium levels. During the last 100 days of the simulation the average network size is 2466 clones with an average connectivity of 12 functional connections. Because the potential repertoire corresponds to 10^4 clones it appears that by repertoire selection 25% of the possible clones participate in the immune network.

Two other novel properties are shown in Figs. 4c and 4d. The rate at which the 2D distributions shown in Fig. 3 change is measured by the "nominal distance" between the patterns. We classify clone sizes into three categories: $B(\mathbf{x}) < \theta_1$, $\theta_1 \leq B(\mathbf{x}) < \theta_2$, and

$B(\mathbf{x}) \geq \theta_2$, and increase the nominal distance between two time steps one unit for each clone that switches to another category. A time plot of the nominal distance is shown in Fig. 4c. Following an early peak, the rate at which the patterns change approaches an equilibrium. Apparently, the actual repertoire is dynamic and continues to change. After a 1000 time steps the nominal distance is still very similar to the one shown in Fig. 4c (not shown). Finally, we counted the number of active clones in each of the four quadrants of the shape space. Time plots are shown in Fig. 4d. The averages during the last 100 time steps are 825, 570, 416, and 570, respectively. In order to test whether this deviates from a random distribution we calculated that $\chi^2 = 145$ for this distribution. With three degrees of freedom, $p < 10^{-4}$, so this is a very significant deviation from a random distribution.

Pattern Formation

The two main parameters that we varied in the bit-string model were the reactivity of the receptors (i.e., the bit-string matching) and the productivity of the bone marrow. In order to study the distribution of clones in the shape space of the bit-string model we computed the Hamming distances between all of the pairs of bit-strings in the network. If the coverage of shape-space is random, the distribution of Hamming distances is expected to be binomial. We showed that the bit-strings tended to be similar to each other, i.e., the Hamming distances were smaller than expected. The deviation increased with the rate of recruitment. We argued that these deviations were due to the formation of clusters of clones with similar bit-strings sharing the same stimulatory field. In the CA the reactivity and the productivity parameters correspond to the size of the neighborhood σ and the probability of recruitment P_R respectively. The effect of these parameters on pattern formation in shape space is studied in Figs. 5 and 6.

The 2D distributions shown in Fig. 5 reveal that the scale of the pattern scales with σ. First, the distance between the black and the gray stripes increases with σ. The explanation is that for large neighnorhoods clones have to be located far from each other to reduce the field strength. Second, the patterns become more global if σ increases. One striking "global" phenomenon which we observed repeatedly (Fig. 5c and 5d) are the long stripes that span the entire shape space. Such lines change position but seem to be anchored around shape $[0, 0]$. For large values of σ (Fig. 5d) the line forms the boundary between a density populated part of shape space and an empty part of shape space. Another global pattern that appears for large values of σ is that large regions in shape space tend to be either black or gray. Thus it seems that the usage of shape space changes from fine grained patterns to large scale clusters as σ increases.

In Fig. 6 we show the 2D distributions of large clones for $P_R = 0.1, 0.2, 0.4$, and 0.8. First, the figure shows that increasing P_R increases the thickness of the lines and the density of populated regions. Thus, areas in which the field is stimulatory are more densily populated. Because this increases the field strength, this implies that the distance between stripes has to increase. (This effect is so small that it is hardly visible.)

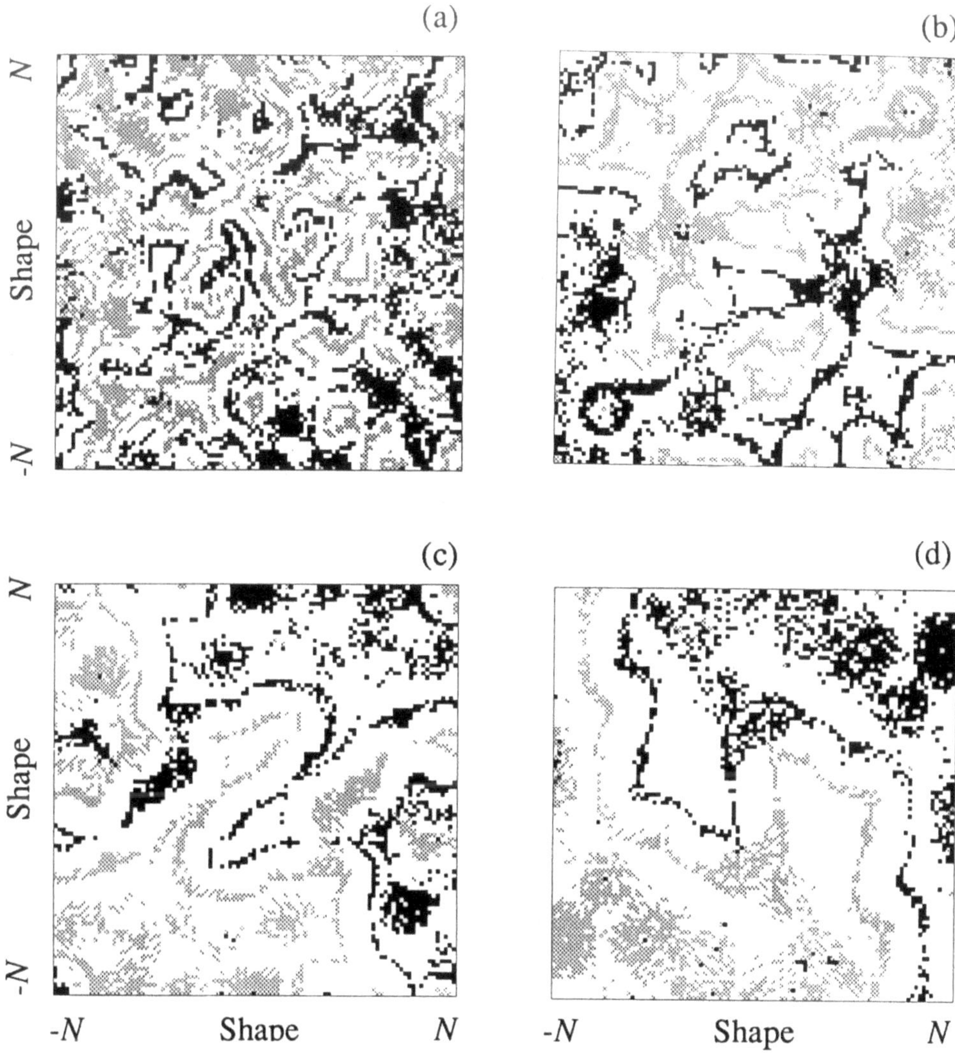

Figure 5. Shape space usage as a function of σ, the neighborhood size. Parameters: $\theta_1 = 5, \theta_2 = 10, \theta_3 = 15, \Delta = 4$, and $P_R = 0.5$. The gray-scales vary between white for $B(\mathbf{x}) = -1$ and black for $B(\mathbf{x}) = \theta_3$. (a) $\sigma = 2$. (b) $\sigma = 3$. (c) $\sigma = 4$. (d) $\sigma = 5$.

A densily populated area in shape-space means that idiotypes tend to be similar. This confirms the results of the bit-string model. The reason these patterns become more pronounced as P_R increases is that P_R determines the rate at which clones can invade areas with stimulatory fields. Second, on a larger scale, the figure shows that increasing the recruitment rate causes the repertoire selection to become more global. In Fig. 6a black and gray lines are distributed all over shape space. In Fig. 6d most clones are

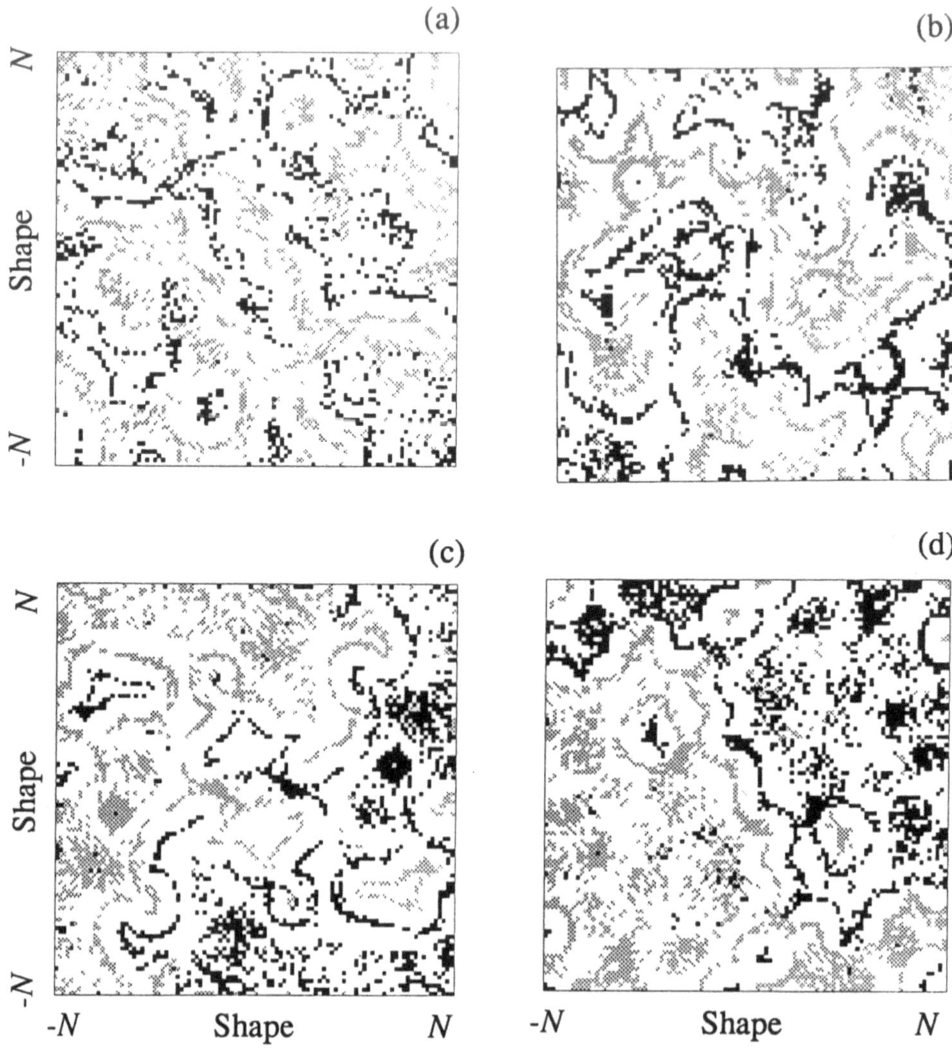

Figure 6. Shape space usage as a function of P_R, the rate of recruitment. Parameters: $\sigma = 3, \theta_1 = 5, \theta_2 = 10, \theta_3 = 15, \Delta = 4$. The gray-scales vary between white for $B(\mathbf{x}) = -1$ and black for $B(\mathbf{x}) = \theta_3$. (a) $P_R = 0.1$. (b) $P_R = 0.2$. (c) $P_R = 0.4$. (d) $P_R = 0.8$.

located in the upper right triangle of shape space. This is similar to what we observed for σ in Fig. 5.

As in Fig. 4d, we counted the number of active clones in each of the four quadrants. The χ^2 of the average distributions is plotted in Fig. 7a,b. This confirms the suggestion made above that the randomness of shape space usage decreases when σ or P_R increase. In Fig. 7c,d we plot the average nominal distance over the last 100 days of two simulations.

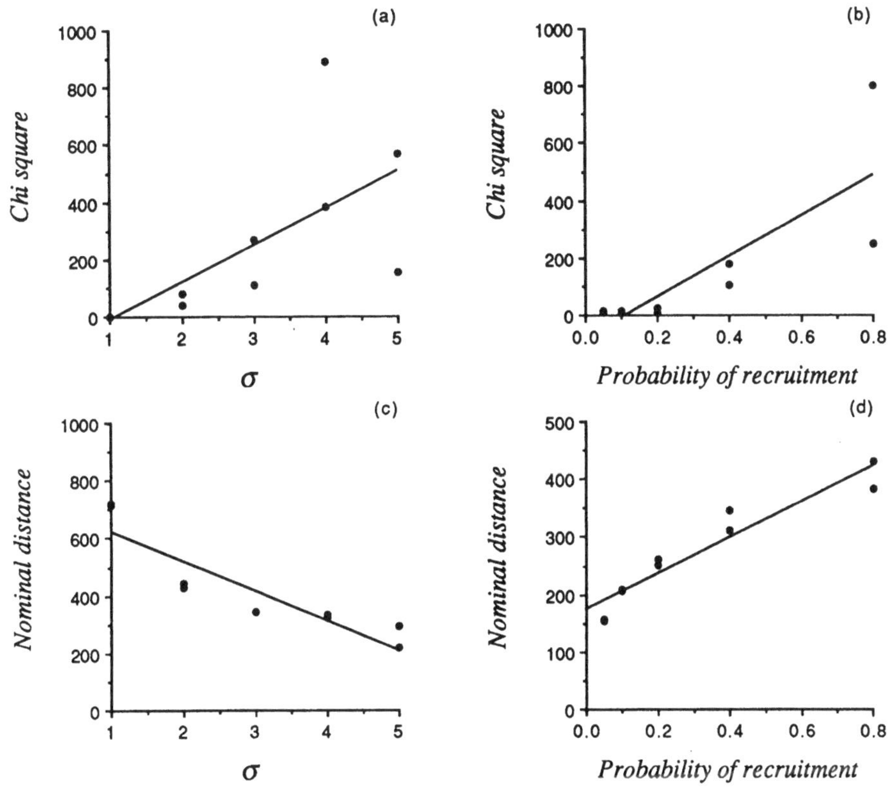

Figure 7. The scale and the rate of change of the shape space patterns. (a) χ^2 as a function of σ. (b) χ^2 as a function of P_R. (c) nominal distance as a function of σ. (d) nominal distance as a function of P_R.

When σ increases the rate of change of the patterns decerases. When P_R increases the rate of change of the patterns increases.

Pattern scale. We conclude that the scale of patterns increase with recruitment in both the bit-string model and in the asynchronous CA. The present results contribute to the understanding of these bit-string results because we now see how an increase in the recruitment decreases the randomness of the selected repertoire. This appears due to two effects. First, stimulatory areas are more rapidly invaded. This gives rise to a cluster of similar shapes. Second, the randomness involved with recruitment may play a role. Because novel clones are supplied at random sites and at random times, the rate of recruitment determines the degree of randomness in these systems. The fact that an increase in the randomness may increase the scale of the ordered regions in a CA has been described for voting rules. In voting rule systems frustration due to annealing (Vichnia, 1986), and/or simple random noise, create larger domains and a larger scale

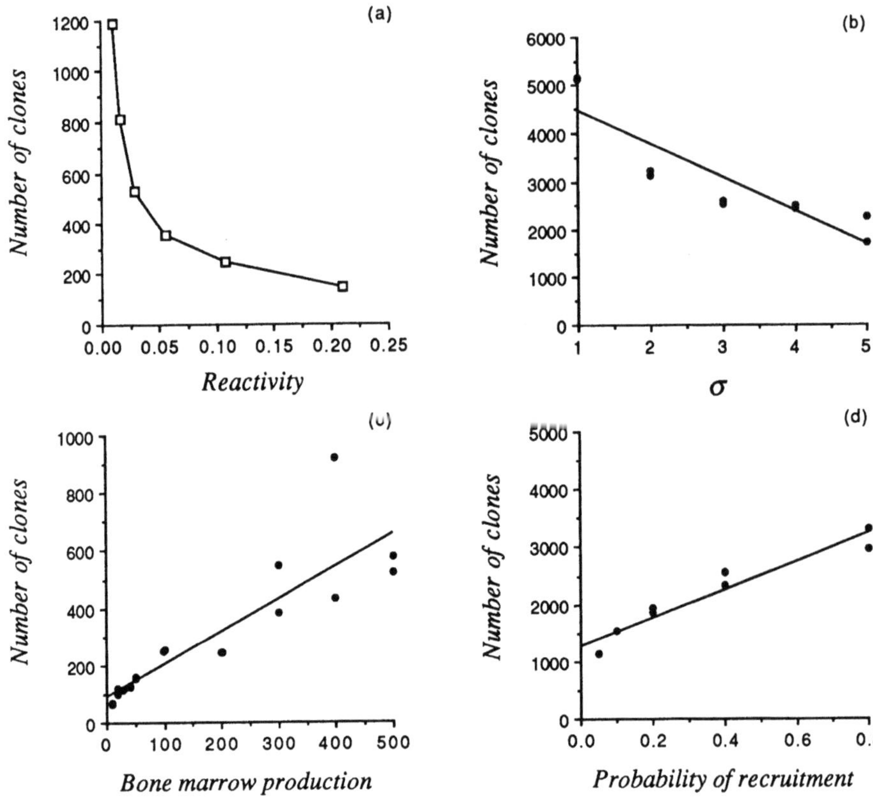

Figure 8. The number of clones in the repertoire. The network size as a function of reactivity in the bit-string model (a), and as a function of reactivity (i.e., σ) in the CA (b). The network size as a function of recruitment in the bit-string model (c) and in the CA (d).

of order (Toffoli & Margolus, 1987). Despite the fact that our transition rules Eqs. (8-10) are not voting rules, the randomness, which keeps boudaries fluid and which allows separate domains to merge, seems to have a similar effect in our immune network.

Size

One of the main question that we addressed in the bit-string model was the number of clones in the repertoire. Since there is no explicit regulation of the number of clones in the model the size of the network was an emergent property which turned out to vary around an equilibrium value. In Fig. 8 we show how the number of clones in the bit-string model varied as a function of the reactivity (Fig. 8a) and the productivity (Fig. 8c).

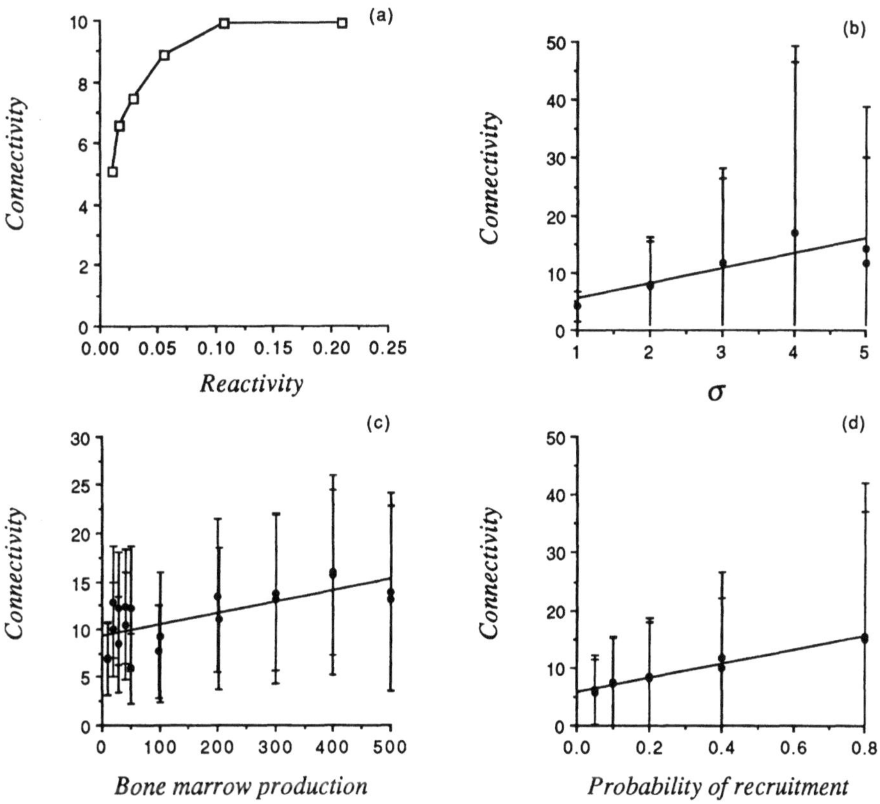

Figure 9. The actual connectivity of the repertoire. The network connectivity as a function of reactivity in the bit-string model (a), and as a function of reactivity (i.e., σ) in the CA (b). The network connectivity as a function of recruitment in the bit-string model (c) and in the CA (d). The error bars indicate the average standard deviation of ten samples.

In Figs. 8b and 8d we show that very similar results are obtained in the CA. Thus, increasing σ (i.e., the reactivity) decreases number of active populations. (Note that graphs in Fig. 8a and 8b would look more similar if the data in Fig. 8b were plotted as a function of the variance σ^2 instead of the standard deviation σ). In Fig. 8b the number of clones in the actual repertoire decreases from 5000 for $\sigma = 1$ to 2000 for $\sigma = 5$. Thus the selection for participation in the actual repertoire becomes stronger if σ or the reactivity increases. Conversely, the number of clones in the actual repertoire increases linearly as a function of the bone marrow production rate and of P_R. Thus, the shape space becomes more densily covered when the rate of recruitment increases. This can be explained by the formation of dense clusters of similar shapes.

Connectivity

Another emergent property of the bit-string model was the connectivity among the clone in the network, which like the network size varied around an equilibrium value. Connectivity was defined as the average number of network connections per clone. The results shown in Fig. 9a and 9c show that the connectivity of the network increases with reactivity and with bone marrow production rate. However, the increase in the connectivity is small compared to range over which the parameters are changed. Thus, we concluded that the connectivity is mainly regulated by selection in the network and is hardly dependent on the reactivity and bone marow production parameters.

In Fig. 9b and 9d we show that similar tendencies are found in the CA. Increasing σ increases the functional connectivity sligthly (Fig. 9b). In Fig. 9b the connectivity increases from 4 for $\sigma = 1$ (i.e., 13 neighbors) to 13 for $\sigma = 5$ (i.e., 385 neighbors). Similarly, the connectivity slightly increases with P_R, see Fig. 9d. The fact that the connectivity hardly depends on the neighborhood size is as counterintuitive as the fact that connectivity hardly depended on the matching probability in the bit-string model. The explanation for this counterintuitive phenomenon is similar for both systems. First, repertoire selection reduces the number of clones in the actual repertoire when clones have too many interactions. In the bit-string paper we discussed how this form of self-organization can be interpreted as a form of self-regulatory completeness (De Boer & Perelson, 1991). Second, because of the pattern formation, clones tend to be more similar which reduces the connectivity.

The error bars in Fig. 9b and 9d indicate the average standard deviation of the connectivity. (Each connectivity point is based upon ten samples, so each error bar is the average of ten standard deviations). The figure shows that the standard deviation increases with σ and P_R. This is another manifestation of the observation that the patterns become more ordered if either σ or P_R increases. Due to the pattern formation some areas are highly connected and some areas are lowly connected. Thus, the distribution of connectivities becomes more variable. Interestingly, in Fig. 9c of the bit-string model one can also see that the standard deviation of the connectivity increases with bone marrow production. This can now also be explained in terms of the increase in the order due to larger neighborhoods or the faster recuitment.

Fields

The histograms in Fig. 10 depict the distribution of field strengths for active B cells (Fig. 10a,b) and for unused sites in the shape space (Fig. 10c,d). The white bars denote the proportion of sites having a small field, i.e., $h(\mathbf{x}) \leq \exp[\theta_1]$, the gray bars denote stimulatory field strengths, i.e., $\exp[\theta_1] < h(\mathbf{x}) < \exp[\theta_2]$, and the striped bars denote suppressive fields, i.e., $h(\mathbf{x}) \geq \exp[\theta_2]$. Thus 45% of the used parts of shape space are stimulated (Fig. 10a,b) and 80% of the unused parts of shape space (Fig. 10c,d) are

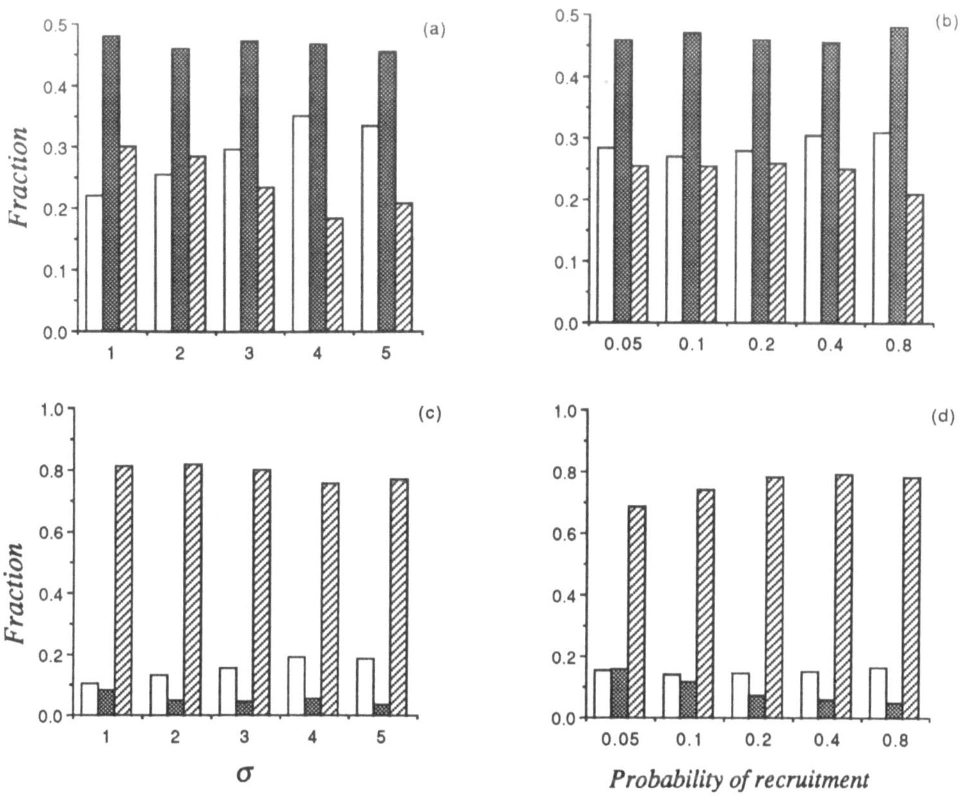

Figure 10. The distribution of field strengths of clones in the actual repertoire (a,b) and of the un-used sites (c,d). The white bars denote the proportion of sites having a small field, i.e., $h(\mathbf{x}) \leq \exp[\theta_1]$, the gray bars denote stimulatory field strengths, i.e., $\exp[\theta_1] < h(\mathbf{x}) < \exp[\theta_2]$, and the striped bars denote suppressive fields, i.e., $h(\mathbf{x}) \geq \exp[\theta_2]$.

suppressed. Both percentages hardly depend on σ or P_R. In the bit-string model we found that the majority of clones in the repertoire were suppressed. Here we find that only 20–30% of the actual repertoire is suppressed. An explanation for this discrepancy might be that suppressed clones are more quickly removed in the CA model than in the bit-string model. In the bit-string model we introduced more than one cell per recruitment event. Thus it takes longer to for the population to disappear by natural decay. Additionally, because in the bit-string model antibodies could be long-lived clones could persist for longer times.

There seems to be a slight effect of the neighborhood size on the ratio of suppressed and non-stimulated clones. Increasing σ increases the proportion of non-stimulated sites, and decreases the proportion of suppressed sites (Fig. 10c). This confirms earlier observations of other shape space models (De Boer *et al.*, 1992a).

Discussion

We have compared the behavior of an asynchronous CA model of the immune network with our previous ordinary diifferential equation/bit-string model. Our most important question was whether the inclusion of a realistic rate of the recruitment of new shapes would affect the results obtained with the bit-string model. Because all results seem to be similar this does not seem to be the case. Although we have been careful to point out how the two models can be derived from each other, it is important to realize how different the two approaches are. The fact that we find similar results for such a wide class of network models suggests that these results are generic.

Apart from the confirmation of our previous results in a more realistic parameter range, the CA model contributes to our understanding of the previous results on pattern formation. This is mainly due to the straightforward two dimensional representation of the CA. First, the 2D grayscale pictures of the CA really show that stimulatory areas become densily populated. This confirmed our interpretation of the same results in the bit-string model. Second, we have agrued that patterns become more ordered because recruitment may play the roles of annealing or randomness in voting rule systems (Toffoli & Margolus, 1988). A similar effect of random recruitment has also been demonstrated to increase the scale of spatial patterns of T and B cells in a CA model of T cell B cell segregation (Hogeweg, 1989). Similar effects may have played a role in the bit-string model and explain the observed deviation from a random distribution of Hamming distances (De Boer & Perelson, 1991).

The CA formalism makes it feasible to study high recruitment rates and very large networks. Thus, the more unrealistic formalism of a 2D CA shape space model has made it feasible to study the immune network for more realistic parameter regimes.

Acknowledgements

We thank Drs. Maarten Boerlijst for many helpful discussions. Portions of this work were performed under the auspices of the U.S. Department of Energy. It was supported in part by NIH Grant RR06555 (A.P.), and by the Santa Fe Institute through their Theoretical Immunology Program.

References

De Boer, R. J. (1988). Symmetric idiotypic networks: connectance and switching, stability, and suppression. In: Theoretical Immunology, Part Two, (A. S. Perelson, ed.), SFI Studies in the Science of Complexity., Vol. III, Addison-Wesley, Redwood City, CA, pp. 265–289.

De Boer R.J. (1991) Recent developments in idiotypic network theory. Neth. J. Med. 39: 254–262.

De Boer, R. J. & Hogeweg, P. (1989a). Memory but no suppression in low-dimensional symmetric idiotypic networks. Bull. Math. Biol. 51: 223–246.

De Boer, R. J. & Hogeweg, P. (1989b). Unreasonable implications of reasonable idiotypic network assumptions. Bull. Math. Biol. 51: 381–408.

De Boer, R. J., Kevrekidis, I. G. & Perelson, A. S. (1990). A simple idiotypic network model with complex dynamics. Chem. Eng. Sci. 45: 2375–2382.

De Boer, R. J. & Perelson, A. S. (1991). Size and connectivity as emergent properties of a developing immune network. J. Theor. Biol. 149: 381–424.

De Boer R. J., Segel L. A. & Perelson A. S. (1992a). Pattern formation in one and two dimensional shape space models of the immune system. J. Theor. Biol. **155**, 295–333.

De Boer R. J., Neumann A. U., Perelson A. S., Segel L. A. & Weisbuch, G. W. (1992b). Recent approaches to immune networks. In: Proceedings First European Biomathematics Conference, (Capassso, V. & Demongeot, P., eds.), Springer, Berlin (in press).

De Boer R. J., Kevrekidis I. G. & Perelson A.S. (1992c) Immune network behavior I: From stationary states to limit cycle oscillations. (submitted).

De Boer R. J., Kevrekidis I. G. & Perelson A.S. (1992d) Immune network behavior II: From oscillations to chaos and stationary states. (submitted).

De Boer R. J. & Van der Laan, J. D. (1992). A cellular automaton model of the immune network. In: Thinking about Biology, (Stein, W.D & Varela, F. J., eds.), Addison-Wesley, Redwood City, CA.

Farmer, J. D., Packard, N. H. & Perelson, A. S. (1986). The immune system, adaptation, and machine learning. Physica D22: 187–204.

Farmer, J. D., Kauffman, S. A., Packard, N. H., & Perelson, A. S. (1987). Adaptive dynamic networks as models for the immune system and autocatalytic sets. Annals New York Acad. Sci. 504: 18–130.

Fish, S. Zenowich, E., Fleming, M. & Manser, T. (1989) Molecular analysis of original antigenic sin. I. Clonal selection, somatic mutation, and isotype switching during a memory B cell response. J. Exp. Med. 170: 1191–1209.

Freitas, A. A., Rocha, B. & Coutinho, A. (1986). Lymphocyte population kinetics in the mouse. J. Immunol. 91: 5–37.

Hogeweg, P. (1989) Local T–T and T–B interactions: a cellular automaton approach. Immunol. Letters 22: 113–122.

Holmberg, D., Andersson, Å., Carlson, L. & Forsgen, S. (1989). Establishment and functional implications of B-cell connectivity. Immunol. Rev. 110: 89–103.

Jerne, N. K. (1974). Towards a network theory of the immune system. Ann. Immunol. (Inst. Pasteur) 125 C: 373–389.

Klinman, N. R. (1972). The mechanism of antigenic stimulation of primary and secondary clonal precursor cells. J. Exp. Med. 136: 241–260.

Klinman, N. R., Press, J. L., Pickard, A. R., Woodland, R. T. & Dewey, A. F. (1974). Biography of the B cell. In: The Immune System, Genes, Receptors, Signals, Sercarz, E. E., Williamson, A. R. & Fox, C. F., eds. Academic Press, New York, pp. 357–365.

Murray, J. D. (1989). Mathematical Biology. Springer-Verlag, Berlin.

Neumann, A. U. & Weisbuch, G. (1992). Window automata analysis of population dynamics in the immune system. Bull. Math. Biol. 54: 21–44.

Neumann, A. U. & Weisbuch, G. (1992). Dynamics and topology of idiotypic networks. Bull. Math. Biol. (in press).

Perelson, A. S. (1984). Some mathematical models of receptor clustering by multivalent ligands. In: Cell Surface Dynamics: Concepts and Models (A. S. Perelson, DeLisi, C. & Wiegel, F. M., eds.), Marcel Dekker, New York, pp. 223–276.

Perelson, A. S. (1988), Toward a realistic model of the immune system. In: Theoretical Immunology, Part Two, SFI Studies in the Sciences of Complexity (A. S. Perelson, ed.), Addison-Wesley, Redwood City, CA, pp. 377-401.

Perelson, A. S. (1989). Immune network theory. Immunol. Rev. 110: 5–36.

Perelson, A. S. & Oster, G. F. (1979) Theoretical studies on clonal selection: minimal antibody repertoire size and reliability of self-non-self discrimination. J. Theor. Biol. 81: 645–670.

Perelson, A. S. & DeLisi, C. (1980). Receptor clustering on a cell surface. I. Theory of receptor cross-linking by ligands bearing two chemically identical functional groups. Math. Biosciences 48: 71–110.

Perelson, A. S. & Segel, L. A. (1991). On the shape-space approach to the immune system: A B cell antibody model. J. Stat. Phys. 63: 1113–1131.

Press, W. H., Flannery, B. P., Teukolsky, S. A. & Vetterling, W. T. (1988). Numerical Recipes in C. The Art of Scientific Computing., Cambridge University Press, Cambridge.

Riley, R. L. & Klinman, N. R. (1986). The affinity threshold for antigenic triggering differs for tolerance susceptible immature precursors vs mature primary B cells. J. Immunol. 136: 3147–3154.

Segel, L. A. & Perelson, A. S. (1988). Computations in shape-space: a new approach to immune network theory. In: Theoretical Immunology (A. S. Perelson, ed.), Part Two, SFI Studies in the Science of Complexity., Vol. III, Addison-Wesley, Redwood City, CA, pp. 321–343.

Segel, L. A. & Perelson, A. S. (1989). Shape-space analysis of immune networks. In: Cell to Cell Signalling: From Experiments to Theoretical Models, (A. Goldbeter, ed.), Academic Press, New York, pp. 273–283.

Stewart, J. & Varela, F. J. (1989). Exploring the meaning of connectivity in the immune network. Immunol. Rev. 110: 37–61.

Stewart, J. & Varela, F. J. (1990). Dynamics of a class of immune networks. II. Oscillatory activity of cellular and humoral components. J. Theor. Biol. 144: 103–115.

Stewart, J. & Varela, F. J. (1991). Morphogenesis in shape-space. Elementary metadynamics in a model of the immune network. J. Theor. Biol.153: 477–498.

Toffoli, T. & Margolus, N. (1987) Cellular Automata Machines. A New Environment for Modeling., MIT Press, Cambridge MA.

Varela, F. J., Coutinho, A., Dupire, B. & Vaz, N. N. (1988). Cognitive networks: immune, neural, and otherwise. In: Theoretical Immunology, Part Two, (A. S. Perelson, ed.), SFI Studies in the Science of Complexity., Vol. III, Addison-Wesley, Redwood City, CA, pp. 359–375.

Varela, F. J. & Stewart, J. (1990). Dynamics of a class of immune networks. I. Global stability of idiotype interactions. J. Theor. Biol. 144: 93–101.

Varela, F. J. & Coutinho, A. (1991). Second generation immune networks. Immunol. Today 12: 159–166.

Vichniac, G. (1986). Simulating physics with cellular automata. Physica 10D: 96–115.

Weisbuch, G. (1990). A shape space approach to the dynamics of the immune system. J. Theor Biol. 143: 507–522.

Weisbuch, G., De Boer, R. J. & Perelson, A. S. (1990). Localized memories in idiotypic networks. J. Theor. Biol. 146: 483–499.

MEMORY B CELLS STABILIZE CYCLES IN A REPRESSIVE NETWORK

Ulrich Behn [1,2], J. Leo van Hemmen [3], Bernhard Sulzer [3]

Institut für Theoretische Physik [1]
Ruhr-Universität Bochum
Universitätsstraße 150
Postfach 102148
W-4630 Bochum 1
Germany

1. INTRODUCTION

In this paper we analyze a model (Behn and van Hemmen, 1989a,b) which is based on the concept of idiotypic networks formed by B lymphocytes (Jerne 1974, 1984; for recent reviews see, e.g., Perelson, 1988, 1989, Varela and Coutinho, 1991) and shows that there is a synergy of two mechanisms for memory.

The first mechanism is —as proposed by Jerne (1974) — that memory is stored "in the symbiotic equilibrium of a cycle that contains an image". The simplest cycle consists of two sorts of antibodies, say Ab_1 and Ab_2, where the 'useful' Ab_1 recognize the antigen Ag and *is* recognized by the 'anti-antibody' Ab_2. The cycle is closed if Ab_1 also recognizes Ab_2 which then act as an *internal image* of the antigen. Since the production of a sort of antibody is stimulated if its paratope recognizes a complementary structure (which may be the epitope of other antibodies or an antigen) and inhibited if its epitope *is* recognized by a complementary structure, the interaction may result in a dynamical equilibrium which provides a reservoir of useful antibodies Ab_1 even in the absence of the antigen Ag — a memory.
The second mechanism is that, after a sufficiently strong primary response against Ag, there appear memory B cells which have a life-time orders of magnitudes longer than 'normal' B cells and which can become activated in a new encounter with Ag. This finally results in a production of antibodies Ab_1.

We show that in the framework of our model for the idiotypic network which includes memory cells the above mentioned cycles are *unstable* unless the corresponding memory cells are present, or, in other words: memory cells stabilize cycles (Behn and van Hemmen, 1989a,b; Behn et al., 1991).

Permanent address: Sektion Physik, Universität Leipzig, Augustusplatz 10, O-7010 Leipzig, Germany
Physik-Department, TU München, James-Franck-Straße, W-8046 Garching bei München, Germany

NATO ASI Series, Vol. H 66
Theoretical and Experimental Insights into Immunology
Edited by A. S. Perelson and G. Weisbuch

A natural 'by-product' of the analysis is a support of Jerne's idea that "the essence of the immune system is the repression of its lymphocytes" (Jerne, 1974). In our model repression is not only important to prevent the system from exploding but is also crucial for the formation of stable cycles. For instance, if stimulation would dominate inhibition, no stable 2-cycles would exist even in the presence of memory cells (Behn, van Hemmen, 1989a,b).

To concentrate on essential mechanisms and not to burden the analysis our model contains a number of simplifications. We do not distinguish between antibodies on the surface of a lymphocyte and those secreted by plasma cells. To describe the dynamics of the antibodies we adopt a bilinear form of the interaction as used by Farmer et al. (1986) which is familiar from the kinetics of chemical reactions. Since there is a clear separation of time scales we decouple in the simplest version of our model the dynamics of the antibodies from that of the memory cells and assume that the latter once appeared after a sufficiently strong primary response — remain constant in number.

In Section 2 we formulate the model and give a qualitative argument that elucidates the mechanism which may lead to a nonzero finite solution of the network equations representing a memory. In Section 3 we analyze the behaviour of small subsystems and show that a 2-cycle interacting with an antigen has three relevant states (the *virgin state*, the *immunized state*, and the state of *chronic infection*) and displays in its dynamics the difference between primary and secondary response. In Section 4 we propose a specific stimulation therapy for a chronic infection. In the last Section we discuss several extensions of the model and directions of further investigations.

2. LOTKA-VOLTERRA NETWORK

We consider a *given set* of constituents of the idiotypic network and describe the dynamics of the number x_i of antibodies of sort i, $i = 1, \ldots, N$, per unit volume interacting with other antibodies and with antigens. Here y_j denotes the number of antigens of sort j per unit volume. In a simple 'mean-field' approach the interaction is bilinear in x_i and y_j (of Lotka-Volterra form) and we obtain (Behn and van Hemmen, 1989a, b)

$$\dot{x}_i = x_i \left(\sum_{j=1}^{N} (m_{ij} - \kappa m_{ji}) x_j - \gamma \right) + d_i \sum_{j=1}^{N} m_{ij} x_j + (d_i + x_i) \sum_{j=N+1}^{N+R} m_{ij} y_j \ , \tag{1}$$

where γ is the inverse life time of noninteracting antibodies and m_{ij} measures the matching between paratope of antibody i with a complementary structure j (on antibody j or antigen j). There is a *stimulation* if paratope i matches a complementary structure and an *inhibition* if epitope i is recognized by a complementary structure. The parameter κ allows for an asymmetry between stimulation and inhibition.
The effect of memory cells is described by d_i which is inserted if the total stimulus per x_i, viz., $\left(\sum_j m_{ij} x_j + \sum_k m_{ik} y_k\right)$, exceeds some threshold μ. Since the lifetime of

memory cells is orders of magnitude longer than that of antibodies we here decouple the dynamics of antibodies and memory cells and assume that the number of memory cells, once appeared, stays constant in time. Without memory cells, $d_i \equiv 0$, we recover the model proposed by Farmer et al. (1986).
More support for the bilinear form of the interaction and the existence of a threshold is given elsewhere (Behn et al., 1991). In passing we note that we have not included a source term on the right hand side of (1) describing a possible influx of new cells of sort i from the bone marrow which would, however, not alter our conclusions.

The dynamics of the antigens y_i, $i = N+1, \ldots, N+R$, is described by

$$\dot{y}_i = y_i\left(\alpha - \sum_{j=1}^{N} m_{ji}x_j\right) , \tag{2}$$

where α, the virulence, is the difference between proliferation rate and the inverse life time; for most relevant antigens α is positive. Eq. (2) allows an exponential growth of antigen during the initial stage of an infection.

We note that the system (1), (2) is invariant against a scaling of the matching parameters $\{m_{ij}\}$ by λ if $\{x_i\}$, $\{y_i\}$, and $\{d_i\}$ are scaled by λ^{-1}. Thus matching and number of constituents per unit volume are measured in arbitrary units choosen for numerical convenience. The time scale is fixed by choosing a value for the life-time of the antibodies of the order of 10...20 days. The matching m_{ij}, here treated as a parameter, could be determined in the frame of a shape space approach (see, e.g., Farmer et al., 1986, Segel, Perelson, 1988) which is beyond the scope of this paper.

In a living organism the list of variables is of course dynamic itself. New sorts of antibodies (lymphocytes) are introduced by the bone marrow and there is an enhanced mutation during the process of clonal selection. We think, however, that the understanding of the dynamics of a given set of constituents is indespensable for extending both theory and simulation to large scale 'living' networks.

We now discuss the qualitative behaviour of system (1) in the absence of antigens by focussing on the dynamics of the total number of antibodies $s = \sum_{k=1}^{N} x_k$,

$$\dot{s} + \gamma s = (1-\kappa) \sum_{j,k=1}^{N} x_j m_{jk} x_k + \sum_{j,k=1}^{N} d_j m_{jk} x_k . \tag{3}$$

In the *virgin state* ($d_j \equiv 0$) the trivial fixed point $s = 0$ (i. e. *all* $x_i \equiv 0$) is stable if repression dominates: $\kappa > 1$. Nonzero fixed points can exist only for $\kappa < 1$ but they *all* are locally unstable (Behn and van Hemmen, 1989b).
The presence of memory B cells (some $d_j \neq 0$) may destabilize the zero fixed point and allow the formation of stable nontrivial fixed points if the $\{d_j\}$ are large enough *and* $\kappa > 1$.
Supposing that all x_i are of the same order of magnitude, say x, we have a competition between the two terms on the right hand side of (3). The first is negative (for $\kappa > 1$) and bilinear in x, the second is positive and linear in x. Thus for large x the first dominates

and prevents the system from exploding, whereas for small x the second dominates and — provided the $\{d_j\}$ are large enough — prevents the system from dying out.
As a result of the competition we have a nonzero finite solution (e.g. a fixed point or a limit cycle) which provides a reservoir of useful antibodies — a memory. This is possible since the organism acquires in the course of its life (due to encounters with antigen or due to interaction between antibodies) a set of nonzero $\{d_j\}$ which means, formally speaking, a symmetry breaking of the virgin state.
We paranthetically note that the above arguments, made on a purely formal level, strongly support Jerne's philosophy (1974) that stresses the crucial role of repression.

3. SMALL SUBSYSTEMS

Since the space of possible paratopes and its complementary structures is high dimensional it is obvious that a good matching between two randomly selected paratopes and epitopes has a very small probability. The formation of a fully connected network is a sort of percolation problem (Perelson, 1989, de Boer, 1989). Here we assume to work in a parameter regime where small subsystems connected by good matching are most probable. The smallest subsystem is a 2-cycle of two sorts of antibodies, x_1 and x_2, which mutually recognize each other and thus, in an equilibrium of stimulation and suppression, provide a reservoir of useful antibodies.

For antibodies x_1 and x_2 with mutual matching $m_{12} = m_{21} = m$ and antigens y which stimulate x_1 with matching $\overline{m}$, in the presence of memory cells $d_1 = d_2 = d$, eqs. (1) and (2) simplify to

$$\dot{x}_1 = x_1\left[(1-\kappa)mx_2 - \gamma\right] + dmx_2 + (d + x_1)\overline{m}y \ , \tag{4}$$

$$\dot{x}_2 = x_2\left[(1-\kappa)mx_1 - \gamma\right] + dmx_1 \ , \tag{5}$$

$$\dot{y} = y(\alpha - \overline{m}x_1) \ . \tag{6}$$

The system (4)-(6) has three fixed points $z^s = (x_1^s, x_2^s, y^s)$,

$$z_1^s = (0,0,0) \ , \tag{7}$$

$$z_2^s = (a,a,0) = \left(\frac{d-\gamma/m}{\kappa-1}\ , \ \frac{d-\gamma/m}{\kappa-1}\ , \ 0\right) \ , \tag{8}$$

$$z_3^s = (a_1,a_2,b) = \left(\frac{\alpha}{\overline{m}}\ , \ \frac{d}{\kappa-1+\gamma\overline{m}/(\alpha m)}\ , \ (a_1-a_2)\frac{md+\gamma}{\overline{m}d+\alpha}\right) \ , \tag{9}$$

which are relevant if their components are nonnegative. There are three parameter regions with, respectively, one, two, or three relevant fixed points: the *virgin state*, the *immunized state*, and the state of *chronic infection*. Figure 1 shows for these states schematically the fixed points and the flows according to a local stability analysis (Behn et al., 1991) .

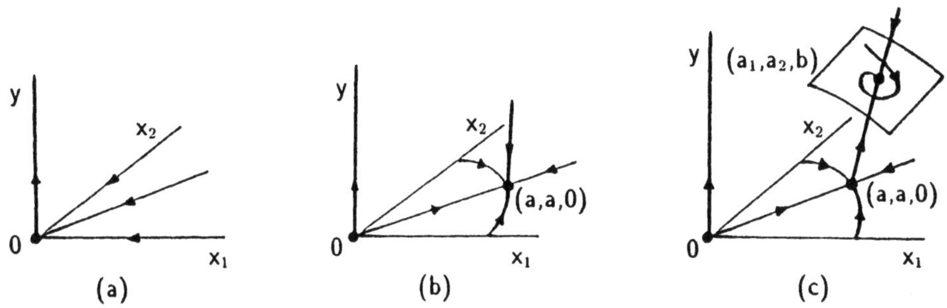

Fig. 1. Fixed points and schematic flows for a 2-cycle of two sorts of antibodies x_1 and x_2 with mutual matching m. The antigens y are recognized by antibodies x_1 with matching $\overline{m}$. For no (or few) memory cells we have the *virgin state* (a). The antibodies x_1 and x_2 die out if they are not additionally stimulated. The *immunized state* (b) needs a minimal amount of memory cells. An infection ends in a healthy state if the antigen is not too virulent. For virulent antigens (or poor matching $\overline{m}$) we have the state of *chronic infection* (c). Further explanation is given in the text.

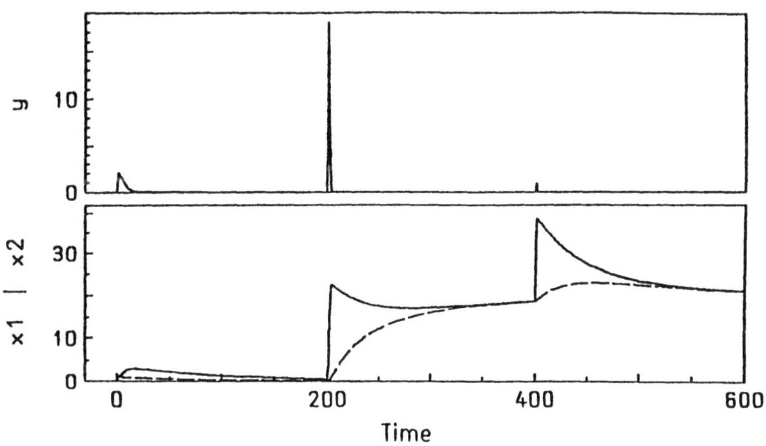

Fig. 2. *Primary* and *secondary response* of a 2-cycle (x_1,solid line; x_2, dashed line) to an antigen y recognized by x_1. After a small subcritical dose $y_{in} = 2$ at $t = 0$ the system relaxes back to the virgin state. At $t = 200$ it is infected with a dose $y_{in} = 20$ strong enough to initiate the appearance of dormant B cells which leads to a stable 2-cycle (*primary response*). The *secondary response* after injection of the same dose $y_{in} = 20$ is much more efficient. ($\alpha = 0.05$, $\gamma = 0.01$, $\kappa = 1.1$, $m = 0.05$, $\overline{m} = 0.03$, $d = 0.4$).

In the *virgin state*, the relevant fixed point $z_1^s = 0$ is stable in the (x_1,x_2) plane if not too many memory cells ($d < \gamma/m$) are present. For $\alpha > 0$ it is unstable in y direction; the virgin state may become infected.

For $d > \gamma/m$ the fixed point $z_1^s = 0$ looses the stability and z_2^s becomes relevant. The fixed point $z_2^s = (a, a, 0)$ describes a healthy *immunized state*, the production of useful antibodies x_1 is maintained even in the absence of antigen due to antibodies x_2 — the *internal image* of the antigen. z_2^s is stable as long as the antibody is not too virulent ($\alpha < \overline{m}a$); an infection is spontanously cured.
For a virulent antigen (or poor matching, or not enough antibodies) we have $\alpha > \overline{m}a$ and z_2^s looses the stability. The fixed point $z_3^s = (a_1, a_2, b)$ becomes relevant describing the state of a *chronic infection* — a 'remis' in the competition between the antigen and the antibodies.
This simplest — and most probable — subsystem exhibits also a reasonable dynamical behaviour. For instance, as shown in Figure 2, it explains the efficiency of the *secondary response*.

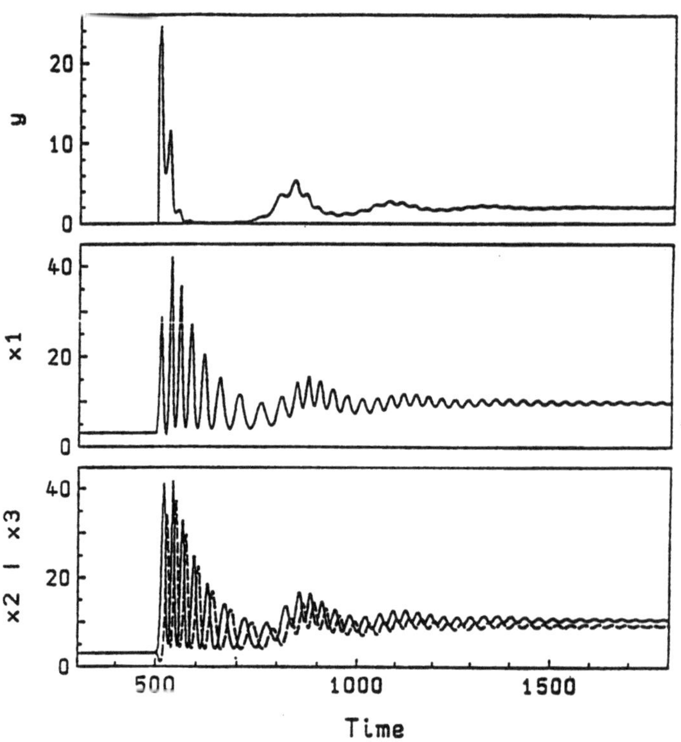

Fig. 3. Response of 3-cycle (x_1, x_2, solid line; x_3, dashed line) on injection of a dose of antigen y_{in} = 20 which is virulent enough leading to a chronic infection. There are the dominant predator-prey oscillations and internal oscillations on a smaller time scale. ($\alpha = 0.1$, $\gamma = 0.002$, $\kappa = 1.1$, $m = \overline{m} = 0.01$, $d = 0.5$).

The response of a 3-cycle on injection of a dose of antigen which leads after typical predator-prey and internal oscillations to a chronic infection is shown in Figure 3. The observation that for a certain period after the injection the antigen seems to be extinct leads us to a possible therapy for a chronic infection.

4. SPECIFIC STIMULATION THERAPY FOR A CHRONIC INFECTION

We consider again a 2-cycle which is now *un*stable against a virulent antigen. After an initial infection the system goes to the state of chronic infection.
The idea of a specific stimulation therapy is to induce a second infection *for the purpose of therapy.* To this end we inject a dose of antigen y_{in} repeatedly, say I times. After this, as shown in Figure 4, the antigen seems — for a while — to disappear before it recovers and the fixed point of chronic infection is reached again.

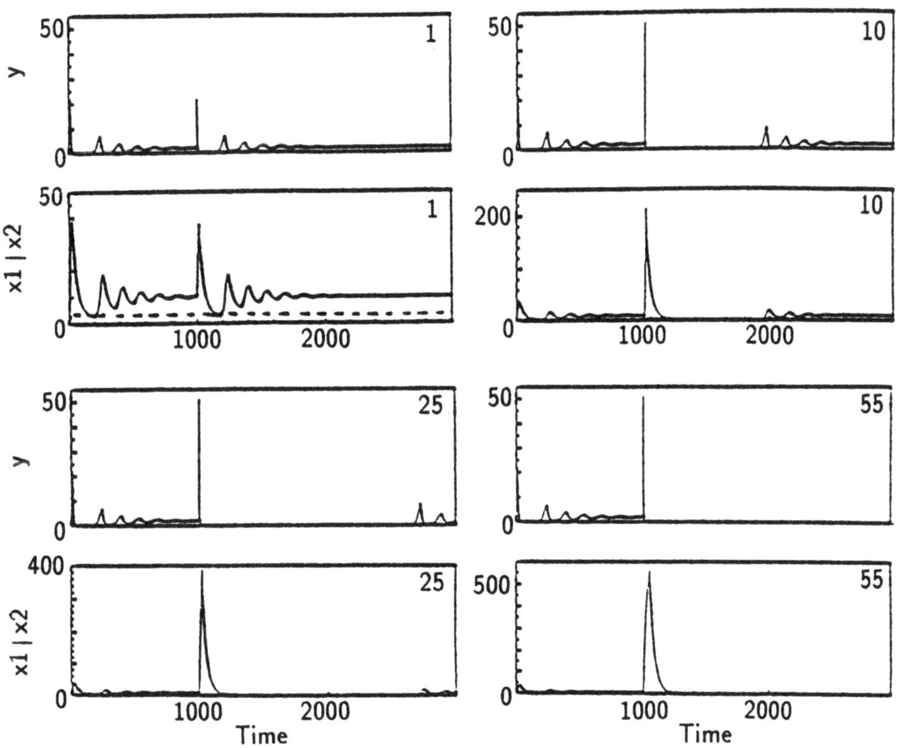

Fig. 4. Specific stimulation therapy. We show the response of a 2-cycle (x_1, solid line; x_2, dotted line) which is unstable against a virulent antigen y stimulating x_1. All parts of the figure have the first infection at $t = 0$ in common after which the system goes to the state of chronic infection (note the different scales at the ordinate). At $t = 1000$ an additional dose of antigen $y_{in} = 20$ is repeatedly injected each time steps over different periods of time I (I=1, 10, 25, 55 as indicated in the right upper corner). After this, the antigen appears nearly extinct before it reappears (latent period). The latent period increases in a nonlinear way with I. Remarkable enough, the antigen increases by only a factor two whereas the antibodies may increase by orders of magnitude. ($\alpha = 0.1$, $\gamma = 0.01$, $\kappa = 1.1$, $m = 0.05$, $\overline{m} = 0.03$, $d = 0.4$).

This phenomenon can be easily understood looking at the flow in the x_1-log(y) plane (x_2 is only weakly varying) as shown in Figure 5. During the injection of antigen Eq. (6) is replaced by $\dot{y} = y(\alpha - \overline{m}x_1) + y_{in}/\Delta\tau$ which means a shift of the fixed point z_3^s. ($\Delta\tau$ is the interval between two injections assumed to be small against characteristic times of the system). The new state is reached asymptotically with increasing injection period I and provides useful antibodies x_1 whose number is an order of magnitude more than that in the state of chronic infection. Once the injections are stopped, the system moves 'freely' on trajectories governed by Eqs. (4)-(6); the 'surplus' antibodies reduce the antigen whose number decreases exponentially. For exponentially small values of y the antigen can be considered extinct.

Since it may be hazardous to apply arbitrarily high total doses of antigen $y_{tot} = y_{in}I/\Delta\tau$ we looked for an optimal distribution of a fixed total dose over the period I (Behn, van

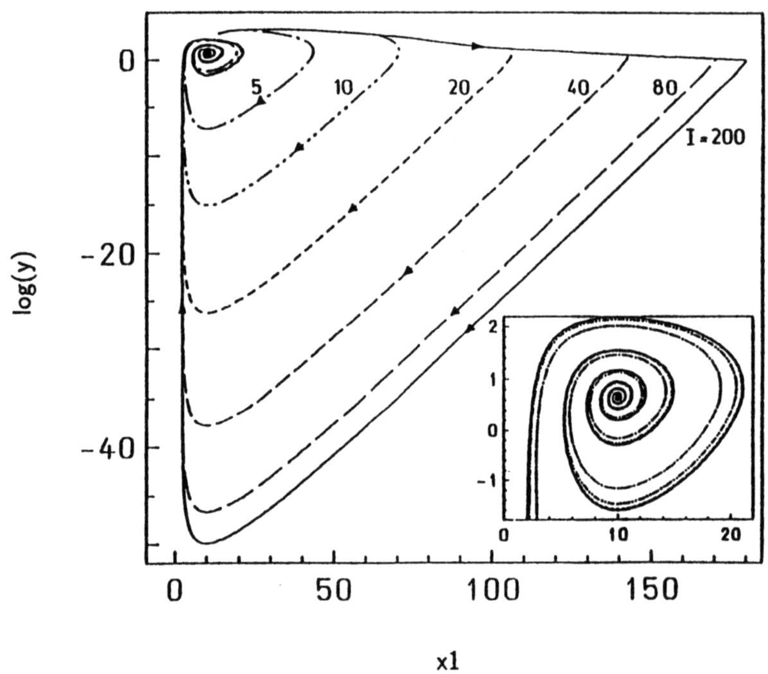

Fig. 5. Flow diagram for the specific stimulation therapy as in Fig. 4. (Since x_2 is nearly constant we show only the projection on the $x_1 - \ln y$ plane). The injections of antigen over a period I drive the system out of the fixed point of chronic infection (left upper corner) towards a new steady state (located in the right upper corner – though not shown) where the number of useful antibodies is orders of magnitude higher. After the injections the system moves on trajectories where the antigen decreases exponentially. For small y the antigen can be considered as extinct (Formally, however, the system returns to the stable fixed point, see also insertion). Parameters are as in Fig. 4 except for $\overline{m} = 0.01$ and $y_{in} = 5.0$.

Hemmen, Sulzer, 1991). We have found that there exists an optimal therapy; it can be favorable to dispense repeatedly a small dose of antigen rather than injecting a large dose once ('homeopathic' principle).

We remark that it is not necessary to use an antigen with its full virulence. The same effect could be reached for a weakened virulence or with a substitute with similar matching that does not reproduces itself at all. The only aim is to provoke a production of useful antibodies above the value of the fixed point of chronic infection.

The specific stimulation therapy, i.e. curing a chronic infection by provoking it to become acute, is well-known in medicine. It was applied, for instance, to tuberculosis before the era of antibiotics and chemotherapy (see, e.g., Alexander-Crespera, 1951). In theoretical immunology there exist models of unspecific stimulation (see, e.g., Marchuk, 1983). Our approach, however, is to the best of our knowledge the first attempt to explain a stimulation therapy in the context of a network theory.
We finally note that a desensibilisation therapy for allergic diseases could be explained in a similar vein.

5. DISCUSSION AND CONCLUDING REMARKS

We have seen that the model exhibits quite a reasonable behaviour — in spite of its crude simplifications. So we could think about $\{x_i\}$ and $\{y_i\}$ as a sort of 'lumped' variables describing the gross behaviour of subsystems the details of which we do not know or are not interested in. We may nevertheless ask whether the properties of the model remain stable if we include more details. Below we briefly discuss several extensions.

Separate treatment of lymphocytes and secreted antibodies. We have analyzed numerically the behaviour of small cycles in a scheme describing separately B cells, secreted antibodies, and antigens in a similar fashion to Eqs. (1) and (2). For reasonable parameter setting we have found (Behn et al., 1991) that the systems still have stable fixed points corresponding, e.g., to a healthy immunized state or to a chronic infection. The trajectories of lymphocytes and secreted antibodies closely follow each other, which — in a sense — justifies not distinguishing between them. (It is, however, just natural that the behaviour may as well become more complex as more details are included, see e.g. de Boer et al. (1990), Perelson (1989)).

Dynamics of memory cells and delay. We have shown that cycles are stabilized by memory cells. What however do the memory cells benefit from the cycles? Why is it not enough to have memory cells? To answer this questions we have to take into account that also memory cells are suspected to have a finite life-time and would die out if not the activity maintained by the cycle would ensure the replacement of memory cells.
In a more sophisticated version of our model we take into account both the dynamics of memory cells and their *finite* lifetime — finite, though much longer than that of 'normal' B cells. In doing so we should bear in mind that memory B cells appear only with a certain delay after the first encounter with the antigen. The delay is the time period

needed by a stimulated lymphocyte to proliferate. Only after several cell divisions the members of the clone differentiate into plasma cells and memory cells. It is further possible that 'new' memory cells could become activated only after some resting time (dormant cells). This would, though, only prolong the delay and does not give a new mechanism. (Therefore we use synonymously the terms dormant cells and memory cells).

In the above context the dynamics of a (symmetric) 2-cycle of antibodies x_1, x_2 with memory cells d_1, d_2 is given by

$$\dot{x}_1 = x_1\left[(1-\kappa)mx_2 - \gamma\right] + m\, d_1\, x_2\,, \tag{10}$$

$$\dot{d}_1(t) = -\delta d_1(t) + \epsilon m x_1(t-\tau)x_2(t-\tau) - m d_1(t)x_2(t)\,, \tag{11}$$

and the corresponding equations for x_2 and d_2. Here δ is the inverse life-time of memory cells, the second and third term on the right hand side of (11) describe the production of memory cells with a delay τ after initial stimulation of normal cells and the 'awakening' of stimulated memory cells, respectively.

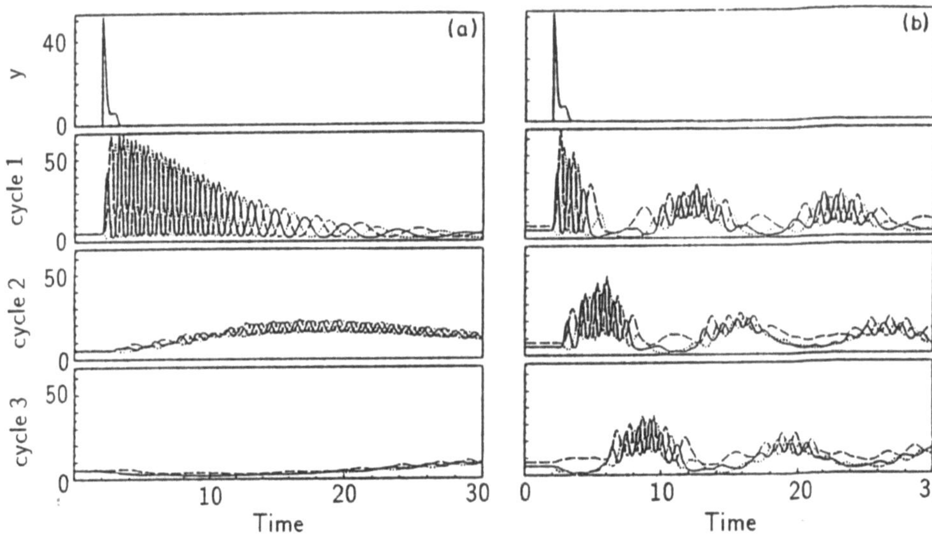

Fig. 6. Weakly coupled cycles. We consider three 3-cycles with intracycle matching $m = 0.2$ for weak intercycle matching $\tilde{m}$ where $\tilde{m}/m = 0.02$ (a), 0.1 (b). Antibody 3 in cycle k is recognized by antibody 1 of cycle $k-1$ ($k = 2,3$). Antibody 1 of cycle 1 matches with $\overline{m} = 0.2$ to the antigen which is injected at $t = 2$ ($y_{in} = 50$). Cycles 2 and 3 show in case (a) only a very modest response with considerable delay whereas in case (b) the time scale is faster and predator-prey oscillations are more pronounced. For $\tilde{m}/m = 1$ (not shown here) one observes a complex multifrequent response which does not allow to seperate the whole system into subsystems; the efficiency in fighting against the antigen decreases. ($\alpha = 1.0$, $\gamma = 0.01$, $\kappa = 1.01$, $d = 0.1$).

For delays within a reasonable interval we have shown analytically (Behn and van Hemmen, unpublished) that the 2-cycle still has a stable fixed point corresponding to a healthy immunized state. This provides an immune memory that last longer than the lifetime of its constituents.

Interaction with the network. Recently it has been argued (Varela and Coutinho, 1991) that only 10...20 percent of an adult's immune system constitutes a network whereas the remaining majority, the 'clonal part', is made of disconnected units responsible for the defense against invadors.
The picture of a 'sea' of weakly connected constituents with strongly connected 'islands' (cycles) and Perelson's gel-sol metaphor (1989) is not too different from these ideas.
To describe such a situation we propose an analytically tractable model which consists of K quasi-independent antibodies interacting via a matching of order m/K. In addition to the network we take single cycles coupling to one of the networks constituents or to the whole network. Here one can also include a noise term. The network then either functions as background or 'swallows' the cycle(s). Figure 6 illustrates that weakly interacting cycles are still well defined entities. There is no localisation/percolation dilemma (cf. Weisbuch et al., 1990).

Innovation and adaptation. After an, preferently analytic, understanding of the behaviour of small subsystems connected by good matching and weakly interacting with other constituents of the network, we can try to describe the dynamics of both the list of variables and parameters by embedding the Lotka-Volterra equations into a hierarchical scheme (cf. Farmer et al., 1986, Behn and van Hemmen, 1989a,b). Within such a scheme one could describe the appearance of new sorts of cells from the bone marrow replacing those which are not stimulated and die out (*innovation*) and the enhanced mutation during the cloning of stimulated cells which generates antibodies centered about the stimulated type (*adaptation*).

We finally mention that it was independently argued by Perelson (1989) that "memory may be carried by both static and dynamic means". Our results give some evidence that there is a synergy of both possibilities.

More details about the foundations of the model and on both analytical and numerical properties, a more extensive discussion of the literature and a more complete list of references may be found elsewhere (Behn et al., 1991).

ACKNOWLEDGMENT

We gratefully acknowledge most valuable discussions with Rob de Boer, Franco Celada, Geoffrey Hoffmann, Marcelline Kaufman, René Lefever, Alan Perelson and Gérard Weisbuch as well as helpful comments by Ingrid Behn. Thanks are due to Adrian Lange for his assistance in preparing the manuscript.

REFERENCES

Alexander-Crespera H (1951) Praktisches Lehrbuch der Tuberkulose, Johann Ambrosius Barth, Leipzig, pp 190-194

Behn U, Hemmen JL van (1989a) On the theory of networks for the immune system. In: Ebeling W, Peschel M (eds) Dynamical Networks, Akademie Verlag, Berlin, pp 162-172

Behn U, Hemmen JL van (1989b) Network description of the immune system: Dormant B cells stabilize cycles. J Stat Phys 56:533-545

Behn U, Hemmen JL van, Sulzer B (1991) Memory in the immune system: Network theory including dormant B cells. Submitted to J theor Biol

Boer RJ de (1989) Extensive percolation in reasonable idiotypic networks. In: Atlan H, Cohen IR (eds) Theories of Immune Networks. Springer, Berlin

Boer RJ de, Kevrekidis I, Perelson AS (1990) A simple idiotypic network model with complex dynamics. Chem Eng Sci 45:2375-2382

Boer RJ de, Perelson AS (1990) Size and connectivity as emergent properties of a developing immune system. J theor Biol 149:381-424

Farmer JD, Packard NH, Perelson AS (1986) The immune systeme, adaption and machine learning. Physica 22D:187-204

Jerne NK (1974) Towards a network theory of the immune system. Ann Immunol (Inst Pasteur) 125C:373-389

Jerne NK (1984) Idiotopic networks and other preconceived ideas. Immunol Rev 79:5-24

Marchuk GI (1983) Mathematical models in immunology. Optimization Software Inc Publications Division, New York

Perelson AS (ed) (1988) Theoretical Immunology, Parts I and II. In: Santa Fe Institute Studies in the Science of Complexity, Vol II and III. Addison-Wesley, Redwood City

Perelson AS (1989) Immune network theory. Immunol Rev 110:5-36

Segel LA, Perelson AS (1988) Computations in shape space. A new approach to immune network theory. In: Perelson AS (ed) Theoretical immunology, part two. Addison-Wesley, New York, pp321-344

Varela FJ, Coutinho A (1991) Second generation immune networks. Immunol Today 12:159-166

Weisbuch G, Boer RJ de, Perelson AS (1990) Localized memories in idiotypic networks. J theor Biol 146:483-499

IV. From Molecular to Cellular Immunology

IMPLICATIONS OF COOPERATIVITY AND DIMENSIONAL COMPLEXITY FOR IMMUNOLOGICAL RECOGNITION

Neil S. Greenspan and Laurence J. N. Cooper
Institute of Pathology
Case Western Reserve University
2085 Adelbert Road
Cleveland, Ohio 44106
United States of America

INTRODUCTION

Concept clarification can play an important facilitating role in the development of theoretical understanding (Mayr). In this chapter, our intent is to explore, and clarify, conceptual issues pertaining to immunological recognition. The first section offers an analysis of the concept of epitope (and related concepts), and the second section reviews our evidence for cooperative binding to multivalent antigen by murine IgG3 antibodies and considers the potential implications of this phenomenon for our understanding of the structural determinants of antibody specificity. We contend that several features of the relationship between antibody and antigen, or more generally, receptor and ligand, defy completely satisfactory description and explanation using standard assumptions, language, and concepts. Specifically, we argue that there are different aspects of epitopeness, including: noncovalent contact, determination of the energy of interaction, and determination of differential energies of interaction between one receptor and two or more ligands. These different dimensions of epitopeness have at times been confounded, and we delineate how the structural correlates of these various aspects of molecular recognition differ. In addition, we identify three key properties of ligand-receptor relationships and discuss the implications of these properties for our conception of epitopes.

NATO ASI Series, Vol. H 66
Theoretical and Experimental Insights into Immunology
Edited by A. S. Perelson and G. Weisbuch

These three properties are: 1) cooperativity -the functional contribution associated with a given structural element is often influenced by the nature of other structural elements (i.e, structure-function correlations are context-dependent), 2) hierarchy - the components of an epitope are likely to vary in the degree to which they contribute to the epitope, and, of equal importance, different hierarchies may be required for different aspects of epitopeness, and 3) multiple levels of resolution - a statement that adequately characterizes a bimolecular interaction at a low level of resolution (e.g., constituent amino acid residues) may not be adequate at a higher level of resolution (e.g., atomic contacts).

THE NATURE OF EPITOPES

The portion of an antigen molecule that is specifically recognized by an antibody combining site, or paratope, is generally referred to as an antigenic determinant, or epitope. One approach to investigating the meaning of a statement is to consider what expressions can be used to replace that statement (Quine). Therefore, we might ask what expression can be used to replace, for example, "Hen egg-white lysozyme expresses the D1.3 epitope." A reasonable answer might be a statement such as, "Hen egg-white lysozyme is bound by monoclonal antibody D1.3 with an affinity of 4.5 x 10^7 M^{-1}" (Harper *et al.*). This superficially inconsequential exchange is revealing, since we have replaced a statement (the first) which tends to foster the belief that an epitope can be defined as a structure independent of the antibody with an expression (the second) which makes clear that identification of an epitope is dependent on the interaction with the defining complementary structure (paratope). Thus, epitope is fundamentally a relational concept (Van Regenmortel).

If one accepts that reference to a given epitope is really a statement characterizing a relationship between two molecules, then it becomes conceivable to imagine that two structurally distinct ligands are able to interact with a given receptor with comparable affinity constants. For example, if an unknown molecule, distinct from hen egg-white lysozyme, was found to bind to the monoclonal antibody D1.3 with an affinity close to $4.5 \times 10^7 M^{-1}$, under the same physical conditions as were applicable to the measurement of the D1.3-lysozyme interaction, then no certain inferences about the structural features of this new ligand for D1.3 could be drawn. Thus, the kind of operational statement that is ultimately required to make reference to an epitope permits degeneracy in the relationship between structure and binding. This argument is not intended to deny a unique structural basis for the interaction of a particular ligand, such as hen egg-white lysozyme, with a particular receptor, such as D1.3. However, it is designed to draw attention to the difficulties of language used in describing empirical observations on ligand-receptor interactions. Stating that our new D1.3 ligand, from above, "expresses the D1.3 epitope," could be expected to lead some readers to unwarranted expectations about the structural relationships between hen egg-white lysozyme and this other ligand for D1.3.

It should be noted, in the context of the above argument, that identical affinity constants can accommodate a range of association and dissociation rate constants, constrained by the fact that the ratio of the two rate constants remains identical. Thus, the above argument can be recast, at what could be regarded as a higher level of resolution, by focusing on kinetic constants as opposed to equilibrium constants.

The occurrence of degeneracy in molecular recognition can be understood if several factors, often given insufficient attention, are considered. First, molecules, especially macromolecules, cannot, in the most general case, be regarded as static structures for questions at high levels of resolution (Karplus and McCammon). Theoretical studies on molecular dynamics suggest that molecular fluctuations are likely to have important influences on many functional activities, including molecular recognition (McCammon and Harvey), and empirical studies provide strong evidence in favor of this contention. For example, recent studies have demonstrated that there are striking conformational changes associated with the binding of cyclosporin A to cyclophilin (Wuthrich *et al.*; Jorgensen), and in the binding of insulin to the insulin receptor (Hua *et al.*). Second, solvent molecules can play pivotal roles in ligand-receptor interactions (Quiocho *et al.*). For example, it is conceivable that structural changes in either ligand or receptor, that eliminate hydrogen bonds between ligand and receptor, can be compensated for by rearrangements of bound water molecules such that one ligand-receptor hydrogen bond is replaced by a pair of ligand-water and water-receptor hydrogen bonds. Third, the exact role of a given atomic group in the ligand or the receptor can vary in subtle ways, so that the same substituent in two slightly different molecules may not contribute equivalently to interactions with the same complementary structure. Examples of cooperative intramolecular interactions have been described for several types of macromolecules, including: antibody (Chien *et al.*), DNA (Lesser *et al.*), and DNA-binding proteins (Jordan and Pabo). In each of these cases, interactions between non-contact residues and contact residues influence the contributions of the contact residues to a ligand-receptor interaction.

In the primary literature pertaining to epitopes, it is unusual to find an explicit statement as to what, in principle, would constitute a complete description of an epitope. Based on the standard definitions in textbooks, a complete definition of an epitope would amount to a list of contact residues (or atoms) with the corresponding spatial coordinates. It has been suggested (Laver *et al.*) that crystallographic analysis of a complex of protein antigen and antibody Fab fragment can provide the information required for such a complete epitope characterization. If one defines a protein epitope as an undifferentiated list of contact amino acids, then methods for the determination of three-dimensional molecular structure, such as crystallography, are currently the only approaches potentially capable of yielding a complete description of an epitope. However, one can argue that even crystallography does not necessarily provide a complete description of an epitope. For instance, the classification of amino acid residues, on antigen or antibody, as contact or non-contact residues involves a subjective or arbitrary element (Getzoff *et al.*; Laver *et al.*). There is no consensus for the exact interatomic distances that should be taken to indicate noncovalent interaction, nor is there a standard rule for including or excluding amino acid residues as a constituents of an epitope if they contact paratopic residues only indirectly through shared hydrogen bonds to an intermediary water molecule. Furthermore, it is highly likely that some contact residues are more critical to the binding energy of the paratope-epitope interaction (Novotny *et al.*). This point is complicated, however, by the possibility that the relative importance of a particular residue, or set of residues, depends on whether one is concerned primarily with the energetics of the particular epitope-paratope interaction under consideration, or with accounting for the differences in the energies of interaction for two different epitopes with the same paratope. Thus, the ideal description of an epitope should be not just a list of contact residues (including spatial information), but a <u>hierarchical list</u>. If one allows that different hierarchies can be applied to the same epitope, depending on whether that epitope is being analyzed in isolation or in comparison with one or more other epitopes, the likelihood that crystallography alone can provide a complete description of a protein epitope diminishes still further. A final limitation of crystallography relates to the flexible, dynamic nature of proteins (Karplus and McCammon), such that the range of conformations through which a given epitope oscillates is not fully captured by standard crystallographic methods (see above).

Based on the above comments it should be clear why there is a lack of correspondence between the standard definition of epitope in textbooks of immunology and the usage of the term in the experimental literature. While the standard definition focuses on contact residues, the majority of studies aimed at characterizing epitopes use methods that can never identify contact residues with certainty. These methods can yield information of great value, however, as there are many reasons for attempting to characterize an epitope. In many instances (e.g., vaccine development), the central issue is to determine what primary structural differences between two molecules account for differential reactivity with a single paratope. Although, in many instances, differential reactivity of two ligands for a given receptor can be attributed to differences in the contact residues associated with the two ligands, and in some cases may solely involve differences in contact residues, it is not necessarily the case that differences in binding are solely due to differences in the identities of contact residues. As already noted, changes in non-contact residues can affect the interactions of contact residues (cooperativity of recognition). Thus, mutational analysis is highly appropriate for some cases of epitope or paratope characterization, and crystallographic information is not always essential (although certainly useful).

In conclusion, there are at least three key aspects of epitopeness: 1) the structural basis for intermolecular contact, 2) the structural basis for the energetics of a single ligand-receptor interaction, and 3) the structural basis for the difference in energies of interaction between, for example, one paratope and two or more epitopes. While X-ray crystallography, and, more recently, nuclear magnetic resonance spectroscopy, can contribute to the determination of macromolecular structure and the direct identification of contact atoms (aspect 1), these techniques alone cannot provide complete information on aspects 2 and 3. Biochemical and genetic methods are necessary, at least in some ultimate sense, for the elucidation of questions pertaining to energetics and specificity. Furthermore, it may not be possible, at least in some cases, to uniquely assign energetic contributions to given atoms or groups of atoms, as elegantly demonstrated for the complex of Eco RI and the cognate DNA sequence (Lesser *et al.*).

INTERMOLECULAR COOPERATIVITY AND ANTIBODY BINDING

Immunologists have recognized for many years that the strength of antibody binding to multivalent antigen (functional affinity) is dependent on the intrinsic affinity (equilibrium association constant characterizing the interaction between one epitope and one paratope) of the relevant variable modules (heavy chain variable domain plus light chain variable domain) and the number of such variable modules that can simultaneously interact with the antigenic surface (Karush). This latter property has been thought to depend primarily on the structural valence of a given immunoglobulin isotype and the segmental flexibility exhibited by that isotype (Crothers and Metzger). For example, an IgG molecule, with two variable modules covalently linked, has a structural valence of two, but in interaction with a multivalent antigen carrying widely-spaced epitopes, only one or the other of the variable modules might be able to bind at a given time, especially for a relatively inflexible IgG subclass. Thus, according to this perspective, IgG antibodies of different subclasses, expressing identical intrinsic affinities (identical variable domains), and varying in segmental flexibility, due to structural differences in the hinge or C_H1 domains (Schneider *et al.*), could exhibit significant differences in binding to multivalent antigen. In spite of this theoretical reason to anticipate the existence of

IgG subclass-associated differences in binding to multivalent antigen, however, there is little empirical evidence to confirm this expectation, and investigators often assume that IgG antibodies expressing identical variable modules but different heavy chain constant domains (different subclasses), will bind identically to multivalent antigen surfaces (Bindon *et al.*).

In contrast, we have described variable module-identical murine IgG antibodies, specific for the cell wall polysaccharide of group A streptococci (GAC) and representing three of the four murine IgG subclasses, that do exhibit significant differences in binding to multivalent antigen (Cooper *et al.*). In our studies, the IgG3 monoclonal antibody (mAb) binds more strongly than IgG1 and IgG2b mAbs of the same specificity (Figure 1).

Differences in segmental flexibility among the mAbs do not appear likely to explain all of the subclass-associated differences in strength or mode of binding. First of all, available measurements of murine IgG segmental flexibility suggest that while IgG2b is significantly more flexible than IgG3, IgG1 is just slightly less flexible than IgG3 (Dangl *et al.*). Thus, if the binding differences are due solely to differences in segmental flexibility, it is necessary to postulate that IgG3 has the optimum degree of flexibility for the antigens used in these studies, and that the hypothetical curve relating binding to flexibility has a very sharp optimum, given the similarity in flexibilities for IgG1 and IgG3.

A stronger argument against the ability of segmental flexibility differences to fully explain the binding differences among our three group A carbohydrate-specific mAbs relates to the ability of the IgG3 mAb, but not the IgG1 or IgG2b mAbs, to bind cooperatively to the bacteria (Figure 2). The operational definition for cooperative binding in this context is the ability of unlabeled group A carbohydrate-specific antibody to

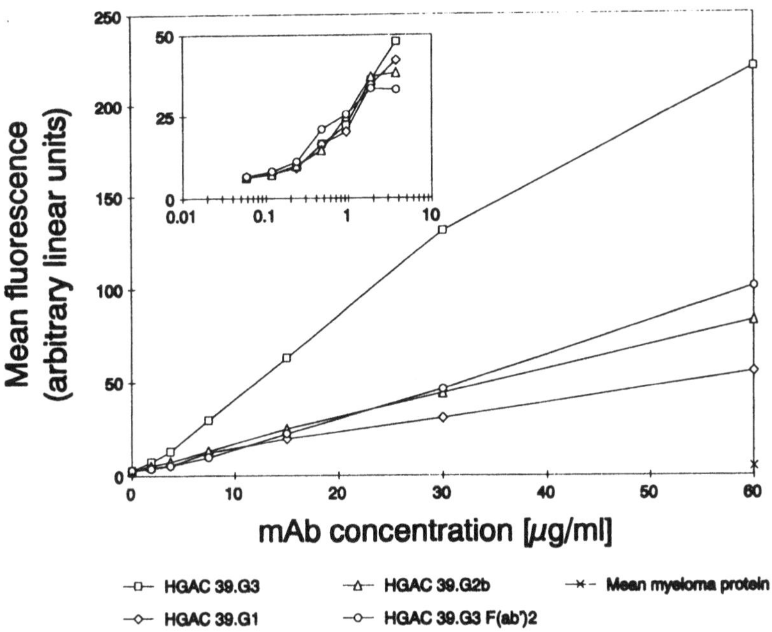

Figure 1. Concentration dependence of mean binding (arbitrary fluorescence units) of group A carbohydrate-specific mAbs to heat-killed, pepsin-treated group A intermediate streptococci or anti-mouse kappa-coated beads (inset) by indirect flow cytometry with FITC-labeled polyclonal goat anti-mouse kappa antibodies, at 37°C. Each point represents the mean fluorescence of approximately 20,000 fluorescent events. The 99% confidence intervals were smaller than the symbols used to plot the data points. Reproduced, slightly modified, with permission from Cooper *et al.* The Journal of Immunology, Figure 1b, 146:2659-2663, 1991.

enhance the binding of radiolabeled group A carbohydrate-specific antibody. As illustrated in Figure 2, the only unlabeled-radiolabeled antibody pair that exhibits substantial enhancement is IgG3-IgG3. While one might be able to explain the relative lack of enhancement for IgG1 or IgG2b at least in part on the basis of non-optimal segmental flexibility, it is not possible to account for the IgG3-IgG3 enhancement phenomenon itself on the basis of an optimum flexibility.

How can we account for the ability of IgG3 mAbs to bind cooperatively to the bacteria? Our working hypothesis is that the phenomenon is the result of intermolecular cooperativity.

In other words, we hypothesize that IgG3 molecules adjacent to each other on the antigen surface engage in noncovalent

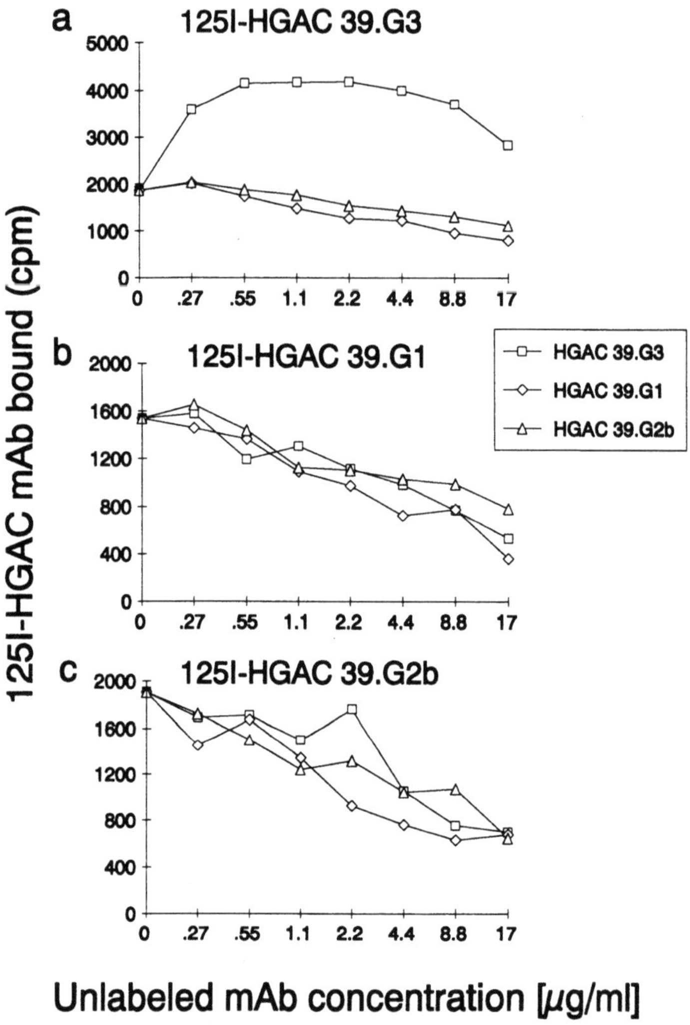

Figure 2. The effect of unlabeled group A carbohydrate-specific mAbs on the binding (cpm) of radiolabeled mAbs to solid-phase heat-killed, pepsin-treated group A intermediate streptococci, at room temperature. Reproduced with permission from Cooper *et al.* The Journal of Immunology, Figure 4, 146:2659-2663, 1991.

interactions with one another, thereby stabilizing the complex of antibody with antigen. It is essentially this same sort of mechanism that is believed to play a crucial role, through protein-protein interactions, in the assembly of transcriptional regulatory complexes (Ptashne; Mitchell and Tjian). Since, unlabeled IgG3-derived $F(ab')_2$ fragments, of appropriate specificity, do not enhance the binding to bacteria of intact radiolabeled IgG3 antibody, and intact unlabeled IgG3 fails to enhance the binding radiolabeled IgG3-derived $F(ab')_2$ fragments (Greenspan *et al.*, 1987; Greenspan *et al.*, 1988; Greenspan *et al.*, 1989), we believe that it is likely that the noncovalent contacts are between Fc regions. In support of these conjectures, there is abundant evidence that IgG3 molecules, and IgG3-derived Fc regions in particular, are self-associating (Grey *et al.*; Abdelmoula *et al.*).

Several questions are raised by the foregoing. First, can these results be incorporated into a biological framework? Second, how do these results extend our conceptions of the mechanistic basis of antibody binding to antigen? Third, what are the implications of the hypothesized cooperative binding mechanism for the relationship between structure (of antibody or antigen) and specificity of binding.

A biological context for our observations is provided by the following points: 1) mice synthesize IgG antibodies predominantly of the IgG3 subclass in response to many bacterial polysaccharides (Perlmutter *et al.*), 2) anti-carbohydrate antibodies typically exhibit low intrinsic affinities (10^3 - 10^6 M^{-1}) in comparison with anti-protein antibodies (10^5 - 10^{10} M^{-1}) (Karush), and 3) bacterial cell wall and capsular polysaccharide antigens typically exhibit high epitope densities (e.g., Schalch *et al.*). Therefore, our primary working hypothesis is that cooperative binding represents an adaptation permitting better binding of low intrinsic affinity antibody to high epitope-density surfaces. Other, not necessarily mutually exclusive, explanations are also conceivable (Greenspan and Cooper, submitted).

As pointed out at the beginning of this section, the traditional view of the relationship between the strength of antibody binding to multivalent antigen and antibody structure does not take into account structural variation in the C_H2 and C_H3 domains (Fc region). In fact, there may be many instances where such variation is of no consequence to antigen binding, but we have now identified one system where such structural differences do significantly influence the interaction with antigen, in the form of bacteria. This result illustrates the principle, discussed in the preceding section of this chapter, that many mechanistic pathways can lead to the same outcome, or that different structural solutions can be found to accommodate a given functional demand. The relevance of this conclusion may extend to applications of antibodies, as well as to understanding the physiology of antibody-mediated immunity in mice.

The essence of the concept of molecular specificity is discrimination, as, for example, by a receptor between two or more ligands. This discrimination can be expressed quantitatively as a ratio of equilibrium association constants characterizing the interactions between the receptor and the respective ligands. Two receptors could be said to express different specificities if they are characterized by different ratios of affinity constants for the same pair of ligands. In our studies on IgG mAbs specific for group A carbohydrate, we have found that the IgG3 mAb binds better than the IgG1 and IgG2b mAbs to a number of different targets, but the magnitude of the difference varies with the target (Cooper *et al.*, manuscript in preparation). The most significant variation among the targets is probably that related to epitope spacing; at lower epitope densities, the magnitude of the IgG3-IgG1/IgG2b difference decreases (Cooper *et al.*, manuscript in preparation). Therefore, in a potentially biologically-relevant sense, the intact IgG3 antibody expresses a different specificity than the

IgG1 and IgG2b antibodies despite the fact that all three antibodies contain identical variable modules.

Two important conclusions follow from the above argument. First, although the variable domains are certainly the chief structural determinants of antibody specificity, they are not necessarily the only structural elements that can influence specificity. Another way of stating this point is that structural alterations in heavy chain constant domains can potentially affect relative functional affinities for different antigens. Second, specificity for antigen can be affected by epitope spacing as well as by epitope structure. Therefore, one might argue that the definition of immunological self can depend not only on what epitopes are expressed but also on what spatial relationships the expressed epitopes bear to one another. Recently, the possibility of protein specificity being dependent on the spacing of specific sites, as opposed to the fine structure of those sites, as been demonstrated for sequence specific DNA-binding proteins involved in the regulation of transcription (Naar *et al.*; Luisi *et al.*).

CONCLUSIONS

Common themes uniting the two preceding sections are: 1) the accepted language and concepts used in discussions of immunological recognition conceal underlying logical and empirical complexity, and 2) the correlation of receptor or ligand structure with the potential for molecular recognition can serve as one example of the much broader, and very fundamental, activity of structure-function correlation for biological macromolecules. An important conclusion based on both sections is that the standard ways of discussing the relationship between molecular structure and molecular interaction create an expectation that there is a one-to-one relationship between structural elements and functional properties. Rigorous analysis demonstrates that frequently, the functional meaning of a given structural element depends, in part, on the structural environment in which it is embedded, and therefore, there is frequently a many-to-many relationship between structure and function. Furthermore, the degree to which cooperative interactions (context-dependent effects) and hierarchical relationships among components of a structure are apparent may depend on the level of resolution of the analysis.

References:

Abdelmoula M, Spertini F, Shibata T, Gyotoku Y, Luzuy S, Lambert P-H, Izui S (1989) IgG3 is the major source of cryoglobulins in mice. J Immunol 143:526-532

Bindon CI, Hale G, Bruggemann M, Waldmann H (1988) Human monoclonal IgG isotypes differ in complement activating function at the level of C4 as well as C1q. J Exp Med 168:127-142

Chien NC, Roberts VA, Giusti NM, Scharff MD, Getzoff ED (1989) Significant structural and functional change of an antigen binding site by a distant amino acid substitution: proposal of a structural mechanism. Proc Natl Acad Sci USA 86:5532-5536

Cooper LJN, Schimenti JC, Glass DD, Greenspan NS (1991) H chain domains influence the strength of binding of IgG for streptococcal group A carbohydrate. J Immunol 146:2659-2663

Crothers DM, Metzger H (1972) The influence of polyvalency on the binding properties of antibodies. Immunochemistry 9:341-357

Dangl JL, Wensel TG, Morrison SL, Stryer L, Herzenberg LA, Oi VT (1988) Segmental flexibility and complement fixation of genetically engineered chimeric human, rabbit and mouse antibodies. EMBO J 7:1989-1994

Getzoff ED, Tainer JA, Lerner RA (1988) The chemistry and mechanism of antibody binding to protein antigens. Adv Immunol 43:1-98

Greenspan NS, Monafo WJ, Davie JM (1987) Interaction of IgG3 anti-streptococcal group A carbohydrate (GAC) antibody with streptococcal group A vaccine: enhancing and inhibiting effects of anti-GAC, anti-isotypic, and anti-idiotypic antibodies. J Immunol 138:285-292

Greenspan NS, Dacek DA, Cooper, LJN (1988) Fc region-dependence of IgG3 anti-streptococcal group A carbohydrate antibody functional affinity. I. The effect of temperature. J Immunol 141:4276-4282

Greenspan NS, Dacek DA, Cooper LJN (1989) Cooperative binding of two antibodies to independent antigens by an Fc-dependent mechanism. FASEB J 3:2203-2207

Grey HM, Hirst JW, Cohn M (1971) A new mouse immunoglobulin: IgG3. J Exp Med 133:289-304

Harper M, Lema F, Boulot G, Poljak RJ (1987) Antigen specificity and cross-reactivity of monoclonal anti-lysozyme antibodies. Molec Immunol 24:97-108

Hua QX, Shoelson SE, Kochoyan M, Weiss MA (1991) Receptor binding redefined by a structural switch in a mutant human insulin. Nature 354:238-241

Jordan SR, Pabo CO (1988) Structure of the lambda complex at 2.5 A resolution: details of the repressor-operator interactions. Science 242:893-899

Jorgensen WL (1991) Rusting of the lock and key model for protein-ligand binding. Science 254:954-955

Karplus M, McCammon JA (1983) Dynamics of proteins: elements and function. Ann Rev Biochem 53:263-300

Karush F (1978) The affinity of antibody: range, variability , and the role of multivalence. In: Litman GW, Good RA (eds) Immunoglobulins, Plenum Publishing Corporation, New York

Laver WG, Air GM, Webster RG, Smith-Gill SJ (1990) Epitopes on protein antigens: misconceptions and realities. Cell 61:553-556

Lesser DR, Kurpiewski MR, Jen-Jacobson L (1990) The energetic basis of specificity in the Eco RI endonuclease-DNA interaction. Science 250:776-786

Luisi BF, Xu WX, Otwinowski Z, Freedman LP, Yamamoto KR, Sigler PB (1991) Crystallographic analysis of the interaction of the glucocorticoid receptor with DNA. Nature 352:497-505

Mayr E (1988) Toward a new philosophy of biology: observations of an evolutionist, The Belknap Press, Cambridge, MA

McCammon JA, Harvey SC (1987) Dynamics of proteins and nucleic acids, Cambridge University Press, Cambridge

Mitchell PJ, Tjian R (1989) Transcriptional regulation in mammalian cells by sequence-specific DNA binding proteins. Science 245:371-378

Naar AM, Boutin J-M, Lipkin SM, Yu VC, Holloway JM, Glass CK, Rosenfeld MG (1991) The orientation and spacing of core DNA-binding motifs dictate selective transcriptional responses to three nuclear receptors. Cell 65:1267-1279

Novotny J, Bruccoleri RE, Saul FA (1989) On the attribution of binding energy in antigen-antibody complexes McPc 603, D1.3, and HyHel-5. Biochemistry 28:4735-4749

Perlmutter RM, Hansburg DH, Briles DE, Nicolotti RA, Davie JM (1978) Subclass restriction of murine anti-carbohydrate antibodies. J Immunol 121:566-572

Ptashne M (1988) How eukaryotic transcriptional activators work. Nature 335:683-689

Quine WV (1987) Quiddities: an intermittently philosophical dictionary, The Belknap Press of Harvard University Press, Cambridge, Massachusetts

Quiocho FA, Wilson DK, Vyas NK (1989) Substrate specificity and affinity of a protein modulated by bound water molecules. Nature 340:404-407

Schalch W, Wright JK, Rodkey LS, Braun DG (1979) Distinct functions of monoclonal IgG antibody depend on antigen-site specificities. J Exp Med 149:923-937

Schneider WP, Wensel TG, Stryer L, Oi VT (1988) Genetically engineered immunoglobulins reveal structural features controlling segmental flexibility. Proc Natl Acad Sci USA 85:2509-2513

Van Regenmortel MHV (1989) Structural and functional approaches to the study of protein antigenicity. Immunol Today 10:266-272

Wuthrich K, von Freyberg B, Weber C, Wider G, Traber R, Widmer H, Braun W (1991) Receptor-induced conformation change of the immunosuppressant cyclosporin A. Science 254:953-954

Shannon Information as a Measure of Amino Acid Diversity

Samuel Litwin
Fox Chase Cancer Center
Philadelphia, PA, USA
and
Rita Jores
Institut Pasteur
Paris, France

Abstract:

The Wu-Kabat (KW) measure of diversity is intended to relate antigen binding structure to residue position. Jores, et al (1) (JAM) generalized KW to resolve hypervariable regions in T-cell receptor β chains. This set the stage for further improvements in diversity measurement. Use of Shannon information improves diversity measurement in several important ways. Several drawbacks to the KW measure, and also the JAM measure are:

1. The sampling variability of these measures is not available, so statistical comparison of diversity among residue sites is not done.

2. Both KW and JAM are unstable in the sense that a single new observation can cause substantial jumps in their values.

3. Neither KW nor JAM can represent the scope of amino acid diversity since they do not account for the proportions of all twenty amino acids.

The Shannon measure (H) addresses 1) greatly reduces 2) and is not subject to 3). We obtain the variance of H along with that of KW and JAM, and show that the coefficient of variation of H is much smaller than that of the other measures.

Knowledge of the variance at each site permits statistical assessment of differences in local diversity. For example, rapidly alternating peaks and valleys resembling noise might indicate structural properties present in a family of sequences. The Shannon measure is capable of being used globally as well as locally because it is additive. That is, if $H(X)$ is the diversity of locus X and $H(Y)$ is that of Y, and if X and Y are independent, then the diversity of X and Y taken together is $H(X) + H(Y)$. As a result, diversity may be assigned not only to a site but also to a family of sequences. Dependent sites require special treatment.

The Shannon measure is obtained by first estimating the multinomial proportions of the 20 amino acids. These estimates, $p_1, \ldots, p_{20}$ are then used to find $-\sum_{i=1}^{20} p_i \log_2(p_i)$. The estimated probabilities lead to a direct assessment of the variance of the Shannon measure and to a Monte Carlo method for finding the variance of any measure dependent on the multinomial. Graphs, in the *KW* format, of the three measures for mouse V_κ chains including 95% confidence bounds are presented. The Shannon measure shows a much broader distribution of diversity than either the *KW* or *JAM* measures for the mouse data. This and its lower noise level may be of importance in relating diversity to protein structure. Its usefullness extends to identification of conserved as well as highly diverse residues.

NATO ASI Series, Vol. H 66
Theoretical and Experimental Insights into Immunology
Edited by A. S. Perelson and G. Weisbuch

The Kabat-Wu, JAM, Shannon, and Nei measures of amino acid diversity:

Wu and Kabat (3) introduced a measure of amino acid diversity in 1970 in an attempt to relate antibody function to structure.

The Kabat-Wu measure (3) is defined as follows:
n = sample size, number of sequences being discussed,
d_1 = the number of distinct amino acids in the sample,
f_i = the number of occurrences of amino acid i, and

$$f_{[1]} = \max_{i=1}^{20} f_i$$

Then,

$$KW = d_1/(f_{[1]}/n)$$

and since $1 \leq d_1 \leq 20$, $1 \leq KW \leq 400$. Wu and Kabat gave their measure in terms of direct observation. They did not think of it as an estimate for a population parameter; however, we cast its limiting value for large samples into that role, namely

$$\lim_{n\to\infty} KW = k/p_{[1]}$$

where k is the number of amino acids in the entire population under consideration, and $p_{[1]}$ is the probability associated with the most prevalent amino acid. If we knew k and $p_{[1]}$ we could say we know KW exactly. Since we do not, the definition should be regarded as an estimate for the true, but unknown, population value.

JAM measure:

Jores et al (1) generalized the Kabat Wu measure in an attempt to sharpen its discrimination for peaks among the residue sites of TCR beta chains. Their measure, like KW, is defined in terms of observable quantities; namely, let
d_2 = The number of amino acids that appear at least twice, and
$f_{[2]}$ = The second highest frequency of an amino acid. Then,

$$JAM = \frac{\binom{d_1}{2} + d_2}{\max\left[\binom{f_{[1]}}{2}, f_{[1]}f_{[2]}\right] / \binom{n}{2}}$$

Again, the population parameter being estimated is associated with the limiting value of their definition, which, in turn is taken as an estimate of this limit. $p_{[2]}$ is defined analogously to $p_{[1]}$.

$$\lim_{n\to\infty} JAM = \frac{\binom{k}{2} + k}{p_{[1]} \max(p_{[1]}, 2p_{[2]})}$$

Nei Measure:

Nei (4) defined a measure of diversity for a different purpose. The Nei measure is the probability that two randomly selected sequences will have different amino acids at a particular site.

$$Nei = 1 - \sum_{i=1}^{20} p_i^2$$

It serves here to introduce the set of multinomial probabilities $p_1, \cdots, p_{20}$ which are the probabilities associated with selection of the twenty amino acids at a particular site. The Nei measure is defined directly in terms of these population parameters. The Kabat Wu measure depends on only one of the multinomial probabilities, the JAM only two. Given any subset of the multinomial parameters, say $p_1, \cdots, p_j, j < 19$, two populations can differ in diversity if their remaining $p_i, j < i \leq 20$, differ. Hence a measure of diversity should take all the p_i into account. An unbiased estimate of the Nei measure is:

$$\hat{N} = 1 - \sum_{i=1}^{20} \frac{k_i(k_i - 1)}{n(n-1)}$$

where k_i is the number of occurrences of the ith amino acid.

Shannon measure:

The Shannon measure, (2), of information, H, or entropy as it is often called, was created to describe the complexity or diversity of a message source. It is defined in terms of the multinomial parameters p_i and, like the Nei measure, it must be estimated from the data.

$$H = -\sum_{i=1}^{20} p_i \log_2 p_i$$

where $0 \leq H \leq 4.322$ bits for 20 categories.

An important property of the Shannon measure is the fact that it can be used to summarize the diversity of sets of residue sites if assignments of amino acids to these sites are independent.

Specifically, if $H(X)$ is the diversity of site X, and $H(Y)$ is that of site Y, then $H(X \otimes Y) = H(X) + H(Y)$ if X and Y are independent. The symbol $\otimes$ means the cartesian product of the sample spaces. This property starts us thinking about the possibility of discussing the diversity of families of proteins, not just structurally motivated local diversity.

Additivity gives the Shannon measure its intuitive appeal. One would like to feel that two books contain twice as much information as one and this is true of the Shannon measure. Background for this measure is discussed in Shannon's book (2).

A good estimate of H, for samples of size $n \geq 100$, is obtained by replacing the p_i by their maximum likelihood estimates k_i/n. Namely:

$$\hat{H} = -\sum_{i=1}^{20} (k_i/n) \log_2(k_i/n)$$

Calculation of the mean and variance of this estimate of H may be reduced to operations requiring only the binomial or trinomial distribution rather than the complete multinomial. By this means the bias of the estimator may be determined, as well as approximate error bounds. Letting $t_i = (k_i/n)\log_2(k_i/n)$ we have:

$$\hat{H} = -\sum_{i=1}^{20} t_i$$

$$E(\hat{H}) = -\sum_{i=1}^{20} E(t_i)$$

and

$$var(\hat{H}) = \sum_{i=1}^{20} var(t_i) + 2\sum_{i=1}^{19} \sum_{j=i+1}^{20} cov(t_i, t_j)$$

The mean and variance of t_i may be obtained by use of the binomial distribution $b(k, n, p_i)$. The covariance of t_i with t_j is found using the trinomial:

$$tri(r, s, p_i, p_j) = \begin{pmatrix} & n & \\ r & s & n-r-s \end{pmatrix} p_i^r p_j^s (1 - p_i - p_j)^{n-r-s}$$

Thus:

$$E(t_i) = \sum_{k_i=0}^{n} t_i b(k_i, n, p_i)$$

with a similar expression for $E(t_i^2)$. Then $var(t_i) = E(t_i^2) - (E(t_i))^2$. Finally:

$$E(t_i t_j) = \sum_{r_i=0}^{n} \sum_{s_j=0}^{n-r_j} t_i t_j \, tri(r_i, s_j, p_i, p_j)$$

and

$$cov(t_i, t_j) = E(t_i t_j) - E(t_i)E(t_j)$$

Rank order disagreement among measures:

The several diversity measures will not agree as to the rank order of diversity of a series of amino acid populations, as the following example shows.

Let S_1, S_2 be sets each containing only two amino acids, and suppose that their population fractions are as in the table:

Set	$p_{[1]}$	$p_{[2]}$	*KW*	*JAM*
S_1	1/2	1/2	4	6
S_2	2/3	1/3	3	6.75

Thus: $KW(S_1) > KW(S_2)$ and $JAM(S_1) < JAM(S_2)$. A similar conclusion can be reached for the Shannon and Nei measures.

A new measure of diversity may cause some distortion in currently held views of diversity. Hence it should have a great deal to recommend it.

Advantages of the Shannon measure:

Shannon information allows direct comparison of two diversities. One can ask if one residue position is twice as diverse as another using this measure. In the case of several independent positions, the total diversity is simply the sum of the separate diversities. No corresponding property exists for any of the other measures discussed here. Advantages of the Shannon measure are summarized below:

1. It uses the information fully, i.e. it depends on the probabilities of all twenty amino acids. The KW and JAM measures do not.

2. It has a very low coefficient of variation compared to the KW or JAM measures.

3. It is additive over independent residue positions. None of the other measures have this property.

4. It is a ratio scale.

5. Simple exact computation of its mean and variance exist. This is not true of the KW or JAM measures.

Behavior of the four measures for a family of multinomial distributions:

All the above diversity measures are invariant to permutations of the multinomial proportions $p_1, \ldots, p_{20}$. Therefore we consider only those multinomial distributions with $p_1 \geq p_2 \geq \cdots \geq p_{20}$. Figure 1 shows a subset of 16 such distributions. These illustrate a set of distributions obtained from the curve $y = (1-x)^{power}$ for $0 \leq x \leq 1$ by taking $p_i = Cy(i/20)$ and adjusting C to normalize the result. In the figure $0 \leq power \leq 30$ and may vary continuously. The resulting set of distributions constitute a representative swath through the space of all multinomial distributions. With *power* as a parameter figure 2 shows the relative values of the four measures over the swath of multinomial distributions. KW and JAM filter out highly diverse distributions, while Nei only changes by a factor of two in this range of the parameter. We include the Nei measure here only for comparison. Since it does not have the additive property of the Shannon measure, we do not consider it further.

A Monte Carlo simulation was used to show the relation of the parameter to the coefficient of variation for each of the four measures. Figure 3 shows that the c.v. is much smaller for H and Nei than for KW or JAM. Thus H will be more useful in determining when two residue sites differ in diversity.

Application to mouse V_κ gene:

A set of 105 mouse V_κ sequences was obtained from (5) and assayed for diversity using KW, JAM, and H. The results are shown in figure 4. The top and middle plots in figure 4 show how the KW and JAM measures single out the peaks of diversity. The figures also depict 95% confidence intervals for the three measures. As a rough measure of significance two sites differ significantly in diversity if their confidence intervals do not overlap. Not much significant change can be seen using either KW or JAM measures, but H shows a fine structure of diversity differences which may be related to structure in the associated antibodies. Comparison with crystallographic structures (not shown) shows that highly conserved sites according to the H measure tend to reside on the inside while exposed ones are on the outside of the antibody molecule. This shows that the H measure is useful for distinguishing conserved as well as diverse sites.

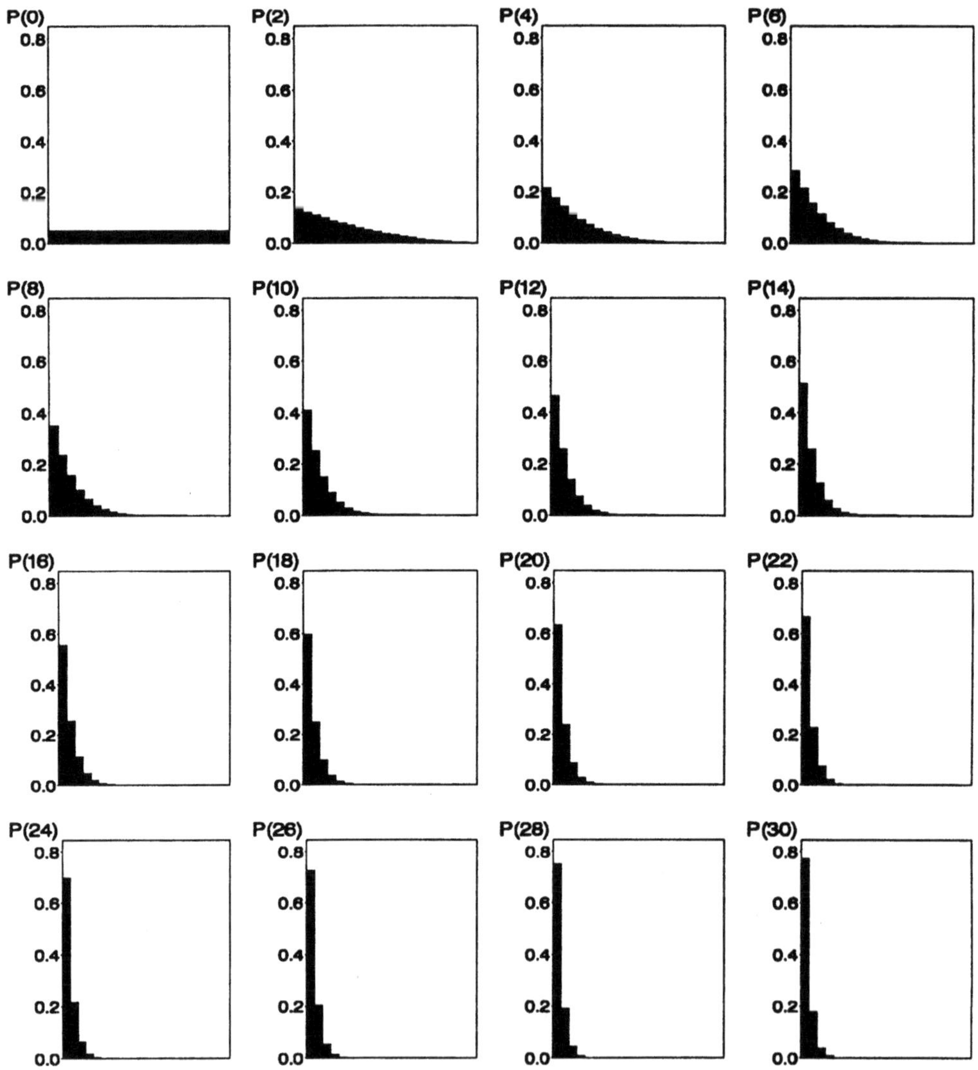

Figure 1: A swath through the space of 20*nomial* distributions induced by the function $y = (1 - x)^{power}$ showing a wide range of diversities. *power* varies continuously in figures 2 and 3.

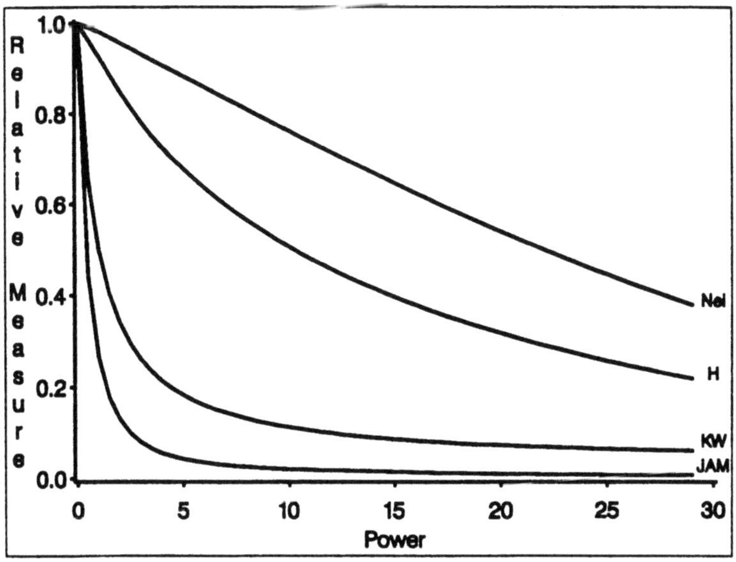

Figure 2: *Nei*, *H*, *KW*, and *JAM* measures plotted relative to their maximum values for the swath of 20*nomials* shown in figure 1. *KW* and *JAM* tend to filter out peaks of diversity while *Nei* only halves over the plotted range of the swath.

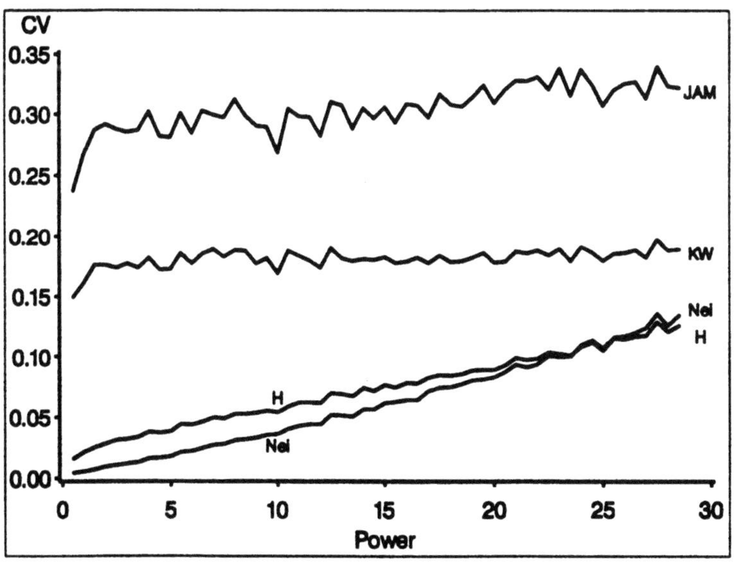

Figure 3: Monte Carlo estimates of the coefficients of variation of the four measures showing that: $CV(JAM) > CV(KW) > CV(H) \simeq CV(Nei)$. Peaks and valleys tend to correspond between curves since the same data sets were used for all four estimates. *H* and *Nei* are clearly better than *KW* and *JAM* from this viewpoint.

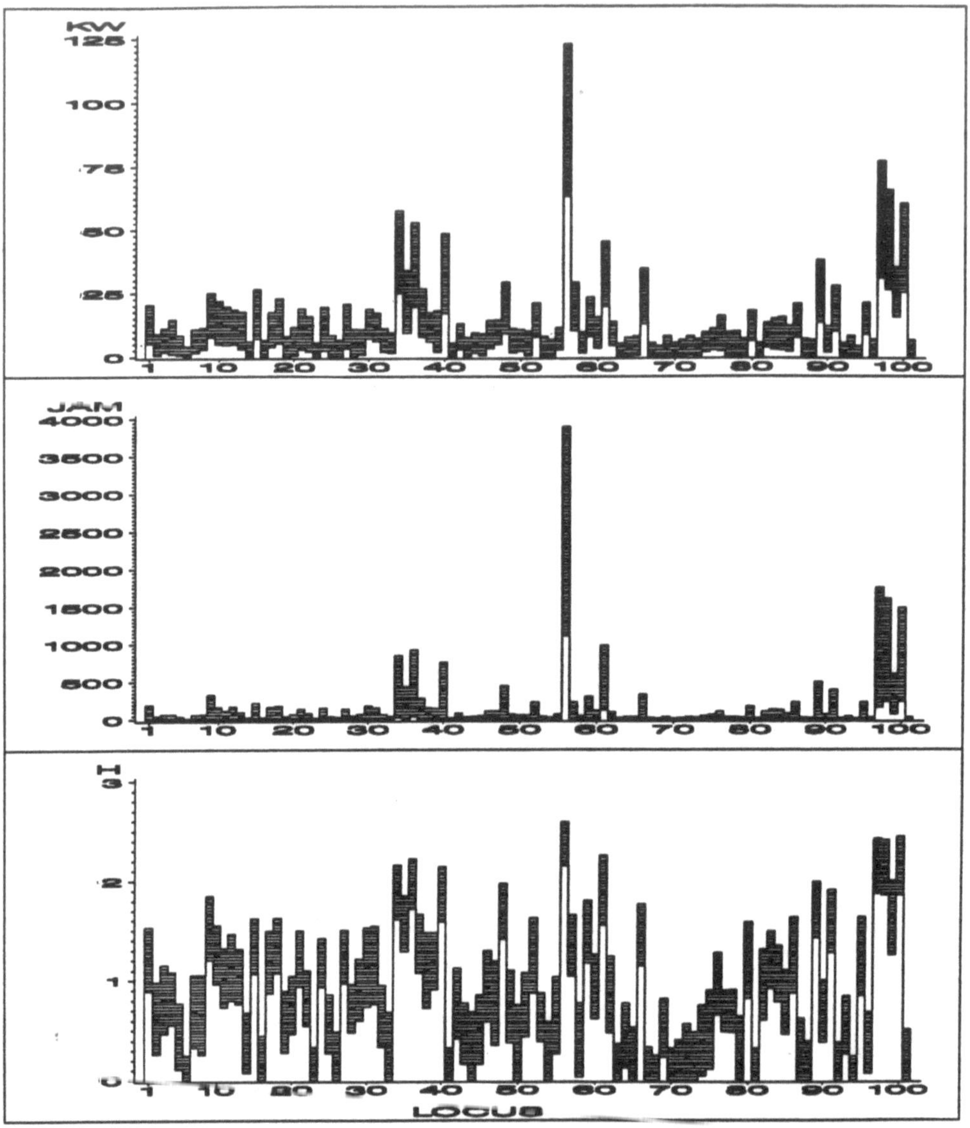

Figure 4: Mouse V_κ gene diversity plotted in *KW* format according the *KW* (top), *JAM* (middle), or *H* (bottom) measures. Dots show the positions of point estimates of diversity while shaded bands indicate the range of 95% confidence intervals. Most residue sites in either *KW* or *JAM* plots are indistinguishable by the simple criterion of having non- overlapping 95% confidence intervals. *H* measure confidence intervals, being much narrower, allow many distinctions to be made, among them indications of conserved as well as hyper-variable residues.

References:

1. Jores, R., Alzari,P.M., and Meo, T. *Resolution of hypervariable regions in T-cell receptor beta chains by a modified Wu-Kabat index of amino acid diversity.* Proc. Natl. Acad. Sci. USA, Vol. 87, Dec. 1990.

2. Shannon, C.E. *The mathematical theory of communication.* University of Illinois Press: Urbana, Illinois, 1949.

3. Wu, T.T. and Kabat, E.A. *An Analysis of the Sequences of the Variable Regions of Bence Jones Proteins and Myeloma Light Chains and Their Implications for Antibody Complementarity.* J. Exp. Med. 132, pp. 211-250, 1970.

4. Nei, M. *Analysis of gene diversity in subdivided populations.* Proc. Natl. Acad. Sci. USA. Vol. 70, No. 12, Part I, pp. 3321-3323, Dec. 1973.

5. Strohal, R., Helmberg, A., Kroemer, G., and Kofler, R. *Mouse* V_κ *gene classification by nucleic acid sequence similarity.* Immunogenetics vol 30:425-493, 1989.

EXPERIMENTAL STRATEGIES, ANTIGENICITY AND INTERPRETATION OF INTERMOLECULAR INTERACTIONS

Frederique Lisacek
Centre de Genetique Moleculaire
CNRS
91190 Gif sur Yvette
France
E-mail: lisacek@frcgm51.bitnet

Abstract

Prediction of antigenic sites in proteins is mainly based on two characteristics: homology between sequences and hydrophobicity indexes. Using these properties, the respective points of view of the antigen and of the antibody are only successively taken into account. Experimental procedure(s) are defined to identify properties of the interaction between the antigen and the antibody. Assuming interactions are expressed in terms of physico-chemical properties of amino acids, is it possible to isolate each contributing factor to binding? What is the basis for an additivity principle? Can common features of epitopes be found to characterize their binding capacity? T-cell epitopes have also been studied. Some results are mentioned, unfortunately raising more questions than providing answers.

Introduction

The molecular basis for antigen-antibody interactions is a key issue of immunology. In order to study experimentally these interactions, one approach is to match the antigen to its most meaningful bits, that is, peptides, representing epitopes [1]. Synthetic peptides as abstractions of whole antigens are being used in two different ways. Within a reductionist approach: peptide sequences are extracted from the native antigen sequence and tested with a known corresponding antibody. Within an artificial approach: peptides can be designed de novo to fit into a given antibody, regardless of their similarity with the native antigen. The definite advantage of such an experimental framework is the speed with which results can be obtained. The controversy about the immunogenicity of synthetic peptides will not be discussed here.

It is assumed that mechanism(s) of recognition in proteins can be, at least partly, explained in terms of amino acid properties. Analysing experimental data supposedly leads to draw predictive principles of antigenicity. It will be shown that, despite of

NATO ASI Series, Vol. H 66
Theoretical and Experimental Insights into Immunology
Edited by A. S. Perelson and G. Weisbuch

a relatively rigorous formalization of the problem, many questions remain unsolved. General, that is, universal features of epitopes probably do not make sense.

1. Theoretical basis

Formal studies of peptides essentially involve searching for patterns in proteins and generating principles to detect such patterns [2,3]. Relationship(s) between amino acid sequences are established through alignment. A pattern is then determined when a number of amino acids can be matched one after another in a set of protein sequences. A sequence can also be considered as one of physico-chemical properties such as hydrophobicity or charge, regularities in the occurrence of these properties can determine another type of pattern, that is, short sequences of physico- chemical characteristics.

Antigenicity has been shown to partly depend on intrinsic factors of epitopes [4]. Accessibility or hydrophobicity profiles of sequences provide some indication of how amino acids are involved in epitope delineation [5,6] but, they have not been used so far, as a very reliable basis for prediction. These properties are only characterizing intramolecular interactions and the influence of intermolecular interactions needs to be assessed. Principles drawn from theoretical calculations [7] combined with X-ray crystallography of protein complexes [8,9,10], provide the main body of data related to protein- protein interactions. Unfortunately, the production of such data is time-consuming.

2. Experimental data

As pointed out earlier, peptides are synthesized to study the specificity of antigen-antibody interactions. Some years ago, Mario Geysen has designed a simultaneous multiple peptide synthesis method. Using this method allows to test thousands of peptide in monoclonal or polyclonal sera, within a short period of time [1]. A variety and quantity of immunological tests have been accumulated for years.

Sets of peptides have been tested in three different situations:

1- Overlapping peptides spanning a whole protein sequences and testing yields two sets of native peptides: reactive versus non reactive.

2- Peptide analogs: each amino acid involved in most epitope sequences is successively replaced by the 19 others and testing yields the set of all acceptable substitutions at each position in the epitope.

3- Artificial repertoire including complete sets of dipeptides or tripeptides.

The first case is the first step in the "reductionist approach", mentioned in the introduction. One can discriminate two sets of peptides on the basis of reactivity. Assuming that this criterion is undoubtfully discriminant -which isn't the case in practise, due to the high rate of noise in data- it is possible to characterize a set of reactive peptides with properties of amino acids which are not shared by non reactive peptides. The problem is then to assess whether such properties are still true for another set of reactive peptides arising from another antigen.

The second case provides a qualitative measure of the antigen-antibody interaction. Assuming non-replaceability is synonymous with intermolecular contact, results have been analyzed in [11].

The third case is related to what is called a "mimotope" by H.M Geysen and colleagues, that is, a synthetic peptide able to mimic the features of an epitope [1]. This case belongs to the context of the previously mentioned "artificial approach". It is focussed on the setting of experimental procedure(s) aimed at broadening the scope of antibody binding.

3. Reductionist approach and data analysis

The first issue is the sequence description. If amino acids are considered as twenty distinct entities, then epitopes are characterized by occurrences and co-occurrences of particular amino acids. Propensity factors for occurrences of amino acids in epitopes were defined in [11].

Otherwise, any physico-chemical properties of amino acids can be used to define a number of classes (from 2 to 20), namely, hydrophobicity, size, number of hydrogen donor or acceptor, etc. Looking for a characterization of epitopes is then looking for regular occurrences of physico-chemical properties.

Given a set of reactive peptides and a set of non reactive peptides, rules are generated to discriminate an epitope from a non reactive peptide (details of the method in [12]). For instance, let us suppose amino acids are clustered according to: hydrophobicity, size, hydrogen donor, hydrogen acceptor. It is assumed that these properties can take discrete values. Thus, amino acids are encoded into binary variables (true or false). In other words, any amino acid is or is not hydrophobic, is or is not large, is or is not a hydrogen donor, ... Each position in the peptide is then specified by a succession of binary variables.

With a set of fifty reactive peptides and one of two hundred non reactive peptides, all of length 8, reactivity is characterized by a set of logical rules of the type: (non hydrophobic in position 1) and (hydrophobic in position 4) (non hydrophobic in position 3) and (large in position 5) and (non large in position 6) (hydrogen donor in position 2) or (hydrophobic in position 3) etc.

Forty to sixty of such rules are generated. They are verified in at least 7020Thresholds are modifiable.

For each tested antigen, a set of rules has been generated from the discrimination of reactive from non reactive peptides, with no significant overlapping of these rules.

Modifying the encoding of amino acids, the above thresholds or other parameters has not yielded any improvement of the generalization of results. Mixing peptides extracted from various sequences has not led to better conclusions (unpublished data).

The possible inadequacy of the methodology, and/or the incompleteness of experimental data, cannot be discarded as explanations for the absence of definite results. But still, couldn't the generalized impossibility to generalize be a telling paradox? Further assumptions need to be made to refine the notion of reactivity and non reactivity, that is, reactivity might depend on some context yet to be specified.

4. The mimotope strategy

4.1 Assumptions

Given an antibody which sequence is usually unknown, step by step construction processes of mimotopes, called "mimotope strategies", have been defined [1,13]. Any of the strategies relies on exhaustivity. Given a building principle, all possible combinations are systematically tested. The obvious restriction is combinatorial explosion.

If the specificity of the antibody is not known, what are the possible starting points, what is a step of the construction process? The length of peptides is apparently the first parameter to be considered. The minimal basis is a dipeptide.

The 400 possible dipeptides can be consistently tested with given antibodies. Best binding dipeptides constitute initial components of synthetically built mimotopes. Dipeptides are extended by adding one amino acid after another, each corresponding to best binding extended tri, tetra, etc, - peptides.

What is the basis for an additivity principle? Is a dipeptide always the conjunction of properties of two amino acids? In other words, how independent are physico-chemical properties of amino acids? As a corollary, is there a minimal set of interactions involved in the binding of a peptide to an antibody?

For example, the relative influence of size with respect to positive charge, can be assessed by gradually shortening the side chain of positively charged amino acids, which happen to be bulky as well. Let us suppose, a peptide containing lysine is reactive to a given antibody. Lysine (which side chain is: C4H8-NH2) is first replaced by ornithine (which side chain is: C3H6-NH2), each time it occurs within a peptide and then by diaminobutyric acid (which side chain is: C2H4-NH2). If in either case, the modified peptide binds the antibody, the dominance of the positive charge is established [14].

To try and answer these questions in general, constituents of peptides are usually not restricted to the twenty natural amino acids. The mimotope strategy was often used with L- and D- isomers of amino acids [1,13,15].

4.2 Discussion

The extension process of dipeptides to longer peptides would be well founded if specificity was gained at each step. In fact, it was shown that peptide volume is a fair justification of how reactive dipeptides are and how this affects the significance of this particular mimotope strategy [14]. Since the specificity of the testing antibody is unknown, it can only be assumed that the ability of an antibody to interact with a peptide relies on either a general property common to all antibodies (similar structure, same function), or a unique property related to its specificity. Specificity implies uniqueness of interaction and is, by definition, true or false. On principle, there are enough possibilities of rank ordering 400 dipeptides according to their binding capacity, that could reflect the specificity of all possible antibodies. If binding were to depend on a specific property of the antibody, this antibody should induce a preference for particular dipeptides. But, volume-equivalent dipeptides of different composition can fit indistinctly in antibody cavities. Such a contradiction suggests that binding does not depend on the specificity of the antibody. Then, dipeptide-antibody interaction is likely to rely on a general property of antibodies.

Requirements for binding differ for dipeptides and tripeptides. Crystallographic data emphasizes the different behavior of amino acids within dipeptides and tripeptides binding to the Bence-Jones dimer [15]. In fact, the weakness of the strategy shows that peptide-antibody specific recognition does not rely on a simple additivity principle and surely requires more than a single dipeptide to start with. Nevertheless, such an experimental approach allows to formally set questions and obtain partial answers.

Combinatorial explosion has been a definite hindrance to the development of mimotope strategies. Recent improvements in experimental techniques offer new potentials for large scale testing [16]. Completely unrelated sequences -DFLEKI and CRFVWC- bind a known antibody with high affinity which confirms the validity of the concept of mimotope.

5. Comments on T-cell epitopes

In recent years, T cell epitope characterization has been facilitated by the use of synthetic peptides. Analysis of data led Rothbard and Taylor [17] to identify a motif common to almost all T cell epitopes known at the time. Earlier, DeLisi and Berzofsky [18] had emphasised the helical propensity of those epitopes. The list has since lengthened, although in a biased way. Indeed, many tests have been performed relying on one or both of the previously suggested properties of epitopes, hence reinforcing them. Other strategies such as MHC binding prediction [19] or the testing of sequences delineated by two cysteines, have also guided the choice of peptides. As a result, interpretations of any collection of data related to T cell epitopes need to be placed in such context.

The length of an epitope is a parameter that has not yet been specified. Are motifs corresponding to signals (beginning or end) in the protein sequences? Are they delineating the core of an epitope? Neither of the published patterns give a precise answer to these questions.

A number of experiments have shown quite recently that provided diverse MHC types are tested, the antigenic regions of the protein cover a large percentage of the whole molecule. In the example of the egg hen lysozyme [20], T-epitopes span two thirds of the sequence, regardless of the MHC type. Epitope prediction is made rather irrelevant when no information on the MHC type is given, which happens to be most of the time. A more complete and detailed interpretation of these results is given in [20]. Introducing the idea of crypticity might refine the validity of a motif. Besides, it is likely that regularity will be observable from the point of view MHC molecules. Supporting evidences in this direction are established in [21].

Conclusion

The study of antigenicity at the molecular level has been tackled experimentally, through the use of synthetic peptides. From ever growing datasets, a major problem is to assess the importance of constraints defining intermolecular interactions.

The existence of a hierarchical organization of constraints in binding has been pointed out in general protein study [22]. Moreover, constraints tend to be context-sensitive. There aren't any reason to believe that such a situation should not apply to the case of antibody-antigen, or of the antigen-MHC-receptor complexes. The interpretation of results is made difficult by the lack of information on the nature of contexts. Prediction is all the more so. A context is for instance, a particular MHC type. Observed regularities in [21] are easier to analyze. Otherwise, the contribution of a formal approach seem to be somehow limited to statements of what not to do. Anyhow, a close link between experimental work and theory allows to slowly clear the way and improve the quality of data.

Acknowledgments

I wish to particularly thank Mario Geysen for inviting me in his laboratory where I have been involved in quite interesting and challenging research activities and surrounded by a very welcoming and competent team.

References

1. H.M Geysen et al. (1987) "Strategies for epitope analysis using peptide synthesis", J. Immun. Meth. 102: 259-274

2. I.A Wilson et al. (1985) "Identical short peptide sequences in unrelated proteins can have different conformations: a testing ground for theories of immune recognition", Proc. Natl. Acad. Sci. USA 82: 5255-5259

3. M.J Rooman, S.J Wodack (1988) "Identification of predictive sequence motifs limited by protein structure data base size", Nature 335: 45-49

4. J.A Berzofsky (1985) "Intrinsic and extrinsic factors in protein antigenic structure", Science 229: 932-940

5. T.P Hopp, K.R Woods (1983) "A computer program for predicting protein antigenic determinants", Molec. Immun. 20: 483-489

6. D.J Barlow et al. (1986) "Continuous and discontinuous protein antigenic determinants", Nature 322: 747-748

7. C. Chothia (1984) "Principles that determine the structure of proteins", Ann. Rev. Biochem. 53: 537-572

8. A.G Amit et al. (1986) "The three-dimensional structure of an antigen-antibody complex at 2.8 Angstroms resolution", Science 233: 747-753

9. P.M Colman et al. (1987) "Three-dimensional structure of a complex of antibody with influenza virus neuraminidase", Nature 326: 358-363

10. E.A Padlan et al. (1989) "Structure of an antibody-antigen complex: crystal structure of the HyHEL-10 Fab-lysozyme complex", Proc. Natl. Acad. Sci. USA 86: 5938-5942

11. H.M Geysen et al. (1988) "Cognitive features of continuous antigenic determinants", J. Molec. Recognition 1, 32-41

12. H. Soldano, J-L. Moisy (1985) "Statistico-syntaxic learning techniques", Biochimie 67: 493-498

13. H.M Geysen et al. (1986) "A priori delineation of a peptide which mimics a discontinuous antigenic determinant", Molec. Immun. 23: 709-715

14. F. Lisacek (1991) "Explanation for the non-specificity of small peptide-antibody interaction", in press.

15. G. Tribbick et al. (1989) "Similar binding properties of peptide ligands for a human immunoglobulin and its light chain dimer", Molec. Immun. 26, 625-635

16. J.K Scott, G.P Smith (1990) "Searching for peptide ligands with an epitope library", Science 249: 386-390

17. J.B Rothbard, W.R Taylor (1988) "A sequence pattern common to T cell epitopes", EMBO 7: 93-100

18. C. DeLisi, J.A Berzofsky (1985) "T cell antigenic sites tend to be amphipathic structures", Proc. Natl. Acad. Sci. USA 82: 7048-7052

19. A. Sette et al. (1989) "Prediction of major histo- compatibility complex binding regions of protein antigens by sequence pattern analysis", Proc. Natl. Acad. Sci. USA 86: 3296-3300

20. E. Sercarz (1991) contribution to the advanced research workshop in Theoretical Immunology, this volume

21. K. Falk et al. (1991) "Allele-specific motifs revealed by sequencing of self-peptides eluted from MHC molecules", Nature 351: 290-296

22. W.A Lim, R.T Sauer (1989) "Alternative packing arrangements in the hydrophobic core of lambda repressor", Nature 339: 31-36

RECEPTOR - LIGAND INTERACTIONS AND DIFFUSION EFFECTS

Stig I. Andersson
Research Group of Global Analysis
Chalmers Science Park
S-412 88 GÖTEBORG, Sweden

1. INTRODUCTION

The objective of this paper is to briefly outline in a non-technical way a mathematical model for some basic interaction mechanisms of importance in cell biology and immunology. We extend the existing analysis of receptor-ligand interactions and diffusion to include general geometric shapes for the cell membrane as well as shape-dependent diffusion mechanisms. Cell membranes are viewed as closed two-dimensional manifolds with a receptor distribution depending on the geometric characteristics of the membrane. Indeed this constitutes a step towards modeling quite realistic situations where the geometry-dependence of internationalization mechanisms could also be studied.

Technically, the main ingredient will be the theory of membranes in its usual differential geometric setting. Differential geometry is the natural framework for the study of global aspects of closed two-dimensional membranes embedded in e.g. a liquid environment R^3.

It will be advocated that membranes are flexible enough devices to serve the two-fold purpose of providing models both for the dynamics of receptor-ligand complexes on cell surfaces and internalization mechanisms (involving mainly local geometry) and for conformational dynamics of proteins (involving mainly global geometry). Despite the very different nature and great difference in length scales of these two phenomena, from a geometric point of view they do share features, conveniently modeled by membranes. This at least for resolutions weaker than the intermolecular length scale.

Mathematically there are two relevant aspects of such membranes; the differential geometric one and the differential topological one. Standard references for these are

NATO ASI Series, Vol. H 66
Theoretical and Experimental Insights into Immunology
Edited by A. S. Perelson and G. Weisbuch

Klingenberg 1977 and Hirsch 1988 (chapter 9), respectively. General information on membranes is provided by Nelson et al 1989. As for conformational dynamics, which will only be mentioned in this article, we refer to McCammon & Harvey 1987, Lipowsky 1991, Finkelstein & Reva 1991, Frauenfelder et al. 1991 for some recent points of view.

Technical details will be kept at a minimum and we shall instead concentrate on the ideas and the interrelation of biological and mathematical aspects of the mechanisms.

The work here originated in theoretical work on the interleukin 2-receptor and its role in T-cell activation. On experimental grounds, two main mechanisms had been proposed for this receptor; the affinity conversion model (ACM) and the preformed heterodimer model (PHM) (cf. the contribution presented by Alan S. Perelson in this volume for background material and a thorough discussion of this question). Mathematical work done by Perelson et al. turned out to favor PHM, whereas similar type of work done by a group in Göteborg seems to point at ACM as the preferred mechanism. During this work the Göteborg group faced serious problems in understanding the implications of some of the chief experimental papers and we failed in understanding the conclusiveness in some papers involving Scatchard plots (cf. Kermode 1989). Therefore, the conjecture evolved that only the coupling of ACM and PHM respectively to a reasonable internalization mechanism would provide means to discriminate between the two hypothesis. Internalization would then also necessarily include a detailed analysis of receptor-ligand kinematics and kinetics of receptor-ligand complexes. Heuristically, such a conjecture, that the difference between ACM and PHM would turn out only in conjunction with an internalization mechanism, has some support in the simple fact that in any experiment, internalization is virtually impossible to block.

It is also important to note that the question of ACM vs. PHM is far from being only an academic one; in e.g. drug-design - and more generally in computer-aided molecular biology - it would be of great significance to have à priori information about the number of epitopes of the subunits constituting a molecule such as the interleukin 2-receptor. Exactly this type of information is likely to be provided through mathematical modeling (c.f. Mayo 1989 for another example of this type).

This program is now under work within the group in Göteborg and is far from finished. Here the focus will be on one ingredient of this program; a mathematical model for the interaction of membrane-bound, lateraly mobile receptors with ligands in a surrounding solvent. The model includes naturally aggregate formation of receptor-ligand complexes and the genesis of coated pits. The new aspect is the inclusion of the geometry-dependence of these mechanisms.

In fact it will be indicated how membrane fluctuations combined with conventional aggregation mechanisms for receptor-ligand complexes could account for the formation of coated pits (and vesicles), modifying earlier mathematical work in this field.

Finally, the framework here is also proper for understanding why it is functional, from a control-theoretic point of view, for a receptor to have a complex structure such as that exhibited by the interleukin 2-receptor. Quite a few receptor systems are homologous to interleukin 2, so it is a general feature. This aspect will however be pursued elsewhere.

2. THE CONVENTIONAL PICTURE

Coated pits are transient structures on the cytoplasmic side of the cell membrane, capable of trapping cell-surface bound receptor-ligand complexes. Through deformation ("adding curvature") coated pits could turn into coated vesicles as a first step towards internalization.

Receptor kinetics is commonly attributed to:

- unaided diffusion
- local flow effects, created e.g. by the formation of coated pits,
- convection (global flow effects),
- ligand-induced dynamics.

Receptors are membrane-bound (integral parts of the cell membrane) elements, free to move laterally in the membrane. The receptor kinetics and internalization have been studied extensively, mainly from the local geometric point of view. We mention just a few of these important results.

Goldstein et al. 1988 studied membrane flow effects in connection with the formation of coated pits. The formation of coated pits gives rise to a radial local flow towards the center of the coated pit. It is shown that unless the lifetime of coated pits is shorter than 10 sec., the effect of this local flow in transporting randomly inserted receptors to the coated pit is negligible. Also, this is not a probable mechanism for keeping receptor-ligand complexes trapped in the coated pit for a time long enough for internalization. On the other hand, convection is a more probable candidate and it is shown that it has considerable effect under typical conditions. DeLisi 1980 and Perelson & DeLisi 1980 undertook early studies of receptor clustering mechanisms including the kinematics of solution phase ligands binding/dissociating to/from cell surface receptors as well as receptor-receptor interaction. Other diffusion aspects were studied in Shoup & Szabo 1982, Abney et al 1989 and the experimental evidence that receptor-mediated endocytosis proceeds through coated pits is presented in Lund et al. 1990 along with a quantitative analysis for hormone-induced internalization.

3. THE GEOMETRIC PICTURE

Biological membranes behave in many respects like two-dimensional fluid membranes, fluctuating in the surrounding solvent. In particular, cell-membranes are well described by closed membranes of this type. For random/fluctuating membranes there are good geometric theories available, where even the many stabilizing features of realistic membranes could be incorporated. Also, although R^2 is good enough for local phenomena, transport phenomena require a global geometric picture for their description. For example, one of the conclusions in Abney et al. 1989 is that locally the diffusion coefficient is determined by Brownian motion whereas the global diffusion coefficient will be strongly geometry-dependent. Also global geometry is needed to describe the fact that receptors are not uniformly reactive. The basic geometric picture we shall adopt here to describe receptor-ligand and receptor-receptor interaction is the following: on the cell-membrane X we consider a random distribution of receptors (essentially viewed as "charges"), attached as integral members of the cell membrane and confined to move on X (laterally mobile). X is here viewed as a two-manifold which is C^∞ and closed. The interaction between this receptor distribution and ligands, combined with cell surface fluctuations may well cause - as will be explained below - coated pits, i.e. local negative curvature deformations of X and lead to vesicle-formation. Before presenting the consequences of this picture, we need some elements of fluid membrane theory.

4. MEMBRANES

The basic constituents of biological membranes are lipid molecules which are arranged in characteristic ways in the membrane dictated by the different degrees of solubility of the various parts of the lipid molecule. The global structure of such a membrane may be very complex from the topological point of view. From the modeling point of view and for resolutions weaker than the intermolecular length scale, the membrane may be viewed as a two-dimensional closed surface immersed into R^3 (fluid membrane freely fluctuating in solvent, and where thickness $\ll$ shape-dimensions). In differential geometric terms such a surface is characterized (modulo equivalences) by the induced metric g_{ij} and the extrinsic curvature K_{ij}. These quantities from the basis of invariants out of which surface models, e.g. Hamiltonians may be constructed (c.f. below).

From the differential topological point of view connected, closed and orientable surfaces are classified according to their genus p, an integer ≥ 0 and related to the Euler characteristic $\chi(X)$; by $\chi(X) = 2 - 2p$. To have genus p means roughly to be a sphere with p handles. As explained in Frauenfelder et al. 1991, one is now starting to gain insight into the interrelation of shape ($\sim$ genus) and biological function, so this aspect should be an important one in the future. It is easy to write down a number of Hamiltonians, based on surface invariants (the algebra spanned by g_{ij}, K_{ij}) and which exhibit a rich conformational dynamics. One of the simplest is given by:

$$H = \int dA\{\mu + \frac{\kappa}{2}(M - M_0)^2 + \overline{\kappa}K\}$$

describing the statistical properties of thin membranes quite well. Here $\kappa, \overline{\kappa}$ are the rigidity and Gaussian rigidity coefficients, respectively, M is the mean curvature, M_0 a spontaneous curvature and K is the standard Gaussian curvature. μ finally is related to a kind of chemical potential. Given such a Hamiltonian, the determination of the shape of the corresponding closed two-membrane is a minimal surface question. So in principle it is the variational problem

$$\delta H = 0 \,,$$

keeping total area and the total volume fixed.

Now, without going into mathematical details, more sophisticated Hamiltonians may be constructed which:

- take into account stabilizing regions, known to exist (cytoskeletons),
- reflect the influence of thermal fluctuations on the shape (minimal surface problem) via the temperature dependence (linear) of the rigidity coefficients κ and $\overline{\kappa}$,
- take into account the effect of say a scalar distribution in the membrane. In our case, such a scalar distribution would be a receptor distribution which couples to membrane variables and is potentially able to interact with ligands and hence in this way influence the global topology.

5. RECEPTOR-LIGAND INTERACTION AND INTERNALIZATION IN A GLOBAL GEOMETRIC FRAMEWORK

In the presence of a receptor distribution $\rho(x_1, x_2)$ in the membrane, coupling to internal curvature degrees of freedom and to exterior ligands in the surrounding solvent, the question of writing down Hamiltonians become tricky. Nevertheless, a number of trial candidates may be constructed, all of them complex and I will not write them down here since we shall not compute anything with them. Studying these Hamiltonians one sees that they share the following property: the spontaneous curvature

$$M_0 \sim |\rho(x_1, x_2)|^2 .$$

Hence the various curvature measures depend directly on the receptor distribution, which in this way influence strongly the global topology (shape). This type of phenomena is sometimes referred to as curvature instability. We have then two aspects of aggregation mechanisms to take into account:

- a fluctuating membrane, the shape of which is directly influenced by thermal fluctuations (via temperature dependence of rigidity coefficients) and by a receptor distribution (curvature instability)

- the dynamics of the receptor distribution $\rho(x_1, x_2)$ on the membrane (closed two-manifold).

The mathematical machinery needed to describe the interplay of these two aspects is somewhat heavy but manageable. To perform calculations in this framework I would need to set up this whole machinery. I believe however that it is more informative if I instead summarize in words the main outcome of such calculations. I shall then describe properties which are somewhat generic in the sense that they are shared by most of these models, to some extent independent of the choice of the actual Hamiltonian! Indeed, I believe strongly in the value of the robustness of the arguments and models used at the present state of knowledge. Qualitative, generic models which do not (essentially) depend on the details of the process are more likely to be useful than more intricate and sensitive ones. Roughly the results may be described as follows: Spontaneous curvature may be created via thermal fluctuations (as well as through collisions, friction with solvent etc), curvature instability mechanisms (internal membrane-bound density distributions) as well as via interaction of the membrane-bound distribution with a ligand distribution in the solvent (extrinsic curvature instability). The influence of these three mechanisms on the global topology, gives rise to a fantastically rich shape-phenomenology, including vesicle formation (negative curvature). As a rule the corresponding variational problem (minimal surface problem) is an extremely difficult one. The dynamics of the membrane-bound distribution on this fluctuating membrane depend on the global topology and the solvent ligand distribution with which it interacts. Technically this dynamics is described essentially by the Hamilton equations of motion, but with the manifold fluctuations built in.

Results

1. The global geometry/topology as induced by the above mechanisms account for pit and vesicle formation.

2. Coated pits and - in a limiting case - coated vesicles could be formed in this way taking into account the receptor distribution dynamics.

3. The formation of coated pits and vesicles is considerably enhanced by global topology fluctuations and by direct ligand-receptor interactions. Local flows become generically important, modifying, for example, the results of Goldstein et al. 1988.

4. Under reasonable assumptions, this mechanism is able to keep receptor-ligand complexes trapped for a sufficient time period to permit internalization.

The full framework for these calculations will be presented elsewhere.

References

Abney J R, Scalettar BA, Owicki JC (1989) Self-diffusion of interacting membrane proteins. Biophys J 55:817-833.

De Lisi C (1980) The biophysics of ligand-receptor interactions. Quart Rev Biophys 13:201-230.

Finkelstein A V, Reva B A (1991) A search for the most stable folds of protein chains. Nature 351:497-500.

Frauenfelder H, Sligar S G, Wolynes P G (1991) The energy landscapes and motions of proteins. Science 254:1598-1601.

Goldstein B, Wofsy C, Echavarria-Heras H (1988) Effect of membrane flow on the capture of receptors by coated pits. Biophy J 53: 405-414.

Hirsch M W (1988) Differential Topology, 3rd ed. Springer, Berlin, Heidelberg, New York.

Kermode JC (1989) The curvilinear Scatchard plot: Experimental artifact or receptor heterogeneity. Biochem Pharamacol 38:2053-2060.

Klingenberg W (1977) Eine Vorlesung über Differentialgeometrie, (Heidelberger Taschenbüch Springer, Berlin, Heidelberg, New York.

Lipowsky R (1991) The conformation of membranes. Nature 349:475.

Lund K A et al. (1990) Quantitative analysis of the endocytic system involved in hormone-induced receptor internalization. J Biol Chem 265: 15713-15723.

Mayo K H et al. (1989) Epidermal growth factor receptor binding is not a simple one-step process. J Biol Chem 264: 17838-17844.

Mc Cammon J A, Harvey SC (1987) Dynamics of Proteins and Nucleic Acids, Cambridge Univ. Press, Cambridge.

Nelson, D, Piran T, Weinberg S (eds) (1989) Statistical Mechanics of Membranes and Surfaces, (Jerusalem Winter School for Theoretical Physics Vol 5), World Scientific, Singapore.

Perelson A S, De Lisi C (1980) Receptor clustering on a cell surface I. Math Biosci 48: 71-110.

Shoup D, Szabo A (1982) Role of diffusion in ligand binding to macromolecules and cell-bound receptors. Biophys J 40:33-39.

V. Theories and Applications to Clinical Immunology

MATHEMATICAL MODELING OF THE IMMUNE RESPONSE DURING ACUTE VIRAL INFECTIONS

G. Bocharov and A. Romanyukha
Institute of Numerical Mathematics, Academy of Sciences
Leninskij prospect 32-A, Moscow, 117334 USSR

1 Introduction

In this paper we present an approach to the quantitative description of the organization and the dynamic properties of the immune response during acute viral infections in man. The study of virus infections is a central aspect of experimental and clinical investigations of the immune response. The real problem in understanding the nature of immunity is related to the difficulty of dealing conceptually with the complexity of the immune system *in vivo* (Doherty, 1986; De Boer & Perelson, 1991; Paul, 1984; Perelson, 1988; Stewart et al, 1989; Weisbuch et al, 1990). Mathematical modeling is a constructive tool of knowledge organization and offers a new direction for theoretical reasoning and analysis.

Immunological mechanisms, which are only one part of the defense mechanisms of vertebrate hosts, nevertheless play an extremely important role in limiting infectious disease. The immune system is built of a very large number of mobile cells with relatively short life times but great proliferative potential, so that the system as a whole is able to deal efficiently with many different and highly variable viruses and bacteria (Zinkernagel et al, 1985). In order to understand what constitutes a functional immune system *in vivo* , the interactions between various lymphocyte subsets during the development of an immune response, the regulation of such a response, and the recall of immunity following subsequent challenge, should all be assessed in quantitative terms for a number of specific viruses.

An important step in this direction is the translation or inversion of immunologic, virologic and clinical data into estimates of the kinetic parameters characterizing virus-sensitive interactions between tissues and the immune system in various compartments - blood, lymph nodes and target tissues. The mathematical model of the antiviral immune response developed by Marchuk et al (1991a) is employed as a phenomenological tool to this end. The key aspect of the adjustment between the model and various data on the immunity to particular infections is the derivation of a consistent data set, the **Generalized Picture** of an infectious disease, which provides a theoretical definition of the structure of the normal course of disease and the antiviral immune response. This approach is employed for mathematical modeling of two viral infections in man: uncomplicated influenza A and acute hepatitis B. The consistency of the model with clearly established experimental facts on passive immunization and vaccination is tested. The balance between immune

NATO ASI Series, Vol. H 66
Theoretical and Experimental Insights into Immunology
Edited by A. S. Perelson and G. Weisbuch

protection and immunopathology is a fundamental problem of theoretical and clinical immunology (Zinkernagel et al, 1985; Zinkernagel et al, 1989). We suggest the way in which the question may be addressed in the model for weakly cytopathic hepatitis B viruses. The relative roles of specific immune mechanisms and non-specific defense mechanisms at different time intervals for highly cytopathic influenza A viruses are analysed using the kinetic characteristics of infection and immune processes offered by the quantitative model of infectious disease.

2 The immune system and viral infections

There are a variety of non-specific mechanisms (interferons (IFN), complement, etc.) that contribute to the limitation of infection within the first 2 or 3 days. However, it is accepted that recovery from primary viral infection is provided by the T-cell response, whereas protection against reinfection by the same viruses depends mostly on pre-existing mucosal IgA or systemic IgG. The basic immune processes involved in acute viral infections are relatively well understood and are briefly summarized below (see Ada & Jones, 1986; Doherty, 1985; Doherty, 1986; Mitchell et al, 1985; Popper & Schaffer, 1986; Zinkernagel et al, 1985).

To initiate infection viruses enter their host and infect the sensitive tissue cells, which results in viral replication and the damage of infected cells. The latter is the basis for clinical illness. All viruses give the immune system a signal on the surface of infected cells or via antigen-presenting cells. Both the T-cell and the B-cell parts of the immune system are specifically activated during viral infection. Induction and triggering of effector T-cells depends on viral antigens being presented by macrophages in lymph nodes, or by tissue macrophages in association with MHC molecules (class I for cytotoxic T lymphocytes, CTL; and class II for helper T-cells). Helper T-cells are necessary for the generation of both the humoral and the CTL mediated immune responses; it is established that different subsets, Th1 and Th2 respectively, provide the corresponding activities. CTLs following clonal expansion and differentiation enter the lymph and localize to tissue sites of virus replication, where they deliver a lytic signal to infected cells before infectious progeny are assembled and leave the cell. Antigen driven activation of particular B-cell clones results in B-cell proliferation and differentiation into mature plasma cells and memory B-cells. Plasma cells produce immunoglobulins which neutralize and remove virus particles from lymph, blood or other body fluids (mucosa).

For acute infectious agents, one does not have to evoke sophisticated suppressor cells or cellular network circuits to regulate an immune response after fast elimination of acute cythopathic viruses (Zinkernagel et al, 1985).

The human lymphatic system consists of bone marrow, thymus and the peripheral lymphoid organs, including the lymph nodes (LN). Blood and lymph are important transport media for immunocompetent cells and antibody molecules. The lymph nodes draining an infected organ are the principal sites of stimulation, proliferation and differentiation of lymphocytes: in the case of acute hepatitis B there is a tree-like system of lymph nodes draining the liver having a total volume of about 50ml; in uncomplicated influenza A virus infection the activation processes are localized in lung-associated lymphoid tissue with a volume of about 12ml. In order to develop estimates of the kinetic rates of cell-cell interactions, it is necessary to identify the precise localization of the immune response during the particular viral infection in question.

The present scheme of immune responses in acute viral infection is simplified and idealized, but it is applicable to all viral pathogens and may be used to derive a general mathematical model of antiviral immune responses.

3 A mathematical model of antiviral immune responses - a framework for the organization of knowledge

The conceptual model of the basic immune processes operative during acute viral infection, with interactions described in terms of positive and negative influences, may be interpreted and organized into a population level mathematical model. The following formulation is not unique, due to the incompleteness of the conceptual model and the lack of regular quantitative principles or specific techniques of adequate mathematical formalization of immunological phenomena. However, previous experience in this field (see Bell,1970; Bell et al, 1978; Perelson, 1988a; De Boer & Perelson, 1991) shows that a central point is a proper balance between the character of the immunological data to be assimilated in a model, and the fine specification of the interactions to be formulated mathematically.

The mathematical model of antiviral immune responses presented here has the following state variables (with t as the independent time variable): free virus population - $V_f(t)$; antigen-presenting cell population - $M_V(t)$; Th1 cell population - $H_E(t)$; Th2 population - $H_B(t)$; CTL population - $E(t)$; B-cell population - $B(t)$; plasma cell population - $P(t)$; antibody population - $F(t)$; virus-infected sensitive tissue cell population - $C_V(t)$; destroyed sensitive tissue cells - $m(t)$.

The interactions between the populations are described in the framework of: the **Clonal Selection Theory** ; the birth-death balances of populations of cells and molecules ; the additive character of interactions and their proportionality to population sizes ; an average fixed CTL-, Th1- and Th2-specificity; and fixed antibody-antigen affinity. It is postulated that the damage of virus-infected target organ cells leads to non-specific weakening of the immune response.

The duration of cell division and differentiation is taken into account. The model is formulated mathematically as a system of delayed differential equations describing the time evolution of the ten state variables (see Marchuk et al 1991a for details):

$$dV_f/dt = uC_V + nb_{CE}C_VE - g_{VF}V_fF - g_{VM}V_f - g_{VC}V_f(C^* - C_V - m) \; ;$$

$$dM_V/dt = g_{MV}M^*V_f - a_MM_V \; ;$$

$$dH_E/dt = b^E{}_H[x(m)r^E{}_HM_V(t-t^E{}_H)H_E(t-t^E{}_H) - M_VH_E] - b_p{}^{H_E}M_VH_EE + a^E{}_H(H^*{}_E - H_E) \; ;$$

$$dH_B/dt = b^B{}_H[x(m)r^B{}_HM_V(t-t^B{}_H)H_B(t-t^B{}_H) - M_VH_B] - b_p{}^{H_B}M_VH_BB + a^B{}_H(H^*{}_B - H_B) \; ;$$

$$dE/dt = b^E{}_p[x(m)r_EM_V(t-t_E)H_E(t-t_E)E(t-t_E) - M_VH_EE] - b_{EC}C_VE + a_E(E^* - E) \; ;$$

$$dB/dt = b_p{}^B[x(m)r_BM_V(t-t_B)H_B(t-t_B)B(t-t_B) - M_VH_BB] + a_B(B^* - B) \; ;$$

$$dP/dt = b_p{}^Px(m)r_PM_V(t-t_P)H_B(t-t_P)B(t-t_P) + a_P(P^* - P) \; ;$$

$$dF/dt = r_fP - g_{FV}FV_f - a_FF \; ;$$

$$dC_V/dt = sV_f(C^* - C_V - m) - b_{CE}C_VE \quad b_mC_V \; ;$$

$$dm/dt = b_{CE}C_VE + b_mC_V - a_mm \; ; \quad x(m) = 1 - m/C^* \; .$$

From the point of view of an immunologist, the model is incomplete as compared to the complexity of a real immune response *in vitro* and therefore *in vivo*. On the other hand from the modeling point of view, the model seems to be too complex as compared to the character of available quantitative data: the model requires identification of about forty-five parameters, and current experimental facts on antiviral immune responses do not seem adequate to the task. To fill the gap we introduce an intermediate level of data organization between the experimental facts and the mathematical model: the Generalized Picture of Infectious Disease.

4 Theoretical description of antiviral immune responses during infectious disease - the Generalized Picture approach

The available immunological and clinical data concern different aspects of acute hepatitis B and influenza A infections: the time course of different antigenic markers and specific antibodies, the variations in subpopulations of peripheral blood lymphocytes and macrophages, the levels of hormones, morphological examinations of the liver and the lung, dynamics of basic clinical indices. By putting forward clear assumptions, it is possible to integrate particular *in vivo* and *in vitro* data sets, including the heterogeneous and partial observations which are available, into a theoretical model of the natural disease course. In other words, it is possible to derive a consistent homogeneous description, in quantitative terms, of the typical kinetics of an antiviral immune response during the infection - the generalized picture (GP) of the acute virus infection.

Definition. The generalized picture of the acute hepatitis B or uncomplicated influenza A virus infection is considered to be the quantitative presentation, in the framework provided by the variables of the mathematical model, of the various immunologic, virologic and pathologic processes occurring in the human organism after virus invasion and resulting in acute infection.

Table 1

Example of the data set comprising the generalized picture of acute hepatitis B (for complete data set and references see Marchuk et al 1991a)

Time (days)	Estimated value of variable	Comments
		Concentration of HBsAg and DNA-containing virus particles (Sensitivity of radio immunoassay test is 10^7-10^8 HBsAg pt ml^{-1})
	V_f	
0	10^0 - 10^6	Infection dose
60	0.5×10^8	Sensitivity level of observation method
70	0.2×10^9	Period of exponential growth
80	0.8×10^9	Onset of non-specific symptoms, beginning of immune reaction
100	0.5×10^{10}	Onset of specific symptoms, period of maximum antigen concentration
120	10^1	Early stage of recovery, disappearance of viral particles
		Concentration of destroyed hepatocytes (proportion of steady-state concentration of liver hepatocytes)
	m	
90	0.8×10^{-2}	Physiological norm
100	0.6×10^{-1}	Onset of specific symptoms
110	0.2	Peak of illness
120	0.9×10^{-1}	Early recovery
130	0.2×10^{-1}	Upper bound of the physiological norm
		Concentration of HBsAg-class II MHC specific Th1 cells in LN (relative to the steady-state concentration in lymph node)
	H_E	
100	0.1×10^2	The present data describe
105	0.1×10^3	proliferative immune response interval
110	0.1×10^4	

The detailed analysis of the GP concept is given in Marchuk et al (1991a). Here we present some of the main statements on which the GP is based:

1. The immune response takes place in the system of LNs draining the site of viral infection;
2. The immune response in the peripheral LNs of animals is a natural biological model of the immune response processes in the system of lymph nodes draining the liver or the lung in man. The experimental data on antigen-induced immune response in the peripheral LNs of animals (changes in lymph node structure, frequency of lymphocytes of a given specificity, cell changes in efferent lymph) can be extrapolated to man;
3. The concentration of foreign antigen determines the beginning of pronounced immune response in LNs in a threshold-like manner;
4. Non-specific symptoms (fever, the absence of appetite, etc.) are the clinical manifestations of the initiation of the antiviral immune response and may be used to coordinate the viral kinetics and macrophage activation.

Part of the GP data set for the case of acute hepatitis B is presented in Table 1. The pragmatic aim of the GP is to assemble and represent the data necessary to obtain estimates of the model parameters corresponding to a particular virus infection.

5 Analysis of immune system activation in terms of kinetic parameters

The approach to parametric analysis is based on taking into account the localization of immune response and disease variables rather than considering the organism as a well-mixed homogeneous reactor. The parameter set of the antiviral immune response model may be divided into two groups. The first consists of parameters for which direct estimates are available on the basis of various observations and data : the steady-state population sizes of Th-cells, CTLs, B-cells, plasma cells and macrophages which are activated in particular virus infections; the half-life times of cells and immunoglobulins; the characteristics of the differentiation processes of T- and B-cells; the rates of production of antibodies by plasma cells, and viruses by infected cells; the antigen-antibody association rate; and so on (see Marchuk et al 1991a). The second group consists of parameters, notably the kinetic rates of cell interactions in sensitive tissues, and of macrophage-lymphocyte cooperation in the LNs compartment, for which no direct estimates are available. To assess these parameters we generalized some results of *in vitro* observations of antigen induced T- and B-cell activation in mixed lymphocyte cultures, thus defining the notions of "time scale" and "sufficient number of cooperating cells" characterizing the realization of a particular process; we then introduced

Table 2

Values of some of the model parameters used in simulation of the immune response in acute hepatitis B and influenza A infections
(for complete parameter set see Marchuk et al, 1991a)

	Physical meaning of parameters	Hepatitis B	Influenza A	Units
M^*	concentration of class I or class II MHC bearing macrophages in LN	10^{-15}	10^{-15}	M
H^*_E	concentration of antigen-specific Th1 cells in LN	10^{-18}	10^{-16}	M
E^*	concentration of antigen-specific precursors of CTLs in LN	10^{-18}	10^{-16}	M
B^*	concentration of antigen-specific B cells in LN	10^{-18}	10^{-17}	M
F^*	concentration of antigen-specific antibodies	8.3×10^{-14}	8.3×10^{-13}	M
C^*	concentration of susceptible cells in a target organ	0.5×10^{-12}	1.7×10^{-11}	M
$b^E{}_H$	rate constant of Th1 cell stimulation	3×10^{16}	2×10^{16}	M
$b^E{}_p$	rate constant of CTL stimulation	5×10^{33}	10^{33}	$M^{-2}day^{-1}$
$b^B{}_p$	rate constant of B cell stimulation resulting in B cell proliferation	8×10^{32}	10^{34}	$M^{-2}day^{-1}$
γ_{VM}	rate constant of macrophage stimulation by virus antigen	9×10^{9}	$2x10^{8}$	$M^{-1}day^{-1}$
σ	rate constant of sensitive tissue cell infection by viruses	2×10^{9}	$2x10^{10}$	$M^{-1}day^{-1}$
b_m	rate constant of virus infected cell damage due virus cytopathicity	0.07	1.5	day^{-1}
ν	rate constant of virus replication and secretion per infected cell	1	510	day^{-1}

simple algebraic models relating these notions to the required estimates of the kinetic rate constants. The approach may be illustrated by one example.

The parameter $b^E{}_H$, referred to as the helper T-cell stimulation rate constant, describes the transition rate of activated Th from the G1-phase to the S-phase of the cell cycle. It is established that the duration of the preparation for division of initially activated Th cells is $t_{H:G1\text{-}S}$ = 14 - 24 hr, and the antigen-mediated *in vitro* proliferation of lymphocytes has been observed to involve direct physical contact with antigen-presenting macrophages, M_V. It is known that during pronounced immune response to foreign antigen, the number of antigen-specific Th cells in LN increases 10^2 - 10^3 times to reach values of 7 x 10^4 - 7 x 10^6 cells ml^{-1}. There are two basic characteristics of the process of Th-cell stimulation: the time $t_{H:G1\text{-}S}$, and the value of $M_V{}^{suf}$, the number of macrophages sufficient to provide the above-specified degree of proliferation of Th cells. The second may be estimated as the total number of macrophages expressing class II MHC molecules in a LN, M^*. Then the admissible range of $b^E{}_H$ may be obtained from the following relationship:

$$b^E{}_H = (t_{H:G1\text{-}S} * M^*)^{-1} = 4 \text{ x } 10^{14} \text{ to } 3 \text{ x } 10^{15} \text{ M}^{-2} \text{ day}^{-1}.$$

The cognate cooperation of Th cells and antigen-presenting macrophages in LN resulting in Th stimulation is realized by the following steps: the collision of M_V and Th, the specific binding of M_V and Th, and intracellular processes of Th cell preparation for division under the action of various signals and soluble factors. The M_V and Th collision rate constant $b_{col}{}^{MH}$ could be evaluated using chemical kinetics techniques under the assumption of LN liquid phase homogeneity. One can obtain the estimate $b_{col}{}^{MH}$ = 10^{15} M^{-2} day^{-1}, which supports the $b^E{}_H$ estimate and may be useful for initial guessing and interpretation of parameter identification results. The list of some parameters for both the acute hepatitis B and the the uncomplicated influenza A infection is presented in Table 2 (see also Marchuk et al, 1991a,b; Romanuykha et al, 1991; Bocharov and Romanyukha).

6 Quantitative mathematical models of the immune response to hepatitis B and influenza A. Fitting and validation

The problem is to provide a quantitative description of an object, given that there is no a priori information on the stochastic nature of the data which characterize that object. This problem is treated by using a mathematical model, and fitting the model to the data. In other words, the problem is to determine the parameter values giving the minimum deviation between a particular model and the corresponding data set.

The GP data sets of both the hepatitis B and the influenza A infections are characterized by large variation in absolute values over the observation interval and by equal significance of the data points independently of their particular values. To measure the goodness of fit between the model and the data we introduced a least-squares function of the relative distance between observation points y_{obs} and the corresponding solution values y(t,a), where a is the vector of model parameters:

$$\Phi(a) = [(1-y(t,a)/y_{obs})^2 + (1-y_{obs}/y(t,a))^2]$$

In a large system with realistic data, the standard situation is that only a few parameters can be identified. The generalized picture data set has a limited number of points (29 for acute HBV infection and 17 for influenza A infection). One needs to utilize available knowledge about the processes described in the model to make the fitting procedure correct. To this end we split the total observation interval into a sequence of smaller subintervals, and associated the kinetics of observed data within each subinterval with the operation of a smaller number of processes from the whole set described in the model. In this way, using information on the natural history of the infectious process and the immune response, we were able to choose reasonable values for the various parameters. Then, using regular fitting procedures, we refined the estimates of about ten parameters to obtain the final parameter sets corresponding to acute hepatitis B and influenza A virus infections. The general limitation of the adopted procedure is that the parameter fit thus obtained is not unique. It is therefore most important to validate the model by testing its predictive ability against experimental situations (see Marchuk et al, 1991b; Bocharov and Romanyukha). The mathematical model of antiviral immune response in acute hepatitis B infection was shown to be quantitatively correct with respect to the following tests:

- the effect of passive immunization on hepatitis B virus (HBV) challenge;
- the effect of vaccination on immune response and resistance to challenge with live HBV (see Fig.1);
- the relationship between the dose of inoculated HBV and duration of the incubation period.

The mathematical model of antiviral immune response in uncomplicated influenza A infection was shown to be quantitatively correct with respect to the effect of passive immunization on virus challenge. A major problem in making the validation is the transformation of experimental data for antigen and immunoglobulin quantities, expressed in the units usually reported (titres, IU, counts per minute, etc), into the direct physical quantities.

7 Extreme forms of viral challenge to the immune system in terms of the mathematical model parameters

The influenza A virus and the hepatitis B virus may be considered as examples of two extreme viral pathogens which threaten the human organism in

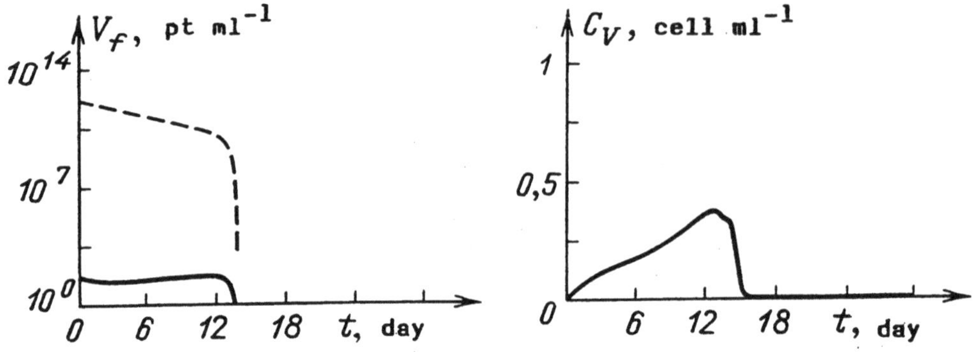

Fig.1 Modeling of vaccination protective effect on HBV challenge. The inoculated HBV dose is 50 HBV l^{-1}. The homeostatic concentration of antibodies to HBsAg is 5×10^{10} IgG molecules l^{-1}.The dose of simultaneously administered HBsAg vaccine is 1.25×10^{14} pt l^{-1}. The time course of vaccine (- - -), viruses(—), and infected cells is shown. The immune response induced by vaccine prevents from hepatitis B infection.

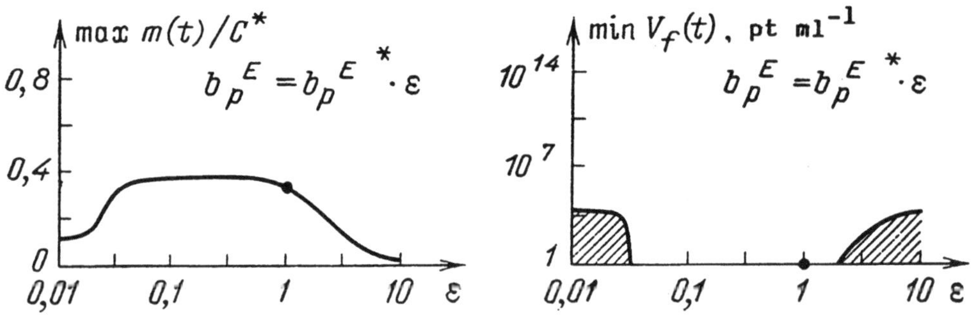

Fig.2 Relationships between CTL reactivity rate parameter b_p^E and disease severity (m_{max}) and outcome (V_{fmin}) characteristics. Dashed domains correspond to chronization of the infection. b_p^{E*} - the basic parameter value given in table 2.

different ways. Influenza A virus is highly cytopathic: the replication rate ν = 500 pt day^{-1}, the infection rate $\sigma = 2 \times 10^{10}$ M^{-1} day^{-1} and the infected cell destruction rate b_m = 1.5 day^{-1}, so there is the need for rapid and efficient elimination by the host. Hepatitis B virus infection is initiated by a weakly cytopathic virus (ν = 1 pt day^{-1}, $\sigma = 2 \times 10^9$ M^{-1} day^{-1}, b_m = 0.07 day^{-1}), and infected hosts do survive independent of the immune response. The majority of infected individuals generate a pronounced primary immune response which eliminates HBV, together with long-lasting protection provided by B-cell memory.

Influenza A infection in adults proceeds in the context of previous antigenic experience (Ada & Jones, 1986; Mitchell et al, 1985). The majority of influenza A induced memory T-cells are cross-reactive, so the frequency of precursor T-cells available for clonal expansion (the steady-state size) is higher (about 10 to 100 fold greater) than that for hepatitis B. The fast character of influenza A infection provides a sharper time scale, thus influencing the way in which clone formation is described in the model: no more than two divisions are considered during the acute phase. The kinetic characteristics of antiviral immunity for the two different virus infections may be assessed by the values of the stimulation-rate parameters for T- and B-cells. It is interesting to note that the stimulation rates of helper T-cells and CTLs turn to be very close for acute infections of influenza A and hepatitis B. This result should be considered in the context of experimental observations that the secondary T-cell response is characterized by kinetics with no appreciable acceleration and no augmentation of the response (Zinkernagel et al, 1985). However the B-cell stimulation rate estimate is 10-fold greater for influenza A.

To assess the contribution of CTL-mediated destruction of virus infected cells to the gravity of the disease (immunopathology) and recovery from infection, we consider the following characteristics of the model solution over the disease interval: minimum $V_f(t)$ - i.e., the degree of virus clearance; and maximum $m(t)$ - i.e. the severity of the disease. Varying (for example) the stimulation rate of CTLs, related to T-cell responsiveness, it may be shown that for HBV infection the immunopathology is necessary if the virus is to be eliminated from the host (Fig. 2). An analysis of the relative growth rates of virus and CTL, based on the model system of equations, gives the following estimates of their ratio:- about 1:1 and 100:1 for acute hepatitis B and influenza A respectively. These numbers make the point that fast non-specific (IFN) mechanism(s) play a vital role in limiting the spread of influenza virus during an initial phase, before the specific immune response is able to bring the infectious process under control.

8 Conclusions

In many fields of pure and applied science, the central issue is the determination of mathematical models that are consistent with both observations and prior knowledge. We have presented a sketch of our experience in developing quantitative mathematical models of antiviral immune responses for certain specific viral infections. The schematic model of the basic immune mechanisms which provide protection of the host is not complete. Nevertheless, it serves to relate many different experimental facts and heterogeneous data from immunology, virology and clinical practice into an integrated and intuitively consistent theoretical description of the immunity to viral infection: we call this the **generalized picture** of particular infectious diseases. The set of model parameters provides a single scale for the analysis and comparison of immune responses during different infections. Moreover, the basic assumptions of the model are explicitly formulated, and hence are open for supplement and refinement from both the theoretical and the experimental points of view.

The principal problem of research is, indeed, the lack of homogeneous data sets: the problems of relating *in vitro* and *in vivo* data, animal and human data, experimental units and the physical numbers of cells and substances, etc. This fundamental problem leads to an uncertainty in our mathematical description which we are not yet able to express in quantitative terms. However, precisely because the basic assumptions of the model are explicitly formulated, the limitations of the single model approach to different infections come into clear focus. Thus, we are able to conclude that a realistic mathematical model of fast virus infection should contain explicitly the non-specific mechanisms contributing to the limitation of the infection at early stages (IFN response). At the same time, we believe that the corresponding correction of the model will not destroy the core descriptive tools - data in the form of the "generalized pictures", and particular sets of parameter values. Refined quantitative multiparameter modeling of immune responses in viral diseases seems to be a logically necessary step in the development of deterministic and descriptive theoretical tools for the analysis of immune reaction mechanisms *in vivo*.

The authors thank Dr. J. Stewart for valuable comments and stimulating discussions.

References

Ada GL, Jones PD (1986) The Immune Response to Influenza Infection. Curr Top Microbiol Immunol 128:1-54

Bell GI (1970) Mathematical Model of Clonal Selection and Antibody Production. J Theor Biol 29:191-232

Bell GI, Perelson AS, Pimbley GH (eds) (1978) Theoretical Immunology. Marcel Dekker, New York and Basel

Bocharov GA, Romanyukha AA Mathematical Model of Antiviral Immune Response. III. Influenza A Virus Infection. (in preparation)
De Boer RJ, Perelson AS (1991) Size and Connectivity as Emergent Properties of a Developing Immune Network. J Theor Biol 149:381-424
Doherty PC (1985) T cells and viral infections. Br Med Bull 41:7-14
Doherty PC (1986) Some problem areas in the interaction between viruses and the immune system. Immunol Cell Biol 65:279-286
Mitchell DM, McMichael AJ, Lamb JR (1985) The immunology of influenza. Br Med Bull 41:80-85
Marchuk GI, Petrov RV, Romanyukha AA, Bocharov GA (1991a) Mathematical Model of Antiviral Immune Response. I. Data Analysis, Generalized Picture Construction and Parameters Evaluation for Hepatitis B. J Theor Biol 151:1-40
Marchuk GI, Romanyukha AA, Bocharov GA (1991b) Mathematical Model of Antiviral Immune Response. II. Parameters Identification for Acute Viral Hepatitis B. J Theor Biol 151:41-70
Paul WE (ed) (1984) Fundamental Immunology. Raven Press, New York
Perelson AS (ed) (1988) Theoretical Immunology, SFI Studies in the Science of Complexity, vol II & III. Addison-Wesley, Redwood, CA
Perelson AS (1988) Towards a realistic model of the immune network. In: Perelson AS (ed) Theoretical Immunology Part Two, SFI Studies in the Science of Complexity, vol III. Addison-Wesley, Redwood, CA. 337-401
Popper H, Schaffer F (eds) (1986) Progress in Liver Disease, vol VIII. Grune & Stratton, Orlando
Romanyukha AA, Bocharov GA, Marchuk GI (1991) Mathematical modeling of infectious diseases. In: Da Prato (ed) Mathematical Aspects of Human Diseases, CNR series on applied mathematics. (in press)
Stewart J, Varela FJ, Coutinho A (1989) The Relationship Between Connectivity and Tolerance as Revealed by Computer Simulation of the Immune Network: Some Lessons for an Understanding of Autoimmunity. J Autoimmun (Suppl) 2:15-23
Weisbuch G, De Boer RJ, Perelson AS (1990) Localized Memories in Idiotypic Networks. J Theor Biol 146:483-499
Zinkernagel RM, Hengartner H, Stitz L (1985) On the role of viruses in the evolution of the immune responses. Br Med Bull 41:92-97

MIMICKING THE STRATEGY OF THE IMMUNE SYSTEM: INSIGHT GAINED FROM MATHEMATICS

Z. Agur, G. Mazor[1] and I. Meilijson[1]
Department of Applied Mathematics and Computer Science
The Weizmann Institute of Science
Rehovot IL-76100, Israel

Abstract

The humoral immune response to antigenic challenge involves a Darwinian process of somatic mutations and selection for B-cells carrying higher affinity antibodies. Insufficient knowledge about this process may hamper the use genetically engineered antibodies for vaccination. The aim of the present work is to circumvent some of the experimental difficulties in the study of affinity maturation by investigating this system mathematically. Using dynamic programming methods we look for mutation rate as a function of clone size, which maximizes the probability that the required antibody structure is generated before the pathogen kills the host. We show analytically that the globally optimal strategy for a fast production of high affinity antibodies is to utilize a step-function mutation rate, i.e., a minimal mutation rate in early stages of the immune response, followed by the maximal possible rate when the proliferating B-cells population size exceeds a given threshold. Based on the methodology we have followed it may be concluded that the good performance of this simple, thrifty, two-stage strategy cannot be improved by evolutionary development of a more complex control of the response. Laboratory experiments are suggested for testing the validity of our theoretical results.

Introduction

Monoclonal antibodies can now be genetically engineered and there seem to be good prospects for future construction of antibodies with any desired structure. Yet, it is not clear that this outstanding progress in biotechnology will be sufficient for generating effective vaccination. The immune system does not operate through a predicted tertiary structure, but, rather, through a continuous mutation and selection process. "Can we imitate this strategy and indeed improve on it?"[12]

---

[1] School of Mathematical Sciences, Raymond and Beverly Sackler Faculty of Exact Sciences, Tel Aviv University, Israel.

NATO ASI Series, Vol. H 66
Theoretical and Experimental Insights into Immunology
Edited by A. S. Perelson and G. Weisbuch

In the animal high affinity antibodies are produced in two stages. The first stage involves the generation of a "preimmune" repertoire of B-cells, expressing predominantly low affinity antibodies. The second stage follows antigenic challenge and involves clonal expansion and further diversification of the expanding clones by a poorly understood process of somatic hypermutation. This process involves an accelerated rate of point mutations along the DNA sequences encoding the variable region of the antibody, in conjunction with a larger proliferation rate of B-cells bearing the higher affinity antibodies[11,5]. It has been suggested that hypermutations occur during any round of antigenic challenge[3,2], whereas another view holds that cells carrying modifications that increase antibody binding affinity are selected only into the memory compartment, to produce an improved secondary response upon re-exposure to the antigen[11]. Involving fast dynamics on different organization levels of the biological system these processes are too intricate to be analyzed by laboratory experiments alone. Indeed, different dynamics of affinity maturation are currently implicated on the basis of the same experimental results[8,9].

In view of the difficulties in the experimental method can one still hope to gain a substantially good understanding of the immune strategy to enable its improvement? An attempt to aid in elucidating the humoral response is made in the present work, which treats affinity maturation as an optimization problem. Using dynamic programming methods we identify the globally optimal strategy for a fast production of high affinity antibodies and show that its properties are similar to those reported for the hypermutation mechanism in the response to hapten 2-phenyl-5-oxazolone[6]. The mathematical analysis is described in detail below, while the biological conclusions are elaborated in the biological literature[1].

Upon pathogenic invasion the humoral immune system is required to produce high affinity antibodies, by modifying a preimmune low affinity antibody, so as to eliminate the pathogen with minimum damage to the body. Pathogen-induced body deterioration dictates an early activation of the mutation process, to minimize the time until a "right" mutation occurs. On the other hand, as the mutation process is random and the chances of generating right mutations are slim, activating hypermutations when the B-cell clone size of moderate affinity antibodies is still small, may cause a random drift to low affinity antibodies, and, as a consequence, a response arrest. The mathematical analysis brought forward below will suggest how these contradicting requirement can be optimally reconciled.

In our model mutations are assumed to occur at random along the DNA sequences that determine the antibody-antigen binding efficacy. Mutation and proliferation occur one at a time, in the locally independent way characteristic of Markov processes. This assumption stipulates that mutations are expressed immediately at the phenotypic level, and not only after the next proliferation[8]. We assume further that only one specific antibody structure, not contained in the preimmune repertoire, is capable of initiating the process of pathogen elimination and that by clonal selection and hypermutation a preimmune antibody structure may be modified to generate the required specific structure. For analytic tractability the epitope and the pathogen-induced host's death rate are taken to be constant and clonal expansion is assumed to involve only B-cells bearing antibodies that are one mutation removed from the required structure. Below we discuss the relaxation of these assumptions.

Methods and Results

Denote the proliferation rate of the proliferating B-cells by λ, the pathogen-induced host's death rate by r, and the DNA length of the binding region by m. We are looking for a mutation rate as a function of clone size, which maximizes the probability of pathogen elimination. Letting I_n and P_n stand for the probability of survival and the mutation rate respectively, when the population size of the proliferating B-cells is n, we obtain the probability that the next event is not the death of the host, as follows

$$\frac{\lambda n + mnP_n}{\lambda n + mnP_n + r}. \tag{1}$$

From (1) we can derive a recursive relation for I_n:

$$\begin{aligned} I_n &= \frac{\lambda n}{\lambda n + mnP_n + r} I_{n+1} + \frac{(m-1)nP_n}{\lambda n + mnP_n + r} I_{n-1} + \frac{nP_n}{\lambda n + mnP_n + r} = \\ &= \frac{n}{\lambda n + mnP_n + r} \left(\lambda I_{n+1} + (m-1)P_n I_{n-1} + P_n \right). \end{aligned} \tag{2}$$

From (2) we obtain:

$$\frac{\partial I_n}{\partial P_n} = n \cdot \frac{(\lambda n + r)(m-1)I_{n-1} - \lambda nmI_{n+1} + \lambda n + r}{(\lambda n + mnP_n + r)^2},$$

whose sign is independent of P_n. It follows by the Howard policy-iteration theorem of dynamic programming[4] that any optimal strategy would use for any size of proliferating B-cell population, either the maximal or the minimal mutation rates.

If there is no limit on how high the mutation rate might be, then the optimal strategy is given by Theorem I:

Theorem I.

Let

$$n_o = \lfloor m \cdot \left(1 - \frac{r}{\lambda}\right) \rfloor$$

then

$$P_n = \begin{cases} 0 & \text{if } n \le n_0; \\ \infty & \text{if } n > n_0; \end{cases}$$

is optimal.

Proof.

Let P_n be a step function of that type, i.e:

$$P_n = \begin{cases} 0 & \text{if } n \leq N; \\ \infty & \text{if } n > N. \end{cases}$$

For $n = N + 1$, the next change is positively a mutation, as the mutation rate is infinite. Therefore

$$I_{N+1} = \frac{m-1}{m} \cdot I_N + \frac{1}{m}. \tag{3}$$

For $n = N$, no mutation can occur, and the next change is either cell proliferation (stepping to $N + 1$) or the host's death. Hence

$$I_N = \frac{\lambda N}{\lambda N + r} \cdot I_{N+1}. \tag{4}$$

From (3) and (4) we obtain

$$I_N = \frac{\lambda N}{\lambda N + r} \cdot \left(\frac{m-1}{m} \cdot I_N + \frac{1}{m} \right).$$

In other words,

$$I_N = \frac{\lambda N}{\lambda N + rm}. \tag{5}$$

As mutation rate is zero from 1 to N, I_1 can be represented in terms of I_N:

$$\begin{aligned} I_1 &= \frac{\lambda}{\lambda + r} \cdot I_2 \\ &= \dots = \frac{\lambda}{\lambda + r} \cdot \frac{2\lambda}{2\lambda + r} \cdot \dots \cdot \frac{(N-1)\lambda}{(N-1)\lambda + r} \cdot I_N \\ &= \frac{\lambda}{\lambda + r} \cdot \frac{2\lambda}{2\lambda + r} \cdot \dots \cdot \frac{(N-1)\lambda}{(N-1)\lambda + r} \cdot \frac{N\lambda}{N\lambda + rm} \\ &= \frac{1}{1 + \frac{r}{\lambda}} \cdot \frac{2}{2 + \frac{r}{\lambda}} \cdot \dots \cdot \frac{N-1}{N - 1 + \frac{r}{\lambda}} \cdot \frac{N}{N + m \cdot \frac{r}{\lambda}} \\ &= \frac{N! \cdot \Gamma\left(1 + \frac{r}{\lambda}\right)}{\left(N + m \cdot \frac{r}{\lambda}\right) \cdot \Gamma\left(N + \frac{r}{\lambda}\right)}. \end{aligned} \tag{6}$$

For N to be optimal with respect to I_1, we need:

$$\frac{N\lambda}{N\lambda + rm} \geq \frac{N\lambda}{N\lambda + r} \cdot \frac{(N+1)\lambda}{(N+1)\lambda + rm} \tag{7}$$

and

$$\frac{(N-1)\lambda}{(N-1)\lambda + r} \cdot \frac{N\lambda}{N\lambda + rm} \geq \frac{(N-1)\lambda}{(N-1)\lambda + rm}. \tag{8}$$

From (7) we get:

$$\frac{N\lambda}{N\lambda+rm} \geq \frac{N\lambda}{N\lambda+r}\cdot\frac{(N+1)\lambda}{(N+1)\lambda+rm} \Longleftrightarrow$$

$$m\left(1-\frac{r}{\lambda}\right) \leq N+1. \tag{9}$$

From (8) we get:

$$\frac{(N-1)\lambda}{(N-1)\lambda+r}\cdot\frac{N\lambda}{N\lambda+rm} \geq \frac{(N-1)\lambda}{(N-1)\lambda+rm} \Longleftrightarrow m\left(1-\frac{r}{\lambda}\right) \geq N. \tag{10}$$

From (9) and (10) we get:

$$N \leq m\left(1-\frac{r}{\lambda}\right) \leq N+1,$$

and thus

$$n_0 = \lfloor m\left(1-\frac{r}{\lambda}\right)\rfloor \tag{11}$$

is always optimal, and if $m\left(1-\frac{r}{\lambda}\right)$ is an integer, then $m\left(1-\frac{r}{\lambda}\right)+1$ is optimal as well.

Note that for n_0 to be larger than zero, λ has to be larger than r, which is a reasonable constraint.

Calculating I_k, when $k > n_0$:

$$I_{n_0+1} = \frac{m-1}{m}I_{n_0} + \frac{1}{m}$$

$$I_{n_0+2} = \frac{m-1}{m}I_{n_0+1} + \frac{1}{m} = \left(\frac{m-1}{m}\right)^2 I_{n_0} + \frac{m-1}{m^2} + \frac{1}{m},$$

and by induction we get:

$$I_{n_0+k} = \left(\frac{m-1}{m}\right)^k I_{n_0} + 1 - \left(\frac{m-1}{m}\right)^k$$

$$= 1 - \left(\frac{m-1}{m}\right)^k \cdot (1 - I_{n_0}). \tag{12}$$

Equation (12) makes sense, as its right-hand side is exactly a unity minus the probability of missing all the k instantaneous mutations, and then failing also with the n_0 B-cells that were left.

For $0-\infty$ step-function P_n, we obtain :

$$I_n = \begin{cases} \frac{(N-1)!}{(n-1)!}\cdot\frac{\Gamma(n+\frac{r}{\lambda})}{\Gamma(N+\frac{r}{\lambda})}\cdot\frac{N}{N+m\cdot\frac{r}{\lambda}} & \text{if } n \leq N \\ 1-\left(\frac{m-1}{m}\right)^{n-N}\cdot\left(1-\frac{N}{N+m\cdot\frac{r}{\lambda}}\right) & \text{if } n > N. \end{cases} \tag{13}$$

In particular, for the optimal n_0, $n_0 = \lfloor m\left(1-\frac{r}{\lambda}\right)\rfloor$, we get:

$$I_{n_0} = \frac{n_0}{n_0 + m\cdot\frac{r}{\lambda}} \approx \frac{m\cdot\left(1-\frac{r}{\lambda}\right)}{m\cdot\left(1-\frac{r}{\lambda}\right)+m\cdot\frac{r}{\lambda}} = 1-\frac{r}{\lambda}.$$

From now on, I refers to the strategy above exclusively.

To prove optimality we have to show that the strategy (P_n), defined by P_n, is indeed better than any other strategy we may use. Using dynamic programming methods, we will show that (P_n) is excessive, i.e., we will check that when $P_n = 0$ an increase in the mutation rate to infinity for the first step, followed by the strategy (P_n), can not be of an advantage, and vice versa. Excessivity is sufficient for optimality because the host's death rate, r, makes this a Discounted Dynamic Programming problem with bounded rewards[4].

For $n > n_0$, we have to check that changing P_n from ∞ to 0 for one step would not be of an advantage.

Lemma I.

$$I_{n_0+k} \geq \frac{(n_0+k)\lambda}{(n_0+k)\lambda + r}\cdot I_{n_0+k+1} \qquad \forall k \in N$$

Proof.

$$I_{n_0+k} \geq \frac{(n_0+k)\lambda}{(n_0+k)\lambda + r}\cdot I_{n_0+k+1} \iff$$
$$n_0 \geq m\left(k\cdot\frac{(m-1)^k}{m^{k+1}-m(m-1)^k}-\frac{r}{\lambda}\right).$$

But $n_0+1 \geq m\,(1-r/\lambda) \Longrightarrow n_0 \geq m\left(1-\frac{1}{m}-\frac{r}{\lambda}\right)$ so it is enough to prove that $1-\frac{1}{m}-\frac{r}{\lambda} \geq k\cdot\frac{(m-1)^k}{m^{k+1}-m(m-1)^k}-\frac{r}{\lambda}$ for all k.

$$1-\frac{1}{m}-\frac{r}{\lambda} \geq k\cdot\frac{(m-1)^k}{m^{k+1}-m(m-1)^k}-\frac{r}{\lambda} \iff$$
$$\left(1+\frac{1}{m-1}\right)^k \geq 1+\frac{k}{m-1},$$

which is well known, and this completes the proof of Lemma I.

For $n \leq n_0$, we have to check that changing P_n from 0 to ∞ for one step would not be of an advantage.

Lemma II.

$$I_n \geq \frac{m-1}{m}\cdot I_{n-1}+\frac{1}{m} \qquad \forall n \leq n_0$$

Proof.

$$I_n \geq \frac{m-1}{m} \cdot I_{n-1} + \frac{1}{m} \Longleftrightarrow$$
$$I_n \geq \frac{(n-1)\lambda + r}{(n-1)\lambda + rm}$$

For $n = n_0$:

$$\frac{n_0\lambda}{n_0\lambda + mr} \geq \frac{(n_0-1)\lambda + r}{(n_0-1)\lambda + mr} \Longleftrightarrow$$
$$n_0 \leq m\left(1 - \frac{r}{\lambda}\right),$$

which agrees with (11).

By induction we get

$$I_n \geq \frac{(n-1)\lambda + r}{(n-1)\lambda + mr} \Longleftrightarrow$$
$$\frac{(n-1)\lambda}{(n-1)\lambda + r} \cdot I_n \geq \frac{(n-1)\lambda}{(n-1)\lambda + mr} \Longleftrightarrow$$
$$I_{n-1} \geq \frac{(n-1)\lambda}{(n-1)\lambda + mr}$$

thus it is enough to prove that

$$\frac{(n-1)\lambda}{(n-1)\lambda + mr} \geq \frac{(n-2)\lambda + r}{(n-2)\lambda + mr}$$

$$\frac{(n-1)\lambda}{(n-1)\lambda + mr} \geq \frac{(n-2)\lambda + r}{(n-2)\lambda + mr} \Longleftrightarrow$$
$$n-1 \leq m\left(1 - \frac{r}{\lambda}\right),$$

which agrees with (11). This completes the proof of lemma II, and therefore completes the proof of the optimality of (P_n). Q.E.D.

However, the globally optimal strategy, defined in Theorem I, is biologically unrealistic since it requires an infinite rate of hypermutation and a real-time control of its onset, depending critically on the number of B-cells: if the rate of hypermutation is very high, the activation of this process causes an immediate decrease in B-cell clone size below the threshold n_0, whereupon hypermutation should instantaneously be switched off, only to be switched on again upon further cellular proliferation. To avoid the need of such a biologically infeasible real-time control of the onset of hypermutation, we will limit the strategies to those that activate the process only once. For this purpose, the maximal mutation rate has to be low enough, relative to the proliferation rate, so that the number of B-cells in the clone will not get a negative trend merely by the activation of hypermutation. Therefore, we assume that mutation rate has as upper bound a value which is slightly smaller than $\frac{\lambda}{m-1}$. For mutation rates below this bound, the B-cell population size tends to increase (until the "decisive" moment of host's death or correct mutation), and thus hypermutations may be activated only once during the infection, without too large a probability of losing all the reacting B-cells. Moreover, putting such a bound on the mutation rate makes it also biologically reasonable, and thus solves both problems.

For biologically realistic values for the parameters m, λ and r, the restriction to these strategies has a very minor effect on the probability of survival. For all examples calculated, the probability of response failure, $1-I_1$, under the best of these biologically feasible policies, increases by no more than one percent of the probability of response failure when using the optimal unbounded mutation rate strategy. Fig 1 shows numerical results of the analysis, for rates close to $\lambda/(m-1)$. The probability of response

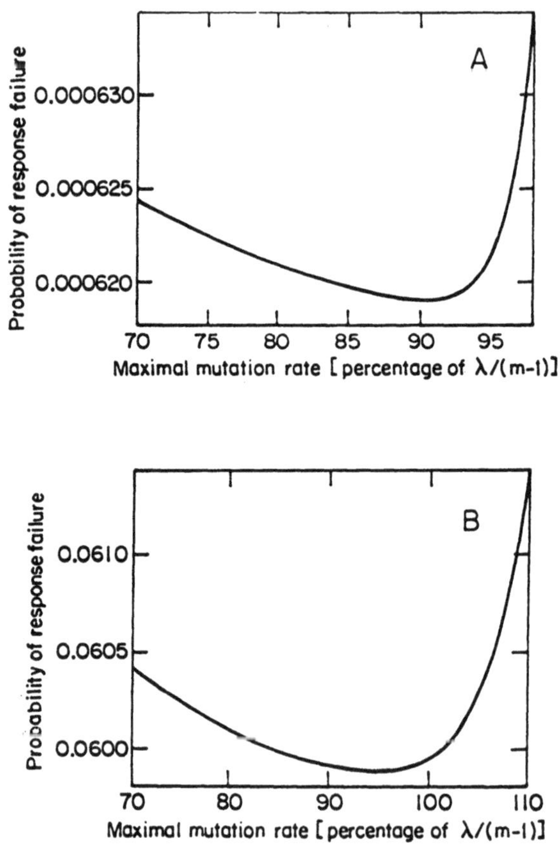

Fig. 1. Probability of response failure $(1-I_1)$ for upper bounds for the mutation rate close to $\frac{\lambda}{m-1}$ and a single activation of the hypermutation mechanism. **A.** $\frac{r}{\lambda}=0.0001$, $m=100bp$; for these parameters the ratio between the minimal probability of response failure under bounded mutation rate (6.19×10^{-4}) and the probability of response failure under the optimal unbounded mutation rate strategy (6.174×10^{-4}) is 1.00236. **B.** $\frac{r}{\lambda}=0.01$, $m=100bp$; for these parameters the ratio between the minimal probability of response failure (0.05988) and the probability of response failure under the optimal unbounded mutation rate strategy (0.05978) is 1.00170.

failure at the bottom of the vertical axis corresponds to the optimal unbounded mutation rate strategy. Note that if mutation rates are decreased below $\approx 95\%$ of $\frac{\lambda}{m-1}$, the probability of response failure increases slowly. In contrast, for upper bounds on mutation rate larger than $\frac{\lambda}{m-1}$, that is, for rates that create a drift to low population sizes, the death probability increases significantly.

Based on the methodology we have followed it may be concluded that the good performance of these simple, thrifty, two-stage strategies cannot be improved by evolutionary development of a more complex control of the response. Altough the analysis presented in this work refers to B-cells bearing antibodies that differ by one mutation from the required structure, we have also shown analytically that the optimality of this "bang-bang" strategy, holds for B-cells carrying antibodies that are more than one mutation removed from the required structure. Analysis of the mutation-selection process operating on different clones of B-cells, under constant or varying antigenic challenge, is under way.

We now compare the analytical results with *in vitro* estimation for the immune response to the hapten 2-phenyl-5-oxazolone[6]. These experiments suggest that around the seventh day of response, the mutation rate is augmented to become $\approx 10^{-3}$ mutations/bp/generation. According to our mathematical prediction, onset of hypermutation should immediately follow the establishment of a clone of size $m\left(1-\frac{r}{\lambda}\right)$, which, for the realistic assumption of r being much smaller than λ, roughly equals m, i.e., a few hundred nucleotides. Hence, the time until the establishment of such a clone, having initially one B-cell of that type, is nine or ten generations $(= log_2 500, log_2 1000)$. As B-cell generation time is now assumed to be roughly 10 hours[7], this corresponds to four days of proliferation. Bearing in mind the 3-day time-lag that precedes clonal expansion in the primary immune response (Milstein, personal communication), our result matches the experimental observations. For a sequence of $\approx 1000bp$ and a hypermutation rate $\approx 10^{-3}$, the probability of generating a mutated B-cell is close to the probability of B-cell proliferation. Thus, also the upper bound for the mutation rate appears to be in agreement with laboratory estimations.

Laboratory experiments are warranted for testing the biological relevance of our theoretical results. Our analysis suggests that a step-function mutation rate with a transition from the minimal to the maximal possible rate of mutation, should practically optimize the immune response at any round of antigenic challenge. This result supports Milstein *et al's* experimental work (see above). Further mathematical work (Agur *et al*, in preparation) provides tools for using the observed distribution of silent mutations in an expanding B-cell clone for checking if hypermutations occur also during the secondary and subsequent responses. The analysis and the computations presented in this work further indicate that the onset of hypermutation is determined by the size of the proliferating B-cell clone. An experimental support to this result may be obtained by modifying B-cell proliferation rate; a higher proliferation rate should result in an earlier onset of hypermutation, and vice versa. Moreover, a corollary to this result is that the time preceding the onset of hypermutation is shorter in later rounds of antigenic challenge. This suggestion can be tested experimentally, once the occurrence of hypermutations in the secondary and further responses is verified. Another conclusion from the present analysis is that the activation of hypermutation before the critical

B-cell clone size is reached may lead to an inefficient response. It may be interesting to check if such a mechanism is involved in the generation of tolerance.

References

1. Agur Z, Mazor G, Meilijson I (1991) Maturation of the humoral immune response as an optimization problem. Proc R Soc London B 245:147–150

2. Berek C, Milstein C (1987) Activation of memory and virgin B cell clones in hyperimmune animals. Immunol Rev 96:23–41

3. Berek C, Griffiths GM, Milstein C (1985) Molecular events during maturation of the immune response to oxazolone. Nature 316:412–418

4. Blackwell D (1965) Discounted dynamic programming. Annals of Math Statistics 36:226–235

5. French DL, Laskov R, Scharff MD (1989) The role of somatic hypermutation in the generation of antibody diversity. Science 244:1152–1157

6. Griffiths GM, Berek C, Kaartinen M, Milstein C (1984) Somatic mutation and the maturation of the immune response to 2-phenyl-oxazolone. Nature 312:271–275

7. Maclennan ICM, Gray D (1986) Antigen driven selection of virgin and memory B-cells. Immunol Rev 91:271–275 (1986)

8. Manser T (1990) The efficiency of antibody affinity maturation: can the rate of B-cell division be limiting? Immunol Today 11:305–307

9. Manser T (1991) Reply. Immunol Today 12:93–94

10. Milstein C (1991) Affinity maturation of antibodies. Immunol Today 12:93

11. Rajewsky K, Forster I, Cumano A (1987) Evolutionary and somatic selection of the antibody repertoire in the mouse. Science 238:1088–1094

12. Winter G, Milstein C (1991) Man-made antibodies. Nature 349:293–299

Rash Theory

Lee A. Segel
Department of Applied Mathematics and Computer Science
The Weizmann Institute of Science
Rehovot IL-76100, Israel

Alan S. Perelson and James M. Hyman
Theoretical Division
Los Alamos National Laboratory
Los Alamos, NM 87545 USA

Sidney N. Klaus, M.D.
Department of Dermatology and Venereology
Hadassah Hospital
Jerusalem IL-91120, Israel

Abstract

A first step is made toward mathematical modeling of pathogen–immune system interactions that might lead to the observed spatial distributions of rashes. Two types of patterns are defined. Type I patterns are stationary and are composed of stripes, networks, or regular arrays of spots. Type II patterns are composed of moving waves of linear, spiral, or target forms. Known pattern formation models (of reaction-diffusion type) are interpreted as relevant to very simple representations of immune-pathogen interactions. Criteria for Type I and Type II patterns are discussed. No observed immune–related patterns can be clearly identified as of Type I, but Type II rash patterns definitely occur, for example in erythema gyratum repens.

Introduction

The study of spatial patterns has interested theoreticians for decades. Trying to find the reason for an observed spatial pattern is a challenging scientific problem, for the existence of such patterns can often provide valuable clues to underlying mechanisms. For example, the immense time and length scales involved make it difficult to investigate the astrophysical phenomena that underlie the formation of galaxies. The fact that these galaxies take on various types of spiral forms provide an important constraint on

NATO ASI Series, Vol. H 66
Theoretical and Experimental Insights into Immunology
Edited by A. S. Perelson and G. Weisbuch

theories (see for example Bertin and Lin, 1985). Any good model must be able to predict this variety of observed spirals. Similarly, a study of the variegated spatial patterns of inflammation, known as rashes, can provide information on the underlying actions of the immune system and pathogens. In fact, the particular pattern of a rash is used to diagnose a variety of diseases. We present here some initial steps in a study of rashes, with emphasis on possible mechanisms of generation of rashes by the interaction of the immune system with pathogenic agents.

There are two principle bases for patterns of rashes. These might be termed *allogenesis* and *autogenesis*, respectively. In allogenesis, the reason for the pattern is, in a sense, separate from the underlying mechanisms of the disease. For example, a rash in shingles is due to the attack of a virus on a nerve. Thus the rash pattern reflects the underlying nerve pattern (Figure 1). By contrast, there are numerous patterns of rashes that seem to have no obvious counterpart in underlying anatomy. They are generated autonomously; hence the term autogenesis. (See Figure 7 below.) It is autogenic patterns that we wish to study.

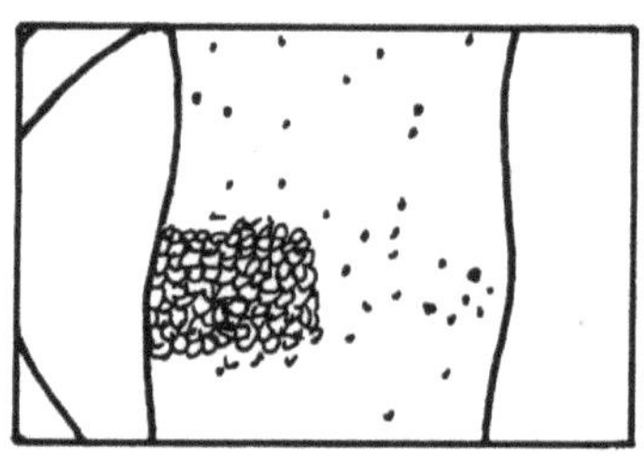

Figure 1. Shingles (herpes zoster): the rash is largely confined to the area above the nerves that are the focus of the viral attack. Sketched from a slide of a patient.

We shall search for pattern formation mechanisms in the framework of what may be called interaction-transport equations. In the present context such equations embody assumptions concerning the nonlinear interaction between various pathogen and host factors and the transport of these quantities in space. Pattern formation has been studied particularly intensively in the subclass of interaction-transport equations that will be of primary interest here, namely reaction-diffusion problems where the interaction is chemical and the spatial transport is via diffusion. However the same general principles of pattern formation occur in broad classes of interaction-transport equations (Levin and Segel, 1985).

Interaction-transport equations can exhibit various types of spatial patterns. One group of such patterns consists of stripes, hexagonal networks, or regular arrays of spots (Figure 2). We term these Type I patterns. Such patterns appear in various areas of fluid mechanics, notably the Bénard-Marangoni convection problem where flow patterns develop in a layer of fluid heated from below. [A recent reference is the paper of Shtilman and Sivashinsky (1991).] Similar patterns appear from so-called Turing (1952) nstabilities in reaction-diffusion equations. It is natural to suspect that Type I patterns ılso may be characteristic of certain rashes. We will shortly pursue this possibility by ormulating a simple model of rash formation that takes the form of reaction-diffusion equations.

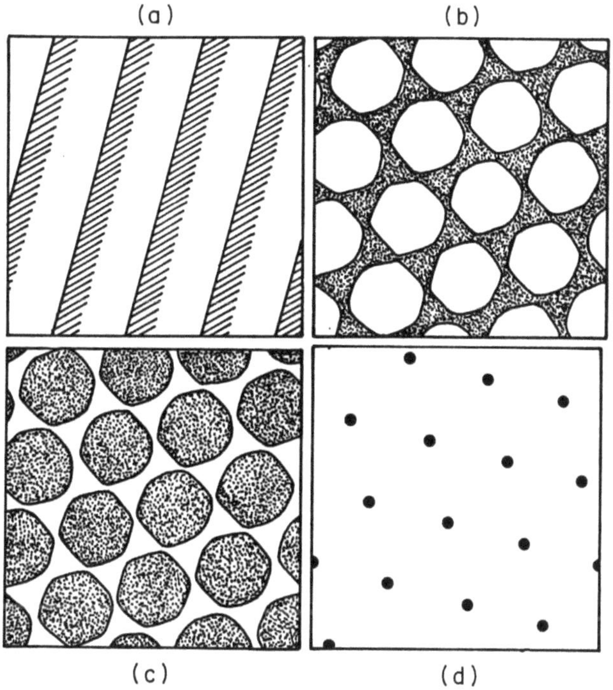

Figure 2. Ideal type I patterns. (a) Stripes. (b) Hexagonal network. (c) Regular array of large spots. The exterior of the spots forms a hexagonal network; patterns (b) and (c) are complementary. (d) Same as (c), except that the spots are smaller.

There is another widespread mechanism of pattern formation that leads to spiral and "target" waves and combinations thereof (Figure 3). See Winfree (1980). Unlike the stationary Type I patterns, *Type II patterns always move.* We shall also consider the possibility of patterns of this type.

Note that Figures 2 and 3 contain idealized versions of Type I and Type II patterns. As we shall discuss further below, in practice one expects to observe various modifications of the ideal situation.

Figure 3. Ideal type II patterns. (a) Two successive snapshots (solid line, dashed line) of a traveling "straight" wave. (b) Snapshot of a spiral wave at a particular instant. The wave rotates in time. Redrawn from Winfree (1972), Fig. 1. (c) Snapshot of centripetally propagating waves, yielding a target pattern. (Except at the center, target and spiral waves are often indistinguishable.) (d) Two interacting sets of centripetal waves. The observed pattern is due to the fact that colliding waves annihilate one another, since after the passage of a wave the medium is temporarily refractory to further excitation.

A model for Type I patterns

We now present a very simple mathematical model that can illustrate possible mechanisms for Type I autogenic pattern formation. This model will be based on two fundamental variables. The first, $V(x,y,t)$, is the concentration of some pathological agent, for example a virus. For simplicity, we shall usually refer to V as a virus concentration. Indeed, viruses are frequently implicated in rash formation. The rashes of measles and chicken pox are common examples. Nevertheless it should be borne in mind that V might represent other pathological agents. (Note that in autoimmune disease, the "pathological agent" may in fact be a normal bodily constituent whose presence stimulates an immune response.) The second variable, $I(x,y,t)$, represents the concentration of some product of the immune system that attacks the pathological agent. In this first model we neglect changes in the direction normal to the skin, so that V and I should usually be thought of as average concentrations in a layer of skin.

The structure of the equations for both V and I is as follows:

rate of change of concentration as a function of time	=	effect of "births"	−	effect of "deaths"	+	net "diffusive" flow to (x, y) from nearby spatial points.

The corresponding equations are

$$\frac{\partial V}{\partial t} = r(V)V - d(I,V) + D_V \left[\frac{\partial^2 V}{\partial x^2} + \frac{\partial^2 V}{\partial y^2} \right], \tag{1a}$$

$$\frac{\partial I}{\partial t} = s(I,V) - n(I)I + D_I \left[\frac{\partial^2 I}{\partial x^2} + \frac{\partial^2 I}{\partial y^2} \right]. \tag{1b}$$

In (1a), $r(V)$ denotes the proliferation rate of the virus, as a function of the level of virus. The function $d(I,V)$ represents the rate of killing of the virus by the immune agent. The diffusion terms represent viral spreading by the infection of adjacent cells, at a rate proportional to the viral diffusivity coefficient D_V. The macroscopic appearance of diffusion arises from simplifying what might be a complicated chain of events and assuming that infection spreads by a random walk from cell to cell.

In equation (1b), the term $s(I,V)$ represents secretion of the immune agent. Both the existing concentration of the immune agent and the amount of virus can influence the secretion rate. In particular, one expects the secretion rate to increase at higher viral levels. If the immune agent is some sort of cell, then $s(I,V)$ may be viewed as a source term corresponding to the rate of appearance of the cells in the skin. The term $n(I)$ represents the "death rate" of I, a measure of the decay of efficacy of the immune agent with time. The final terms in the equation represent the spatial diffusion, with diffusivity D_I, of the immune agent. This would be ordinary chemical diffusion if I were a chemical, and would be caused by random motion if I represented a cell.

In constructing this simple model, we have not been specific about the entities involved. Such a tactic has both advantages and disadvantages. Disadvantageous is the fact that without somewhat precise specification of the constituents of the model, it is hard to design an experiment to test the theory. There is the compensating advantage that a general theory might reveal broad principles that operate in a variety of examples to generate the observed spatial patterns. If a measure of success is achieved with a simple and general model, one can later investigate other models that are more realistic and complex.

We will identify the presence of a rash with a sufficiently high level of I. As in psoriasis, for example, I may be a cytokine that induces abnormally high proliferation of skin cells (probably by increasing the fraction of cells that leave G_0 to proliferate) and also promotes abnormal adherence of dead cells to the skin. Alternatively, I could be a factor that increases blood vessel permeability, as in cutaneous angiitis. Note that by taking a single "immune variable" we almost certainly are lumping several factors: several different immune-related agents may have the functions of attacking V, affecting skin cell proliferation, modifying capillary permeability etc. Moreover, the intensity of a rash at a given time may well depend on the past history of I, not just its present value. Indeed, it is conceivable that the intensity of the rash is better correlated with the level of V than with the level of I.

For definiteness we take the following fully specified pair of equations for further study.

$$\frac{\partial V}{\partial t} = (1+kV)V - m^2 IV + \frac{\partial^2 V}{\partial x^2} + \frac{\partial^2 V}{\partial y^2}, \qquad \frac{\partial I}{\partial t} = IV - I^2 + \theta^2 \left[\frac{\partial^2 I}{\partial x^2} + \frac{\partial^2 I}{\partial y^2}\right]. \quad (2a, b)$$

Here k, m, and θ are constants. These equations can be regarded as a special case of (1). They also can be obtained from a more general version of (1) through the introduction of dimensionless variables. In particular, the parameter θ^2 can be thought of as the ratio of the diffusivities D_I/D_V. (See Appendix.)

We will examine the above equations in the case that k is positive. Thus, for the range of populations under consideration, an increase in the viral population increases the viral reproduction rate. As discussed below, such an autocatalytic effect seems essential for Type I pattern formation. We expect that at higher values of the viral population, the viral birth rate will decrease rather than increase. However, we shall assume for the moment that the viral population level remains sufficiently small that the ultimate decrease in proliferation rate can be ignored. Shortly we will modify the growth term so that this restriction will not be needed.

In (2), the interaction of virus and immune factor—whether for viral destruction [Equation (2a)] or immune stimulation [Equation (2b)]—is assumed to follow mass action kinetics and therefore to be proportional to the product IV. The decay of the immune factor is assumed to be dominated by some sort of cooperative effect that can be modeled by a term proportional to I^2. We shall see that an autocatalytic suppressive effect of this nature is another necessity for Type I pattern formation.

Reaction-diffusion theory (see, for example, Segel and Jackson, 1972) assists one to conjecture that the system (2) constitutes an example of a reasonable model for describing the major factors that might induce Type I rash patterns. The particular form of the model was selected to be identical with a model previously derived and analyzed in the context of an investigation of pattern formation in ecology (Segel and Levin, 1976). This has the advantage that existing fairly extensive calculations can be exploited. These calculations concerned a one dimensional version of the model, wherein y-dependence was ignored. The calculations showed that if and only if certain conditions on the parameters hold, there is a stable uniform solution to the equations. That is, no pattern is expected to arise, since I and V are at constant values throughout the skin. However under other conditions, the uniform state is unstable: inevitable small perturbations will begin to grow exponentially in time. A nonlinear analysis indicates that the perturbation growth eventually slows and a new steady state results. This teady state is spatially nonuniform. It consists of essentially sinusoidal increases and lecreases of V and I about their mean values. Thus a striped pattern may be generated. An outline of the relevant results found by Segel and Levin (1976) is presented in Appendix A.]

We have analyzed the two-dimensional system (2) numerically. In our initial computer simulations we noted that at certain points the viral population seemed to be growing without bound. To curb this tendency we modified equation (2a) to

$$\frac{\partial V}{\partial t} = \left(\frac{1+kV}{1+\epsilon V^2}\right)V - m^2 IV + \frac{\partial^2 V}{\partial x^2} + \frac{\partial^2 V}{\partial y^2}. \tag{2c}$$

The parameter ϵ was taken to be small enough so that $\epsilon\overline{V}^2 \ll 1$, where $\overline{V}$ is the uniform steady state value of V. Thus the various linear and nonlinear stability analyses, which were done for (2a) and which are relevant for $V \approx \overline{V}$, are essentially unaffected by this change. However the new per capita proliferation rate $(1+kV)/(1+\epsilon V^2)$ now decreases to zero for large V. As expected, "blow up" of V no longer occured with the modified model. In a one dimensional analysis, the analytical results of Segel and Levin (1976) were reproduced in detail. This gives confidence in the computer simulations. Upon extending the numerical calculations to two dimensions, we found as expected that a two dimensional perturbation gives way to a one dimensional array of stripes (Figure 4).

Figure 4. Striped pattern obtained by numerical simulation of (2b) and (2c). Parameters as in (A13). (a) Contour plots of viral level V at $t = 0$. $[7 \le V \le 11.6$, *contour interval* $= 0.2]$. *(b) Immune factor I at* $t = 0$. $[6 \le I \le 7.9$, *interval* $= 0.1]$. *(c) V at* $t = 20$ $[7.4 \le V \le 8.25$, *interval* $= 0.05]$. *(d) I at* $t = 20$ $(7 \le I \le 8.02$, *interval* $= 0.02)$. *In (c) and (d), stripes of high and low values are indicated by H and L, respectively. Calculations were carried out over the square* $-12.7 \le x,\ y \le 12.7$. *The computer plots, however, here and below, are slightly rectangular, and only the lower half of the rectangle is shown. The concomitant distortion of the pattern is not significant for our purposes.*

A number of comments should be made. (a) Initial conditions were chosen arbitrarily. It is seen in these and other simulations that (as is the case in numerous examples of pattern formation), after a transient the initial conditions have no essential effect on the patterns. (b) Although their initial conditions are different, V and I soon end up essentially in phase. This is expected for models such as ours that fulfill conditions (6) given below (R. Dillon, P.K. Maini and H.G. Othmer, unpublished. Cited by Benson, Sherratt and Maini, 1992). (c) There is little difference between the depicted patterns at $t = 20$ and the patterns at $t = 10$ and $t = 40$. That is, the pattern is already quite steady at $t = 10$. (d) The boundaries have little effect on the patterns. (e) High values of V or of I occur in horizontal bands, in areas designated by H in Figures 4(c) and 4(d). Stripes of rash centered at these high areas would range in width from zero to complete coverage of the area depending on whether the threshold value for rash formation was greater than the maximal value of I, somewhat less than this value, or even less than the minimal value of I.

Analogy with other examples of pattern formation (Segel and Stuart, 1962) suggests that if the parameter m is diminished, there might be a chance for another type of pattern. Indeed, as is shown in Figure 5, this expectation is borne out. When $t = 1$ the effect of the initial conditions has already essentially disappeared, and a striped pattern has been generated. As time goes on, however, the pattern becomes more and more spotlike. However even after 80 time units, the pattern is far from exhibiting the hexagonal symmetry that one would expect at steady state from general arguments in the theory of pattern formation. As is seen from Appendix 1, our time unit is of the magnitude of the viral doubling time.

Figure 5. Viral levels V in simulations as in Figure 3, with the same initial conditions, except that the parameter a is reduced from 0.61 to 0.52. Note the development of an array of spots. (a) $t = 1$ $[17 \le V \le 25.5$, *interval* $= 0.5]$. *(b)* $t = 10$ $[22.8 \le V \le 28.8$, *interval* $= 0.3]$. *(c)* $t = 40$ $[18.4 \le V \le 32.8$, *interval* $= 0.8]$. *(d)* $t = 80$ $[15 \le V \le 35$, *interval* $= 1.0]$.

The fact that considerable time might elapse before the pattern approaches a steady ymmetrical state has implications for the type of rash patterns that one would antici- ate if models such as the one under investigation indeed capture important aspects of he biological situation. It could well be that supercritical values of I will trigger a rash f they persist more than a certain time. This condition might be fulfilled at a time that s small compared to the time it takes for a symmetrical pattern to evolve. If so, then he observed rash pattern would be highly asymmetrical.

In a manner that is conventional in the theory of pattern formation, we have ap- roached the generation of Type I patterns by examining the stability of a uniform tate of pathological agent V and immune product I. This can be regarded as merely a nathematical device: when patterns deviate only slightly from the uniform state we can mploy the methods of nonlinear stability theory. Another possible justification of our pproach arises from the speculation that rash patterns might suddenly appear owing o the slow change of a parameter and the consequent destabilization of the uniform tate.

To illustrate a possible scenario, let us focus attention on the parameter $m^2 \equiv \beta/\delta$. s explained more fully in the Appendix, m^2 is proportional to a coefficient β that neasures the effectiveness of the immune system in killing virus. It turns out (Segel nd Levin, 1976) that the uniform state switches from being stable to being unstable vhen $f(\theta)$, defined in (A6), switches from positive to negative. A little manipulation hows that the uniform state is unstable if and only if $m < m_c$, where the *critical value* f m is given by

$$m_c = \frac{1}{2}[\theta k + \theta^{-1}]. \tag{3}$$

ne can imagine the conflict between the viruses and the immune system slowly shifting n favor of the viruses, resulting in a decrease in the parameter $m \equiv \sqrt{\beta/\delta}$ via a decrease n the immune system effectiveness factor β. If this process resulted in a decrease of n below its critical value m_c a rash would rapidly appear. Reversal of the decrease in n could cause the rash to disappear again. [Technical remark: In the mathematical nalysis m is regarded as a constant. However the analysis will remain approximately rue if m varies slowly.]

model for Type II patterns

We turn now to a study of Type II patterns. As with Type I pattern, the general rinciples can be illustrated on a variety of models. We again select model equations that ave already been studied so that the broad features of our approach can be delineated vithout the considerable work that is required to analyze even a fairly simple system of onlinear partial differential equations.

We employ as our prototypical model for Type II patterns a pair of reaction- liffusion equations that are essentially those put forward by Tyson and Fife (1980) as a highly simplified model of the pattern-forming Belousov-Zhabotinski chemical reaction. In the form employed by Keener and Tyson (1986) these equations can be written

$$\frac{\partial V}{\partial t} = \frac{1}{\epsilon}V(1-V) - \frac{f}{\epsilon}I\frac{V-q}{V+q} + D_V\left[\frac{\partial^2 V}{\partial x^2} + \frac{\partial^2 V}{\partial y^2}\right], \tag{4a}$$

$$\frac{\partial I}{\partial t} = V - I + D_I\left[\frac{\partial^2 I}{\partial x^2} + \frac{\partial^2 I}{\partial y^2}\right]. \tag{4b}$$

Comparing with the general form (1) we note that the proliferation rate $r(V)$ is proportional to $1 - V$ and thus represents the standard "logistic" assumption that per capita proliferation decreases linearly as the population increases. Destruction of virus is assumed to increase linearly with immune factor I. We shall assume that V is initially everywhere greater than the small constant q ($q \ll 1$). If this is the case, V will always remain greater than q. Thus the dependence of viral destruction on V is reasonable; the rate is a monotonically increasing and saturating function of V. In (4b), the supply of immune factor I is assumed to be proportional to the viral level. In the absence of such a supply, the amount of I decays exponentially.

Equations (4) have been extensively analyzed. When coefficients are constant, the solution takes the form of a rotating spiral [Figure 3(b)]. For very small ϵ and $f = 1$ the spiral seems to undergo rigid rotation around a fixed tip. For larger values of ϵ, however, the tip meanders in regular or irregular patterns (Jahnke and Winfree, 1991).

To appreciate the nature of the various solutions to (4) we must employ the concept of *excitability*. Excitability is said to occur in situations where solutions always tend to a steady state, but superthreshold perturbations ignite a large transient response. This response eventually fades away as the solution returns to the steady state. Subthreshold perturbations disappear rapidly. Following a response, time must elapse before another response can be excited (refractory period in recovering media).

In the spatially independent case ($D_V = D_I = 0$), the system (4) exhibits excitability when $f > 1 + \sqrt{2}$ but is not much bigger, or when $f < \frac{1}{2}$ but not much smaller. Oscillations occur when $\frac{1}{2} < f < 1 + \sqrt{2}$, (Tyson 1977). Target patterns can be anticipated if an oscillatory patch is surrounded by excitable medium. Indeed, suppose that the coefficient f varies slowly in space such that the kinetics of (4) with diffusion neglected is oscillatory near the origin and excitable further from the origin. Then it can be shown that a succession of waves radiate out from the vicinity of the origin, forming a target pattern (Tyson, 1985).

In summary, model (4) exhibits Type II structures.

Rash patterns in vivo

Having formulated representative models for Type I and Type II patterns, we now naturally inquire as to whether such patterns of rashes are known. Disappointingly, no Type I inflammatory patterns seem to occur: there seem to be no instances of more or less regular, stationary, stripes of inflammation nor of reasonably regular arrays of inflamed spots or reticular lattices. "An exception that proves the rule" is the reticular network depicted in Figure 6. This is a Type I pattern but it is almost certainly not directly due to interaction between a pathogen and the immune system. Rather the pattern stems from stresses generated by an abnormal drying of the skin and concomitant cracking. The phenomenon is similar to the cracking observed in mud flats. But in one sense this example does illustrate the type of phenomenon we have in mind, for the same combination of weak nonlinear effects and symmetry considerations that underlies hexagonal patterns in Bénard convection and in reaction diffusion equations may also be responsible for the appearance of reticular patterns that are sometimes seen in mud flats. A useful general survey of two dimensional patterns is that of Weaire and Rivier (1984). Thus Figure 6 probably should be regarded as a Type I pattern, but not one generated directly by a pathogen-immune system interaction. It is not easy to supply references that might be relevant to a study of skin cracking. A recent paper that may be helpful is Delannay and Warren's (1991) study of the array of parallel cracks induced by the uniaxial straining of a brittle film that is bonded to a ductile substrate. Their approach might be extended to yield a reticular pattern when the applied strain is of a more general type. The multilayer geometry that they consider is reminiscent of the structure of skin.

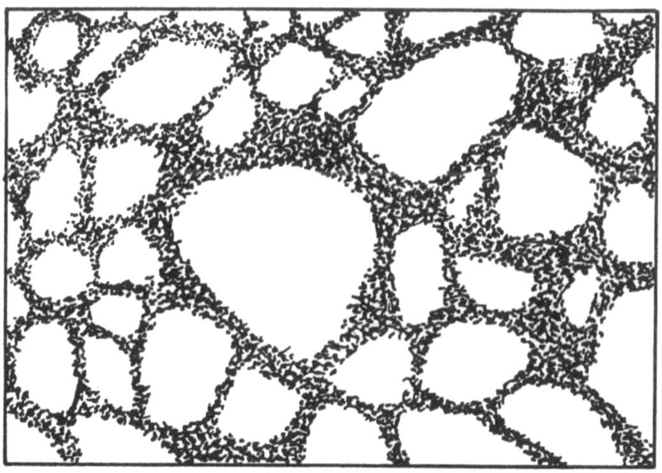

Figure 6. A reticular network of scales in ichthyosis vulgaris. Redrawn from a portion of Figure 8 in Ferrandez (1987).

Patterns very reminiscent of the prototypes in Figure 3, particularly gyrate or figurate erythemas, are known to dermatologists. They have been described as possessing a "distinct erythematous border which may be annular, arcuate, gyrate, polycyclic, serpiginous or reticulate in overall pattern. This border typically demonstrates relatively rapid expansion or migration over a period of hours or days... A hypersensitivity reaction to either exogenous or endogenous antigen has been postulated to occur in all four entities" (Willis, 1978). Figure 7 is a sketch of the rapidly moving rash pattern (often a centimeter per day—Willis, 1978) on the trunk of a patient with the most spectacular of the erythemas, erythema gyratum repens or EGR (Burgdorf and Goltz, 1987). These authors mention that EGR almost always is associated with an internal malignancy.

Figure 7. "Wood grain" rash pattern on the trunk of a sufferer from erythema gyratum repens. The lines correspond to raised scaly erythematous lesions. Sketched from Burgdorf and Goltz (1987), Figure 88-3.

Because of their characteristic general shapes, because they invariably are composed of moving bands, and because their (changing) location seems to bear no essential relation to underlying physiological structures, there seems little doubt that the erythemas constitute examples of autogenic Type II patterns. Although solid evidence seems lacking, clinicians tend to adhere to the hypothesis that these structures are typically the product of activities of the immune system.

Type II patterns are also found in the fungal disease tinea imbricata. Patterns in fungi have been studied to some extent, e.g. by Edelstein and Segel (1983), but these are not of central interest here.

Discussion

We set out to unearth possible mechanisms for the distribution of rashes whose primary cause can be traced to operations of the immune system. Our discussion was based on two classes of patterns, Type I and Type II, that are known from general theoretical investigations of pattern formation. In searching for instances of these patterns in dermatology, we were led to broaden our purview. Type I and II patterns of skin abnormalities can in fact be traced to at least three influences— mechanical, congenital and immune.

Figure 6 showed one example of a Type I (reticular) pattern, whose fundamental cause is "mud cracks" appearing in abnormally dry skin. Another example of mechanically generated pattern is the parallel stripes of adolescent striae that sometimes appear on the backs of weightlifters (Levine and Goolami 1986, Figure 165). Congenital abnormalities can lead to various pigmentation patterns of linear, spiral, or spotted character. Study of such patterns is akin to research on zebra stripes and leopard spots. See Murray (1989) for a comprehensive summary of theoretical progress in this area, employing underlying models that are formally similar to those employed here.

This paper is primarily concerned with inflammatory rashes, whose basis is the interaction of a pathogen with the immune system. We have seen that Type I patterns of such rashes seem not to exist, but that EGR constitutes an excellent example of a Type II pattern.

In a brief letter to the editor, Moore (1982) called attention to the "striking morphological similarity of EGR to chemical waves seen in the Belousov-Zhabotinski reaction." We have taken this observation further here (at first without knowledge of Moore's observation) by examining a simple model for the genesis of rash patterns that is of the same class as models for the Belousov-Zhabotinski reaction and other pattern forming mechanisms. We now demonstrate how our examination of mathematical models can perhaps shed light on the observation that Type I inflammatory patterns seem much rarer than Type II patterns. We will be able to isolate a major feature that is expected to characterize the underlying cause of the observed Type II patterning.

What are the underlying requirements for reaction-diffusion models to yield, respectively, Type I and II patterns? The form of the diffusion terms is fixed, so the question really is, what kinds of underlying homogeneous kinetics, symbolized in general by the equations

$$\frac{\partial V}{\partial t} = f(V,I), \quad \frac{\partial I}{\partial t} = g(V,I), \tag{5a, b}$$

yield Type I and Type II patterns?

There are two main qualitative requirements for Type I patterns (Turing 1952, Segel and Jackson, 1972). One is that *at steady state* the signs of the partial derivatives

$$\begin{bmatrix} \partial f/\partial V & \partial f/\partial I \\ \partial g/\partial V & \partial g/\partial I \end{bmatrix} \tag{6a}$$

must essentially be of the form

$$\begin{bmatrix} + & - \\ + & - \end{bmatrix}. \tag{6b}$$

The requirements $\partial f/\partial I < 0$ and $\partial g/\partial V > 0$ of (6) are intrinsic to our virus-immune system model. The first states that the immune factor tends to decrease virus, while the second states that the presence of virus tends to boost the production of the immune factor. (Type I patterns can also appear if $\partial f/\partial I > 0$ and $\partial g/\partial V < 0$, but such conditions are immunologically infeasible.) The requirements $\partial f/\partial V > 0$, $\partial g/\partial I < 0$ imply that at steady state the virus V must stimulate its own production while the immune factor I must inhibit its own production. Hence V is called the ***activator*** variable and I the ***inhibitor***. We note that for the example (2), at steady state [when $f(V,I) = g(V,I) = 0$], $\partial f/\partial V = kV$. Hence if $\partial f/\partial V > 0$ then $k > 0$, yielding the postulated autocatalytic viral (activator) growth. Similarly, if the autocatalytic term I^2 in (2b) were replaced by a more conventional term proportional to I, then $\partial g/\partial I$ would vanish at steady state.

A second general requirement for a Type I pattern is that activation be short range and inhibition long range, i.e., that $D_V > D_I$. If I is a chemical that diffuses, and V is a "virus" that "diffuses" by randomly spreading from cell to cell, it is certainly reasonable to hypothesize that inhibition is relatively long range compared to activation.

The final requirement for Type I pattern is a quantitative relation, exemplified by $m < m_c$, that yields instability of the uniform state. It is difficult to make *a priori* estimates for the likelihood of such a requirement being fulfilled.

One can imagine mechanisms that might provide the required positive feedback of virus on its own proliferation on the one hand, and the required negative feedback of the immune factor on *its* production. For example, one can envision the possibility of increasing viral proliferation at higher viral levels through the agency of helper viruses (Berns and Bohenzky, 1987). Nonetheless the fact that Type I patterns seem not to occur indicates that even if the required feedbacks exist they are not sufficiently strong, or that for some other reason the instability condition is not satisfied.

When Type II patterns are found it is very often the case that one of the variables is excitable, a concept introduced in our remarks concerning Equation (4). Technically speaking, if the nullcline $\partial V/\partial t = 0$ is N-shaped, then V is usually excitable for suitable parameter choices. In more general terms, the existence of the N-shaped nullcline means that for a range of I, $\partial V/\partial t$ switches from negative to positive and back to negative as V increases.

Excitability requires positive feedback of a quantity on itself when a threshold value of that quantity is exceeded, and values of the quantity are less than an upper threshold. Thus superthreshold stimulation leads to an explosive burst of growth in the amount of the excitable variable. Excitability occurs in many biological contexts. Perhaps the most important of these is associated with propagation of electrical signals by voltage-dependent channels—leading to traveling waves in nerves. Heart muscle is an excitable medium that exhibits spiral and other waves. Another example is spiral waves in aggregating D. discoidium amoebae, where positive feedback is provided by cAMP-induced cAMP secretion. (Winfree, 1980).

Although very common, excitability is not necessary for Type II patterns, which also occur in oscillatory media (reviewed by Tyson, 1985). Continual oscillations appear when the "forces" that counteract disturbances are too strong, and repeatedly draw the system well past rest conditions. Waves of activity can travel through oscillatory media if appropriate transport mechanisms, such as diffusion, propagate disturbances to adjacent spatial locations.

It appears that suitable excitability or oscillation occurs in the context of skin diseases, yielding the observed Type II patterns.

We have asserted that Type I patterns seem never to occur. It could be, however, that the appearance of spots in a number of common diseases is not random, as it seems at first sight, but rather exhibits an order that is consistent with a Type I pattern-forming mechanism. This leads us to suggest that it would be worthwhile for dermatologists to make serial photographs of the emergence of the spots and to examine the pattern for randomness by various statistical procedures (Ripley, 1988). If clear instances of non-randomness were found, it would be worthwhile for theorists to determine what predictions of spot emergence patterns would be generated by Type I models.

More generally, dermatologists interested in forwarding the study of rash patterns would provide invaluable data by collecting timed sequences of photographs that showed the temporal development of rash patterns, whether the distribution appeared "random" or organized.

The value of investigating the temporal behavior of rash distributions can be appreciated by contemplating various relationships concerning waves in excitable media that have been derived by theorists. One of these is that for spiral waves

$$\frac{\text{wavelength} \cdot \text{wave propagation speed } v}{\text{diffusivity } D} = Q, \tag{7}$$

where Q is a number that is typically about 100 (See Winfree 1991). Here and below, D refers to the diffusion of the excitable entity—in our case, the "virus". (For Type II patterns, the diffusivity of the other entity is not a critical parameter.)

As cited for example in the review of Tyson and Keener (1988) excitable media support solitary waves whose speed V is related to the diffusivity D and the rate of exponential growth k of the excitable factor by the approximate "Luther Equation"

$$V \approx 2\sqrt{kD}. \tag{8}$$

Another way to interpret k is that the doubling time of the excitable factor, following a superthreshold stimulus is $0.7/k$. For spiral waves, it appears that if the time scales for changes in I and V are not too different Luther's equation still offers a rough approximation (Winfree, 1991).

For spiral waves an accurate relationship is known between the velocity v of a plane wave in the medium and the actual velocity v_n, normal to the spiral. For sufficiently small curvature κ this local velocity is given by (Tyson and Keener, 1988)

$$v_n = v + D\kappa. \tag{9}$$

[The curvature κ at a point is the reciprocal of the radius of curvature R ($\kappa = 1/R$), where R is the radius of the circle that best fits the curve at that point (the tangent circle)].

As was already done a decade ago by Winfree (unpublished: private communication) one can employ (7) and (8) to obtain some interesting order of magnitude estimates. Consider EGR, with its patterns whose wavelength is of magnitude one cm, and which move one cm/day. From (7), this yields a diffusion coefficient of 0.01 $\mathrm{cm}^2/\mathrm{day}$, assuming that EGR is associated with spiral waves. In a time $t = 1$ day, the "virus" would diffuse $\sqrt{Dt} = 1$ mm, i.e. the virus would spread randomly across some tens of cells. Given $v = 1$ cm/day and $D = 0.01$ $\mathrm{cm}^2/\mathrm{day}$, we obtain from (8) the estimate $k^{-1} \approx 10^{-2}$ days ≈ 15 min for the time scale of viral doubling. These rough estimates of D and k^{-1} seem feasible, and give a clue to the biological processes that might underly EGR.

Our study of Type I and Type II patterns is just one step in the analysis of rash distributions. An example of an additional question of considerable interest is the genesis of apparently random patterns—which could be called Type Zero. Another matter concerning rashes that is worthy of future investigation concerns the configuration of individual lesions.

Acknowledgements. D. Kriegel, J.D. Murray and E. Winocour provided valuable advice . J. Tyson and A. Winfree were particularly helpful, providing both tactical and strategic assistance. Portions of this work were done under the auspices of the U.S. Dept. of Energy, partially supported by NIH grant AI28433 (ASP, LAS). Partial support was also received from US-Israel Binational Science Foundation grant 89-146.

Appendix: Calculations concerned with system (2)

To see more clearly the origin of the equations in (2), let us consider a special case of (1). To emphasize that they are dimensional, we shall append an asterisk to our previous notations for the various variables. The model we shall consider is

$$\frac{\partial V^*}{\partial t^*} = V^*(\rho + \alpha V^* - \beta E^*) + D_V\left[\frac{\partial^2 V^*}{\partial (x^*)^2} + \frac{\partial^2 V^*}{\partial (y^*)^2}\right],$$

$$\frac{\partial I^*}{\partial t^*} = I^*(\gamma V^* - \delta I^*) + D_I\left[\frac{\partial^2 I^*}{\partial (x^*)^2} + \frac{\partial^2 I^*}{\partial (y^*)^2}\right].$$

Upon introducing the dimensionless variables

$$V = \frac{V^*}{\rho/\delta}, \quad I = \frac{I^*}{\rho/\gamma}, \quad t = \frac{t^*}{\rho^{-1}}, \quad x = \frac{x^*}{(D_V/\rho)^{1/2}}, \quad y = \frac{y^*}{(D_V/\rho)^{1/2}},$$

we obtain equations (2), with dimensionless parameters

$$k = \alpha/\gamma, \quad m^2 = \beta/\delta, \quad \theta^2 = D_I/D_V.$$

The system (2) has a uniform steady state solution $V(x,y,t) \equiv \overline{V}$, $I(x,y,t) = \overline{I}$, where

$$\overline{V} = \overline{I} = p^{-2}, \quad p^2 \equiv m^2 - k. \tag{A1}$$

We assume that

$$m^2 > k > 0 \tag{A2}$$

so that the concentrations $\overline{V}$ and $\overline{I}$ will be positive. (From now on, it is convenient to write our results in terms of the parameters m and p rather than m and k). For completeness we note that the system (2) also possesses the steady state $V = I = 0$. Because of the autocatalytic effect ($k > 0$), this state is always unstable.

Linear stability theory requires consideration of the small perturbations

$$V' = V - \overline{V}, \quad I' = I - \overline{I}. \tag{A3}$$

Let $\boldsymbol{V}'$ denote a vector with components (V', I'). A wide class of perturbations can be formed by superposing sinusoidal disturbances. Hence it is fruitful to investigate disturbances of the form

$$\boldsymbol{V}' = \epsilon e^{\sigma t} \boldsymbol{C}_1 \cos qx \tag{A4}$$

where the small parameter ϵ measures the size of the perturbation. $\boldsymbol{C}_1$ is a constant vector, and the constant spatial period of the disturbance is given by $2\pi/q$.

We wish the uniform state to be stable to spatially uniform perturbations ($q = 0$). This turns out to require

$$m^2 - p^2 < 1. \tag{A5}$$

When $q > 0$, the $\theta - q$ plane is divided into two regions, respectively where $\sigma > 0$ (instability) and $\sigma < 0$ (stability). The boundary of this area, where $\sigma = 0$, is given by $h(\theta) = 0$, where

$$h(\theta) \equiv \theta^2 q^4 + p^{-2}[1 - (m^2 - p^2)\theta^2]q^2 + p^{-2}. \tag{A6}$$

The curve's minimum is at the point (θ_c, q_c), where

$$\theta_c = (m - p)^{-1}, \quad q_c^2 = (m - p)/p. \tag{A7}$$

For θ near θ_c

$$\sigma = \overline{\sigma}(\theta - \theta_c) + O(\theta - \theta_c)^2, \quad \text{where } \overline{\sigma} = \frac{2(m-p)^2}{p[1 - (m-p)^2]}. \tag{A8}$$

Segel and Levin (1976) reformulated more standard approaches to nonlinear stability theory by simultaneously employing successive approximations and multiple time scales. They showed that the "most dangerous mode", when $q = q_c$ could be represented by

$$\boldsymbol{V}' = \epsilon A \boldsymbol{C}_1 \cos q_c x + \epsilon^2 A^2 [\boldsymbol{C}_0 + \boldsymbol{C}_2 \cos 2q_c x] + O(\epsilon^3). \tag{A9}$$

The analysis assumes that $|\epsilon| \ll 1$, where

$$\epsilon^2 = \theta - \theta_c. \tag{A10}$$

The amplitude $A(t)$ turns out to satisfy

$$\frac{\partial A}{\partial t} = \epsilon^2[\overline{\sigma} A - \gamma A^3]. \tag{A11}$$

Lengthy calculations show that γ is a positive number given by

$$\gamma = \frac{p^4(30p^2 - 22mp - 23m^2)}{36m(m-p)[(m-p)^2 - 1]} . \tag{A12}$$

If nonlinear terms in (A11) are neglected, $A \sim \exp(\sigma t)$ where σ is given by (A8). Thus the results of linear theory are recovered. According to (A11), after an initial period of exponential growth the amplitude A approaches the constant value $\sqrt{\overline{\sigma}/\gamma}$.

In our numerical work, we employed the parameters

$$m = 0.61, \quad \theta = 4, \quad k = 0.25, \quad \epsilon = 10^{-4}. \tag{A13}$$

Hence the steady state (for $\epsilon = 0$) is given by

$$p^{-2} = 8, \quad \theta_c = 3.9, \quad 2\pi/q_c = 0.73, \quad \sigma = 0.055 \quad \gamma = 0.028.$$

The magnitude of the sinusoidal perturbation at steady state, $\sqrt{\sigma/\gamma}$, is 1.4. To obtain the spot patterns we reduced m to 0.52.

References

Benson DL, Sherratt JA, Maini PK (1992) Diffusion driven instability in an inhomogeneous domain. Bull. Math. Biol., in press

Bertin G, Lin CC (1985) Formation and maintenance of spiral structure in galaxies. In: van Woerden H, et al (eds) The Milky Way Galaxy. Internat. Astronom. Union Symp. 106. D. Riedel, Dordrecht, p 513–532

Burgdorf WHC, Goltz RW (1987) Figurate erythemas. In: Fitzpatrick TB, Eisen AZ (eds) Dermatology in General Medicine, 3rd edn. McGraw Hill, New York, p 1010–1017

Berns, KI, Bohenzky, RA (1987) Adeno-associated viruses: an update. Adv. Virus Res. 32:243–306

Delannay, F, Warren, P (1991) On crack interaction and crack density in strain-induced cracking of brittle films on ductile substrates. Acta Metall. Mater. 39(6):1061–1072

Edelstein L, Segel LA (1983) Growth and metabolism in mycelial fungi. J. Theoret. Biol. 104:187–210

Ferrandez C (1987) Clinical Atlas of Dermatology. Yearbook Medical Publishers Inc, Chicago

Jahnke W, Winfree AT (1991) A survey of spiral-wave behaviors in the oregonator model. Internat. J. Bifurcation Chaos 1(2):445–466

Keener JP, Tyson JJ (1986) Spiral waves in the Belousov-Zhabotinskii reaction. Physica D 21:307–324

Levin SA, Segel LA (1985) Pattern generation in space and aspect by interaction and redistribution. SIAM Review 27:45–67

Levine GM, Goolamali, SK (1986) Diagnostic Picture Tests in Dermatology. Wolfe Medical Publishers Ltd, Ipswich

Moore, HJ (1982) Does the pattern of erythema gyratum repens depend on a reaction-diffusion system? British J. Dermatology 107:723

Murray, DJ (1989) Mathematical Biology, Springer, Berlin

Ripley, BD (1988) Statistical Inference for Spatial Processes. Cambridge University Press, Cambridge

Segel LA, Jackson, JL (1972) Dissipative structure: an explanation and an ecological example. J. Theoret. Biol. 37:545–559

Segel LA, Levin, SA (1976) Applications of nonlinear stability theory to the study of the effects of dispersion on predator-prey interactions. In: Piccirelli R (ed) Selected Topics in Statistical Mechanics and Biophysics. Amer. Inst. of Physics Symp. 27, p 123–152

Segel LA, Stuart JT (1962) On the question of the preferred mode in cellular thermal convection. J. Fluid Mechanics 13:289–306

Shtilman L, Sivashinsky G (1991) Hexagonal structure of large-scale Marangoni convection. Physica D 52:477–488

Turing A (1952) The chemical basis for morphogenesis. Phil. Trans. Roy. Soc. B 237:37–72

Tyson JJ (1977) Analytic representation of oscillations, excitability and traveling waves in a realistic model of the Belousov-Zhabotinskii reaction. J. Chem. Phys. 16:905–915

Tyson JJ (1985) A quantitative account of oscillations, bistability, and traveling waves in the Belousov-Zhabotinskii reaction. In: Field RJ, Burger M (eds) Oscillations and Traveling Waves in Chemical Systems. John Wiley & Sons, Inc, New York, p 93–144

Tyson JJ and Fife PC (1980) Target patterns in a realistic model of the Belousov-Zhabotinskii reaction. J. Chem. Phys. 73:2224–2237

Tyson JJ and Keener JP (1988) Singular perturbation theory of traveling waves in excitable media (a review). Physica D 32:327–361

Weaire D and Rivier N (1984) Soap, cells, and statistics–random patterns in two dimensions. Contemp. Phys. 25:59–99.

Willis WF (1978) The gyrate erythemas. Int. J. Dermatol. 17:698–702

Winfree A (1972) Spiral waves of chemical activity. Science 175:634–636

Winfree A (1980) The Geometry of Biological Time. Springer, Berlin

Winfree A (1991) Varieties of spiral wave behavior: an experimentalist's approach to the theory of excitable media. Chaos 1:303–334

ON SECOND SYMMETRY AND AIDS

Geoffrey W. Hoffmann and Tracy A. Kion
Departments of Microbiology and Physics
University of British Columbia
Vancouver, B.C.
Canada, V6T 1Z3

ABSTRACT

We present a model of the relationships between various classes of antibodies that are present in alloimmune sera. The model is a generalization of "second symmetry", the idea that the total set of antibodies in an A anti-B serum is complementary to the total set of antibodies in the converse B anti-A serum. We discuss the idea of a major axis of shape space in the immune system. We finally briefly review an idiotypic network model of AIDS pathogenesis and some supporting data from our laboratory. An exciting recent result is that injections of 10 nanogram amounts of gp120 or p24 of HIV can prolong the survival of MRL-*lpr*/*lpr* mice, a model for lupus. In view of the fact that there are many similarities between lupus and AIDS, a similar approach may prevent AIDS.

Antibodies in Alloantisera

If two strains of mice, let us call them A and B, are cross-immunized with lymphocytes (A is immunized with B lymphocytes, and B with A lymphocytes), they each make three kinds of MHC-related antibodies, as shown in Table 1 for the A anti-B serum. Most familiar are the conventional anti-foreign MHC antibodies, namely A anti-(B MHC) and B anti-(A MHC). Secondly, there are anti-anti-self antibodies that are made

NATO ASI Series, Vol. H 66
Theoretical and Experimental Insights into Immunology
Edited by A. S. Perelson and G. Weisbuch

Table 1. MHC-related antibodies in an A anti-B alloimmune serum,

1. Conventional anti-class I MHC and anti-class II MHC antibodies
   A anti-(B MHC)
   = Anti-foreign

2. MHC-image antibodies
   A anti-(B anti-A)
   = Anti-anti-self

3. Anti-I-J^B
   Postulated to be anti-(MHC-image of B)
   = Anti-(foreign MHC-image)

against the receptors of the injected lymphocytes that recognize the host. The anti-foreign antibodies in an A anti-B serum have V regions that are complementary to the anti-anti-self antibodies of a B anti-A serum and vice versa (Fig. 1). The anti-anti-self antibodies are MHC-image antibodies, and they can be detected in an inhibition of cytotoxicity assay (Hoffmann et al., 1986) or in an ELISA assay (T. A. Kion, unpublished). The anti-anti-self antibodies in one antiserum interact with anti-foreign antibodies in the complementary antiserum. The third kind of antibodies are anti-I-J antibodies, which we discuss later. For now we will focus on just the conventional anti-foreign and anti-anti-self antibodies.

Extensive absorption of an alloimmune serum with fixed lymphocytes from the immunizing strain, such that all the cytotoxic anti-foreign antibodies are removed, fails to remove the MHC-image antibodies. This has been shown with both inhibition of cytotoxicity (Hoffmann et al., 1986) and the ELISA assay (T. A. Kion, unpublished). Anti-(host MHC) V regions in the naive absorbing lymphocytes may therefore be present at only a very low concentration. This may be the case partly because of the dynamic nature of immunization using viable lymphocytes. During the immune response two population distributions change in an interdependent way; firstly the population distribution in the inocculum lymphocytes

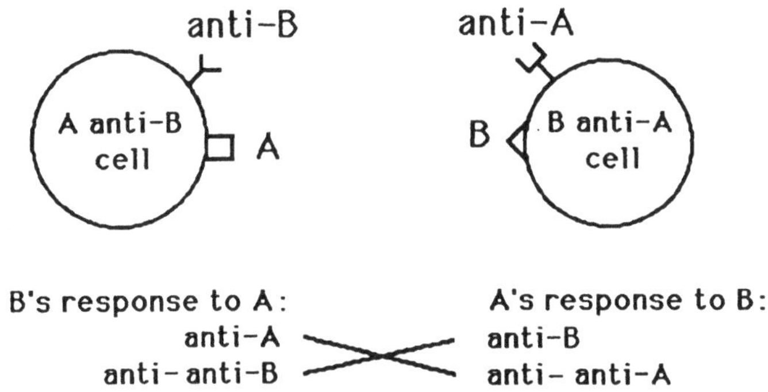

Fig. 1. The immunogenic structures on the surface of lymphocytes of strain A that are injected into a mouse of strain B include the A strain MHC antigens and anti-B receptors of the A strain lymphocytes that recognize and are stimulated by the B strain MHC antigens. If A and B strain mice are cross-immunized, the anti-foreign antibodies made in A are complementary to the anti-anti-self antibodies made in B and the anti-foreign antibodies made in B are complementary to the anti-anti-self antibodies made in A (Hoffmann et al., 1986).

whose V regions constitute the antigen, and secondly the population distribution of the host clones that recognize the foreign idiotypes. Hence V regions that are initially present at a very low level may eventually become the dominant antigen, so that absorption of antibody using naive immunogen may be inefficient.

It is probable that anti-MHC T cell receptors are the main idiotypic antigens for the production of anti-anti-self antibodies, rather than the analogous B cell receptors, because T cells within the injected cells make a more vigorous response to the allogeneic host MHC than do B cells. The fact that the anti-anti-self antibodies produced nevertheless interact with the anti-(foreign MHC) *antibody* V regions in the converse antiserum would follow from the anti-anti-self antibodies having V regions that are a good imitation of self MHC, that is, they are MHC-image antibodies.

We find that there is a lot of cross-reactivity of MHC-image antibodies in their interactions with various anti-MHC antibodies (Hoffmann et al., 1986; Kion and Hoffmann, work in progress). Their specificity can most clearly be defined in terms of what they do not react with. In particular, the A anti-(B anti-A) antibodies (that is, anti-anti-self) do not interact with the A anti-B antibodies (anti-foreign) in the same serum, but they do react with many other anti-foreign allo-antibodies besides their nominal complement, namely B anti-A. This is a case of "hole specificity". The absence of any A anti-(A anti-B) antibodies in the antiidiotypic set can be most simply ascribed to the cytotoxic effect of the A anti-B (i.e. anti-foreign) antibodies eliminating the subset of clones that could make antibodies specific for themselves.

While helper T cells are selected to be anti-class II MHC, we have argued that suppressor T cells have V regions with complementarity to the helper T cell idiotypes (Hoffmann, 1988). The suppressors would then have MHC-image V regions. The famous I-J determinants on suppressor T cells (Tada et al., 1976; Murphy et al, 1976) can most simply be understood as MHC-image determinants (Hoffmann, 1988).

A third kind of antibody that can be detected in alloimmune sera is anti-I-J. Anti-I-J antibodies are in fact the basis of the definition of I-J. If these antibodies are made against the V regions of allogeneic MHC-image suppressor T cell idiotypes, they would be anti-(foreign-MHC-image) antibodies. This idea leads to the picture shown in Fig. 2, which is a model for the relationships between six classes of antibodies present in the pair of complementary alloimmune sera, A anti-B and B anti-A. This model is a generalization of "second symmetry" (Hoffmann et al., 1986), namely the idea that total set of antibodies in an A anti-B serum is complementary to the total set of antibodies in the converse serum, B anti-A.

The model predicts that anti-I-J antibodies should bind to MHC-image determinants on anti-anti-self antibody V regions, and we have preliminary mouse data showing that this is so.

MHC-image and anti-(MHC-image) antibodies play a role in our AIDS model of pathogenesis (below). It has been reported that anti-I-J antibodies react with a subset of CD8 bearing human lymphocytes in man (Lehner et al., 1984), consistent with the idea that there is cross-reactivity between MHC-image determinants on suppressor T cell V regions in mouse and man. Hence anti-I-J antibodies may be a powerful tool for analysing alloimmunity and HIV induced pathogenesis in man.

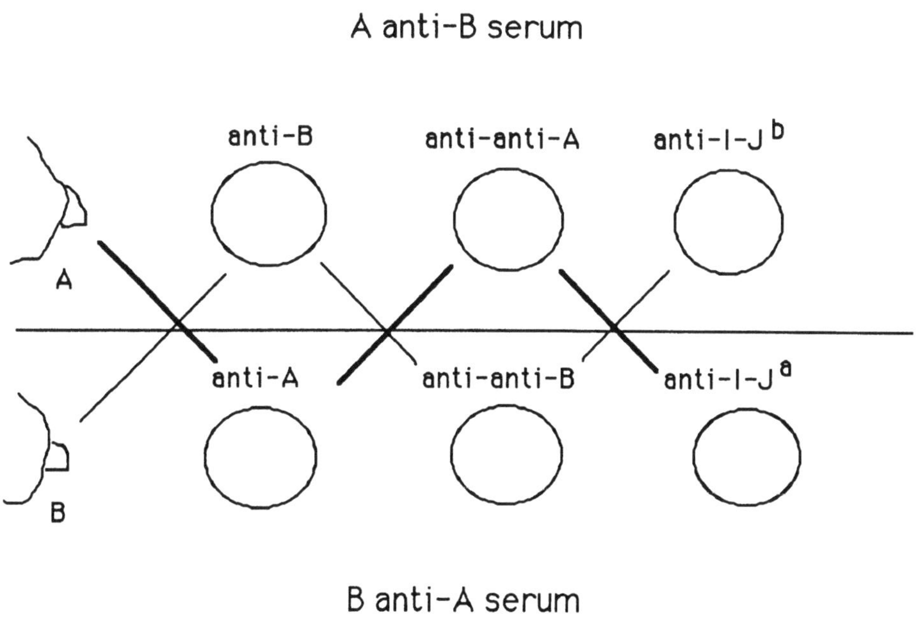

Fig. 2. *Model of the antibodies in two complementary alloimmune sera, A anti-B and B anti-A. Each component in one serum is complementary to one or more components in the other serum.*

A "major axis" of shape space: MHC - anti-MHC - MHC-image, etc.

The immune system is not a system in which all shapes are equal. Immunologists discovered a long time ago that MHC shapes are special for the regulation of the immune system. The importance of anti-MHC shapes was emphasized by the finding that T cells were found to have dual specificity, namely for both an antigen and MHC determinants. Helper T cells are weakly anti-class II MHC, cytotoxic T cells are typically weakly anti-class I MHC, and we have suggested that I-J bearing suppressor T cells have MHC-image determinants (Hoffmann, 1988). It may be useful to discuss this in the context of shape space, a concept that has been developed by Perelson and Segel (1988). We suggest that there is a major axis in the shape space of each immune system, encompassing MHC, anti-MHC, MHC-image, and including anti-MHC-image at least in alloimmune individuals. The V regions of B cells, on the other hand, are not *known* to have any special relationship to this major axis, but since B cells are regulated by T cells, it would be surprising if there were not a secondary effect of the major axis on the B cell repertoire. We have suggested that B cells (in particular IgG producing cells) may, as a class, have V regions that are at least weakly MHC-image, by being positively selected to recognize helper T cell idiotypes that are anti-MHC (Hoffmann et al., 1991). According to our model of I-J, normal animals have MHC-image T cells that play a central role in stabilizing the immune system (Hoffmann et al., 1988). Below we discuss an autoimmune disease in MRL-*lpr*/*lpr* mice that might be viewed as a process in which B cells (and antibodies) encroach to a greater than normal degree on a region of shape space (MHC-image) that is more heavily occupied by T cells than B cells in healthy individuals.

AIDS pathogenesis

Autoimmunity ideas (Ziegler and Stites, 1986; Andrieu et al., 1986; Shearer, 1986; Martinez-A. et al., 1988; Habeshaw and Dalgleish, 1989; Kopelman and Zolla-Pazna, 1988), together with the similarities between AIDS and graft versus host disease (Shearer, 1983), idiotypic network ideas (Jerne, 1974; Hoffmann, 1975, 1988) and the presence of MHC-image antibodies in alloantisera (Hoffmann et al., 1986), lead to an explicit model of AIDS pathogenesis that involves allogeneic lymphocytes as a cofactor (Hoffmann, 1988; Hoffmann et al., 1991). HIV components appear to have MHC-image properties, so an anti-MHC-image component of the immune response to HIV may synergize with the MHC-image component of the immune response to allogeneic lymphocytes in a way that destabilizes the system. Four lines of evidence from our laboratory support or are consistent with this idea.

Anti-collagen antibodies in homosexuals

In a cross-sectional study of sera from homosexuals we detected anti-collagen antibodies in all 17 of the homosexual AIDS patients' sera that we tested (Grant et al., 1989). We also saw these antibodies in 32% of HIV^- homosexuals (n = 44) and in 66% of HIV^+ homosexuals (n = 24), in contrast with 0% of normal heterosexual controls (n = 8). A tentative connection between these results and our model is made on the basis of the fact that anti-collagen is found also in graft versus host disease, a form of alloimmunity that has similarities with AIDS (Shearer, 1983). Anti-collagen may be a marker for alloimmunity, and hence a convenient marker for the presence of this putative AIDS cofactor. This interpretation predicts a longitudinal study should reveal a correlation between the presence of anti-collagen and the rate of progression to AIDS.

MHC-image and anti-HIV antibodies in alloimmune mice

We found that alloimmune mice, but not normal, unimmunized mice, contain both MHC-image antibodies (Hoffmann et al., 1986) and anti-HIV antibodies, namely anti-gp120 and anti-p24 (Kion and Hoffmann, 1991). These mice have not been exposed to HIV or HIV components. We interpret the presence of anti-p24 in these sera in the light evidence for several HIV components, namely gp120, gp41 and Nef having similarities with MHC, and speculate that p24 may be another member in this class. Perhaps there is evolutionary pressure on HIV components to have MHC-image features. Then the anti-gp120 and anti-p24 antibodies in alloimmune sera would be anti-MHC-image.

We classify the anti-gp120 and anti-p24 antibodies as anti-MHC-image rather than anti-MHC because they are not readily absorbed out of the sera by the cells used to immunize the mice (T. A. K. and G. W. H., unpublished), in contrast to anti-MHC antibodies. In this respect these putative anti-MHC-image antibodies are like MHC-image antibodies, which are also not readily absorbed out of the serum by the immunizing cells (Hoffmann et al., 1986). As discussed above for MHC-image antibodies, this may be due to the antigenic T cell idiotypes on the immunogen being present at a very low concentration in naive lymphocyte populations.

MHC-image and anti-MHC-image antibodies in autoimmune mice

We found that the same two kinds of immunity that were postulated to be important in AIDS, namely MHC-image and anti-HIV (putative anti-MHC-image) are present in the MRL-*lpr/lpr* mouse, an autoimmune strain that is a model for *systemic lupus erythematosus* or SLE (Kion and Hoffmann, 1991). These mice too have not been exposed to HIV or HIV components. We detect mainly anti-gp120 of HIV in the sera, with anti-p24 being present in only occasional mice. The presence of both the MHC-image and presumptive anti-MHC-image antibodies in these mice means that the model we formulated for AIDS may be relevant also for *lupus* pathogenesis. In this case there is no obvious external stimulus that triggers the disease, although the possible involvement of a murine retrovirus cannot be excluded. The disease may be simply due to initial conditions (in particular the initial conditions of the N dimensional clonal population vector, where N is the number of clones) being such that a critical threshold in the clonal population space is exceeded. In particular, the system may simply have an initial condition and an initial trajectory in the N-dimensional idiotypic space that leads to elevated levels of MHC-image and anti-MHC-image antibodies. Such an explanation involving a fairly deterministic trajectory could account for the fairly narrow range in the life expectancy (with a mean of about 6 months) that characterizes MRL-*lpr/lpr* mice. It also suggests that a very small perturbation in the initial conditions may suffice to deflect the trajectory away from the undesired region of the N-dimensional clonal space.

Enhancement of longevity of autoimmune mice using low doses of HIV components.

We recently obtained the exciting result that injections with very low doses of certain HIV components, namely gp120 or p24, cause the prolongation of the lives of MRL-*lpr/lpr mice* (Kion and Hoffmann, submitted for publication). The idea of this experiment was to suppress the anti-gp120 immunity seen in these mice, consonant with the idea that these antibodies play a role in pathogenesis. Groups of mice were injected twice (at 3 and 4 weeks of age) with 10 ng of either gp120 or p24, while control mice received injections of phosphate buffered saline or no injections. Control mice died at an average age of about 6 months, while mice in both the p24 and the gp120 injected groups survived to an average age of greater than 10 months.

The injections also caused a decrease (relative to controls) in the amount of anti-gp120 antibodies and anti-collagen antibodies seen four weeks after the injections. That decrease was transient, and the levels of the antibodies were back at close to the control levels eight weeks after the injections.

We could interpret this experiment in the somewhat vague terms of changing the trajectory of an N-dimensional vector, as discussed above. Alternatively, and more concretely in the context of our model of the T cell repertoire (Hoffmann, 1988), gp120 and p24 may be acting as MHC-image substances and stimulating helper T cells, which then stimulate MHC-image suppressor T cells. One might expect that such low level stimulation would not induce immunity along the MHC-image-anti-MHC-image axis, but could instead stabilize the mutual stimulation between anti-MHC helper T cells and MHC-image suppressor T cells.

ACKNOWLEDGEMENTS

This work is funded by the National Health Research and Development Program of Health and Welfare Canada and the Natural Sciences and Engineering research Council of Canada.

REFERENCES

Andrieu, J. M., Even, P. and Venet., A. (1986). AIDS and related syndromes as a viral-induced autoimmune disease of the immune system: an anti-MHC II disorder. Therapeutic implications. *AIDS Research* **2**, 163-174.

Grant, M.D., M.S. Weaver, C. Tsoukas, and G.W. Hoffmann. (1989). Distribution of antibodies against denatured collagen in AIDS risk groups and homosexuals AIDS patients suggests a link between autoimmunity and the immunopathogenesis of AIDS. *J. Immunol..* **144**, 1241-1250 (1990).

Gunther and Hoffmann (1982) Qualitative dynamics of a network model of regulation of the immune system: A rational for the IgM to IgG switch. *J. theoret. Biol.* **94**, 815-855.

Habeshaw, J. and Dalgleish, A (1989). The relevance of HIV env/CD4 interactions to the pathogenesis of Acquired Immune Deficiency Syndrome. *J. AIDS* **2**, 457-468.

Hoffmann, G. W. (1975) A theory of regulation and self-nonself discrimination in an immune network. *Eur. J. Immunol.* **5**, 638-647.

Hoffmann, G. W. (1978) Incorporation of a nonspecific T cell dependent helper factor into a network theory of the regulation of the immune response. *Theoretical Immunology* , G.I. Bell, A.S. Perelson and G.H. Pimbley (eds.) Marcel Dekker, N.Y., (1978) pp. 571-602.

Hoffmann, G. W., A. Cooper-Willis, and M. Chow (1986). A new symmetry: A anti-B is anti-(B anti-A), and reverse enhancement. *J. Immunol.* **137**, 61-68.

Hoffmann, G. W., Kion, T. A., Forsyth, R. B., Soga, K. G. and Cooper-Willis, A. (1988). The N-dimensional network. In *Theoretical Immunology, Part 2,* A. S. Perelson, (ed.), pp. 291-319, Addison-Wesley, Redwood City CA. **Note:** An improved version of the N-dimensional mathematical model of the above paper, together with a new one-dimensional model is presented in "Modeling the immune system network in two dimensions, N dimensions and one dimension" by G. W. Hoffmann, M. J. Lyons, D. J. Mathewson, M. J. Paetkau, and K. G. Soga, (1992) in *Proceedings of the International Seminar on Modeling and Signal Processing of Cellular Systems,* (E. P. Hofer, Ed.) in press.

Hoffmann, G. W. (1988). On I-J, a network centre pole and AIDS. In *The Semiotics of Cellular Communication in the Immune System,* E. Sercarz, F.

Celada, N.A. Mitchison and T. Tada (eds.), pp. 257-271, Springer-Verlag, New York.

Hoffmann, G. W., Kion, T. A. and Grant, M. D. (1991) An idiotypic network model of AIDS pathogenesis. *Proc. Nat. Acad. Sci. (USA)* **88**, 3060-3064.

Hoffmann, G. W. and Tufaro, F. (1989) Serological Distance Coefficients, *Immunol. Letters,* **22**, 83-90, 1989.

Jerne, N. K. (1974). Towards a network theory of the immune system. *Ann. Immunol. (Inst. Pasteur)* **125C**, 373-389.

Kion, T. A. and G. W. Hoffmann. (1991). Anti-HIV and anti-anti-MHC antibodies in alloimmune and autoimmune mice. *Science,* **235**, 1138-1140.

Kopelman, R. G. and S. Zolla-Pazner. (1988). Association of human immunodeficiency virus infection and autoimmune phenomena. *Am. J. Med.* **84**, 82.

Lehner, T., Brines, R., Jones, T., Avery, J. (1984). Detection of cross-reacting murine I-J like determinants on a human subset of $T8^+$ binding, presenting and contrasuppressor cells. *Clin. exp. Immunol.* **58**, 410-419.

Martinez A.C., M. A. R. Marcos, A. de la Hera, C. Marquez, J. M. Alonso, M. L. Toribio, and A. Coutinho. (1988). Immunological consequences of HIV infection: advantage of being low responder casts doubts on vaccine development. *Lancet,* Feb. 27.

Murphy, D. B., Herzenberg, L. A., Okumura, K., Herzenberg, L. A., and McDevitt, H. O. (1976) A new I subregion (I-J) marked by a locus (Ia-4) controlling surface determinants on suppressor T lymphocytes. *J. Exp. Med.,* **144**, 699-712.

Segel, L. A. and Perelson, A. S. (1988) Computations in Shape Space: A new approach to immune network theory. In *Theoretical Immunology, Part 2,* A. S. Perelson, (ed.), pp. 321-343, Addison-Wesley, Redwood City CA.

Shearer, G. M. (1986). AIDS: An autoimmune pathologic model for the destruction of a subset of helper T lymphocytes. *Mount Sinai J. Med.* **53**, 609-615.

Shearer, G. M. (1983). Allogeneic leukocytes as a possible factor in induction of AIDS in homosexual men. *N. Engl. J. Med.* **308**, 223-224.

Tada, T., Taniguchi, M. and David, C. S. (1976) Properties of the antigen-specific T cell factor in the regulation of antibody response of the mouse. IV. Special subregion assignment of the gene(s) that codes for the suppressive T cell factor in the H-2 histocompatibility complex. *J. Exp. Med.*, **144**, 713-725.

Ziegler, J. L. and D. P. Stites. (1986). Hypothesis: AIDS is an autoimmune disease directed at the immune system and triggered by a lymphotropic retrovirus. *Clin. Immunol. and Immunopath.* **41**, 305-313.

SUPERANTIGENS, ALLOREACTIVITY, IMMUNOLOGIC TOLERANCE, AND AIDS: A UNIFIED HYPOTHESIS

Haynes W. Sheppard, Ph.D.
Michael S. Ascher, M.D., F.A.C.P.
Viral and Rickettsial Disease Laboratory,
Division of Laboratories,
California Department of Health Services,
2151 Berkeley Way
Berkeley, CA 94704, USA

ABSTRACT

We present a hypothesis of T cell activation and clonal dynamics which unifies thymic selection, the pathogenesis of AIDS, the effects of superantigens, and the response to alloantigens. These apparently disparate phenomena are integrated under the concept that T cell activation involves synergy between signals at the α-β T cell receptor (TCR) ("signal-1") and costimulatory signals at CD4/CD8 ("signal-2") which together exceed a threshold intensity. Combinations of signals with appropriate intensity and duration lead to clonal expansion while continuous or abnormally intense signals lead to clonal deletion by shifting the population dynamics toward excess programmed cell death. Clonal deletion occurs in the thymus when self-reactive cells are trapped in the presence of "above-threshold" signal combinations. Superantigens and alloantigens deliver the equivalent of signal-1 but have polyclonal effects because they interact with multiple TCR's. In our view, HIV can effects all helper T cells because it delivers an abnormally intense costimulatory signal-2 through CD4 leading eventually to wide spread clonal deletion. The effects of HIV are gradual because peripheral T cells are difficult to tolerize, and because the effects of HIV require the concomitant delivery of signal-1 by antigen.

This synthesis of HIV immunopathogenesis and tolerance has profound implications in both fields: management of AIDS should include control of inappropriate activation and studies of tolerance should exploit the natural experiment of AIDS in the development of co-stimulatory signals which might induce antigen directed clonal deletion and true adult tolerance.

NATO ASI Series, Vol. H 66
Theoretical and Experimental Insights into Immunology
Edited by A. S. Perelson and G. Weisbuch

INTRODUCTION

The AIDS epidemic has occurred during a period of many major breakthroughs in the understanding of immune system structure and function (particularly in the T cell compartment). The T cell receptor has been cloned and characterized[27], the presentation of peptide antigens by MHC was discovered[79], the functional importance of other T cell receptors (particularly CD4 and CD8) has been established[87], superantigens have provided an experimental model for tolerance[1], and the powerful tool of transgenic mice has provided insight into the molecular mechanisms of thymic selection and T cell activation[39,93]. However this very progress has led to a fundamental paradox; namely that positive and negative thymic selection and peripheral T cell activation all appear to be mediated through the same cell surface receptors[13]. In our view, the resolution of this paradox requires a new conceptual framework which can account for these conflicting results.

Similarly, there has been phenomenally rapid progress in the identification and characterization of the human immunodeficiency virus (HIV). However, the initial hypothesis, namely that the virus causes AIDS through direct killing of helper T cells, has been contradicted by a 10 year disease process, a largely effective immune response to the virus, and an extreme paucity of infected cells[7]. These findings have led one prominent scientist to conclude that HIV can not be the cause of AIDS[18] despite unassailable epidemiologic evidence that it is[63]. Resolution of this paradox may also benefit from a new conceptual framework which places something other than direct cytopathic effects at the heart of the HIV mediated disease process.

In considering these two paradoxes, we noticed that some manifestations of HIV infection resemble events that occur in the thymus during the development of T cell tolerance. These include functional anergy[45], lymphocyte activation[21], selective loss of certain T cells[58], and high levels of programmed lymphocyte death through apoptosis[54]. This leads us to propose that AIDS and tolerance (i.e. negative selection) occur through a common mechanism and that the interaction of the virus with CD4, an integral component of the T cell receptor complex (TCRC), provides important insights into both phenomena. In addition, the conceptual framework that integrates AIDS and tolerance also provides explanations for the profound effects of superantigens and for the polyclonal nature of immune responses to alloantigens.

DISCUSSION

Recent studies demonstrate that T cells receive two types of receptor mediated signals. The first occurs through interaction of the α-β T cell receptor (TCR) with an antigenic peptide presented in the context of an MHC molecule ("signal-1"). The second, nonspecific costimulatory signal is provided by other self molecule(s), expressed on accessory cells, which interact with other T cell receptors ("signal-2")[55]. Several molecules have been suggested as receptors for signal-2[72] including recent studies assigning this function to CD4 and CD8[6,19,85]. Paradoxically, the coordinate delivery of both signals results in activation of peripheral lymphocytes but clonal deletion of thymocytes. In contrast, partial or aberrant signals induce paralysis in both cell populations which can often be reversed by providing additional signals[36,82]. Such clonal anergy has been proposed as a second mechanism for the induction of self tolerance[4,35,72].

These findings suggest to us that lymphocyte activation (i.e. proliferation and the expression of effector functions) is an all-or-none phenomena but is initiated by at least two receptor mediated signals which are biochemically additive or synergistic and must cooperate to exceed a critical intensity threshold[33,82]. Thus, the coordinate delivery of signal-1 and signal-2 normally exceeds the threshold while each signal alone is insufficient to activate T cells. Clonal anergy can be viewed as the consequence of "below-threshold" signals which do not induce activation but nevertheless engage sufficient number receptors and induce biochemical events that temporarily prevent the coordinate delivery of other above-threshold signals. A period of rest is required to reset biochemical pathways and receptor expression before responsiveness is recovered.

We have suggested that the immunopathologic consequences of intense T cell activation suggest that one purpose for this complex mechanism is to limit the destructive potential of T cell responses to the tissue sites where antigens are being processed and presented in the "correct" form[7]. As Lewis Thomas has asserted, immune effector mechanisms "are so powerful, and involve so many different defense mechanisms, that we are in more danger from them than from the invaders."[84]. Thus, reversible anergy in T cells that receive "inappropriate" signals may be a safeguard against such unrestricted antigen activation rather than a mechanism normally used for the induction of self tolerance. Furthermore, the maintenance of self tolerance through clonal anergy might be expected to result in a high frequency of autoimmunity.

The key concept that integrates HIV pathogenesis and other immunologic phenomena is that the fate of a T cell population is determined by the overall intensity and/or duration of the activation signals it receives. Unlike most other tissues, lymphoid cells undergo rapid proliferation and functional differentiation in response to receptor mediated signals. Most of the progeny of such activated cells undergo active or "programmed" death through a poorly understood mechanism characterized by apoptosis[3,40,70,78,90,92]. A small number of cells return to a resting state but retain the capacity to repeat the response cycle, i.e. serve as memory cells. This homeostatic mechanism prevents the net expansion or contraction of the immune system.

The nature and kinetics of T cell memory is an area of great uncertainty and controversy[2,11,73]. Experiments showing life-long survival of resting "memory" T cells are contradicted by experiments showing rapid decay of responses in the absence of antigen[24]. Part of this problem is in the definition of the term "memory" which is used by some to indicate the ability to make a secondary and anemnestic response to an antigen. A broader and more fundamental definition of memory, discussed in this NATO workshop, is the continued ability to make a responses at all (i.e. clonal survival). The three alternatives are that; 1) memory cells are very long lived, 2) there is continued maturation of new cells through the thymus in an adult, 3) or memory is maintained through proliferation of clones that cannot be replaced. The key issue in AIDS is the source of the so-called replacement of cells that die. It is our contention that replacement occurs through the progeny of clonal proliferation. If replacement by thymic maturation occurs at all, it is at a low level, is random with respect to antigen specificity, and cannot compensate for significant loss of T cells. Therefore, clonal deletion occurs when the homeostatic regulation of clonal proliferation is disrupted so that no progeny survive. Others have attached significance to memory cells in AIDS pathogenesis, but from the perspective of preferential killing by virus[71].

We have suggested that escape from signals is necessary for the return to the resting state and that high intensity or chronic signals will result in clonal deletion[8]. The regular cycles of stimulation and rest that are required to maintain antigen-specific T cell lines may be an in vitro parallel of this phenomena[89]. Thus the net elimination of T cells (i.e. self reactive cells in the thymus or $CD4^+$ T cells in HIV infection) can be explained by a shift in the dynamics of the same activation process that causes positive selection of non-self reactive thymocytes and effective immune responses in mature lymphocytes.

AIDS as immune system activation

We have previously proposed that the key element in HIV induced immunodeficiency is enhanced immune activation, mediated by the interaction of gp120 with CD4 which delivers a high-affinity costimulatory signal[7,8]. In the course of "normal" antigen-mediated activation, this unique signal-2 nonspecifically enhances T cell responses leading to the clinical manifestations of HIV disease such as cachexia, lymphadenopathy, and non-specific autoimmune phenomena. More importantly, we proposed that HIV infection causes $CD4^+$ T cell loss by shifting clonal dynamics toward excess programmed cell death (PCD). Others have proposed that PCD is an abnormal pathway, induced by "aberrant" HIV mediated signals[5]. We view PCD as an integral part of normal clonal dynamics which is enhanced by HIV signals resulting in an improper balance between proliferation and cell death[75]. We have termed the generalized immune activation "panergy" and the loss of immunologic memory through clonal deletion "imnesia"[7].

When our hypothesis was first proposed, it was based on two observations, that CD4 was the virus receptor[50], and that HIV infection was associated with many clinical and laboratory signs of generalized immune activation. At the time, CD4 was thought to be a cellular adhesion receptor with no established signal transduction capacity[20]. Since then the critical role of CD4 mediated signals in T cell activation[12,14,23,51,53,56,77,87] and in the induction of antigen specific tolerance has been demonstrated[4,25,37,57,83,94].

The ability of HIV to transduce signals through CD4 has also been demonstrated[41,56,59,60]. However, some studies have failed to find signal transduction[29,44,61]. The overall effect of HIV on lymphocyte activation have been reported to be both positive[16,43,52,62,66] negative[26,28,42,68,76]. There may be several reasons for this. First, receptor ligation can produce opposite outcomes depending on the nature of the ligand, the timing and coordination of signal delivery, the degree of crosslinking, and the condition of the T cell population. For example, costimulation through CD4 is dependent on its physical proximity to CD3 which is itself a function of prior activation[46,67]. Second, many in vitro lymphocyte activation experiments do not assess clonal dynamics but compare the results of thymidine incorporation at a fixed time after signal delivery. Such results can be dramatically effected by changes in clonal dynamics which result in the delay or acceleration of the programmed death phase in a response cycle. Finally, many of these experiments have been performed with a variety of native and recombinant gp120 preparations which may vary dramatically in their structure and purity. A recent report of gp120-induced cytokine production showed dramatic differences in the activity of recombinant gp120 produced in different host/vector systems[15].

The in vitro effects of other CD4 ligands also show a similar mixture of negative and positive effects on T cell activation[4,6,12,14,23,51,53,55,56,77,87]. In this context, negative effects are often interpreted as the blocking of a positive signal[76] and the overall consensus is that CD4 mediated signals participate positively in T cell activation[87]. In AIDS, the reverse is true, and the interaction of gp120 with CD4 is generally seen as negative because HIV disease is generally seen as "immune suppression". This is in spite of almost identical sets of data in the two different settings. Despite this controversy over the *in vitro* results, the overwhelming *in vivo* evidence for the association between signs of generalized immune activation and progressive HIV disease argues that the primary *in vivo* result of the gp120/CD4 signal transduction is stimulation and not suppression.

Tolerance

In this context, a simple explanation for negative thymic selection is that self-reactive thymocytes are trapped in the presence of signal-1 and signal-2, cannot rest, and are driven to clonal deletion. Non-self reactive thymocytes receive signal-2 but signal-1 is below-threshold, reducing activation to a level that permits maturation and migration to the periphery (I.E. positive selection)[88]. One problem with this interpretation is that a peripheral immune response to antigen, which also provides both signals, might be expected to cause clonal deletion. The fact that peripheral tolerance is very difficult to induce suggests that mature lymphocytes, unlike self-reactive thymocytes, can escape signals through elimination of the antigen or migration away from the local site of infection. In addition, activation signals may have a less intense effect on mature T cells so that escape from above-threshold signals is more easily achieved[64,80,85].

This model of HIV pathogenesis suggests the possibility that acute antigen stimulation in the context of HIV infection could result in antigen directed adult tolerance. Anecdotal support for this possibility, can be found in an intriguing case report of a liver transplant, coincident with HIV infection, which was successful despite the withdrawal of immunosuppressive therapy[86].

Superantigens

Superantigens such as Mls-1 are the most intensively studied experimental models for the induction of tolerance. A superantigen interacts directly with a substantial proportion of TCR beta-chain variable regions (Vβ) and, if present during fetal development, causes deletion of the T cell clones which bear those Vβ molecules[38,49]. When the superantigen is not endogenous, the corresponding T cell clones are retained and adult exposure to superantigen induces polyclonal activation of these cells. In other words, a superantigen is an aberrant and broadly reactive form of signal-1 which, in the context of CD4 or CD8 costimulation, leads to clonal deletion in the thymus or activation in the periphery. The requirement for costimulatory signals in superantigen activation is variable and may be a function of whether the superantigen can deliver an above-threshold signal by itself[65,74]. In this model of T cell activation, superantigens are simply Vβ binding molecules that deliver a Vβ-specific signal which is equivalent in biochemical intensity to a specific combination of self-MHC and antigenic peptide.

HIV as superantigen ?

Since Mls-1 was recently identified as a retroviral gene product[48] it has been suggested that HIV might also encode a Vβ binding superantigen[1,32,34,91]. However, a superantigen would be expected to cause deletion of a specific Vβ family rather than general depletion of $CD4^+$ T cells[47] or random holes in the repertoire[32]. In fact, the natural history of HIV disease is far more consistent with the effects of a costimulatory signal. An aberrant signal-2 would be a less efficient tolerogen, requiring concomitant signal-1 (provided by foreign antigen presentation), but would have more generalized effects than a superantigen because all helper T cells bear CD4. This would account for both the breadth of HIV induced immunodeficiency and the length of the incubation period between HIV infection and AIDS. Of particular relevance to this issue is the demonstration that MAIDS, a lymphoproliferative/ immunodeficiency disease of mice[9,30] is caused by a virus with typical superantigen properties[31]. Thus, the precedent for an "immunodeficiency" disease due strictly to excess activation signals is well established by this model. As would be anticipated from our hypothesis, this disease is associated with profound activation and proliferation but no general T cell loss.

Alloreactivity

This view of superantigens may also explain the paradoxically intense response to alloantigens[10,17,69]. Studies of thymic selection demonstrate that positive selection requires physical interaction of the TCR with MHC[88]. Thus MHC is, in a sense, a V-β binding protein with superantigen potential and thymic selection would result in a repertoire with only those TCR's which recognize self-MHC with below-threshold intensity. A T cell repertoire selected by self MHC would contain many clones which would recognize foreign MHC as an above-threshold signal-1, resulting in a superantigen-like response. This view predicts that self MHC has a profound effect on selection of the non-self repertoire and may explain some of the variable patterns of V-β expression in different genetic backgrounds[81].

One caveat in this interpretation is the phenomena of MHC restriction in which foreign MHC is unable to present antigen to T cells which have been primed in the context of self-MHC. If the combination of antigenic peptide with MHC generally increases the intensity of signal-1 then foreign-MHC would be expected to present antigen even better than self-MHC. An explanation may be found in the recent observation that protein antigens contain a limited number of peptides that can be processed and combined with MHC and that the subset of peptides that are "presentable" rather than "cryptic" is different in different genetic backgrounds[22]. Therefore MHC restriction may be more a function of antigen processing than differences in the interaction between MHC and TCR. In other words, foreign antigen presenting cells may be able to process and present peptides which were processed and presented in the context of self-MHC.

In summary, we have presented a hypothesis in which thymic selection, superantigen action, allogeneic reactivity and AIDS pathogenesis are seen as variable outcomes of one continuous process, immune activation. The particular fate of cells in each case is determined by the intensity and duration of a cooperative set of signals acting on T cells. The ultimate fate of a clone is determined by the number of progeny that survive each round of proliferation and programmed cell death. Both the antigen signal-1 and the accessory molecule-mediated signal-2 have variable intensity, and the combination is what determines the fate of an exposed T cell population. The paradoxical features of T cell activation and AIDS are unified under this spectrum of activation. Strategies designed to address issues in each area should take advantage of information derived from the other.

REFERENCES

1. Acha-Orbea H, Palmer E. Mls - a retrovirus exploits the immune system. Immunol Today 1991;12:356-361.

2. Agus DB, Surh CD, Sprent J. Reentry of T cells to the adult thymus is restricted to activated T cells. J Exp Med 1991;173:1039-1046.

3. Alles A, Alley K, Barrett JC, et al. Apoptosis: a general comment. FASEB J 1991;5:2127-2128.

4. Alters SE, Shizuru JA, Ackerman J, Grossman D, Seydel KB, Fathman CG. Anti-CD4 mediates clonal anergy during transplantation tolerance induction. J Exp Med 1991;173:491-494.

5. Ameisen JC, Capron A. Cell dysfunction and depletion in AIDS: the programmed cell death hypothesis. Immunol Today 1991;12:102-105.

6. Anderson P, Blue ML, Morimoto C, Schlossman F. Cross-linking of T3 (CD3) with T4 (CD4) enhances the proliferation of resting T lymphocytes. J Immunol 1987;139:678.

7. Ascher MS, Sheppard HW. AIDS as immune system activation II: the panergic imnesia hypothesis. JAIDS 1990;3:177-191.

8. Ascher MS, Sheppard HW. AIDS as immune system activation: a model for pathogenesis. Clin Exp Immunol 1988;73:165-167.

9. Aziz DC, Hanna Z, Jolicoeur P. Severe immunodeficiency disease induced by a defective murine leukaemia virus. Nature 1989;338:505-508.

10. Benoist C, Mathis D. Demystification of the alloresponse. Curr Biol 1991;1:143-144.

11. Beverley PCL. Is T-cell memory maintained by crossreactive stimulation? Immunol Today 1991;11:203-205.

12. Biddison WE, Rao PE, Talle MA, Goldstein G, Shaw S. Possible involvement of the T4 molecule in T cell recognition of class II HLA antigens. J Exp Med 1990;159:783-797.

13. Blackman M, Kappler J, Marrack P. The role of the T cell receptor in positive and negative selection of developing T cells. Science 1990;248:1335-1341.

14. Carrel S, Moretta A, Pantaleo G, Tambussi G, Isler P, Perussia B, Cerottini JC. Stimulation and proliferation of CD4+ peripheral blood T lymphocytes induced by an anti-CD4 monoclonal antibody. Eur J Immunol 1988;18:333-339.

15. Clouse KA, Cosentino LM, Weih KA, et al. The HIV-1 gp120 envelope protein has the intrinsic capacity to stimulate monokine secretion. J Immunol 1991;147:2892-2901.

16. Clouse KA, Robbins PB, Fernie B, Ostrove JM, Fauci AS. Viral antigen stimulation of the production of human monokines capable of regulating HIV1 expression. J Immunology 1989:143;470-475.

17. Demotz S, Sette A, Sakaguchi K, Buchner R, Appella E, Grey HM. Self-peptide requirement for class II major histocompatibility complex allorecognition. Proc Natl Acad Sci USA 1991;88:8730-8734.

18. Duesberg PH. Human immunodeficiency virus and acquired immunodeficiency syndrome: correlation but not causation? Proc Natl Acad Sci 1989;86:755-762.

19. Emmrich F, Strittmatter U, Eichmann K. Synergism in the activation of human CD8 T cells by cross-linking the T-cell receptor complex with the CD8 differentiation antigen. Proc Natl Acad Sci USA 1986;83:8298.

20. Fleischer B, Schrezenmeier H. Do CD4 or CD8 molecules provide a regulatory signal in T-cell activation? Immunol Today 1988;9:132-134.

21. Fuchs D, Hausen A, Hoefler E, et al. Activated T cells in addition to LAV/HTLV-III infection: a necessary precondition for development of AIDS. Cancr Det Prev Suppl 1987;1:583-587.

22. Gammon G, Klotz J, Ando D, Sercarz EE. The T cell repertoire to a multideterminant antigen. Clonal heterogeneity of the T cell response, variation between syngeneic individuals, and in vitro selection of T cell specificities. J Immunol 1990;144:1571-1577.

23. Gilliland LK, Teh HS, Uckun FM, Norris NA, Teh SJ, Schieven GL, Ledbetter JA. CD4 and CD8 are positive regulators of T cell receptor signal transduction in early T cell differentiation. J Immunol 1991;146:1759-1765.

24. Gray D, Matzinger P. T cell memory is short-lived in the absence of antigen. J Exp Med 1991;174:969-974.

25. Grusby MJ, Johnson RS, Papaioannou VE, Glimcher LH. Depletion of CD4+ T cells in major histocompatibility complex class II-deficient mice. Science 1991;253:1417-1420.

26. Gupta S, Vayuvegula B. Human immunodeficiency virus-associated changes in signal transduction. J Clin Imm 1987;7:486-489.

27. Hedrick SM, Cohen D, Nielsen E, Davis M. Isolation of cDNA clones encoding T cell specific membrane-associated proteins. Nature 1984;308:149.

28. Hofmann B, Nishanian P, Baldwin RL, Insixiengmay P, Nel A, Fahey JL. HIV inhibits the early steps of lymphocyte activation, including initiation of inositol phospholipid metabolism. J Immunol 1990;145:3699-3705.

29. Horak ID, Popovic M, Horak EM, Lucas PJ, Gress RE, June CH, Bolen JB. No T-cell tyrosine protein kinase signalling or calcium mobilization after CD4 association with HIV-1 or HIV-1 gp120. Nature 1990;348:557-560.

30. Huang M, Simard C, Jolicoeur P. Immunodeficiency and clonal growth of target cells induced by helper-free defective retrovirus. Science 1989;246:1614-1617.

31. Hugin AW, Vacchio MS, Morse HC. A virus-encoded "superantigen" in a retrovirus-induced immunodeficiency syndrome of mice. Science 1991;252:424-427.

32. Imberti L, Sottini A, Bettinardi A, Puoti M, Primi D. Selective depletion in HIV infection of T cells that bear specific T cell receptor V-beta sequences. Science 1991;254:860-862.

33. Janeway CA. The co-receptor function of CD4. Sem in Immunol 1991;3:153-160.

34. Janeway C. Mls: makes a little sense. Nature 1991;349:459-461.

35. Jenkins MK, Pardoll DM, Mizuguchi J, Chused TM, Schwartz RH. Molecular events in the induction of a nonresponsive state in interleukin-2 producing helper T-lymphocyte clones. Proc Natl Acad Sci USA 1987;84:5409.

36. Jenkins MK, Ashwell JD, Schwartz RH. Allogenic non-T spleen cells restore the responsiveness of normal T cell clones stimulated with antigen and chemically modified antigen-presenting cells. J Immunol 1988;140:3324-3330.

37. Julius M, Newell K, Maroun C, Haughn L. Functional consequences of CD4-TCR/CD3 interactions. Sem in Immunol 1991;3:161-166.

38. Kappler JW, Staerz U, White J, Marrack PC. Self-tolerance eliminates T cells specific for Mls-modified products of the major histocompatibility complex. Nature 1988;332:35-40.

39. Kaye J, Hsu ML, Sauron ME, Jameson SC, Gascoigne NRJ, Hedrick SM. Selective development of CD4+ T cells in transgenic mice expressing a class II MHC-restricted antigen receptor. Nature 1989;341:746-749.

40. Kerr JFR, Wyllie AH, Currie AR. Apoptosis: a basic biological phenomena with wide-ranging implications in tissue kinetics. Br J Cancer 1972;26:239-257.

41. Kornfeld H, Cruikshank WW, Pyle SW, Berman JS, Center DM. Lymphocyte activation by HIV-1 envelope glycoprotein. Nature 1988;335:445-448.

42. Krowka J, Stites D, Mills J, et al. Effects of interleukin 2 and large envelope glycoprotein (gp120) of human immunodeficiency virus (HIV) on lymphocyte proliferative responses to cytomegalovirus. Clin Exp Immunol 1988;72:

43. Kupcu Z, Barrett N, Dorner F, Gallo RC, Eibl MM, Mannhalter JW. Effect of HIV on antigen-induced T cell proliferation. Abstract #T.C.P. 38, V Int Conf on AIDs, 1989, Montreal, Canada.

44. Lamarre D, Capon DJ, Karp DR, Gregory T, Long EO, Sekaly RP. Class II MHC molecules and the HIV gp120 envelope protein interact with functionally distinct regions of the CD4 molecule. EMBO J 1989;8:3271-3277.

45. Lane HC, Depper JM, Greene WC, Whalen G, Waldmann TA, Fauci AS. Qualitative analysis of immune function in patients with the acquired immunodeficiency syndrome. New Eng J Med 1985;313:79-84.

46. Ledbetter JA, June CH, Rabinovitch S, Grossman A, Tsu T, Imboden JB. Signal transduction through CD4 receptors: stimulatory vs. inhibitory activity is regulated by CD4 proximity to the CD3/T cell receptor. Eur J Immunol 1988;18:525-532.

47. Marrack P. Superantigens and immune system design. Clin Imm Spectrum 1991;3:8-10.

48. Marrack P, Kushnir E, Kappler J. A maternally inherited superantigen encoded by a mammary tumour virus. Nature 1991;349:524-526.

49. McDonald HR, Schneider R, Lees RK, Howe RC, Acha-Orbea H, Festenstein H, Zinkernagel RM. T-cell receptor V-beta use predicts reactivity and tolerance to Mls-1a encoded antigens. Nature 1988;332:40-45.

50. McDougal JS, Klatzmann DR, Maddon PJ. CD4-gp120 interactions. Curr Opin Immunol 1991;3:552-558.

51. Miceli MC, Parnes JR. The roles of CD4 and CD8 in T cell activation. Sem in Immunol 1991;3:133-141.

52. Mittler RS, Hoffmann MK. Synergism between HIV gp120 and gp120-specific antibody in blocking human T cell activation. Science 1989;245:1380-1382.

53. Mittler RS, Rankin BM, Kiener PA. Physical associations between CD45 and CD4 or CD8 occur as late activation events in antigen receptor-stimulated human T cells. J Immunol 1991;147:3434-3440.

54. Montagnier L, Guetard D, Rame V, Olivier R, Adams M. Virological and immunological factors of AIDS pathogenesis. Quatr Col Cent Gardes 1989; :11-17.

55. Mueller DL, Jenkins MK, Schwartz RH. Clonal expansion versus functional clonal inactivation: a costimulatory signalling pathway determines the outcome of T cell receptor occupancy. Ann Rev Immunol 1989;7:445-480.

56. Neudorf SML, Jones MM, McCarthy BM, Harmony JAK, Choi EM. The CD4 molecule transmits biochemical information important in the regulation of T lymphocyte activity. Cell Immunol 1990;125:301-314.

57. Newell MK. Haughn LJ, Maroun CR, Julius MH. Death of mature T cells by separate ligation of CD4 and the T-cell receptor for antigen. Nature 1990;347:286-289.

58. Nicholson JKA, McDougal JS, Jaffe HW, et al. Exposure to human T-lymphotropic virus type III/ lymphadenopathy-associated virus and immunologic abnormalities in asymptomatic homosexual men. Ann Intern Med 1985;103:37-42.

59. Nye KE, Knox KA, Pinching AJ. Lymphocytes from HIV-infected individuals show aberrant inositol polyphosphate metabolism which reverses after zidovudine therapy. AIDS 1991;5:413-417.

60. Nye KE, Pinching AJ. HIV infection of H9 lymphblastoid cells chronically activates the inositol polyphosphate pathway. AIDS 1990;4:41-45.

61. Orloff GM, Kennedy MS, Dawson C, McDougal S. HIV-1 binding to CD4 T cells does not induce a Ca2+ influx or lead to activation of protein kinases. AIDS Res Hum Retrovir 1991;7:587-593.

62. Pinching AJ. HIV/AIDS pathogenesis and treatment: new twists and turns. Curr Opin Immunol 1991;3:537-542.

63. Pinching AJ, Jeffries DJ, Harris JRW, Swirsky D, Weber JN. HIV and AIDS. Nature 1990;347:324.

64. Pircher H, Rohrer UH, Moskophidis D, Zinkernagel RM, Hengartner H. Lower receptor avidity required for thymic clonal deletion than for effector T-cell function. Nature 1991;351:482-485.

65. Quaratino S, Murison G, Knyba RE, Verhoef A, Londei M. Human CD4- CD8-alpha-beta+ T cells express a functional T cell receptor and can be activated by superantigens. J Immunol 1991;147:3319-3323.

66. Rieckmann P, Poli G, Fox CH, Kehrl JH, Fauci AS. Recombinant gp120 specifically enhances tumor necrosis factor-alpha production and Ig secretion in B lymphocytes from HIV-infected individuals but not from seronegative donors. J Immunol 1991;147:2922-2927.

67. Rivas A, Takada S, Koide J, Sonderstrup-MacDevitt G, Engleman E. CD4 molecules are associated with the antigen receptor complex on activated but not resting T cells. J Immunol 1988;140:2912-2918.

68. Rosenstein Y, Burakoff SJ, Herrmann SH. HIV-gp120 can block CD4-class II MHC-mediated adhesion. J Immunol 1990;144:526-531.

69. Rotzschke O, Falk K, Faath S, Rammensee HG. On the nature of peptides involved in T cell alloreactivity. J Exp Med 1991;174:1059-1071.

70. Russell JH, White CL, Loh DY, Meleedy-Rey P. Receptor-stimulated death pathway is opened by antigen in mature T cells. Immunology 1991;88:2151-2155.

71. Schnittman SM, Lane HC, Greenhouse J, Justement JS, Baseler M, Fauci AS. Preferential infection of CD4+ memory T cells by human immunodeficiency virus type 1: evidence for a role in the selective T-cell functional defects observed in infected individuals. Proc Natl Acad Sci USA 1990;87:6058-6062.

72. Schwartz RH. A cell culture model for T lymphocyte clonal anergy. Science 1990;248:1349-1356.

73. Scott S, Pandolfi F, Kurnick JT. Fibroblasts mediate T cell survival: a proposed mechanism for retention of primed T cells. J Exp Med 1990;172:1873-1876.

74. Sekaly RP, Croteau G, Bowman M, Scholl P, Burakoff S, Geha RS. The CD4 molecule is not always required for the T cell response to bacterial enterotoxins. J Exp Med 1991;173:367-371.

75. Sheppard HW, Ascher MS. AIDS and programmed cell death. Immunol Today 1999;12:423.

76. Silberman SL, Goldman SJ, Mitchell DB, et al. The interaction of CD4 with HIV-1 gp120. Sem in Imm 1991;3:187-192.

77. Sleckman BP, Rosenstein Y, Igras VE, Greenstein JL, Burakoff SJ. Glycolipid-anchored form of CD4 increases intercellular adhesion but is unable to enhance T cell activation. J Immunol 1991;147:428-431.

78. Smith CA, Williams GT, Kingston R, Jenkinson EJ, Owen JJT. Antibodies to CD3/T-cell receptor complex induce death by apoptosis in immature T cells in thymic cultures. Nature 1989;337:181-184.

79. Solbach W, Moll H, Rollinghoff M. Lymphocytes play the music but the macrophage calls the tune. Immunol Today 1991;12:4-6.

80. Speiser DE, Chvatchko Y, Zinkernagel RM, MacDonald HR. Distinct fates of self-specific T cells developing in irradiation bone marrow chimeras: clonal deletion, clonal anergy, or in vitro responsiveness to self-Mls-1a controlled by hemopoietic cells in the thymus. J Exp Med 1990;172:1305-1314.

81. Sprent J, Gao EK, Webb SR. T cell reactivity to MHC molecules: immunity versus tolerance. Science 1990;248:1357-1363.

82. St.-Pierre Y, Watts TH. Characterization of the signaling function of MHC class II molecules during antigen presentation by B cells. J Immunol 1991;147:2875-2882.

83. Teh HS, Garvin AM, Forbush KA, Carlow DA, Davis CB, Littman DR, Perlmutter RM. Participation of CD4 coreceptor molecules in T-cell repertoire selection. Nature 1991;349:241-243.

84. Thomas L. In: Lives of a Cell, 1974. Viking Press.

85. Turka LA, Linsley PS, Paine R, Schieven GL, Thompson CB, Ledbetter JA. Signal transduction via CD4, CD8, and CD28 in mature and immature thymocytes. J Immunol 1991;146:1428-1436.

86. Vanhems P, Bresson-Hadni S, Vuitton DA, Miguet JP, Gillet M, Lab M, Brechot C. Long-term survival without immunosuppression in HIV-positive liver-graft recipient. Lancet 1990;337:126.

87. Veillette A. Introduction: the functions of CD4 and D8. Sem in Immunol 1991;3:131-132.

88. von Boehmer H, Kisielow P. Self-nonself discrimination by T cells. Science 1990;248:1369-1373.

89. Weyand CM, Goronzy J, Fathman CG. Human T cell clones unsed to define functional epitopes on HLA class II molecules. Proc Nat Acad Sci USA 1986;83:762-768.

90. Wyllie AH, Kerr JFR, Currie AR. Cell death: the significance of apoptosis. Int Rev Cytol 1980;68:251-306.

91. Yagi J, Rath S, Janeway CA. Control of T cell responses to staphylococcal enterotoxins by stimulator cell MHC class II polymorphism. J Immunol 1991;147:1398-1405.

92. Zacharchuk CM, Mercep M, Chakraborti PK, Simons SS, Ashwell JD. Programmed T lymphocyte death. J Immunol 1990;145:4037-4045.

93. Zhou P, Anderson GD, Savarirayan S, Inoko H, David CS. Thymic deletion of V-beta11+, V-beta5+ T cells in H-2E negative, HLA-DQbeta+ single transgenic mice. J Immunol 1991;146:854-859.

94. Zuniga-Pflucker JC, Jones LA, Chin T, Kruisbeek. CD4 and CD8 act as co-receptors during thymic selection of the T cell repertoire. Sem in Immunol 1991;3:167-175.

PARADOXICAL EFFECTS OF SUPPRESSOR T CELLS IN ADJUVANT ARTHRITIS : NEURAL NETWORK ANALYSIS

Henri ATLAN*

Dpt of Medical Biophysics
Hadassah University Hospital
Jerusalem, Israel
and
Irun R. COHEN

Dpt of Cell Biology
The Weizmann Institute
Rehovot, Israel

I. Introduction

A double paradox has been found by N. Karin and I.R. Cohen(Cohen 1989, Karin & Cohen, in preparation) in their results from recent studies on Adjuvant Arthritis commonly used as an experimental model of auto-immune disease. The purpose of this work was to elucidate the mechanisms whereby T cell vaccination can prevent the dysregulation of a preexisting regulatory network, which seems to accompany the onset of the disease.

Previous work (Cohen & Atlan 1989, Atlan & Hoffer-Snyder 1989) had shown that this network was made of at least two couples of regulatory cells, helper and suppressor, antigen specific and anti-idiotypic T cells (Figure 1). Under the combined effects of these regulatory cells, the effector T cells responding to the self-antigen (e.g. a joint cartilage protein in AA, or Myelin Basic Protein (MBP) in Experimental Auto-immune Encephalomyelitis (EAE)) are normally maintained in a resting state. While the antigen specific regulatory cells are active in naïve individuals and prevent a spontaneous onset of the disease, the anti-idiotypic cells are activated upon experimental

NATO ASI Series, Vol. H 66
Theoretical and Experimental Insights into Immunology
Edited by A. S. Perelson and G. Weisbuch

activation of the effector cells in the presence of an adequate adjuvant.

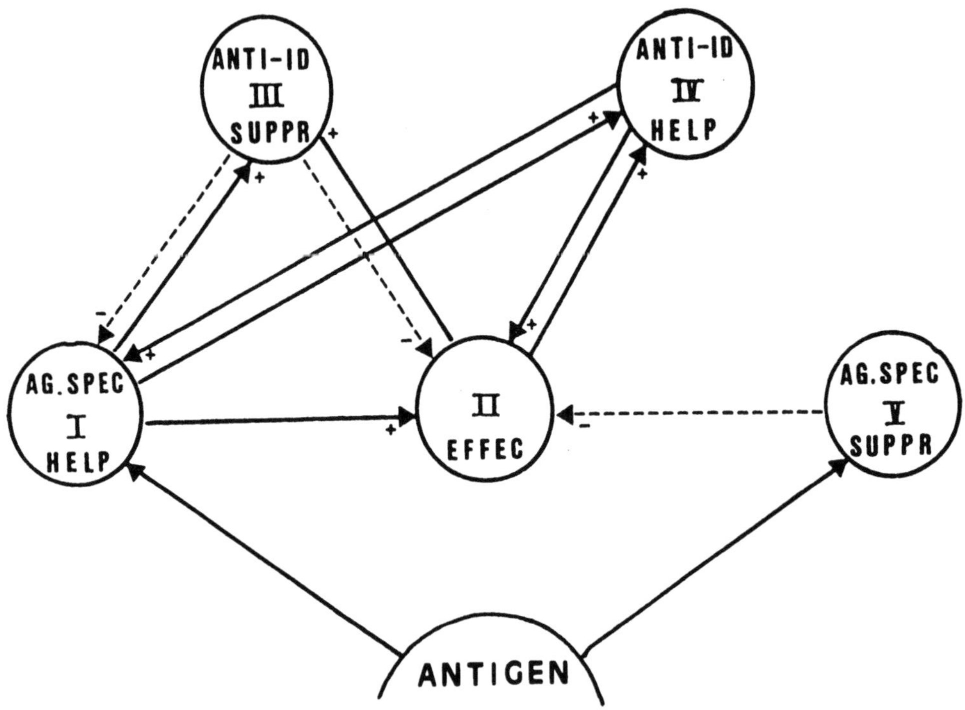

Figure 1

Thus, the onset of the disease depends upon whether a permanent and stable activation of the effector cells is maintained, or it is prevented by the cooperative effect of the regulatory cells.

In our previous publications (Weisbuch & Atlan 1988, Cohen & Atlan 1989), the interactions between these five kinds of T cell populations were represented by a small five unit neural network. Each unit in the network represented a cell population which could be set either in a resting 0 state, or in an activated 1 state. Later on (Atlan & Hoffer-Snyder 1989) , we made the model more sophisticated by allowing three states for each cell population, 0,1,2, where state 2 applied only to a fully activated and proliferating cell population. This allowed us, among other things, to distinguish between an activating effect of the effector cells (unit II in the network) on other cell populations which respond to them (such as the anti-idiotypic units III and IV) when they are in state 1, and their killing effect on the same cells when they are in state 2 (see fig.1).

Assuming a given set of connections between the cell populations, reasonable enough to represent plausible interactions between them, it is easy to compute attractors of the network in the form of steady states or short cycles. Steady states, where the effector cells are in a resting state or in state 1, represent tolerant, non diseased states. Those where effector cells are in state 2, represent auto-immunopathic states.

In our previous work, we concluded, among others, that the onset of the auto-immune disease had to be related to a deficiency, either genetic or induced, in the suppressive effect of one or the two regulatory suppressor T cells. One of the two paradoxes reported by Karin and Cohen forces us to revise the model, since, contrary to the prediction, they consistently found a positive correlation between the onset and severity of AA and an increased suppression of effector cell response to their antigen in vitro, induced by spleen cells from the diseased animals.

The aim of this paper is to present a revised model of the regulatory network which would account for this strange finding. In agreement with the data on the non specific nature of these suppressor cells (see further), the new model has to incorporate a sixth population of cells, different from both the antigen specific and anti-idiotypic suppressor cells accounted for by the previous five unit network. The network analysis will show that the seemingly paradoxical effect of these suppressor cells can be explained by their overall, non specific activity, which leads, under some conditions, to a dominant contra-suppressive effect responsible for the disease.

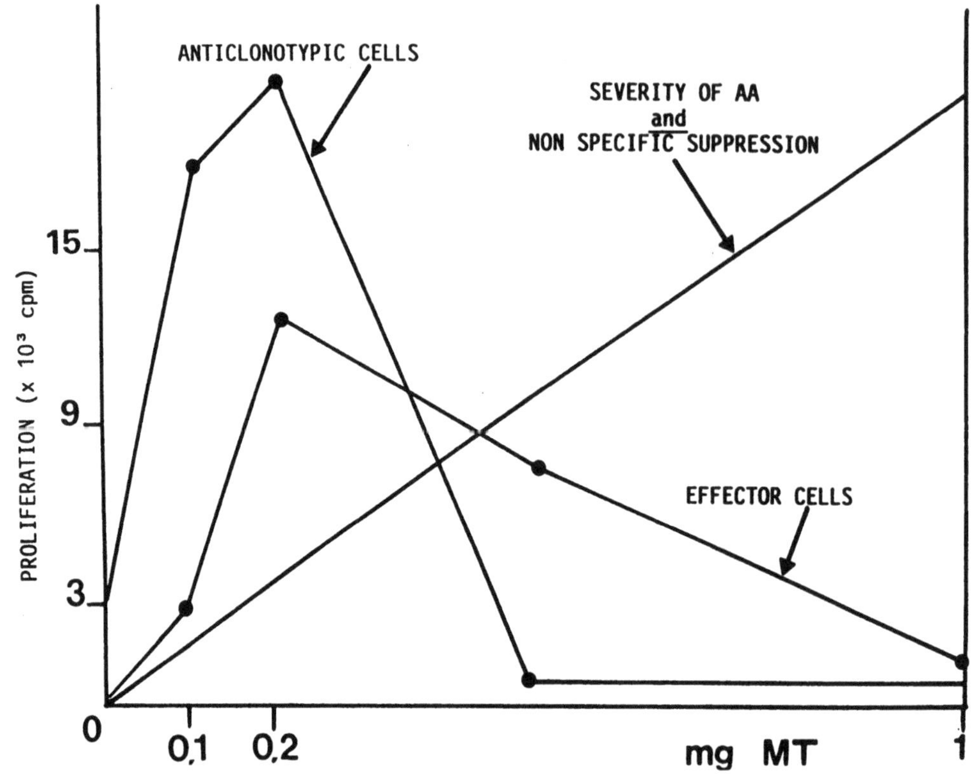

Figure 2 . Main experimental results (Cohen 1989, Karin & Cohen) accounted for by the models of figures 3,4,5,6) : i) positive correlation of non specific suppression with the dose of MT and the severity of AA ; ii) negative correlation of non specific suppression with anti-clonotypic regulatory cell activity and lymph node response to antigen (hsp65).

II.Schematic summary of the relevant experimental findings.

Mycobacterium Tuberculosis (MT) works as an adjuvant, but is able to induce by itself auto-immune AA in rats because of a documented cross-reactive T cell response between one of its proteins (hsp65), and a joint cartilage protein (Van Eden & al. 1985, Van Eden & al. 1988). Lewis rats were inoculated in the foot pad with various amounts of MT from 0.1 to 1.0 mg. The response was scored as no disease, mild AA and severe AA. An obvious correlation was found between the dose of MT and the severity of the disease.

II.1 The T cell suppression paradox

A similar correlation was found between the dose of MT and a suppressive activity of spleen cells from the diseased animals on the proliferative response of effector T cell clones to the antigen. Moreover, among the groups of animals which received the same amount of MT and presented some variability in the severity of the induced disease, the more severe the disease, the higher T cell suppression. The T cell suppression developed in diseased rats was measured in vitro as a reduction of the proliferative response of known effector CD4+ T cell clones to MT or to hsp65 in the presence of X irradiated spleen cells isolated from either AA or naïve rats.

This spleen cell suppression was found to be mediated by CD8+ T cells, and composed of a specific and a non specific part. The specific suppression reduces the proliferation of clones responding to MT only. The non specific reduces the proliferation of other clones responding to different antigens, non relevant to AA. In other words,
a non specific suppressive effect is manifest not only on effector cells responsible for AA, but also on control effector cells responsible for non related different auto-immune disease, such as anti-MBP T cell clones capable to induce EAE.

The non specific spleen cell suppression was found to be relatively more important with the high doses of MT which induced the most severe forms of AA. In addition, the specific suppression was found very early (a few days) after injection of MT, even before any effector cell activation could be detected in the draining lymph nodes of the animal. This finding is an indication of the possible existence of a preformed network of antigen- mediated spleen cell suppression, where the suppressive effect is triggered even before the effector cells respond to their antigen. On the other hand, the non specific spleen cell suppression was developed at a later stage.

II.2. The effector cell paradox

The second paradox concerns the proliferative activity of the effector T cells from draining lymph nodes, considered to be directly responsible for the disease.

As seen in figure 2, whereas this activity is maximum for the moderate doses of MT which produced mild forms of AA, it falls down to almost zero after injection of high doses of MT which produced the most severe forms of AA ! One can speculate on the efficient cause of the disease in those severely diseased animals where the effector cell activity in the lymph nodes is nil. One cannot rule out the possibility that all the active effector cells are present in the joints at this stage. Nevertheless, it is clear that the effector cell activation in the lymph nodes is suppressed, while the maximum spleen cell suppression is correlated with a state of severe auto-immune disease.

II.3 The anti-clonotypic regulatory cells

The MT induced activity of anti-clonotypic cells (anti-anti-hsp65) was measured by the proliferative response of lymph node cells to X irradiated resting effector T cell clones (anti-hsp65), after inoculation of MT.

An early response was found, again before any anti-MT effector cell activity could be detected, indicating the existence of the proformed regulatory network. This anti-clonotypic activity was found to be mediated not only by anti-idiotypic cells, but also by so-called anti-ergotypic cells, i.e. cells responding to activated T cells irrespective of their specificity (Lohse & al. 1989).

Figure 2 shows how this anti-anti-MT activity varies with the dose of MT. A basic activity, already detectable in the absence of MT, raises with low doses of MT which produce mild AA, and decreases in severe forms of disease induced by high doses of MT.

II.4 T cell vaccination

As a reminder, effector T cells which were cloned from the lymph nodes of diseased animals were characterized by i) their response to a self-antigen like a nonapeptide of the joint cartilage (clone A2b) or to the cross-reactive mycobacterial antigen hsp65 (clone M1), and ii) their capability to induce active arthritis when transferred to naïve rats after in vitro activation. Now, the same clones transferred to naïve rats after con-A activation and X ray irradiation produce a vaccination, in the sense that further inoculation of an active pathogenic amount of effector cells, or of MT, does not induce the disease (Cohen 1989). Similarily, after induction of AA by a pathogenic dose of effector cells or of MT, if the animals recover and do not die, they are vaccinated against further induction of the disease.

In such vaccinated animals, the spleen cell suppression and the anti-anti-hsp65 activity of lymph node cells induced by inoculation of different amounts of MT was measured (Karin & Cohen, in preparation). While no disease was observed, as expected, after inoculation of MT, no spleen cell suppression was observed, and a strong anti-anti-hsp65 response of lymph node cells was observed. In other words, T cell vaccination induces a reversal of the balance between cell suppression and anti-clonotypic regulatory T cell activity.

III The network model

III.1 Before vaccination

The most relevant data reported here can be summarized very schematically as in figure 2.

The plain experimental data suggest that very high doses of MT induce a suppressive activity on both the effector cells and the anti-clonotypic regulatory cells. As mentioned above, this spleen cell suppression , both specific and non specific, was found to be mediated by CD8+ T cells. Since this suppressive activity is mostly non specific for high doses of MT, it must be distinguished from the antigen specific suppressor cell population accounted for in our previous five unit network model. In essence, the main finding concerning the regulation of auto-immunity in this set of experiments, is the existence of a balance between a high non specific T cell suppressor activity and a low anti-clonotypic cell activity in severe AA, reversed into the opposite state of low suppressive activity and high anti-clonotypic cell activity in mild AA.

All this can be represented by adding a sixth T cell population (unit VI) to our previous five unit network (fig.3). This unit is connected by suppressive links to the regulatory cells (and possibly to the effector cells as in the alternative model of section III.3), and thus completes the preformed network, but it is normally inactive in the absence of MT.

As in our previous model, unit I represents both antigen specific helpers and the antigen presenting cell system. That is why it is activated by both the self-antigen and the bacterial adjuvant. As previously, all the units are assigned three possible states, 0,1,2. However, if we want our model to account for the experimental data mentioned above, units I and VI must have at least four states (0,1,2,3) since they represent populations of cells which react directly to, at least, four ranges of MT doses.

As previously, and for the above mentioned reason, unit II (effector cells) sends a negative output (-1) when it is in state 2. Otherwise, all the ouputs are equal to the states of the unit from which they originate, and the connection weights are +1 for activating connections (full arrows) and -1 for suppressive ones (dotted arrows).

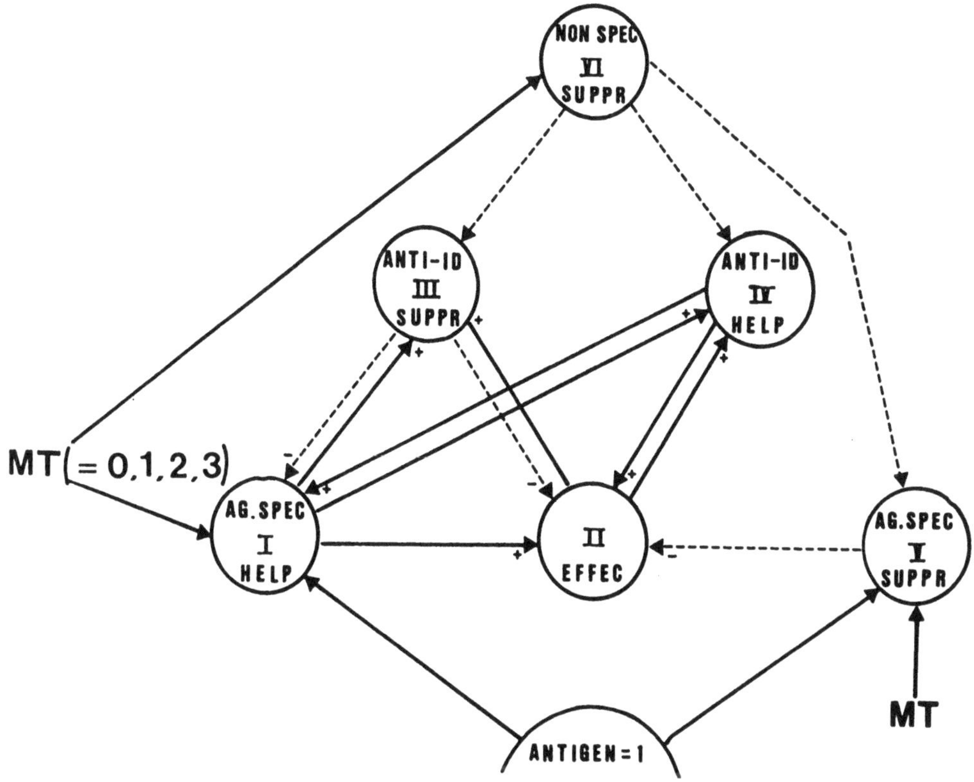

Figure 3 . Model 1. The laws corresponding to the connections have the general form :

$$\text{if } \Sigma\, w_{ij}\, S(j) + IN > \Theta_{ik}\,,\ NS(i) = \Theta_{ik}+1$$

$$\text{if } \Sigma\, w_{ij}\, S(j) + IN < \Theta ik\,,\ NS(i) = \Theta_{ik}\,.$$

w_{ij} is the weight of the connection from automaton j to i. w_{ij} = +1 for activating connections(full arrows) and = -1 for suppressive connections (dotted arrows).
S is the state of an automaton at time t and NS is the next state of an automaton, i.e. at time t+1 . S and NS $\epsilon \{0,1,...,n_i-1\}$ where n_i is the number of states of automaton i $\Theta_{ik} \epsilon \{0,1,...,n_i-2\}$ is a threshold value . Since the maximum value of NS(i) is n_i-1, if, in the first inequality, the sum $\Sigma > n_i-1$, NS(i) = n_i-1.
IN is an external input coming from MT and/or the antigen .External inputs from MT to I and VI equal 0,1,2, or 3.The external input from MT to automaton V equals 1 for all values of MT larger than 0. The external input from the antigen equals 1.

In the particular instance of the network connections in this figure, with *total* external input to I, V and VI being respectively 3, 2 and 2, the laws become:

if -S(III) + S(IV) + 3 > Θ NS(I) = Θ + 1

if S(I) - S(III) + S(IV) - S(V) > Θ NS(II) = Θ + 1

if S(I) + S(II) - S(VI) > Θ NS(III) = Θ +1

if S(I) + S(II) - S(VI) > Θ NS(IV) = Θ + 1

if -S(VI) + 2 > Θ NS(V) = Θ + 1

if +2 > Θ NS(VI) = Θ + 1 (This latter law means that VI is always in state 2

... / ...

Under these conditions, the dynamics of the network are computed, and the attractors are looked for when different values of the external inputs are presented to units I and VI, representing the four ranges of doses, 0,1,2,3, for MT .

As in our previous model, unit V receives an input +1 from the antigen in the absence of MT, and an input +2 in the presence of MT.

The computation is performed by random updating which provides the most general results and is the most plausible, in the absence of any assumption on a specific order of updating, either synchronous or serial, as mentioned in our previous works (Cohen & Atlan 1989).

The results of the computations are shown in table 1. They clearly exhibit the above-mentioned balance of activity between the non specific T suppressor cells (unit VI) and the regulatory anti-clonotypic cells (units III and IV) . In particular, a high dose of MT, represented by an input 2 on VI (and an input 3 on I), produces as a single attractor the state 320002. This state corresponds to a severe form of the disease (effector cells II in state 2), with high suppressor activity (VI in state 2) and no activity of the regulatory cells. This is a result of a suppressive activity of VI on the regulatory suppressive activity of III and V, strong enough to over-compensate the activating effect of the helper units I and IV.Thus, the disease appears as a result of a contra-suppressive effect of the non specific T suppressor cells, activated by high doses of MT.

---------------------------Figure 3 (contin.) since it receives a total external input equal to 2 , and no connection from any other automaton. Of course, the situation will be different in the networks of figures 4 and 6 where it receives other connections.)

Automata I and VI have four states 0,1,2,3 and their threshold values Θ equal 0,1,2. Automata II to V have three states 0,1,2 and their threshold values equal 0,1.

As explained in the text, the state of an automaton j is sent as input to every automaton i to which it sends a connection, except for II which sends an output -1 when in state 2.

The computation, whatever being the order of updating, leads in this particular instance to the attractor 320002, meaning that automaton I is in state 3, II in state 2, III, IV and V in state 0, and VI in state 2. The different attractors obtained for different values of MT are shown in Table1.

As to the second paradox, a possible model is suggested by the result of the simulation when a still higher activity (state 3) is induced on unit VI by a very high dose of MT. As shown in table 1, this is not enough to produce the paradoxical suppression of the effector cell activity since the attractor is now 320003. Some modification of the connections is necessary. Of course, in the absence of experimental data, one can imagine many possibilities. A simple one would be to assume an extension of the suppressive effect of VI to the effector cells II directly, or else to the antigen presenting system (i.e. to I), an indication of which can be found, although in a different system, in Ben Nun & Cohen (1982). Then, the state 000003 becomes the only attractor of the network. This result would account for the effect of the highest doses of MT, where high non specific suppressive activity is associated with no activity of both the regulatory and the effector cells in the lymph nodes, in spite of the animals being severely diseased.

As will be clarified further, this modification in the connectivity may be seen as an early and partial vaccination effect, still insufficient and occuring too late to cure the active disease.

III.2 Network representation of T cell vaccination

Within the framework of the theory of neural network computation, vaccination is interpreted as a learning phenomenon by a second order dynamics which modify the connections. As a result, the network stabilizes in new attractors representing non diseased states, even when it receives the external immunogenic inputs which led it to diseased attractors beforehand.

In the present situation, we assume that a partial hebbian learning rule can be applied so that the connections in the pre-existing , non activated, network are modified by the state of activity of the units after they have been activated. When unit VI is stabilized in an activated state by exposition to a given dose of MT, it triggers new reactivities

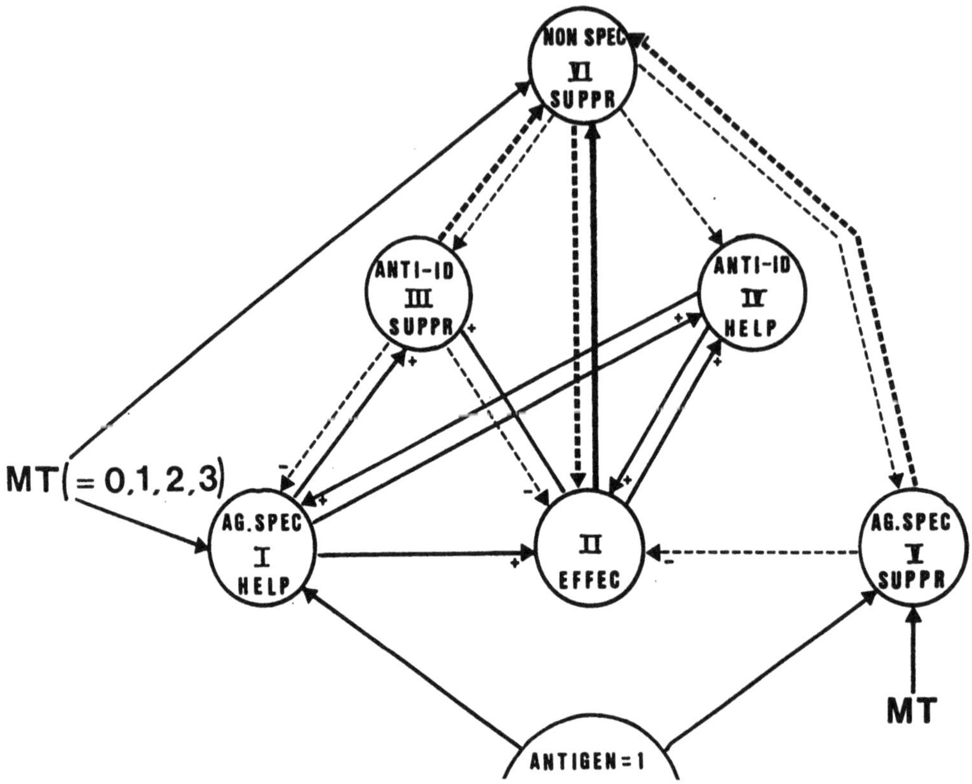

Figure 4. Model 1 after vaccination . The accentuated arrows are new connections created by the learning process which represents the effects of the vaccination.

from the suppressive units III and V, and new interactions with the effector unit II. Thus, the network structure is changed as in figure 4, with new connections from III to VI, V to VI, and between II and VI. The new attractors of the network after learning are shown in table 1. As expected from the above data on the vaccinated state, one can see that high doses of MT now produce non diseased attractors, characterized by low non specific suppressor cell (VI) activity and high regulatory cell (III,IV,V) activity. In addition, the effector cells are never stabilized in their fully active state 2.

III.3 An alternative model

Figure 5 represents a slightly different model which produces similar results. In this model, we assume now an additional direct

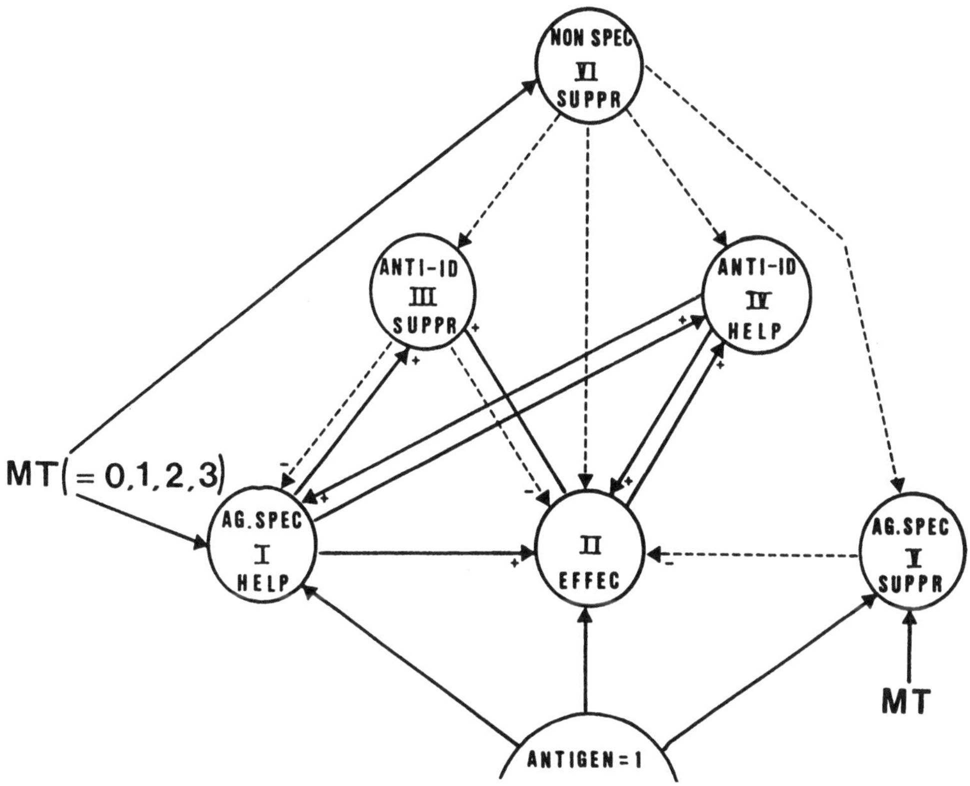

Figure 5. Model 2 . See text.

external input +1 from the antigen to unit II. This may represent a genetically determined direct sensitivity of the effector cells to the self-antigen. On the other hand, contrary to the first model, a direct suppressive connection from VI to II is assumed to be part of the preformed network, before any learning could take place.

As can be seen in table 1, the results are the same, but the assumptions on the changes in the connections related to learning must be slightly different as far as the connections to, and from unit II are concerned (see figure 6 and table 1). In particular, one has to assume a

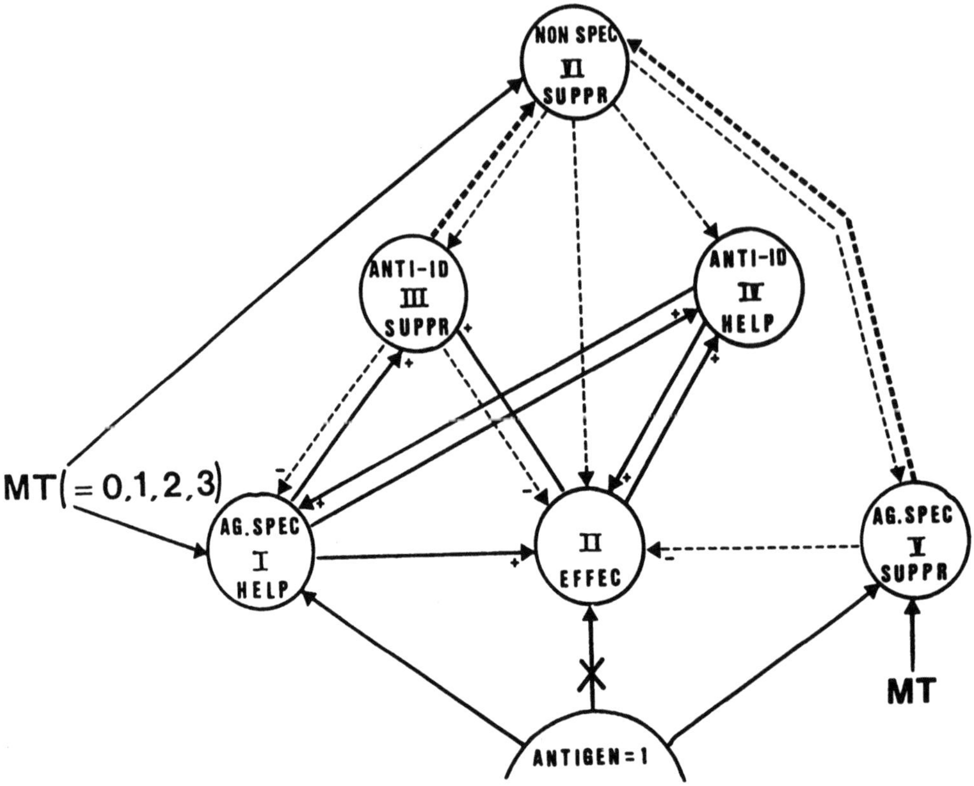

Figure 6 . Model 2 after vaccination . See text.

suppression of the direct responsiveness of the effector cells to the self-antigen, i.e. a suppression of the external input to unit II, as a result of the previous activation of the network.

In both models, the new connection from II to VI assumed in the learning process is not critical. The attractors are the same without it, except for the highest input value +3 ot MT which produces the two attractors 312220 and 300003, instead of the four shown in table 1.

However, the new connections to VI from III and V seem to be critical to explain the effects of vaccination. Their biological counterpart could be tested experimentally.

TABLE 1

MT DOSE	TOTAL EXTERNAL INPUT TO I	TOTAL EXTERNAL INPUT TO VI	ATTRACTORS I	II	III	IV	V	VI
0	1	0	1	0	1	1	1	0
1	2	1	2	1	2	2	1	1
2	3	2	3	2	0	0	0	2
3	3	3	3	2	0	0	0	3
3	3	3**	0	0	0	0	0	3

** MODEL 1 : ADDITONAL CONNECTION VI----→II OR VI----→I OR BOTH

MODEL 2 : SUPPRESSED CONNECTION +1——→II

AFTER LEARNING ("VACCINATION")

MT DOSE	TOTAL EXTERNAL INPUT TO I	TOTAL EXTERNAL INPUT TO VI	I	II	III	IV	V	VI
0	1	0	1	0	1	1	1	0
1	2	1	2	0	2	2	2	0
2	3	2	3	1	2	2	2	0
3	3	3	{3	1	2	2	2	0
			3	1	2	2	1	1
			3	1	2	2	0	2
			3	0	0	0	0	3

TOTAL EXTERNAL INPUT TO V = 1 IF MT = 0
= 2 otherwise

IV Conclusion

As was noticed previously (Atlan 1989), this kind of neural network models has a basic flaw, namely they are not unique. As we have just seen, several different models can account for the same observed facts. However, they provide some insight into the plausibility of mechanisms which may look paradoxical and counter-intuitive. In the present situation, the decisive role of a non specific immunosuppression in the onset of auto-immunity appears very clearly as a suppression of specific regulatory cells, later extended to other, non specific, cells, and even to the effector cells themselves.

* Present address: Service de biophysique, Faculté de Médecine Broussais-Hotel Dieu, 15 rue de l'Ecole de Médecine, 75006, Paris, France.
Aknowledgement: part of the work was supported by an INSERM (Paris) grant.

REFERENCES

Atlan H (1989) Automata Network Theories in Immunology: Their Utility and their Underdetermination.*Bull Mathem Biol* 51: 2 247-253.

Atlan H, Hoffer-Snyder S (1989) Simulation of the Immune Cellular Response by Small Neural Networks. In : Atlan H, Cohen I R (eds) *Theories of Immune Networks,* Springer, Berlin Heidelberg NewYork, p 85-98.

Ben Nun A, Cohen I R (1982) Experimental autoimmune encephalomyelitis (EAE) at the level of cytotoxic T lymphocytes in guinea pigs. *Eur J Immunol* 12 : 709-713

Cohen I R (1989) T Cell Vaccination and Suppression of Autoimmune Disease. *Progress in Immunology* VII , Proceed. 7th Intnl. Congress of Immunology (Eds. F. Melchers & al.), Springer, Berlin, p 867-873.

Cohen I R, Atlan H (1989) Network Regulation of Autoimmunity: An Automaton Model. *J Autoimm* 2: 613-625.

Karin N, Cohen I R (in preparation) Paradoxical T Cell Suppression in the pathogenesis of Adjuvant Arthritis.

Lohse A W, Mor F, Karin N, Cohen I R (1989)Control of Experimental Autoimmune Encephalomyelitis by T Cells responding to Activated T Cells. *Science* 244: 820-822

Van Eden W, Holoshitz J, Nevo Z, Frenkel A, Klajman A,Cohen I R (1985) Arthritis induced by a T lymphocyte clone that responds to Mycobacterium tuberculosis and to cartilage proteoglycans. *Proc Natl Acad Sci USA* 82: 5117-5120

Van Eden W, Thole J E R, Van der Zee R, Noordij A, Embden J D A, Hensen E J, Cohen I R (1988) Cloning of the mycobacterial epitope recognized by T lymphocytes in adjuvant arthritis. *Nature* 331:171-173

Weisbuch G, Atlan H (1988) Control of the Immune Response. *J. Phys. A* 21: L189-192.

IDIOTYPIC REGULATION IN EXPERIMENTAL AUTOIMMUNE THYROIDITIS (EAT)

H. Tang, C. Bédin, B. Texier and J. Charreire
INSERM U.283
Hôpital Cochin
27, rue du Fg St-Jacques
75674 Paris Cedex 14
France

The expression of antibody (Ab) idiotypes (id) by T lymphocytes has been long described in functional assays (Binz et al., 1976; Eichmann et al., 1978; Coutinho & Meo, 1983) and has been taken to suggest VH gene expression in both T and B lymphocytes. When T cell receptor (TCR) proteins (Allison et al., 1982; Meuer et al., 1983) and structural genes (Hedrick et al., 1984; Saito et al., 1984) were identified, the expression of Ab idiotopes by T lymphocytes was reinterpreted as resulting from powerful mechanisms occurring during the somatic selection of TCR on the basis of ids from Abs. Consequently, a functional idiotypic network was supposed to be responsible for this expression (Pereira et al., 1989). Simultaneously, experiments (Ertl et al., 1982; Sikora & Levy, 1980; Sikora et al., 1982; Singhai & Levy, 1987; Martinez-A. et al., 1987a, 1987b) were reported, describing such phenomenon. However, two of the investigators considered that they were aleatory expressions of T cell ids (Martinez-A. et al., 1987a, 1987b).

In a recent work (Roubaty et al., 1990), we demonstrated the induction of protection to experimental autoimmune thyroiditis (EAT) when cloned inactivated cytotoxic T cells (HTC2), specific for a thyroiditogenic epitope from thyroglobulin (Tg) (Remy et al., 1989) were injected into syngeneic recipients three weeks before immunization by Tg. Because on day 0 prior immunization by Tg we detected in sera from HTC2-inoculated animals, anti-Tg and their anti-idiotypic autoantibodies (A-Abs), Ab2-beta, we hypothesized the protection could occur through an idiotypic network (Jerne, 1974; Jerne et al., 1982), where anti-Tg A-Abs would be restricted to the same specificity as the protective T cell clone: the pathogenic epitope of the Tg molecule (Salamero et al., 1987).

The aim of this work was to characterize the *in vivo* A-Ab production after immunization with HTC2 cells and to understand the intermediate step(s) leading to this production.

NATO ASI Series, Vol. H 66
Theoretical and Experimental Insights into Immunology
Edited by A. S. Perelson and G. Weisbuch

MATERIALS AND METHODS.

1. *Animals.* Six to 8 week-old female CBA/J (H-2^k) mice were provided by the CSEAL-CNRS (Orléans-la-Source, France).

2. *Antigens.* Porcine Tg was from Sigma (St-Louis, MO). The purification of < 10 kDa Tg tryptic fragments was performed as follows: 20 mg of Tg were incubated with 1 mg of trypsin-TPCK (Worthington, Freehold, NJ) for 4 h at 37°C in 0.2 M NH4 HCO_3, pH 8.6. Trypsin activity was stopped by the addition of 10 mM (final concentration) of phenylmethylsulfonyl fluoride (Sigma). < 10 kDa Tg tryptic fragments were selected after preparative electrophoresis. For that purpose, trypsinized Tg was applied onto 3 mm slab gel and submitted to SDS-PAGE (0.1% SDS-12% polyacrylamide). Then 1 cm lane of the outer gel was cut and stained with Coomassie Blue to outline the gel region containing < 10 kDa Tg tryptic fragments. Their electroelution was done in 1% acetic acid solution using the Biotrap apparatus (Schleicher & Schuell, Dassel, West Germany).

3. *Preparation of F(ab')2 fragments from monoclonal anti-Tg A-Ab (anti-Tg mA-Ab).*

3.1. *Anti-Tg mA-Ab 3B8G9* (IgG2a, kappa) was produced by fusion of Tg-immunized CBA/J spleen cells with the BALB/c myeloma P3X63 Ag 8.653. Its specificity for < 10 kDa Tg tryptic fragments inducing EAT and its affinity (Ka = 4 nM) for binding to Tg were previously demonstrated (Salamero et al., 1987).

3.2. *F(ab')2 fragments* from 3B8G9 were prepared according to the method described by the enzyme supplier. Briefly, 0.125 ml immobilized pepsin (Pierce Chemical Co., CA, USA) was added to 5 mg 3B8G9 in 1.0 ml of 20 mM sodium acetate buffer (pH 4.5) and incubated for 2 h at 37°C under rocking. Then 3.0 ml of 10 mM Tris-HCl pH 7.5 were added, mixed and centrifuged. Supernatant was passed through protein-A Sepharose (Pharmacia, Uppsala, Sweden) column previously equilibrated with 10 mM Tris-HCl solution pH 7.5. The eluted F(ab')2 fragments were collected and their quality verified in SDS-PAGE using the Phast-System (Pharmacia).

4. *T cell hybridomas.* The cloned Tg-specific cytotoxic T cell hybridoma (HTC2) was previously described (Remy et al., 1989). Briefly it was shown to induce lysis of syngeneic-thyroid epithelial cells or of macrophages pulsed with the EAT inducer < 10 kDa Tg tryptic fragments in the context of class I antigens of major histocompatibility complex. HTC2 cells were maintained in RPMI 1640 (Gibco-BRL, Cergy-Pontoise, France) supplemented with 5% heat-inactivated fetal calf serum, 100 U/ml penicillin and 100 µg/ml streptomycin (complete medium) in the absence of antigen. As control, the fusion partner BW 5147 ($H\text{-}2^k$) was used and the clone 148, a non cytotoxic hybridoma specific for an undefined Tg epitope, absent from the < 10 kDa Tg tryptic fragments.

5. *Immunization of animals with T cell hybridomas.* CBA/J mice were i.p. inoculated with irradiated (2500 Rads) T cell hybridomas. Before irradiation, the ability of HTC2 cells to induce specific lysis of Tg or < 10 kDa Tg tryptic fragments pulsed syngenic macrophages was verified. Each experimental group included at least three mice. On days specified in the result section, animals were bled, sacrified and their thyroids collected for histopathological studies. As control, six naive CBA/J mice were bled and killed on day 0 prior to T cell inoculation; their individual sera gave the control o.d. substracted from experimental o.d.. In some experiments, HTC2 cells were fixed with glutaraldehyde (Sigma, 200 µl for 50×10^6 cells, 30 min, 4°C) and washed extensively in phosphate-buffered saline (PBS) before inoculation into mice.

6. *Detection of A-Abs by enzyme-linked immunosorbent assay (ELISA).*

6.1. *Autoantibodies to Tg.* Briefly, 96-well polyvinyl microtiter plates (Costar No. 3590, Cambridge, MA) were coated overnight at 4°C with 100 µl of 100 µg/ml porcine Tg in carbonate buffer, pH = 9.6. After two washes in PBS-0.05% Tween-20 (T-20), the free protein binding sites were blocked by the addition of 1.5% BSA (2 h, 37°C). Then 100 µl of 1:100 diluted sera from individual mice were incubated (2 h, 37°C) before wells were extensively washed with PBS-T20. Alkaline phosphatase-conjugated goat anti-mouse immunoglobulin G (GAM-IgG) (Sigma) 1:250 diluted was added as second antibody before the substrate (p-nitrophenylphosphate disodium) was deposited. Measurement with a Titertek multiskan spectrophotometer was performed 30 min. later, at 405 nm and results were expressed as Δ absorbance (experimental absorbance - control absorbance).

6.2. *A-Abs to < 10 kDa Tg tryptic fragments* were similarly determined on plates coated with 100 µl of < 10 kDa Tg tryptic fragments at the concentration of 10 µg/ml.

6.3. *Anti-id A-Abs* were determined on plates coated with 100 µl of 2.5 µg/ml F(ab')2 fragments from 3B8G9 anti-Tg mA-Ab, and murine sera were deposited (100 µl, diluted 1:100, 2 h, 37°C). Then < 10 kDa Tg tryptic fragments were added (100 µl, 5 µg/ml, 2 h, 37°C) for displacement of anti-Tg A-Abs bound inside the antigen binding sites. Lastly biotinylated anti-mouse IgG (Fc) (Sigma, 50 µl, 1:5000), streptavidine-coupled to alkaline phosphatase (Sigma, 50 µl, 1:2000) and enzyme substrate were sequentially added. After each reagent, extensive washes with PBS-T20 were done. In the same experiment, incubation with the < 10 kDa Tg tryptic fragments was omitted to allow the detection of total anti-id A-Abs. Ab2-beta were considered as the difference between absorbances without and with < 10 kDa Tg tryptic fragments.

7. *Histopathological studies of cryo-preserved sections.* Thyroid glands were collected in liquid nitrogen and 5 µm sections made under cryostat (ref. 2800 frigocut, Reichert-Jung, Nussloch, W. Germany). Fixation (10 min) in acetone and three washes in PBS were done before addition (20 µl, 30 min at room temperature) of specific reagents. They were monoclonal anti-thy 1-1 Ab (clone 19XE5, Becton-Dickinson, Mountain View, CA, USA) or monoclonal anti-thy 1-2 Ab (clone 30H12, Becton-Dickinson). Then staining was performed with fluorescein-conjugated GAM Ig (Sigma, 1:500, 30 min). Tissues were washed three times in PBS before ten sections per reagent and per thyroid gland were blindly observed under fluorescent microscope (Reichert-Jung).

RESULTS.

1. *Production of A-Abs after immunization of mice with HTC2 cells.* The production and the kinetics of HTC2-induced A-Ab production were studied. For that purpose, CBA/J mice were injected i.p. with 5 x 10^6 mitomycin-inactivated HTC2 cells, bled and killed every five days, until day 30 post-immunization. As shown on *Table 1*, anti-Tg and anti-idiotypic A-Abs to F(ab')2 fragments of 3B8G9, one anti-Tg mA-Ab specific for one EAT inducer epitope were evaluated. These A-Abs were detectable as early as day 5 in murine sera; then their levels regularly increased to reach maximum values on day 20 post-inoculation and decreased.

Surprisingly, augmentation in the numbers of immunizing T cells was not followed by similar increases in the levels of A-Abs (data not shown).

Table 1: Kinetic study of anti-Tg and anti-idiotypic A-Ab to F(ab')2 fragments of 3B8G9 in sera from CBA/J mice injected with 5 x 10^6 inactivated Tg-specific T cell hybridoma (3 mice per group, mean ± SEM).

Days post-HTC2 inoculation	Autoantibodies to Tg	F(ab')2 from 3B8G9
5	252 ± 30*	93 ± 12
10	373 ± 43	142 ± 49
15	387 ± 45	143 ± 31
20	614 ± 76	168 ± 36
25	379 ± 52	132 ± 56
30	334 ± 71	136 ± 14

* Δ absorbance at 405 nm resulting for each serum of mean absorbance minus mean day 0 absorbance (115 for anti-Tg and 128 for anti-id).

2. *HTC2-induced anti-Tg A-Abs are specific for one Tg epitope located in the < 10 kDa Tg tryptic fragment as the epitope recognized by HTC2*. The large size (660 kDa Mr) of the Tg molecule led to consider this molecule as a vast continuum of antigenic determinants. Therefore, the reactivity towards Tg detected after immunization of CBA/J mice with HTC2 cells could be aleatory and directed to epitope(s) different from those exhibited by the immunizing T cells. Because we demonstrated that the Tg epitope recognized by HTC2 cells is included in < 10 kDa Tg tryptic fragments inducing EAT, we investigated if HTC2-induced anti-Tg A-Abs also share this specificity. As shown on *Table 2*, when sera from animals immunized with 1 x 10^6 inactivated HTC2 cells were tested in ELISA for their binding to < 10 kDa Tg tryptic fragments, only sera from HTC2 immunized mice were positive. More precisely, A-Abs to < 10 kDa Tg tryptic fragments were never detectable in sera from CBA/J mice immunized with another Tg specific T cell hybridoma, 148, even when murine sera were tested at a dilution of 1:10 (data not shown).

Table 2: HTC2-induced anti-Tg A-Abs share the same specificity as HTC2 for the < 10 kDa Tg tryptic fragments (3 mice per group, mean ± SEM).

T cells inoculated	A-Abs to	Days post-T cell inoculation (1×10^6) 5	10	15	20	25	30
HTC2	Tg	249±66*	407±48	410±69	511±34	404±10	320±60
	<10 kDa	266±24	249±16	318±35	320±64	236±14	226±46
148	Tg	ND	ND	102±12	123±54	83±7	ND
	<10 kDa	ND	ND	17±2	21±2	19±2	ND

* Δ absorbance at 405 nm.

3. *Mechanisms of anti-Tg A-Ab production after immunization with HTC2 cells.* The detection of anti-Tg A-Abs in murine sera from animals injected three weeks earlier with HTC2 cells led us to hypothesize two mechanisms to explain their production. First, we supposed that cytotoxic T cells could migrate to the recipient's thyroid gland, where they could lyse the thyroid epithelial cells. During this procedure, fragments from murine Tg under synthesis would be released and the production of anti-Tg A-Abs induced. To answer these questions, histopathological studies of thyroid glands and evaluation of A-Abs to Tg or to < 10 kDa Tg tryptic fragments after inoculation of fixed HTC2 were done.

3a. *Histopathological studies of thyroid glands from T cell-inoculated mice.* Ten serial cryo-preserved sections per thyroid gland were blindly examined on days 5, 10, 15 and 20 post-inoculation of 3×10^6 HTC2 or BW 5147 T cells. Then thyroid gland sections were incubated with either anti-thy 1-1 or anti-thy 1-2 mAb before specific staining. In this study, we took advantage of the expression of thy 1-1 Ag by HTC2 and BW 5147 T cells to discriminate injected and recipient's T cells infiltrating the thyroid glands since recipient's T cells would bear thy 1-2 antigen. As shown on *Table 3*, whereas T lymphocytes were never found in the thyroid glands from animals injected with 3×10^6 BW 5147 T cells, thy 1-1 or thy 1-2 lymphocytes were detected in the thyroid glands from only HTC2-inoculated mice. However, the number of T cells infiltrating the thyroid glands and the origin of thêta Ag they expressed varied with the time after HTC2 inoculation. More precisely, thy 1-1 cells were regularly detected on day 5 post-HTC2 inoculation, while they were not on day 20 (13 versus 2 on days 5 and 20 respectively). In turn, very few thy 1-2 cells were detected on day 5 post-HTC2 inoculation while they were always on day 20 (1 versus 10 on days 5 and 20 respectively). It must be noted that the increase in the amount of inoculated HTC2 cells was not followed by a similar increase of infiltrating T cells in the recipient's thyroid glands.

Table 3: Analysis of T cells infiltrating the thyroid glands from T cell inoculated CBA/J mice (3 mice per group).

Thy-1 antigen detected	T cell inoculated (3×10^6)	Days post-T cell injection + 5	+ 10	+ 15	+ 20
thy 1-1	HTC2	13*	1	3	3
	BW 5147	0	1	1	1
thy 1-2	HTC2	1	4	4	11
	BW 5147	0	0	1	0

* Total number of T lymphocytes counted in the 3 thyroid glands.

3b. *Levels of A-Abs in mice inoculated with glutaraldehyde fixed HTC2 cells.* CBA/J mice were injected with 5×10^6 glutaraldehyde-fixed HTC2 cells. Then levels of A-Abs to Tg, to < 10 kDa Tg tryptic fragments and to F(ab')2 fragments from 3B8G9 were individually evaluated and compared to those of control animals which received unfixed HTC2-inactivated cells. This evaluation was made every five days until day 20 post-injection. As shown on *Table 4*, A-Abs were always higher in sera provided by mice which received unfixed HTC2 cells than in those provided by mice inoculated with the glutaraldehyde-fixed HTC2 cells. For example, the percentages of responses obtained on day 20 with fixed HTC2 cells represented 31, 39 and 33% of the counterpart responses obtained with unfixed HTC2 cells. Because we verified that fixed HTC2 cells were unable to induce lysis of syngeneic macrophages pulsed with Tg, we concluded that anti-Tg A-Abs generated after inoculation of fixed HTC2 cells result from the stimulation of the idiotypic network (Roubaty et al., 1990) whereas those generated after inoculation of unfixed HTC2 cells result from both the release of syngeneic Tg after lysis of syngeneic cells by HTC2 cells, and stimulation of the idiotypic network.

Table 4: Kinetics in ELISA of A-Ab productions to Tg, to < 10 kDa Tg tryptic fragments and to F(ab')2 fragments from 3B8G9 in sera from mice inoculated with 5 x 10^6 inactivated HTC2 cells - glutaraldehyde-fixed or not (mean ± SEM of 4-5 mice).

A-Abs to	Fixation of HTC2 cells	Days post-HTC2 cell inoculation 5	10	15	20
Tg	0	208±12*	321±31	424±32	503±43
	+	101±8	116±20	154±20	156±21
		(49)**	(36)	(36)	(31)
< 10 kDa Tg tryptic fragments	0	76±10	150±15	162±34	214±21
	+	29±2	42±5	63±11	83±18
		(38)	(28)	(39)	(39)
F(ab')2 fragments from 3B8G9	0	162±15	345±39	365±21	302±27
	+	16±7	80±10	143±26	101±15
		(10)	(23)	(39)	(33)

* Δ absorbance at 405 nm evaluated in individual mouse sera resulting of mean absorbance on experimental day minus mean absorbance on day 0 (182 for Tg, 70 for < 10 kDa Tg tryptic fragments and 206 for F(ab')2 fragments).
( )** Percentage of counterpart value obtained with unfixed HTC2 cells.

DISCUSSION.

In the present work, we studied the *in vivo* production of anti-Tg A-Abs after immunization of CBA/J mice with a fully syngeneic cloned cytotoxic T cell hybridoma specific for Tg and more precisely for < 10 kDa Tg tryptic fragments inducing EAT.

The rationale of this work originated from our previous experiments (Roubaty et al., 1990) demonstrating that in the absence of immunization by Tg, injection of CBA/J mice with inactivated HTC2 cells is followed, 20 days later, by anti-Tg A-Ab production. Furthermore, we observed that the HTC2 injected animals were protected from contracting EAT after immunization by Tg. The characteristics of anti-Tg A-Ab production were determined in terms of kinetics, dose of inactivated T cells used for injection and existence or not of a relationship between the fine specificity of the immunizing T cells and those of the A-Abs produced.

Kinetic studies demonstrate that T cell-induced anti-Tg A-Ab production resulted from an active immunization procedure. One surprising observation was the lack of dose-effect response between numbers of immunizing HTC2 cells and A-Ab responses. No clear explanation can be proposed except that a very small subset of lymphoid cells is concerned, which is totally stimulated by 1×10^6 inactivated cells. In that respect, it must be noted that in experimental autoimmune encephalomyelitis (EAE), protection was observed when only 10^2 irradiated antigen-specific T cells were inoculated into recipients (Cohen, 1986).

In addition to the unexpected anti-Tg A-Ab production consecutive to inoculation of T cells, the fact that among the two Tg-specific T cell hybridomas, the production of anti-Tg A-Abs specific for a determinant located in the < 10 kDa Tg tryptic fragments occurred only in animals which received HTC2 cells and that immunizing HTC2 cells also share this specificity is surprising. Before we conclude that when a given T cell is injected into animals, it induced A-Abs expressing the same specificity, even if such a coexistence has already been reported for B and T cell responses towards a peptide from pigeon cytochrome C (Hannum et al., 1985), we have to take into consideration that < 10 kDa is large enough to contain several epitopes, at least one recognized by HTC2 cell and another one by A-Ab to Tg. This question needs to be further investigated by determination of amino-acid sequences forming epitopes recognized by A-Ab to Tg and/or by HTC2 cells.

In a last series of experiments, we investigated the mechanisms responsible for T cell-induced anti-Tg A-Ab production. In our initial work describing anti-Tg A-Ab production consecutive to HTC2 cell inoculation into CBA/J mice, two mechanisms were envisioned. Either cytotoxic T cells were migrating to the recipient's thyroid gland where they lyse thyroid cells and therefore allow release of native or fragmented murine Tg; or the Tg complementary structure of the TCR borne by HTC2 cells generates A-Abs bearing the internal image of the nominal Ag which would stimulate anti-Tg A-Ab production.

In favor of the first hypothesis are the early detection of the thy 1-1 positive HTC2 cells in the thyroid glands from mice injected with HTC2 cells and the decreased levels of A-Abs to Tg, to < 10 kDa Tg and to F(ab')2 fragments from 3B8G9 when mice received glutaraldehyde-fixed HTC2 cells. If one considers that after HTC2 T cell inactivation, approximately 35, 20, 5 and 0% cells were still alive on days 1, 3, 5 and 7 post-irradiation respectively, and that their capacity of syngeneic cytotoxicity towards Tg-pulsed macrophages is sub-normal on day 1, while it is abolished by 2/3 on day 3 post-irradiation (data not shown), therefore, this hypothesis is highly probable. In this scheme, Tg peptidic fragments, under the process of Tg synthesis in the thyroid epithelial cells, would be released, including one immunodominant < 10 kDa Tg tryptic fragments usually never seen by lymphoid cells. Nevertheless, the apparent restricted specificity of HTC2-induced anti-Tg A-Abs is surprising, taking into account the numerous potential epitopes borne by the Tg molecule. In fact, anti-Tg A-Abs specific for multiple epitopes should have been detected, as it is observed after the immunization with Tg. However, it cannot be excluded that A-Abs to other Tg epitopes are generated after HTC2 T cell inoculation. They could be directed towards conformational structures of the Tg molecule and therefore, undetectable in our tests.

In an other hand, the second hypothesis cannot be ruled out. Therefore, to determine its relevance, kinetic studies of anti-id A-Ab production in sera from mice inoculated with fixed HTC2 cells were conducted. The detection of anti-id A-Abs, Ab2-beta representing the internal image of nominal antigen, as soon as day 5 post-inoculation of glutaraldehyde fixed HTC2 cells, indicated that this mechanism also interfere in the production of anti-Tg A-Abs.

Altogether, our experiments confirm the production of anti-Tg A-Abs after inoculation of inactivated HTC2 cells. They also demonstrate the restricted specificity of HTC2-induced anti-Tg A-Abs for < 10 kDa Tg tryptic fragments inducing EAT, a specificity also shared by the immunizing HTC2 cells. Concerning the mechanisms responsible for their production, two pathways are hypothesized and shown on *Figure 1*: first, early *in situ* destruction by inactivated cytotoxic T cells of recipient's thyroid cells resulting in the release of one immunodominant Tg fragment and A-Ab synthesis; second, production of A-Abs to HTC2-TCR, structurally similar to anti-idiotypic A-Abs to Tg, further inducing *in vivo* anti-Tg A-Abs. The respective part of these simultaneous mechanisms is under investigation in our laboratory.

ACKNOWLEDGMENTS.

We thank Immunotech (Marseille, France) for Dr H. Tang support, Mr. J.C. Mussard for histopathological studies and J. Decaix for her excellent secretarial work.

REFERENCES.

Allison JP, McIntyre BW, Bloch D (1982) Tumor-specific antigen of murine T-lymphoma defined with monoclonal antibody. J Immunol 129: 2293-2300.

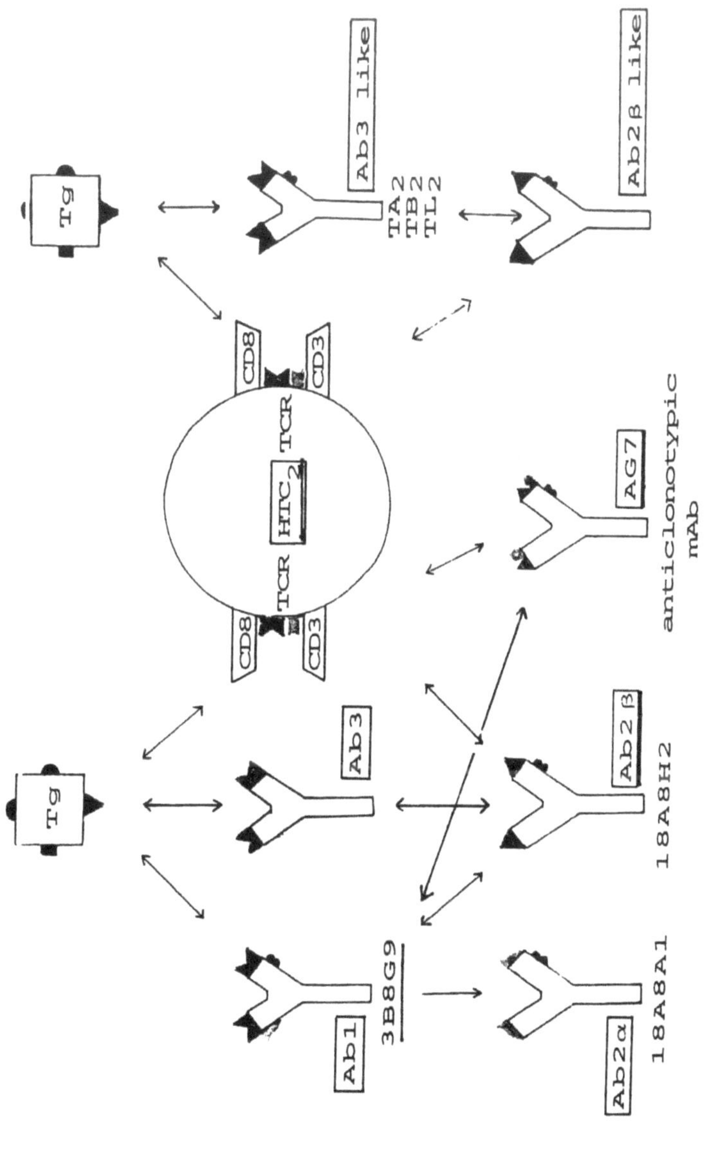

Figure 1: Idiotypic regulation in EAT.

Binz H, Wigzell H, Bazin H (1976) T-cell idiotypes are linked to immunoglobulin heavy chain genes. Nature 264: 639-642.
Charreire J (1989) Immune mechanisms in autoimmune thyroiditis. Adv Immunol 46: 263-334.
Cohen IR (1986) Regulation of autoimmune disease physiological and therapeutic. Immunol Rev 94: 5-21.

Coutinho A, Meo T (1983) Immunoglobulin gene expression by T lymphocytes. Scand J Immunol 18: 79-100.
Dighiero G, Lymberi P, Mazié JC, Rouyre S, Butler-Browne GS, Whalen RG, Avrameas S (1983) Murine hybridomas secreting natural monoclonal antibodies reacting with self antigens. J Immunol 131: 2267-2272.
Eichmann K, Falk I, Rajewsky K (1978) Recognition of idiotypes in lymphocyte interaction. II. Antigen-independent cooperation between T and B lymphocytes that possess similar and complementary idiotypes. Eur J Immunol 8: 853-857.
Ertl HCJ, Greene MI, Noseworthy JH, Fields BN, Nepom JT, Spriggs DR, Finberg RW (1982) Identification of idiotypic receptors on reovirus-specific cytolytic T cells. Proc Natl Acad Sci USA 79: 7479-7483.
Hannum CH, Matis LA, Schwartz RH, Margoliash E (1985) The B10.A mouse B cell response to pigeon cytochrome c is directed against the same area of the protein that is recognized by B10.A T cells in association with the E^kbeta:E^kalpha Ia molecule. J Immunol 135: 3314-3322.
Hedrick SM, Cohen DI, Nielsen EA, Davis MM (1984) Isolation of cDNA clones encoding T cell-specific membrane-associated proteins. Nature 308: 149-153.
Jerne NK (1974) Towards a network theory of the immune system. Ann Immunol (Paris) (Inst Pasteur) 125C: 373-389.
Jerne NK, Roland J, Cazenave PA (1982) Recurrent idiotopes and internal images. EMBO J 1: 243-247.
Martinez-A C, Coutinho A, Bandeira A, de la Hera A, Toribio ML, Marcos MAR, Pereira P (1987a) Sporadic idiotypic cross-reactivities between antibodies and T helper cells: one example of aleatory expression of T cell idiotypes. J Mol Cell Immunol 3: 21-28.
Martinez-A C, Pereira P, de la Hera A, Toribio M, Marcos MAR, Marquez C, Coutinho A (1987b) A common idiotope on T cell receptors and antibodies expressed in the absence of network selection. Eur J Immunol 17: 1391-1394.
Meuer SC, Fitzgerald KA, Hussey RE, Hodgdon JC, Schlossman SF, Reinherz EL (1983) Clonotypic structures involved in antigen-specific human T cell function. Relationship to the T3 molecular complex. J Exp Med 157: 705-719.
Pereira P, Bandeira A, Coutinho A, Marcos MA, Toribio M, Martinez-A C (1989) V-region connectivity T cell repertoires. Ann Rev Immunol 7: 209-249.
Remy JJ, Texier B, Chiocchia G, Charreire J (1989) Characteristics of cytotoxic thyroglobulin-specific T cell hybridomas. J Immunol 142: 1129-1133.
Roubaty C, Bédin C, Charreire J (1990) Prevention of experimental autoimmune thyroiditis through the anti-idiotypic network. J Immunol 144: 2167-2172.
Saitò H, Kranz DM, Takagaki Y, Hayday AC, Eisen HN, Tonegawa S (1984) A third rearranged and expressed gene in a clone of cytotoxic T lymphocytes. Nature 312: 36-40.
Salamero J, Remy JJ, Michel-Béchet M, Charreire J (1987) Experimental autoimmune thyroiditis induced by a 5-10 kDa tryptic fragment from porcine thyroglobulin. Eur J Immunol 17: 843-848.
Sikora LKJ, Levy JG (1980) Genetic control of the immune response to ferredoxin: linkage and mapping of T cell proliferation and antibody production genes to the MHC of mice. J Immunol 124: 2615-2619.
Sikora L., Weaver M, Levy JG (1982) The use of unideterminant fragments of ferredoxin in the genetic mapping of determinant specificity of the immune response. Mol Immunol 19: 693-703.
Singhai R, Levy JG (1987) Isolation of a T-cell clone that reacts with both antigen and anti-idiotype: Evidence for anti-idiotype as internal image for antigen at the T-cell level. Proc Natl Acad Sci USA 84: 3836-3840.

VI. Repertoire Selection

The Secrets of Crypticity and its Relationship to Medical Problems: Autoimmune Disease, Transplantation and Cancer

Gilles Benichou
Vatsala Bhardwaj
Stephen P. Schoenberger
Eli E. Sercarz

Department of Microbiology and Molecular Genetics
5304 Life Sciences Bldg.
University of California, Los Angeles,
405 Hilgard Ave., California 90024-1489.

Abstract

The cryptic self represents those poorly presented antigenic determinants that exist within the organism, against which the T cell repertoire has not been rendered tolerant. Occasional availability of certain cryptic determinants may have a detrimental consequence and cause autoimmunity. On the other hand, cryptic determinants may be able to be harnessed to perform worthwhile tasks such as raising anti-tumor responsiveness.

Definitions and Concepts

It is abundantly clear that autoimmune diseases are only the most obvious signs of the autoaggressive potential of the immune system. Autoreactive clones can regularly be isolated from normal individuals, supporting the view that a broad regulatory control must exist to defend homeostasis. In order to gain the knowledge necessary to conquer such diseases as arthritis, diabetes and multiple sclerosis, it will be important to define those determinants that are the targets of autoimmune activity.

The questions that can be asked are whether these determinants are a particular subset of those on self-proteins? What proportion are they of the whole? What is the nature of the process that regulates responsiveness to these determinants? Each of these questions is still under investigation, but we would like to attempt a somewhat new perspective on these issues, based upon the difference between dominant and "recessive" (subdominant and cryptic) determinants on self-antigens.

NATO ASI Series, Vol. H 66
Theoretical and Experimental Insights into Immunology
Edited by A. S. Perelson and G. Weisbuch

The differentiation between D (dominant), S (subdominant) and C (cryptic) determinants can be defined either after injection of the native protein or of each of the three categories of peptide. (See Table 1)

When a protein antigen is used to prime an animal, and a suspension of lymphocytes from the draining lymph nodes are removed 10 days later for culture in the presence of peptides derived from the antigen, the levels of resulting proliferative responses can be represented as in Figure 1-left. Dominant determinants are defined as those that give a strong response, subdominant determinants a weak response, and cryptic determinants, no response.

Cryptic determinants can exhibit the same range of binding affinities for the MHC as do dominant determinants and also can elicit equally strong T cell responses (see Figure 1-(central)). Therefore, the reason for crypticity doesn't lie in the lack of binding affinity for the MHC or a "hole-in-the-T-cell-repertoire".

What then are some explanations for crypticity? These can be grouped as in Table 2 into three general categories. Each of these possibilities has been implicated for one reason or another in accounting for the failure of a particular determinant, when part of the native, folded antigen, to induce a T cell proliferative response. In this article, we will focus on the competitive aspects of dominance and crypticity.

TABLE 1

DEFINITIONS OF DOMINANCE, SUBDOMINANCE, AND CRYPTICITY

IN VIVO IMMUNIZATION WITH:				
NATIVE PROTEIN		PEPTIDES		
Type of priming:	Response in vitro to peptide Ags:	Type of peptide priming:	Response in vitro to protein Ag:	Response in vitro to peptide Ags:
DSC (Native protein)	D - Strong S - Weak C - None	D S C	Strong Weak None	D Strong S Strong C Strong

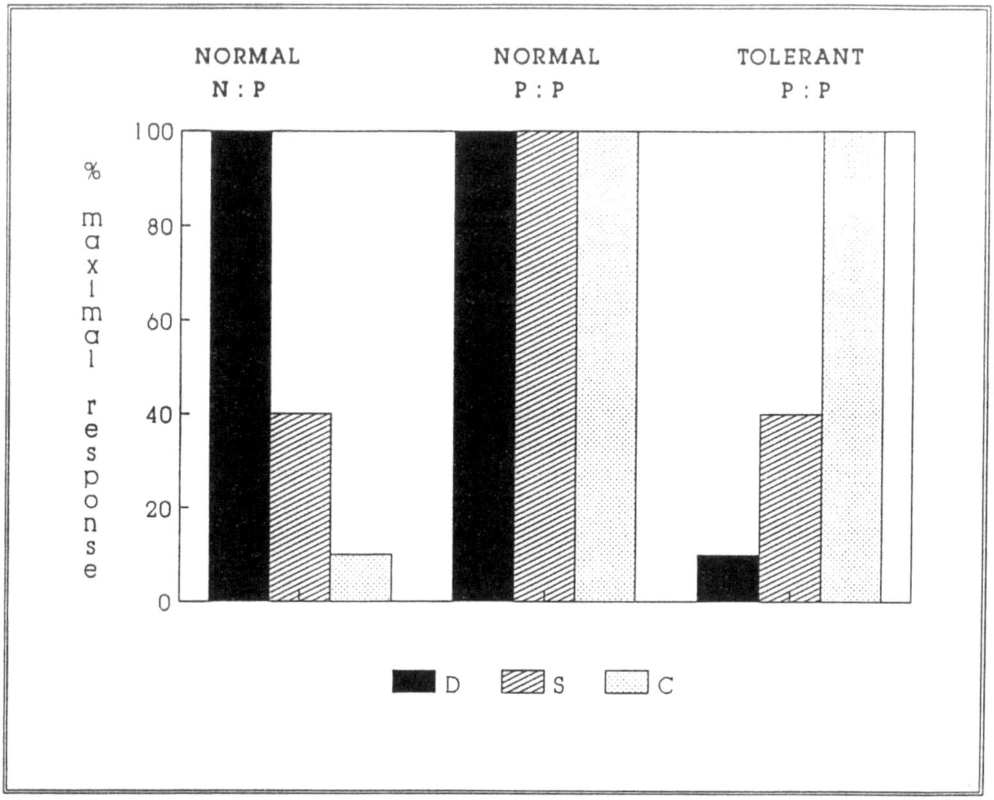

Figure 1A-(left panel)--The in vitro T cell proliferative response in lymph node cells, primed by the injection of native protein (N) and challenged in vitro with dominant (D), subdominant (S) or cryptic (C) peptides (P).

Figure 1B-(central panel)--The mice were primed with peptides, and challenged in vitro with peptides.

Figure 1C-(right panel)--Mice made tolerant in birth or adulthood, are primed with peptide and challenged in vitro with peptide. All in vivo immunizations were in complete Freund's adjuvant.

Differential priming and tolerization by dominant and cryptic determinants

It is now widely accepted that self-proteins are continuously processed and presented as peptides in association with self-MHC molecules for interaction with T cells. Recognition of self-proteins has profound effects, both positive and negative , on the development of the T cell repertoire. However, tolerance must be considered at the level of particular determinants, as animals are tolerant to dominant determinants on a self-protein but not to its cryptic determinants. Experimental support for the hypothesis that T cells that recognize poorly presented or cryptic self-determinants can escape tolerance has been obtained in an experimental model of tolerance (Gammon & Sercarz, 1989). Adult mice which are tolerized to hen eggwhite lysozyme (HEL) by intravenous injection of the antigen in saline do not respond to further challenge with the whole protein or peptides containing dominant determinant regions. However, these HEL-tolerized mice are capable of mounting a potent lymph node T cell proliferative response upon immunization with peptides containing cryptic determinants. The resulting pattern of response in tolerant mice as shown in Fig. 1-(right) is precisely the opposite of that in the unprimed mice in Fig 1-(left).

TABLE 2

REASONS FOR CRYPTICITY

<u>Processing</u>:

1. ***Indolent processing*** - The native protein has certain regions that are very slowly unfolded and processed to disclose the determinant, or regions that fail to be transported.
2. ***Excessive processing*** - determinant is destroyed by proteolytic enzymes.

<u>Competition</u>:

3. ***Agretope competition*** - Other agretopes, with higher affinity for the same MHC molecule, either on the same antigen or from other molecules, compete favorably.
4. ***Determinant capture*** - A downstream determinant on the <u>same</u> molecule binds to an MHC molecule, removing the rest of the attached antigen from the possibility of binding to other MHC molecules.

<u>Hindrance at the binding site</u>:

5. ***Ag--MHC*** - Hindrance at the binding site from association between determinant and MHC molecules.
6. ***TcR--Ag/MHC*** - Hindrance of the TcR from association with MHC-Ag complex.

It was therefore of interest to convert HEL into a self antigen to study the hierarchy of presentation of "self-determinants" during tolerance induction in the thymus. Experiments in HEL transgenic mice were therefore performed to examine this point (Cibotti *et al.*,1992). In fact, H-2^d mice transgenic for HEL were non-responsive to the dominant determinant for this haplotype, p106-116/E^d. Moreover, subdominant and cryptic determinants could induce a response, indicating that the responsive T cells had not been rendered tolerant (Cibotti *et al.*,1992). These results were an indication that the hierarchy of self determinants in the developing thymus was very similar to that in the periphery of normal non-transgenic animals, to the same assemblage of determinants.

To study a bona-fide self antigen, it was then decided to examine class I and II MHC molecules themselves, since they clearly are present in the developing thymus. We found $CD4^+$ T cell proliferative responses could also be raised towards certain polymorphic regions of self-MHC class I and class II molecules. Two self-class I peptides, L^d 61-85 and D^d 61-85 (representing, in large part, their $\alpha 1$ helices) were shown to bind with high affinity to self-class II restriction elements. Interestingly, only the peptide Ld 61-85 was capable of inducing in vivo lymph node T cell proliferation upon immunization of syngeneic B10.A mice. D^d 61-85 failed to stimulate a T cell response, presumably because tolerance had been induced during thymic selection, while L^d 61-85, in spite of its high affinity for self Ia molecules, was not presented efficiently enough to ensure T cell tolerance, owing to the low level of expression of L^d molecules as compared to D^d (Benichou *et al.*, 1990).

In support of this view, some preliminary results indicate that upregulation of the L^d molecule by γ-interferon can actually induce the presentation of the cryptic self-determinant and trigger an autoreactive T cell line directed against L^d 61-80 in the absence of peptides (unpublished data). This indicates that any combination of circumstances which affects the stability of the hierarchy of expression of self-determinants may render visible the cryptic self-determinants, as well as trigger the autoreactive T cells which have not been eliminated during development and initiate an autoimmune process.

Dominant self and cryptic self

All of these results indicate that tolerance must be considered at the level of particular determinants: *the dominant self-determinants* which are efficiently processed and presented to induce tolerance, comprise the "dominant self", while the *cryptic self-determinants,* which fail to make an impact on the T cell repertoire owing to incomplete processing and presentation in the thymus during development (Gammon *et al*, 1991), comprise the "cryptic self". We can estimate from our results with lysozyme determinants in three MHC haplotypes that cryptic-subdominant self is roughly 4 times as large as the dominant self! (See Table 3) The stark realization that there is an excess of sequestered determinants on each self molecule leads to the view that autoimmune disease is directed against determinants which in the normal state are cryptic. The hierarchy of determinants within a self-protein will directly relate to the maintenance of tolerance and prevention of autoimmune pathogenesis. Intramolecular competition between different determinants in a self-sequence determines their relative dominance. Within the continuum from crypticity to dominance of particular self determinants lies the threshold for in-vivo stimulation of autoreactive T lymphocytes. Modification of this threshold, by changing the efficiency of antigen presentation, or factors related to the competition, or the average avidity of reactive T cells, etc, should affect the maintenance of self-tolerance. We tried to address the issue of hierarchy and intramolecular competition by a series of studies on intramolecular competition, described in the subsequent sections.

TABLE 3

DOMINANT VS CRYPTIC DETERMINANTS IN AUTOIMMUNITY

1. There are "sequestered determinants" on each self-molecule, as there are on each foreign antigen.

2. There are many more cryptic and subdominant determinants than dominant determinants. (Roughly 2C:2S:1D)

3. T cells directed against originally cryptic self-determinants are responsible for autoimmune disease.

4. Some T cell memory may be induced to particular cryptic determinants: these T cells may play a special role in autoimmunity.

Intramolecular competition: even cryptic regions have sub-hierarchies.

Subdominant regions of a molecule that is sequestered may also reveal a submolecular hierarchy of displayed determinants, when it becomes unveiled through processing. For example, it was possible that more than one determinant existed within the region L^d 61-85. Therefore, we tested two other peptides, L^d 61-75 and L^d 73-85 corresponding to its N and C terminal portions. Neither of them was capable of recalling in vitro lymph node (LN) T cell responses from mice immunized with L^d 61-85. However, the peptide L^d 65-80 recalled such T cell responses efficiently indicating that the major determinant was located in the central part of the self-peptide. Upon immunization with either L^d 61-75 or L^d 73-85, LN T cell proliferation could be detected in response to the immunizing peptide but neither to the full length sequence L^d 61-85 nor the truncated L^d 65-80. (This fits the criteria indicated in Table 1 describing cryptic determinants.) Accordingly, the self class I peptide L^d 61-85 contains three distinct and overlapping T cell determinants. The dominant determinant, L^d 65-80, stimulates T cells upon immunization with the whole molecule while two cryptic determinants (L^d 61-75 and Ld 73-85) can be detected only after immunization with the peptides themselves. These data show that a hierarchy of T cell determinants can exist within a circumscribed region of a self-molecule consisting of overlapping T cell determinants, which compete for stimulating in vivo autoreactive T cell responses (Benichou *et al.* , manuscript in preparation).

Simultaneous induction of tolerance/priming by dominant/cryptic self-determinants

What T cells are induced specific for the dominant or cryptic determinants on the same peptide during tolerance induction? Adult B10.A mice were tolerized with the self-peptide L^d 61-80. 300 ug of the peptide L^d 61-80 was injected intravenously in saline and 12 days later, the tolerized animals were immunized in the hind foot pads with L^d 61-80 in the presence of CFA. These mice were completely tolerant to the dominant determinant in the draining popliteal lymph nodes in that they could not mount a T cell response against L^d 61-80 or L^d 65-80 that were shown previously to be very strong immunogens. However, upon immunization of the tolerized mice with the same peptide L^d 61-80, a strong T cell proliferative response was recorded in the spleen, directed towards the cryptic L^d 61-75 (Benichou *et al.*, manuscript in preparation). Therefore, we conclude that while tolerance occurred to the dominant self-determinant, no tolerance was induced to the cryptic self-determinant. In fact, the response to 61-75 was so enhanced that it is reasonable to believe that these T cells were actually primed by the tolerizing injection. Such a cadre of primed autoreactive T cells may present a special problem in autoimmune disease.

Intramolecular competition: Subjugation of previously dominant determinants

In an entirely different system, we have been able to study the effect of placing dominant determinants in apposition with other determinants: surprising and unexpected relationships ensue (Bhardwaj *et al.*, 1992). On immunization of B10.PL and PL/J (H-2^u) mice with mouse myelin basic protein, the proliferative T cell response was shown to be highly restricted to an immunodominant N-terminal peptide, N-acetylated Ac1-9 or Ac1-11 (Zamvil *et al.*, 1986), in association with the I-A^u molecule. In order to investigate the mechanism of immunodominance of this determinant and especially competition with attached downstream determinants, we constructed chimeric peptides (CP) in which Ac1-11, or its substituted analog Ac1-11 (Ala4), was linked covalently to 35-47 of MBP, an I-E^u restricted determinant. The latter peptide was synthesized because previous studies (Wraith *et al.*, 1989) had shown that replacement of lys-4 by ala-4 greatly enhanced the peptide's ability to bind to the I-A^u molecule. It was anticipated that in this variant, binding by the I-A^u component would be favored over the I-E^u component in the competition between Ac1-11 and 35-47.

TABLE 4

SUBJUGATION OF PREVIOUSLY DOMINANT DETERMINANTS

IN-VITRO CHALLENGE	IMMUNIZATION WITH[a] Ac1-11:35-47	IMMUNIZATION WITH[a] Ac1-11(Ala4):35-47
	MEAN OF STIMULATION RATIOS	
Ac1-11:35-47 / Ac1-11	22.7	18.3
Ac1-11:35-47 / 35-47	20.4	10.5
Ac1-11(A4):35-47 / Ac1-11	11.8	17.8
Ac1-11(A4):35-47 / 35-47	14.0	10.0

[a] Eight mice in each group were immunized with the chimeric peptides in complete Freund's adjuvant, and their draining lymph nodes were cultured *in vitro* for 5 days and pulsed on the last day with [^{3}H] thymidine. The values represent the mean of the 8 individual ratios found in each animal.

However, following immunization of B10.PL and (SJLxB10.PL)F1 mice with Ac1-11:35-47, responses to the parental determinants, both Ac1-11 and 35-47, were "subjugated" by the response to neodeterminants within the CP (Table 4). Unique chimeric determinants within Ac1-11:35-47 or Ac1-11(Ala4):35-47 conclusively took over the lead position in the hierarchy of dominance following immunization with the CPs, with preferences of even 20-fold. This result was surprising since previously it had been believed that Ac1-11 was a dominant determinant within most contexts. Furthermore examination of the fine specificity of T cell hybridomas raised by the chimeric molecule showed that none of the 11 hybridomas examined reacted to either Ac1-9 or 35-47: *all* were directed to junctional neodeterminants.

Despite this exclusive pattern of immunogenicity, the full length CP was also immunologically active *in vitro*, in that it could stimulate both Ac1-9 and 35-47 specific T cell clones. These results indicate that the enhanced immunogenicity of certain determinants is probably due to the competition for stimulation of T cells superimposed upon a hierarchy of competitive binding to the MHC among different CP-component determinants. There are probably several ways the full length CP can bind to the MHC, and what is finally displayed for the TcR may be quite different depending on how the binding to MHC initially occurs. These results illustrate that depending on the nearest neighbor flanking residues which frame a particular determinant, the T cells activated by a medium length 25-30-mer peptide are quite unpredictable. The "subjugation" phenomenon suggests that it would be impossible to predict a dominant constituent in a long peptide from its previous history as part of the native molecule: there may be "local rules".

Induction of autoimmune disease via T cell determinant mimicry.

In the monitoring of self-nonself discrimination, the tolerance mechanisms -- clonal deletion, clonal anergy, active suppression, idiotypic control, determinant sequestration -maintain a tolerable level of self reactivity. With a large population of T cells directed against cryptic self determinants, it could be predicted that induction of cross-reactive T cells might occasionally occur upon viral or microbial infection. Dominant determinants on the infecting microbe may initially induce T cells which then cross-react with related self-determinants. The essential point, in fact, is that the cross-reactive determinants must belong to the cryptic self, and thereby would address a T cell repertoire that had not been tolerized. Of course, related determinants on 2 molecules may b3e either dominant or cryptic, and as we have pointed out, this may be largely dependent on flanking sequences.

After excitation of the ambient T cell clones, propagation of disease will have to depend on the continued stimulation of self-reactive cells, and the availability of the cryptic cross-reactive determinant at the lesion site. Although this may not be a frequent occurrence, several circumstances might supervene that could contribute to continued propagation of disease; these are presented in Table 5.

Surprisingly, although mimicry has usually been relegated to those situations where considerable identity exists in the eliciting and the self-antigen, recent work with derivatives of the chimeric molecule mentioned above indicate that mimicry can occur without significant sequence identity (Bhardwaj *et al.*, manuscript in preparation). Thus, the heart of the previously mentioned chimeric peptide, 7-11:35-44 (see Table 4), can substitute for Ac1-9 in inducing almost all T cell clones raised against the latter peptide! These findings indicate that mimicry in the induction of disease, because of its unpredictability, could occur more frequently than has been expected.

TABLE 5

PROPAGATION OF AUTOIMMUNE DISEASE

1. Upregulation of MHC and processing: T cell secretion of IFN-γ, TNF, and/or macrophage activating factor, increases display of MHC molecules and perhaps upregulates processing mechanisms.

2. In memory T cells, if higher affinity for Ag-MHC, also display an increase in the quality and quantity of adhesion molecules; they may also induce complementary adhesion molecules on partner APCs.

3. The inflammatory site: Chemotactic factors and lymphokines secreted by T cells and APC bring more cells to sites of inflammation and enhance their activities.

4. Spreading of specificity: The autoantigen starts to be processed more effectively and previously cryptic determinants can recruit additional T cells specific for this molecule and other autoantigens in the inflamed area.

Role of cryptic determinants in allorecognition.

The introduction into a recipient of cells bearing foreign MHC molecules at their surface results in a potent immunological reaction in which donor cells are rapidly and specifically killed and the graft destroyed (Snell, 1981; Dausset, 1981). The problem of allorecognition, despite the fact that it does not represent a physiological situation (with the exception of the fetal-maternal relationship), is nevertheless of great clinical importance. It also represents an important model for understanding the regulation of the immune response. Although the structure that is being recognized by T cells on a "conventional" antigen has been largely elucidated, the structure which is recognized on the allo-MHC molecule by the TCR is still controversial. Moreover, it is well established that during the inductive phase of the alloresponse, the frequency of precursor T cells capable of recognizing the allo-MHC molecule is extremely high for reasons which are still obscure (Matzinger & Bevan, 1977; Kourilsky & Claverie, 1989).

One of the models which has been proposed to explain the polyclonality of the primary alloresponse hypothesizes that self- (recipient) or allo- (donor) derived peptides can interact with the native allo-MHC molecules to create a series of new determinants recognized by T cells (Pierres *et al.*, 1989). This would seem to contradict the dogma of self-restriction but in fact, it is believed that the association of the allo-MHC with certain self-peptides mimics an association between the self-MHC itself with the same or different self-peptides. The allo-MHC, as a new restriction element, should bind a large variety of self-peptides and such associations of allo-MHC + self-peptides might represent combinations to which the immune system has never been previously exposed, for which tolerance has not been induced. Associations would be expected between available, dominant self-peptides that can combine with the allo-MHC molecule to create a new series of determinants. In some cases, these potential determinants will not generate a response because of mimicry with self-MHC-dominant self-peptides, against which the organism is tolerant. Presumably, the portion of the allogeneic response using allo-MHC as restriction element will fortuitously tie into the cross-reactive repertoire that recognizes self MHC-foreign peptide, and has already experienced negative selection. In other cases, the new allo complexes will address that repertoire which was not negatively selected, specific for self-MHC-cryptic self combinations, a repertoire that should be largely intact. In addition, some of the allo-response will be directed against allo-MHC-determinants bound to self-MHC molecules (Benichou *et al.*, 1992).

Crypticity and cancer: A possible source of new targets for therapeutic intervention.

Although self reactivity can lead to severe pathological consequences for the host, the existence in the adult of potentially autoreactive T lymphocytes also raises the possibility that certain subsets of these could be useful in an antitumor immune response. It is becoming clear from experimental systems that some spontaneous "non-immunogenic" neoplasms (the majority of human cancers) actually do express tumor-specific antigens against which a cytotoxic immune response can be directed (Van Pel *et al.*, 1983; Van Der Bruggen *et al.*, 1991). In murine systems these tumor-specific transplantation antigens (TSTA) appear able to function as targets for an effective antitumor immune response. Some of these antigens have been identified in tumors that failed to grow in syngeneic hosts only after introduction of immunogenic variants of tumor cells made through a variety of treatments such as mutagenesis, infection with virus, or transfection with lymphokine genes (reviewed in Melief & Kast, 1991). The existence in tumor bearing hosts of T cells that are potentially able to recognize these antigens indicates that, although these determinants were cryptic to the unprimed immune system, they became more immunodominant in the immune response that followed vaccination with immunogenic variants of the parent tumor. This shift in response induction could be a result of an increase in the average affinity of available antigen-specific T lymphocyte receptors or the lower target determinant density required for secondary stimulation of primed T lymphocytes. The genes encoding two tumor rejection antigens recognized in-vitro by host cytotoxic lymphocytes (CTL) have been identified and are found to encode proteins whose level of expression in normal cells is low or absent (Van Den Eynde *et al,*, 1991; Van Der Bruggen *et al.*, 1991). This observation suggests that identifying cryptic self determinants whose level of presentation on tumor cells differs sufficiently from that in normal cells may provide potential targets for immunotherapy via expansion of antigen specific T lymphocytes in a specific autoimmune response directed primarily against previously cryptic determinants on tumor cells.

The targets for such a response would best be chosen from those oncoproteins that are inextricably associated with maintaining the neoplastic phenotype, such that antigen loss under the selective pressure of immunotherapy would result in loss of transformation (Hellstrom & Hellstrom, 1989). One class of potential targets include proto-oncoproteins, such as ras and p53, in which point mutations have been shown to be associated with a variety of cancers (Bos, 1988; Hollstein *et al.*, 1991). Although several groups have recently investigated the possibility of using point-mutated ras as a target for immunotherapy, using peptides that contain the altered residues to induce effector T lymphocytes, this approach need not be limited to determinants that contain mutated residues (Jung & Schluesener, 1991; Peace *et al.*, 1991). Point mutations in oncoproteins might possibly result in an alteration in the pathways of processing of the entire protein and thereby increase the efficiency of presentation of previously poorly presented determinants. This could result in new MHC binding peptides that could serve as targets for T lymphocytes directed against the novel combination of MHC and self-peptide found only on the tumor cell.

Another class of potential targets might be found among the oncoproteins, such as members of the myc family, that are found to be overexpressed in a variety of cancers (reviewed in Bishop, 1991). The heightened level of expression of a previously cryptic oncoprotein determinant should influence its presentation efficiency and this can be used to discriminate tumor cells from normal cells in an antitumor response.

The examples mentioned of two tumor-specific rejection antigens that have been identified as normal genes but whose products are only detectable in tumor cells, is functionally equivalent to the situations described above in which the density of the complex of MHC and self peptide recognized by specific T lymphocytes reaches a physiologically relevant threshold for T lymphocyte stimulation only in the target cell. By choosing the appropriate target determinants and by increasing the frequency of available reactive T lymphocytes, through adoptive transfer of in-vitro generated peptide-specific clones or through vaccination with peptide, it should be possible to engineer an anti-tumor immune response that recognizes the pathologic determinant density of self-peptide/MHC found on tumor cells while sparing normal cells in which the target self determinant is inefficiently presented and remains cryptic.

References

Benichou, G., Takizawa, P.A., Ho, P.T., Killion, C.C., Olson, C.A. McMillan, M. & Sercarz, E. (1990) Immunogenicity and tolerogenicity of self-MHC peptides. J. Exp. Med. 172: 1341-1346.

Benichou, G., Takizawa, P.A., Olson, C.A., McMillan, M. & Sercarz, E.E. (1992) Donor MHC peptides are presented by recipient MHC molecules during graft rejection. J. Exp. Med. 175: 305-308.

Benichou, G., Fedoseyeva, E., Olson, C.A., Geysen, M., McMillan, M. & Sercarz, E. Intramolecular competition between self-peptides and regulation of tolerance induction. (Manuscript in preparation)

Bhardwaj, V., Kumar, V., Geysen, H.M. & Sercarz, E.E. (1992) Subjugation of dominant immunogenic determinants within a chimeric peptide. (Submitted)

Bhardwaj, V., Kumar, V. & Sercarz, E. Molecular mimicry in the absence of protein sequence homology. (Manuscript in preparation.)

Bishop, J. M. (1991) Molecular themes in oncogenesis. Cell 64: 235.

Bos, J. L. (1988) The ras gene family and human carcinogegesis. Mutat. Res. 195(3): 255.

Cibotti, R., Kanellopoulos, J., Cabaniois, J-P., Halle-Panenko, O., Kosmatopoulos, K., Sercarz, E. & Kourilsky, P. (1992) Tolerance to a self-protein involves its immunodominant but not subdominant determinants. PNAS (in press).

Dausset, J. (1981) The major histocompatibility complex in man. Science 213: 1469.

Gammon, G. & Sercarz, E. (1989) How some T-cells escape tolerance induction. Nature 342: 183-185.

Gammon, G., Sercarz, E. & Benichou, G. (1991) The specificity of the autoreactive T cell repertoire : The dominant self and the cryptic self. Immunology Today 12: 193-195.

Hellstrom, K. E. & Hellstrom, I. (1989) Oncogene-associated tumor antigens as targets for immunotherapy. FASEB Journal 3(6): 1715-1722.

Hollstein, M., Sidransky, D., Vogelstein, B. & Harris, C.C. (1991) P53 mutations in human cancers. Science 253(5015): 49-53.

Jung, S. & Schleusener, H. J. (1991) Human T lymphocytes recognize a peptide of single point-mutated, oncogenic ras proteins. J. Exp. Med. 173: 273-276.

Kourilsky, P. & Claverie, J.M. (1989) MHC restriction, alloreactivity, and thymic education: a common link? Cell. 56: 327-329.

Matzinger, P. & Bevan, M. (1977) Why do so many lymphocytes respond to major histocompatibility antigens? Cell. Immunol. 29: 1.

Melief, C. J. M. & Kast W.M. (1991) Lessons from T cell responses to virus-induced tumors for cancer eradication in general. Cancer Surveys (in press).

Peace, D. J., Chen, W., Nelson, H. & Cheever, M.A. (1991) T cell recognition of transforming proteins encoded by mutated ras proto-oncogenes. J. Immunology 146(6): 2059-2065.

Pierres, M., Marchetto, S., Naquet, P.M., Landais, D., Peccoud, J., Benoist, C. & Mathis, D. (1989) I-A alpha polymorphic residues that determine alloreactive T cell recognition. J. Exp. Med. 169: 1655.

Snell, G.D. (1981) Studies in histocompatibility. Science 213: 172.

Van Den Eynde, B., Lethe, B., Van Pel, A., De Plaen, E. & Boon, T. (1991) The gene coding for a major tumor rejection antigen of tumor P815 is identical to the normal gene of syngeneic DBA/2 mice. J. Exp. Med. 173: 1373-1384.

Van Der Bruggen, P., Traversari, C., Chomez, P., Lurquin, C., De Plaen, Van Den Eynde, B., Knuth, A. & Boon T. (1991) A gene encoding an antigen recognized by cytoloytic T lymphocytes on a human melanoma. Science 254(5038): 1643- 1647.

Van Pel, A., Vessiere, F. & Boon, T. (1983) Protection against two spontaneous mouse leukemias conferred by immunogenic variants obtained by mutagenesis. J. Exp. Med 157: 992-2001

Wraith, D.C., Smilek, D.E., Mitchell, D.J., Steinman, L. & McDevitt, H.O. (1989) Antigen recognition in autoimmune encephalomyelitis and the potential for peptide mediated immunotheraypy Cell 59: 247-255

Zamvil, S.S., Mitchell, D.J., Lee, N.E., Moore, A.C., Kitamura, K., Steinman, L. & Rothbard, J.B. (1986) T-cell epitope of the autoantigen myelin basic protein that induces encephalomyelitis. Nature 324: 258-260.

Role of Cryptic Self Germline Encoded Ig Epitopes in the Selection of the T cell Repertoire

Jian-Jun Chen, Srinivas-Venkatesh Kaveri
and Heinz Kohler
San Diego Regional Cancer Center
San Diego, CA 92121

KEYWORDS / ABSTRACT : self-Ig / T cell repertoire selection / cryptic epitope

Although complexes of self peptides and MHC are believed to play a role in T cell repertoire selection, it is unknown whether hypervariable Ig peptides are involved. Murine T cell lines and clones were generated against a hypervariable region peptide derived from a self germline encoded S107 V1 antibody. These peptide-specific T cells do not respond to the intact S107 antibody, even when presented by a S107 B cell transfectant. The role of such cryptic epitopes in highly polymorphic self antigens in the T cell repertoire selection is discussed.

It is now widely accepted that the repertoire expressed by peripheral T cells is the result of two selective processes that occur during intrathymic development (Marrack & Kappler 1988). Positive selection expands cells able to recognize foreign peptides associated with MHC molecules, and negative selection eliminates cells reactive to self peptides presented by MHC.

---

* This work was supported by grant AG 04180.

NATO ASI Series, Vol. H 66
Theoretical and Experimental Insights into Immunology
Edited by A. S. Perelson and G. Weisbuch

An interesting question remains, however, of how T cells are positively selected with specificity for non-self antigens in the absence of such antigens in the thymus. Evidence exists supporting the concept that self peptides are involved in positive selection (Nikolic-Zugic & Bevan, 1990; Koga, et al. 1989). For example, autoreactive T cells against self-molecules such as hemoglobin exist in the adult T cell repertoire, suggesting that incomplete tolerance to self-antigens may enrich the T cell repertoire (Schild, et al. 1990). If induction of incomplete tolerance to self peptides is a key mechanism for enlarging the anti-non-self repertoire, a considerably larger diversity of self peptides would be required. Such a high degree of diversity exists in the polymorphism of immunoglobulins, which is accessible on B cells in the fetal thymus (Miyama-Inaba, et al. 1988). To investigate the possible role of immunoglobulin self peptides in the acquisition of the T cell repertoire we studied the T cell response to polymorphic structures of a germline encoded antibody in syngeneic mice.

A 24-residue peptide of the CDR2/FRIII region of the heavy chain of T15, representing the self binding locus (Kang, et al. 1988), was used as the T cell antigen. Because of the evolutionary conserved nature of this polymorphic immunoglobulin structure as a target for natural anti-peptide antibodies (Kaveri, et al. 1990), we investigated whether this V_H region peptide is also recognized by self T cells.

The synthesis and purification of Vh50-73 peptides has been described previously (Kang, et al.1988). Balb/c mice were immunized with 50 ug of T15H(50-73) peptide one time in incomplete Freund's adjavant (IFA) and lymph node cells were harvested seven days later. Immunized mice showed strong proliferative responses to the peptide; however, no response was seen against intact T15 (IgA,k). This was surprising, as even when the concentration of intact T15 was 15.12 uM, which is 4 times higher than the effective dose (3.7 uM) for the peptide, it did not stimulate lymph node cells of peptide primed mice (see Fig. 1).

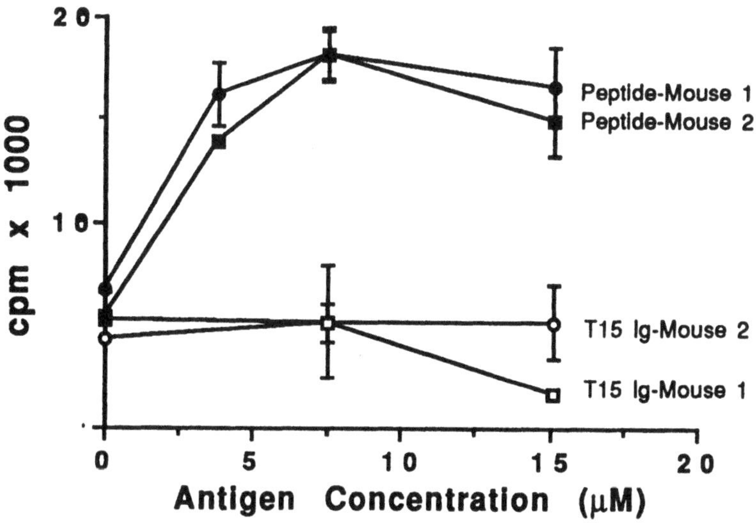

Fig.1 Inability of intact T15 Ig to stimulate T15H(50-73) primed lymphocytes in primary response. Balb/c mice were immunized with 50 ug of T15 H(50-73) in I.F.A. and lymph node cells were collected 7 days later. 5 x 10^5/ well of lymphocytes were stimulated with different doses of T15 H(50-73) peptides or intact T15 Ig. The proliferation assays with [3H] thymidine were carried out as described (Saeki, et al. 1989). After a 48-h period in culture, proliferaton was assessed by 16- to 18- h exposure to 1uCi of [^{3}H]thymidine. Results are expressed as mean cpm+/- SD of duplicate cultures.

To investigate further the failure of the T15 peptide to induce T cells responding to the intact T15 Ig, we developed T cells lines from peptide immunized mice.

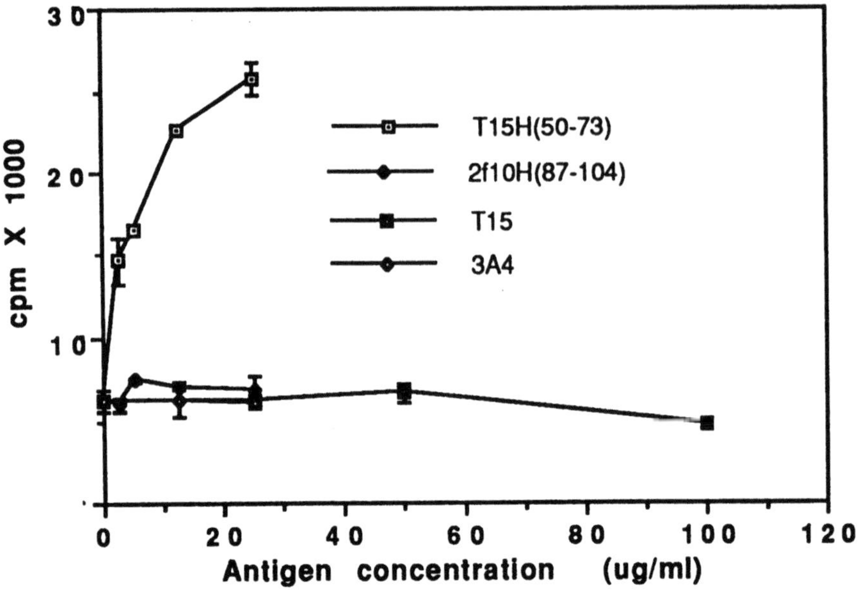

Fig. 2a. Antigen specificity of of T15H(50-73)-specific T cell line. Rested T cells (2 X 10^4 cells/well) were stimulated with T15H 50-73 peptide, 2F10-H3 peptide, T15 Ig and 3A4 Ig in the presence of irradiated syngeneic spleen cells (5X10^5 cells/well). The proliferation assays with [3H] thymidine were carried out as described in Figure 1. The peptide specific T cells were generated by immunizing Balb/c mice with 50 ug of T15H(50-73) peptide in I.F.A. subcutaneously. The draining lymph node cells were harvested 7 days later and stimulated every 10 days with APC and 50 ug/ml of T15 peptide. The peptide T15H 50-73 (ASRNKANDYTTEYSASVKGRFIVS) and 2F10H(87-104) (SAVYFCARSGARATSGDY) were synthesized as described
(Kang, et al.1988).

As seen in Fig. 2a, the line responded only to the T15 peptide and not to a control peptide consisting of a hypervariable region from a unrelated murine monoclonal antibody 2F10 (Raychaudhuri, et al. 1990). Again, no response was seen against the intact T15 Ig and to another unrelated mouse monoclonal antibody, 3A4 (Raychaudhuri, et al. 1986).

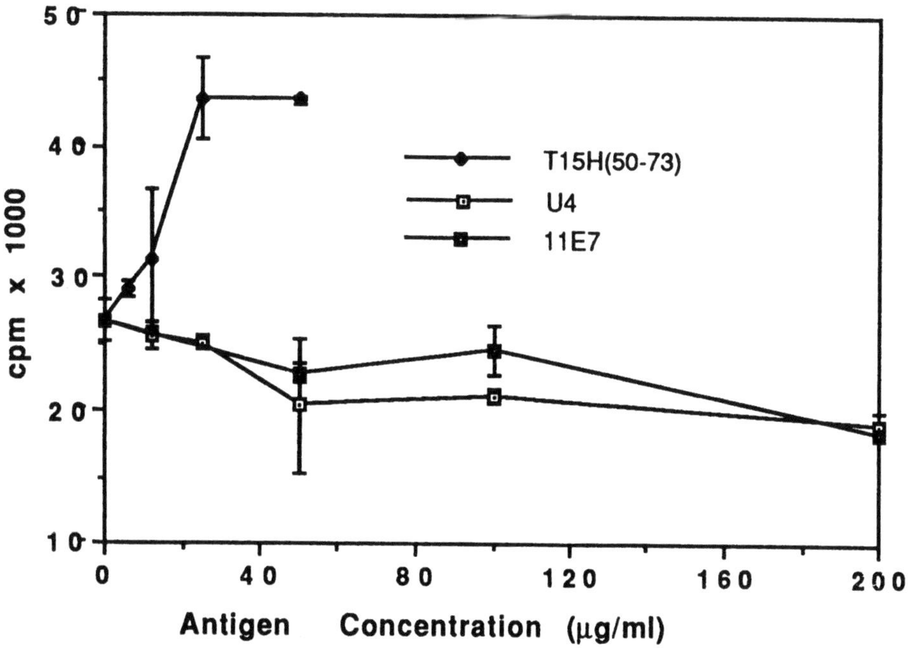

Fig. 2b. Specificity of of T15H(50-73)-specific T cell line. Rested T cells (2 X 10^4 cells/well) were stimulated with T15H 50-73 peptide, 11E7 Ig, and U4 Ig in the presence of irradiated syngeneic spleen cells (5X10^5 cells/well). The proliferation assays with [3H] thymidine were carried out as described in Fig 1.

Considering that IgM but not IgA is the immunoglobulin isotype expressed by B cells in early development, we tested the immune response of peptide specific T cells to 11E7 (IgM,k), which has the T15 germline structure (Kang, et al. 1987). 11E7 was not able to induce the proliferation of peptide specific T cells. To see if the lack of T15 recognition is related to the PC epitope specificity of T15, a non PC binding S107 mutant, U4 (IgA, k; Behar, et al. 1989), was tested in the T cell proliferation assay. Figure 2b shows no response to U4 nor to 11E7.

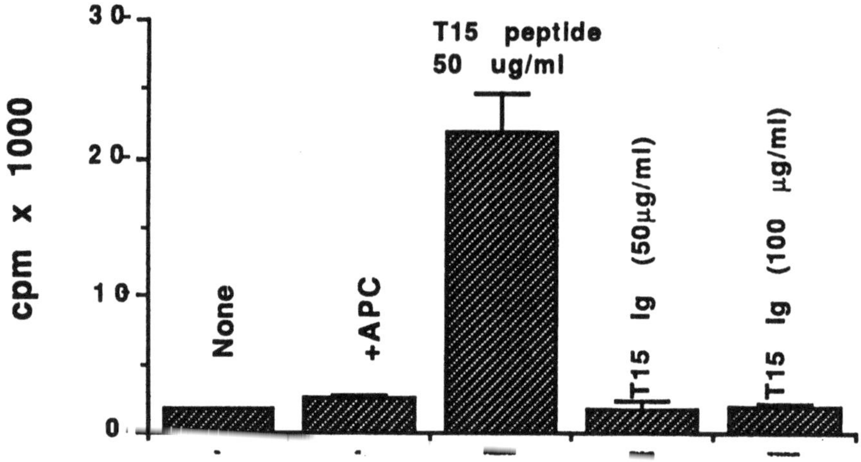

Fig. 3. Inability of T15 transfectant BCG-3R-Id to present endogenous Ig. T15H(50-73)-specific T cell line (2×10^4 cells/well) were cocultured with mitomycin pretreated BCG-3R-Id (2×10^5 cells/well). Different antigens or 5×10^6 irradiated (3300 R) syngeneic spleen cells were added as indicated. The BCG-3R-Id cells were pretreated with 100 ug/ml of mitomycin (Sigma) at 37° C for 1.5 hours and washed with RPMI 3x.

Because endogenous Igs are efficiently presented by B cells, we used the T15 transfectant BCG-3R-Id (IgM, k; Ward, et al. 1988) as a presenting cell. As seen in Fig. 3 the T15 transfectant presented the exogenous T15 peptide but was unable to present the endogenous and intact exogenous T15 Ig despite the expectation that the transfectant should have been a powerful presenter of endogenous immunoglobulin (Weiss & Bogen, 1989).

In order to further characterize these T cells, we established a T15 peptide-specific T cell clone, 3-1, and analyzed the MHC restriction. Clone 3-1 responded to the T15H(50-73) peptide in a dose-dependent manner (Fig. 4) but only when presented by cells from H-2^d haplotype mice (Balb/c, DBA/2 and the congeneic Cal 20 strain). Cells from non-H-2^d mice (A/J and C57BL/6) were not able to present.

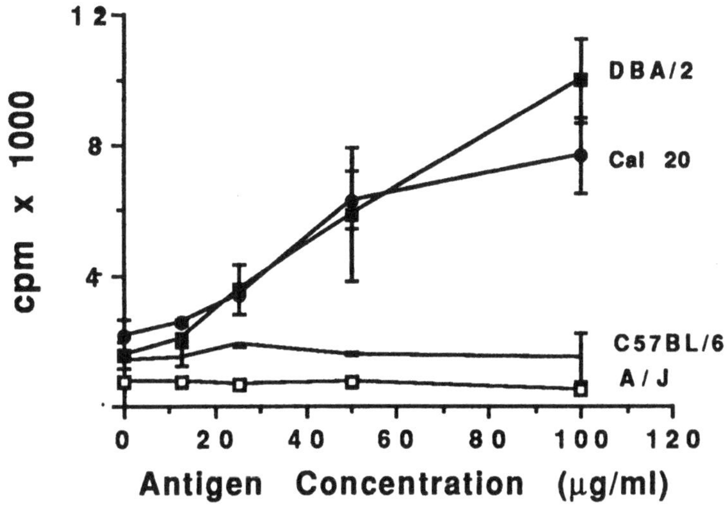

Fig.4. MHC restriction for T cell activation. T15H(50-73) specific T cell clone 3-1 ($2X10^4$ cells/well) were restimulated with different doses of T15 peptide in the presence of irradiated spleen cells ($5X10^5$cells/well) from different strains of mouse. The proliferation assay were carried out as described in Fig.1.

We then addressed the question whether the T15H(50-73) peptide requires processing by antigen-presenting cells. Antigen presenting cells were first treated with chloroquine and then tested for their ability to present the T15H(50-73) peptide. After chloroquine treatment, the APCs were still capable of presenting the peptide to the T cells (Fig. 5). As a control, treatment of APCs with chloroquine abolished the response of the 3A4-specific T cell clone 8-23 (Saeki, et al. 1989) to 3A4 Ig.

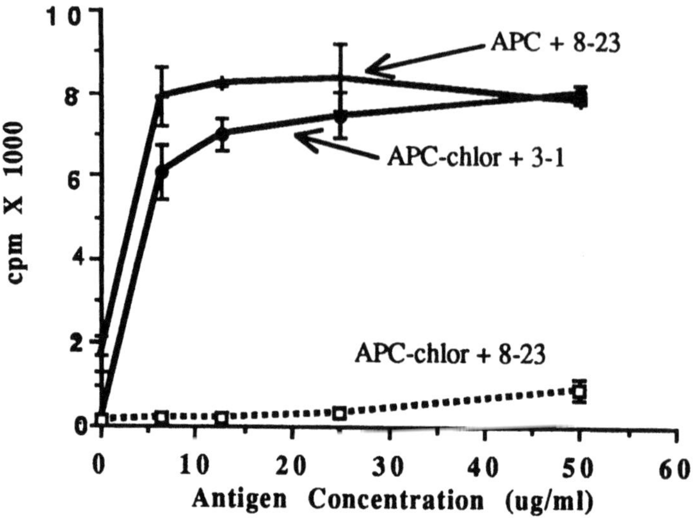

Fig. 5. Chloroquine resistant processing. T15H (50-73) specific T cell clone 3-1 or 3A4 specific T cell clone 8-23 ($2X10^4$ cells/well) were stimulated with different doses of T15 peptide or 3A4 Ig in the presence of either normal irradiated APC or chloroquine pretreated irradiated APC ($5X10^5$ cells/ well). "APC" indicates normal irradiated APC. "APC-chlor " indicates APC pretreated with chloroquine as described (Saeki , et al. 1989).

These results suggest that either this peptide does not require processing or that the enzymes responsible for the processing are not impaired by chloroquine treatment (chloroquine resistant processing). To distinguish between these two possible mechanisms, we used fixed APCs for presentation of the T15H(50-73) peptide to clone 3-1.

In Table I the proliferative response of a representative experiment using fixed APCs is shown. Fixing the APCs abolishes the presentation of the peptide to the clone 3-1. However, if APCs are incubated with peptide prior to fixation, the 3-1 clone responded. Negative controls included T cells alone or fixed preincubated APCs alone, or T cells mixed with the fixed APCs which had not been preincubated with the peptide. These results strongly suggest that processing of the T15H(50-73) peptide is necessary for effective presentation. Interestingly, no response was observed to two subpeptides (T15H50-61, T15H61-73) suggesting that the processing of the 50-73 peptide must preserve the central region of the peptide (data not shown). Further dissection of the precise location of this epitope requires a systemic survey using overlaping peptides.

Table 1 Peptide processing is required for 3-1 clone response

Cells	CPM
T cell	97 +/- 36*
Fixed preincubated APC	238 +/- 14
T cell + APC	401 +/- 15
T cell + fixed APC + peptide #	< 250
T cell + fixed preincubated APC **	19,825 + 350

* mean +/- SD

\# Different doses of T15H (50-73) peptide (12.5, 25 , 50, 100, 200 µg/ml)were tested.

** APC was preincubated with peptide (50µg/ml) before being fixed with 0.15% paraformadehyde.

These results clearly demonstrated that the repertoire of normal mice contains T cells for a self peptide derived from the polymorphic region of a germline encoded immunoglobulin. However, the self Ig peptide is cryptic in the context of the intact Ig structure. The role of cryptic T cell peptides in tolerance has been discussed by Sercarz and colleagues (Benichou & Sercarz, 1990; Benichou, et al. 1990) who have described a cryptic T cell repertoire against self-MHC peptides. They proposed that antigens in immunologically privileged sites as well as antigens processed inefficiently are cryptic self epitopes (Gammon, et al. 1991). We have extended these observations to self peptides derived from the polymorphic regions (CDR) of antibodies. This data raises the possibility that the hypervariable region of self immunoglobulin may contribute to the acquisition of the T cell repertoire in ontogeny. Earlier observations in neonatally anti-mu suppressed mice have shown that the development of the T cell repertoire is influenced by B cells (Martinez-A 1984 &1985). The self peptide derived from Ig gene product combined with the self-MHC may be presented directly by thymic B cells (Rudensky, et al. 1990) or by other thymic APCs and thereby positively select the T cell repertoire in the absence of foreign peptides. If immunoglobulins are involved in thymic education, one would ask why self Ig peptides are able to be presented in the thymus but are cryptic in adulthood? Aberrant Ig gene products with different flanking regions found in fetal tissues (Gallagher & Miller, 1988; Reth & Alt, 1984) may be processed in the thymus differently from that of intact Ig. The other possibility is that thymus-specific proteases process the self peptides in a manner that would be nonphysiologic elsewhere in the immune system.

Alternatively, the cryptic behavior may be due to peptide binding to MHC molecules on thymus epithelium which may not be the same as found elsewhere (Marrack & Kappler, 1989). The role of cryptic self Ig in the thymus would be to increase the TCR diversity in the absence of foreign antigens without activating an adverse autoimmune repertoire. However, creating the T cell repertoire for cryptic self-Ig plants the seeds for pathologic autoreactivity. Ly-1 B cells implicated in autoantibody production is the major phenotype of thymic B cells (Miyama-Inaba, 1988) . Persistence of peripheral Ly-1 B cells may lead to an autoimmune response via help of T cells expanded by the cryptic epitopes. Hence, self Ig peptides may have a double-sided role in the immune system.

REFERENCES

Behar SM, Corbet S,Diamond B,Scharff MD (1989) The molecular origin of anti-DNA antibodies. International Reviews of Immunology, 5:23-42.
Benichou G,Sercarz E (1990) The presentation of self-peptides: tolerance and competition. International Reviews of Immunology 6:75-88.
Benichou G,Takizawa PA, Ho PT, Killion CC, Olson CA, McMillan M, Sercarz EE (1990) Immunogenicity and tolerogenicity of self-major histocompatibility complex peptides. Journal of Experimental Medicine 172:1341-6.
Gallagher PF, Miller JF (1988) Immunoglobulin gene expression is a normal differentiation event in embryonic thymocytes. European Journal of Immunology 18:183-6.
Gammon G, Sercarz EE, Benichou G (1991) The dominant self and the cryptic self: shaping the autoreactive T-cell repertoire. Immunology Today 12:193-5.
Kang CY, Brunck TK, Kieber-Emmons T, Blalock JE, Kohler H (1988) Inhibition of self-binding antibodies (autobodies) by a VH-derived peptide. Science 240:1034-6.
Kang CY, Cheng HL, Rudikoff S, Kohler H (1987) Idiotypic self binding of a dominant germline idiotype (T15). Autobody activity is affected by antibody valency. Journal of Experimental Medicine 165:1332-43.
Kaveri SV, Kang CY, Kohler H (1990) Natural mouse and human antibodies bind to a peptide derived from a germline VH chain. Evidence for evolutionary conserved self-binding locus. Journal of Immunology 145:4207-13.
Koga T, Wand-Wurttenberger A, DeBruyn J, Munk ME, Schoel B, Kaufmann SH(1989) T cells against a bacterial heat shock protein recognize stressed macrophages. Science 245:1112-5.
Marrack P, Kappler J. (1988) The T-cell repertoire for antigen and MHC. Immunology Today 9:308-15.
Marrack P, McCormack J, Kappler J (1989) Presentation of antigen, foreign major histocompatibility complex proteins and self by thymus cortical epithelium. Nature 338:503-5.
Martinez-A C, Pereira P, Bernabe R, Bandeira A, Larsson EL, Cazenave PA, and Coutinho A (1984) Internal complementarities in the immune system: Regulation of the expression of helper T-cell idiotypes. Proceedings of the National Academy of Sciences of the United States of America 81:4520-4523.
Martinez-A C, Bernabe RR, Hera ADL, Pereira P, Cazenave PA, Coutinho A (1985) Establishment of idiotypic helper T-cell repertoires early in life. Nature 317:721-723.
Miyama-Inaba M, Kuma S, Inaba K, Ogata H, Iwai H, Yasumizu R, Muramatsu S, Steinman RM, Ikehara S (1988) Unusual phenotype of B cells in the thymus of normal mice. Journal of Experimental Medicine 168:811-6.
Nikolic-Zugic J, Bevan MJ (1990) Role of self-peptide in positively selecting the T cell repertoire. Nature 344:65-67
Raychaudhuri S, Kang CY, Kaveri SV, Kieber-Emmons T, Kohler H (1990) Tumor idiotype vaccines. VII. Analysis and correlation of structural, idiotypic, and biologic properties of protective and nonprotective Ab2. Journal of Immunology 145:760-7.
Raychaudhuri S, Saeki Y, Fuji H. and Kohler H (1986) Tumor specific idiotype vaccine. I. Generation and characterization of internal image tumor antigen. J. Immunol 137:1743
Reth M, and Alt FW (1984) Novel immunoglobulin heavy chains are roduced from DJ_H gene segment rearrangements in lymphoid cells. Nature 312: 418-423.
Rudensky AY, Mazel SM, Yurin VL (1990) Presentation of endogenous immunoglobulin determinant to immunoglobulin-recognizing T cell clones by the thymic cells. European Journal of Immunology 20:2235-9.
Saeki Y, Chen JJ, Shi LF, Raychaudhuri S, Kohler H (1989) Characterization of "regulatory" idiotope-specific T cell clones to a monoclonal anti-idiotypic antibody mimicking a tumor-associated antigen (TAA). Journal of Immunology 142:1046-52.

Schild H, Rotzschke O, Kalbacher H, Rammensee HG (1990) Limit of T cell tolerance to self proteins by peptide presentation. Science 247:1587-9.

Ward RE, McNamara-Ward M, Webb CF, Altman D, Lim PL, Tucker PW, Kohler H (1988) Regulation of an idiotype+ B cell lymphoma. Effects of antigen and anti-idiotopic antibodies on proliferation and Ig secretion. Journal of Immunology 141:340-6.

Weiss S, Bogen B (1989) B-lymphoma cells process and present their endogenous immunoglobulin to major histocompatibility complex-restricted T cells. Proceedings of the National Academy of Sciences of the United States of America 86:282-6.

Acknowledgement: BCG- 3R-Id transfectant was a kind gift of Dr. Philip W. Tucker, University of Texas, Dallas.

T CELL REPERTOIRE IN AUTOREACTIVE IMMUNOGLOBULIN-TRANSGENIC MICE

M. Zöller

Unité d'Immunobiologie

Institut Pasteur

Rue du Dr. Roux

75724 Paris CEDEX 15

France

SUMMARY

Female BALB/c, which were mated with immunoglobulin (anti-TNP IgM, Sp6)-transgenic male BALB/c mice, received intravenous injections of trinitrophenyl (TNP) coupled to mouse IgG1 during pregnancy to examine the possibility of induction of tolerance in immunoglobulin-transgenic mice, when the immunoglobulin is specific for a soluble "self antigen", i.e. TNP-IgG1. In addition, a possible function of "autoreactive" B cells in selection or deletion of autoreactive T cells was evaluated.

The number of B cells was slightly decreased in non-transgenic (NTG) and transgenic (TG) litters of TNP-IgG1-treated mice. In the early postnatal period, the majority of remaining TNP-specific B cells was in an activated stage. But, responsiveness was severely impaired after renewed contact with the nominal "self" antigen in the postnatal period.

NATO ASI Series, Vol. H 66
Theoretical and Experimental Insights into Immunology
Edited by A. S. Perelson and G. Weisbuch

In addition, during the early postnatal period of hyperreactivity "autoreactive" B cells could present nominal antigen as well as their own idiotype. The capacity of antigen presentation was considerably restrained during the consecutive period of partial anergy, especially after renewed contact with the nominal "self" antigen. Furthermore, despite of the potential to present antigen, addition of activated "autoreactive" B cells to T cell cultures revealed a suppressive effect. The dichotomie between selection and suppression of self- and idiotype-specific T cells by "autoreactive" B cells supports the hypothesis that tolerance towards non-MHC self antigens may be sustained by naturally activated autoreactive B cells.

INTRODUCTION

It is still a matter of debate, where T cells become tolerized for non-MHC self antigens (Bevan, 1977, Zinkernagel et al.,1987, Marrack et al.,1988a, Taylor and Rose,1988, Sprent et al.,1988, Matzinger and Guerder,1989). While it is known that self-reactive B cells constitute the majority of the neonatal B cell repertoire (Dighiero,G. et al,1985), autoimmune phenomena being avoided by a high degree of idiotypic connectivity between naturally activated B cells (Holmberg,D. et al.,1984), similar information with respect to the T cell repertoire is still missing. We have proposed recently that intrathymic B cells may actually be involved in selection of a primary T cell repertoire with specificity for self MHC plus nominal antigen by presenting mainly their own idiotypes, but nominal antigen as well (Zöller,1990a,1990b). As far as nominal self antigens are concerned, these should be presented by intrathymic macrophages and/or dendritic cells, too (Blackman et al,1988, McDonald,1988, Marrack et al., 1988b), and it could be expected that T cells positively selected by B cells may succumb negative selection as it is known for T cells with high avidity for self MHC antigens (Jenkinson et al.,1985, Kisielow,P. et al.,1988). "Idiotype-specific" T cells would not necessarily be eliminated by macrophages and/or dendritic cells, except the majority of B cells expressing one particular idiotype would be activated and free immunoglobulin would be available to be presented by macrophages and dendritic cells.

Taking these assumptions into account, indications whether intrathymic B cells are involved in establishment and selection of the T cell repertoire may be obtained from immunoglobulin-transgenic mice, especially when B cells expressing the transgene are specific for a non-MHC-self antigen. Evidence will be presented supporting the hypothesis that autoreactive B cells are involved not only in selection of the T cell repertoire but also in downregulation of autoreactive T cells depending on their stage of activation.

MATERIAL AND METHODS

Mice: BALB/c mice and Sp6-transgenic BALB/c mice (Rusconi and Köhler,1985, Köhler and Shulman,1980), kindly provided by G.Köhler, Max-Planck Institute for Immunobiology, freiburg, were bred under specific pathogen-free conditions in the animal facilities of the German Cancer Research Center. Where indicated, female BALB/c mice, mated with male BALB/c or male Sp6-transgenic BALB/c mice, received 200µg TNP-IgG1, intravenously, twice per week, throughout pregnancy and the foster period. NTG and TG litters were immunized, intraperitoneally, with $2x10^8$ TNP-sheep red blood cells (SRBC) or 1µg TNP-lipopolysaccharide (LPS), 4 and 3 days, respectively, before testing. Coupling of TNP to ovalbumin (OVA), SRBC and LPS was done as described elsewhere (Zöller and Andrighetto,1984).

Fluorescence analysis: The percentage of surface IgM- and Sp6-positive cells was determined with by fluorescence activated cell scanning using the following reagents: FITC-anti-mouse µ and affinity-purified 20-5 (10µg/ml), an IgG1 monoclonal antibody recognizing the idiotype of Sp6, as first antibody followed by FITC-anti-IgG1 as second antibody.

Effector cells, cultures and assay procedures: Thmic cells (TC) and spleen cells (SC) were obtained by pressing the organs through fine gauze and washing the cells in Hank's balanced salt solution. Bone marrow cells (BMC) were flushed out of the femur. T cells were purified by depletion of macrophages by two rounds of adherence (1 h each) on plastic. The remaining fraction was panned on anti-mouse Ig-coated plates (Wysocki and Sato,1978). The non-adherent fraction contained <1% Ig-positive cells and >95% were Thy1.2-positive. Where indicated, the T cell population was in addition depleted of cells expressing MHC class II antigens by treatment with an anti-MHC class II monoclonal antibody (17 227, kindly provided by G.Hämmerling, Institute for Immunology and Genetics, German Cancer Research Center, Heidelberg, FRG) and rabbit complement. The adherent fraction was regained with a rubber policeman and was used as source of B cells, >90% of which stained with FITC-anti mouse Ig.

B cell frequencies were determined under limiting dilution (LD) conditions, diluting SC or TC from $1x10^5$ or $2x10^6$ to $1.56x10^3$or $3.12x10^4$ cells per 200µl RPMI 1640, supplemented with 10^{-3}M HEPES buffer, $5x10^{-5}$M 2-mercaptoethanol, glutamine, 10% fetal calf serum and 10% IL-2 containing supernatant of rat SC cultures. As filler cells $2x10^4$ irradiated SC were added. B cells were activated by 20ng/ml TNP-LPS. For every dilution 24 replicates were set up. Cells were cultured for 3 days.

Frequencies of proliferating T cells were determined by diluting TC or SC from 51200-400 cells/well in RPMI 1640, supplemented with 10^{-3}M HEPES buffer, $5x10^{-5}$M 2-mercaptoethanol, glutamine and 2% mouse serum. T cells were activated by addition of 4µg TNP-OVA or 2µg Sp6, purified by 2 passages over a Concanavalin A Sepharose column. After 3 days, 0.5µCi of ^{3}H-thymidine was added for an additional period of 12h. Plates were harvested with a Titertek harvester and ^{3}H-thymidine uptake was determined in a ß-counter.

In bulk cultures, B cells were titrated in triplicates from $5x10^3$-$0.3x10^3$ cells/well. SC, depleted of B cells, or T cells depleted of MHC class II antigen-positive cells were added at a concentration of $1x10^5$ cells/well. Antigen concentrations and determination of ^{3}H-thymidine uptake were described above.

For the plaque assay (Jerne and Nordin,1963), freshly harvested TC and SC from 3-5 mice per group were diluted to $1x10^7$ and $1x10^6$ cells/100µl, setting up duplöicates for every sample. LD cultures were centrifuged, the medium was decanted and 100µl of Hank's balanced salt solution was added to every well. To 100µl cell suspension, 25µl of RBC (1:3 dilution) and 25µl guinea pig complement (1:5 dilution) were added. Suspensions were mixed with 300µl agar (45°C), containing 0.75mg/ml DEAE-cellulose and were poured into Petri dishes. Plaque forming cells (PFC) were counted after an incubation period of 3-4 h at 37°C in a humidified atmosphere.

Frequencies were calculated according to the formula F_0 (fraction of non-responding cultures) = e^{-u}, where u = c/w (number of c cells distributed in w wells) (Lefkovits and Waldmann,1979). In each experiment lymphocytes from three to five mice were pooled. Experiments were repeated two to three times, the recorded frequencies represent the mean of the repeats.

RESULTS

Influence of prenatal antigen-treatment on B cell development:

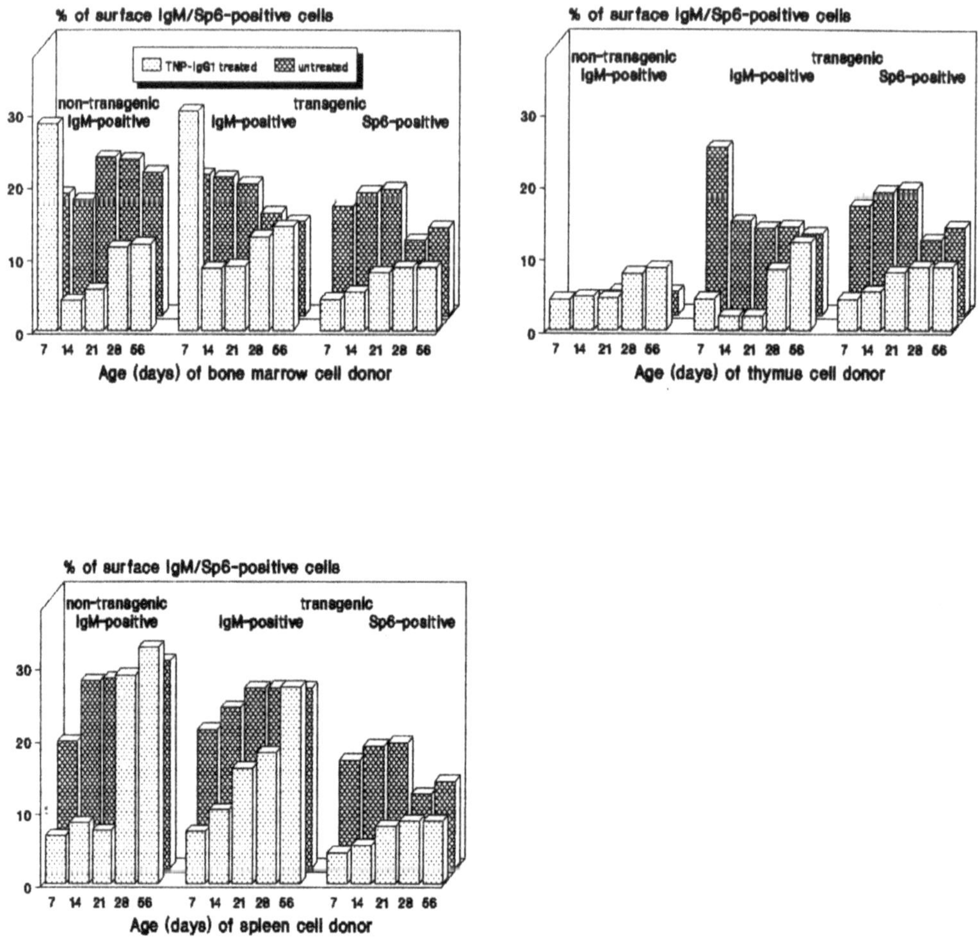

Figure 1. Influence of prenatal antigen-treatment on B cell development in NTG and TG mice: The percentage of surface IgM- and Sp6-positive cells in bone marrow, thymus and spleen of NTG and TG litters, which were untreated or prenatally treated with TNP-IgG1 (see Material and Methods) was determined during postnatal development by FAC-scanning.

Pregnant BALB/c mice, mated with NTG or anti-TNP IgM (Sp6)-TG male BALB/c mice were treated throughout the gestational and foster period with intravenous injections of TNP-IgG1 to evaluate the possible influence of a non-MHC self antigen on the developement of (transgene-expressing) B cells. In bone marrow, thymus and spleen, especially of TG mice, the number of surface IgM-positive cells was decreased during postnatal maturation. Although the number of surface IgM-positive cells returned towards normal levels at the age of around 8 weeks, in TG litters, the number of B cells expressing the transgene remained slightly reduced (Figure 1).

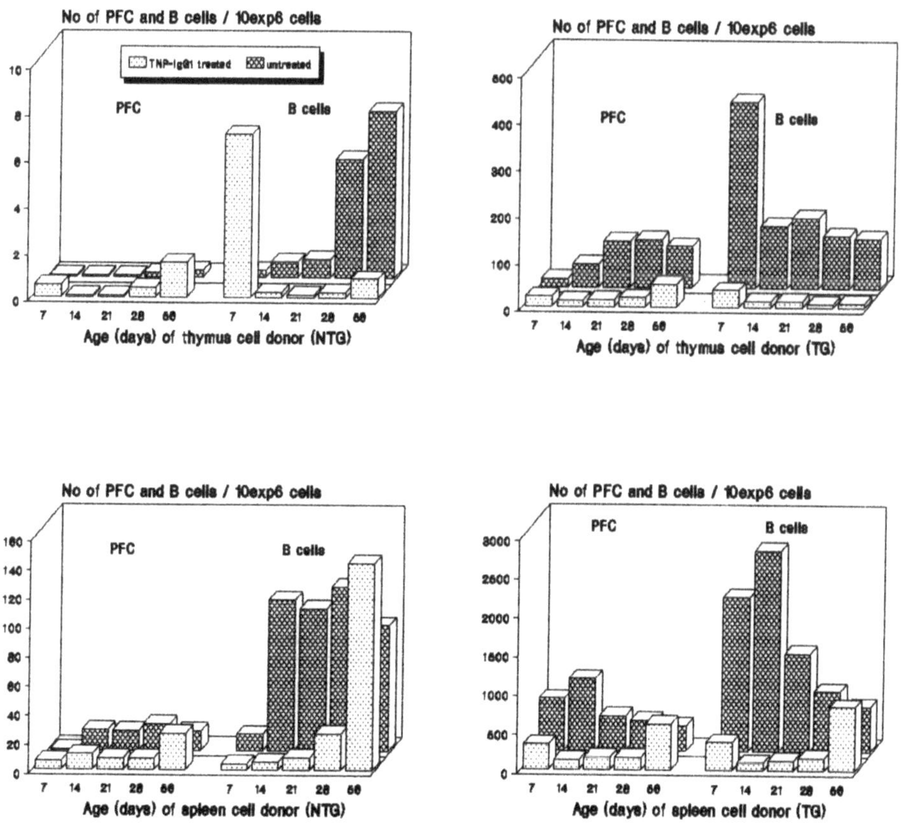

Figure 2. Evaluation of antibody secreting and resting TNP-specific B cells in prenatally antigen-treated NTG and TG mice: The number of anti-TNP IgM secreting B cells was determined throughout postnatal development in thymus and spleen of NTG

and Sp6-TG litters, which were prenatally untreated or treated with TNp-IgG1. The number of PFC was derived from freshly harvested organs, the number of B cells represents frequencies as determined under limiting dilution conditions after in vitro activation with TNP-LPS (see Material and Methods).

At the age of 1 to 4 weeks, the majority of B cells in thymus and spleen of transgenic mice were activated and secreted the transgenic immunoglobulin. A comparison of the number of PFC from freshly harvested organs with the number of B cells as evaluated under limiting dilution conditions revealed that in thymus and spleen of transgenic mice up to 50% of TNP-specific B cells secreted antibody, while in non-transgenic littermates the comparable figures were 4% - 15%. By prenatal treatment of non-transgenic and transgenic mice with nominal antigen, a higher percentage of B cells was driven into the activated stage of antibody secretion (Figure 2 and Table 1).

Table 1

Ratio of non-secreting versus secreting B cells in prenatally untreated and TNP-IgG1-treated mice.

		TNP-specific B cells / TNP-specific PFC			
Donor:		NTG		TG	
Prenatal treatment:		none	TNP-IgG1	none	TNP-IgG1
Organ	Age				
Thymus	7d	3.2	14.0	20.7	1.7
	14d	7.1	2.5	2.6	1.0
	21d	7.8	1.4	1.5	0.8
	28d	25.4	0.4	1.1	0.4
	56d	23.9	0.5	1.2	0.2
Spleen	7d	14.0	0.7	3.0	1.1
	14d	7.5	0.5	2.8	0.8
	21d	7.4	1.1	2.8	0.8
	28d	6.2	3.1	2.0	1.0
	56d	6.4	5.8	1.8	1.4

It is concluded that by prenatal treatment with nominal antigen B cells, thus becoming autoreactive, were driven into a more activated stage. This accounted for NTG as well as TG mice.

B cell activation in prenatally antigen-treated, immunoglobulin-transgenic mice:

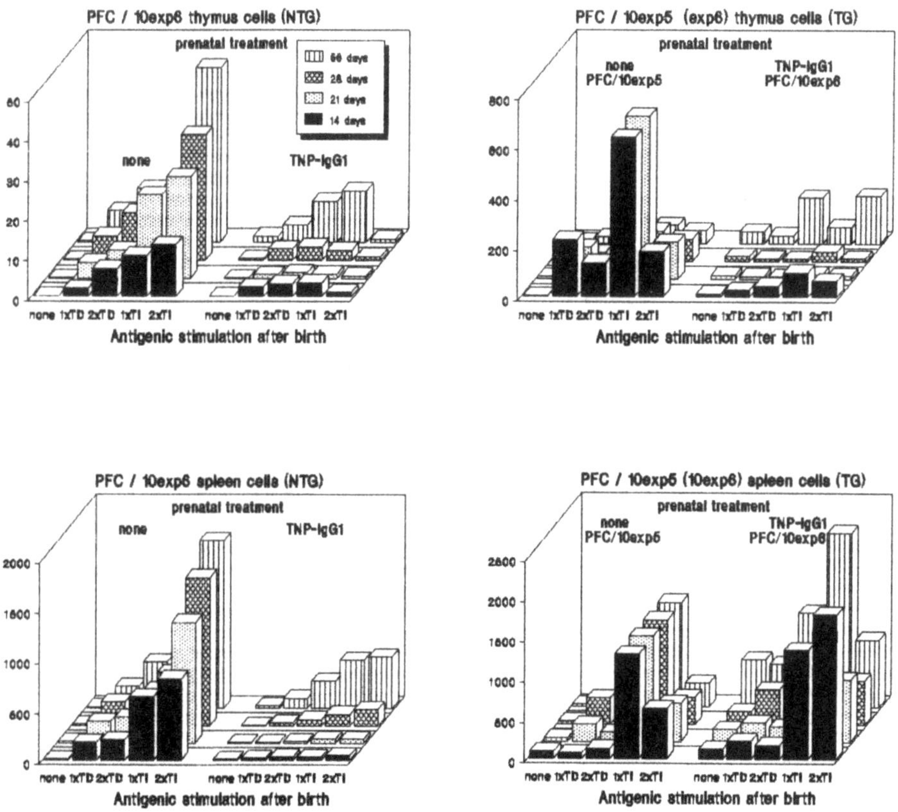

Figure 3. Expansion of B cells after stimulation with T dependent and T independent antigen: The number of anti-TNP IgM secreting B cells was evaluated in thymus and spleen of prenatally untreated and TNP-IgG1-treated NTG and TG litters, which received one or two intraperitoneal injections of TNP-SRBC (TD) or TNP-LPS (TI). PFC were determined 4 and 3 days, respectively, after the last injection.

Table 2

Recruitment of plaque forming cells after antigenic stimulation of prenatally untreated and TNP-IgG1-treated mice.

Organ	Donor	Prenatal Treatment	Age	Ratio of TNP-specific PFC unstimulated:	1xTD	2xTD	1xTI	2xTI
Thymus	NTG	none	14d	1:	20.0	70.0	103.1	130.4
			21d		40.1	69.8	210.0	256.2
			28d		30.8	60.4	45.5	157.8
			56d		26.5	44.8	15.7	146.5
		TNP-IgG1	14d		30.1	36.4	42.9	13.8
			21d		12.5	18.1	13.8	8.8
			28d		7.0	8.5	6.3	2.5
			56d		2.9	6.9	8.7	0.5
	TG	none	14d		43.9	26.4	122.9	34.6
			21d		12.5	6.4	64.9	14.6
			28d		3.4	4.3	11.6	8.6
			56d		3.0	1.2	8.1	5.6
		TNP-IgG1	14d		2.3	3.5	7.8	5.0
			21d		0.8	0.7	1.2	0.9
			28d		0.6	0.5	1.9	0.7
			56d		0.7	3.8	1.4	3.9
Spleen	NTG	none	14d	1:	13.0	14.7	45.5	58.5
			21d		16.3	19.4	39.8	90.9
			28d		12.7	15.3	29.0	80.8
			56d		16.2	34.3	39.3	123.4
		TNP-IgG1	14d		1.7	3.2	3.0	4.2
			21d		1.8	4.3	10.1	10.0
			28d		3.8	8.0	14.7	21.1
			56d		4.1	11.0	19.4	20.9
	TG	none	14d		0.7	1.4	14.0	6.8
			21d		4.8	2.6	29.2	10.6
			28d		8.3	3.0	33.2	8.5
			56d		2.4	2.1	40.3	9.1
		TNP-IgG1	14d		1.8	1.4	10.9	14.6
			21d		1.5	1.2	6.7	4.9
			28d		2.7	1.6	7.3	3.4
			56d		0.9	2.0	3.7	1.4

Despite the fact that prenatal treatment with nominal antigen did not result in anergy of "autoreactive" (transgenic or non-transgenic) B cells at birth, recruitment of antibody-secreting B cells after antigenic stimulation was severely impaired (Figure 3 and Table 2). While in the spleen of 4 weeks old NTG litters the number of PFC increased from 18 to 278 and 1471 by repeated stimulation with TD and TI antigens, respectively and in thymus and spleen of TG mice the number of PFC secreting the immunoglobulin transgene was at least expanded after priming with nominal antigen, in prenatally antigen-treated NTG and TG mice antigenic stimulation lead in most instances to a decrease of the number of PFC. This was especially noted until the age of 4 weeks. The finding was confirmed evaluating after in vivo stimulation the numbers of B cells under limiting dilution conditions in thymus and spleen of NTG and TG, prenatally antigen-treated mice. After repeated in vivo stimulation, a significant decrease in the number of B cells, which could be induced to secrete anti-TNP antibodies, was observed (data not shown).

Thus, prenatal antigen treatment resulted in activation of part of the B cells expressing the transgene. The remaining, non-activated B cells apparently were in an anergic stage. If B cells are indeed involved in selection of the T cell repertoire, a comparison of TNP- and Sp6-specific T cells in prenatally antigen-treated, NTG and TG mice should reveal significant differences.

Antigen- and idiotype-specific T cells in transgenic and prenatally antigen-treated, non-transgenic and transgenic mice:

In the spleen of non-transgenic mice, the frequency of T cells proliferating in response to TNP increased slightly during postnatal development and was significantly increased after

Table 3A

Frequency of proliferating T cells.

		Frequency of proliferating T cells / 10exp6 lymphocytes			
Antigenic stimulus in vitro:		TNP-OA			
Prenatal treatment:		none		TNP-IgG1	
Stimulation in vivo:		none	TNP-SRBC	none	TNP-SRBC
Organ	Donor				
Thymus	NTG, 7d	6.3	10.9	55.6	3.8
	NTG,14d	12.9	35.7	11.7	2.7
	NTG,28d	2.6	8.1	5.4	1.8
	NTG,56d	1.6	5.5	2.5	3.3
	TG, 7d	195.8	134.5	270.3	55.3
	TG, 14d	59.4	43.5	27.2	13.7
	TG, 28d	43.5	38.6	15.7	10.3
	TG, 56d	28.8	22.1	10.6	20.9
Spleen	NTG, 7d	36.1	113.6	170.3	30.6
	NTG,14d	65.4	220.5	98.4	11.5
	NTG,28d	102.4	370.2	35.6	13.7
	NTG,56d	155.6	520.2	24.3	30.2
	TG, 7d	195.8	113.5	413.6	95.1
	TG, 14d	215.8	165.6	89.3	46.1
	TG, 28d	113.8	92.9	20.5	15.2
	TG, 56d	51.5	35.3	145.0	29.1
Antigenic stimulus in vitro:		Sp6			
Thymus	NTG, 7d	117.4	211.2	263.9	101.2
	NTG,14d	56.4	107.1	49.2	18.2
	NTG,28d	19.7	67.1	10.7	6.7
	NTG,56d	1.8	5.7	2.8	5.4
	TG, 7d	258.8	219.3	393.7	113.1
	TG, 14d	116.4	100.4	83.2	28.2
	TG, 28d	60.9	49.3	22.5	8.0
	TG, 56d	14.0	10.1	8.2	10.6
Spleen	NTG, 7d	133.3	242.6	382.2	141.7
	NTG,14d	72.6	149.0	57.9	34.8
	NTG,28d	55.0	115.2	26.7	13.1
	NTG,56d	28.9	49.5	14.8	20.5
	TG, 7d	342.3	335.2	417.1	166.3
	TG, 14d	106.5	77.5	61.3	25.1
	TG, 28d	85.2	71.4	25.7	13.7
	TG, 56d	22.7	16.6	18.1	26.6

Table 3B

Expansion of TNP- and Sp6-specific proliferating T cells after antigenic stimulation.

		Expansion rate of proliferating T cells after stimulation			
Antigenic stimulus in vitro:		TNP-OA		Sp6	
Prenatal treatment:		none	TNP-IgG1	none	TNP-IgG1
Organ	Donor				
Thymus	NTG, 7d	1.73	0.69	1.80	0.38
	NTG,14d	2.77	0.23	1.90	0.37
	NTG,28d	3.12	0.34	3.41	0.63
	NTG,56d	3.44	1.32	3.17	1.93
	TG, 7d	0.69	0.20	0.85	0.29
	TG, 14d	0.73	0.50	0.86	0.34
	TG, 28d	0.89	0.66	0.81	0.36
	TG, 56d	0.77	1.97	0.72	1.29
Spleen	NTG, 7d	3.15	0.18	1.82	0.37
	NTG,14d	3.37	0.12	2.05	0.60
	NTG,28d	3.62	0.38	2.09	0.49
	NTG,56d	3.34	1.24	1.71	1.39
	TG, 7d	0.58	0.23	0.98	0.40
	TG, 14d	0.77	0.52	0.73	0.41
	TG, 28d	0.82	0.74	0.84	0.53
	TG, 56d	0.69	0.65	0.73	1.47

		Ratio of proliferating T cells TNP-IgG1-treated : prenatally untreated mice			
Antigenic stimulus in vitro:		TNP-OA		Sp6	
Antigen in vivo:		none	TNP-SRBC	none	TNP-SRBC
Organ	Donor				
Thymus	NTG, 7d	8.83	3.50	2.25	0.48
	NTG,14d	0.91	0.07	0.87	0.17
	NTG,28d	2.08	0.23	0.54	0.10
	NTG,56d	1.56	0.60	1.56	0.95
	TG, 7d	1.38	0.41	1.52	0.52
	TG, 14d	0.46	0.31	0.71	0.28
	TG, 28d	0.36	0.27	0.37	0.16
	TG, 56d	0.37	0.95	0.59	1.05
Spleen	NTG, 7d	4.72	0.27	2.87	0.58
	NTG,14d	1.50	0.05	0.80	0.23
	NTG,28d	0.35	0.04	0.49	0.11
	NTG,56d	0.16	0.06	0.51	0.41
	TG, 7d	2.11	0.84	1.22	0.50
	TG, 14d	0.41	0.28	0.58	0.32
	TG, 28d	0.18	0.16	0.30	0.19
	TG, 56d	0.87	0.82	0.80	1.60

antigenic stimulation. In the thymus, TNP-specific T cell proliferation could only be detected after antigenic stimulation. Although the frequency of Sp6-specific T cells decreased during postnatal maturation, an augmentation was noted after antigenic stimulation. In Sp6-transgenic mice, TNP- and Sp6-specific proliferation was seen in the thymus and the spleen, but antigenic stimulation did not result in an increase of the number of T cells responding either to TNP-OVA or to Sp6. In prenatally antigen-treated NTG and TG mice, the number of T cells proliferating in response to TNP or Sp6 was increased early after birth, but decreased during postnatal developement. In addition, after antigenic stimulation, proliferation was severely impaired, i.e. the number of thymic and splenic T cells proliferating in response to TNP or Sp6 was significantly decreased (Table 3A and 3B). At the age of 8 weeks, a tendency towards recovery of T cell responsiveness was noted in prenatally antigen-treated, NTG and TG mice.

Since frequencies and responsiveness of TNP- as well as Sp6-specific T cells, indeed, were altered in NTG and TG mice prenatally treated with TNP, it finally was evaluated whether B cells present non-MHC self antigens including their own idiotype, and whether their capacity to present antigen depends on the stage of activation.

Presentation of nominal antigen and idiotypes by B cells:

The hypothesis was evaluated by coculturing T cells, which were depleted of adherent cells, B cells and any remaining MHC class II-positive cells with purified B cells from 1 week old mice evaluating the proliferative response in the presence of nominal antigen (TNP) or the idiotype (Sp6). Thymic and splenic B cells from control mice present nominal antigen mainly after antigenic stimulation. B cells from naive transgenic mice presented nominal antigen and Sp6, they partially lost the capacity to present TNP as well as Sp6 after antigenic stimulation. B cells from 1 week old prenatally antigen-treated transgenic mice did present nominal antigen and their own idiotype. They lost the capacity to present antigen as well as idiotype during postnatal maturation and the capacity to present could not be recovered by antigenic stimulation (data not shown). In fact, even thymic and splenic B cells of 1 week old mice, which were treated with antigen at the age of 4 days, did no longer initiate proliferation of T cells in response to TNP or Sp6 (Figure 4).

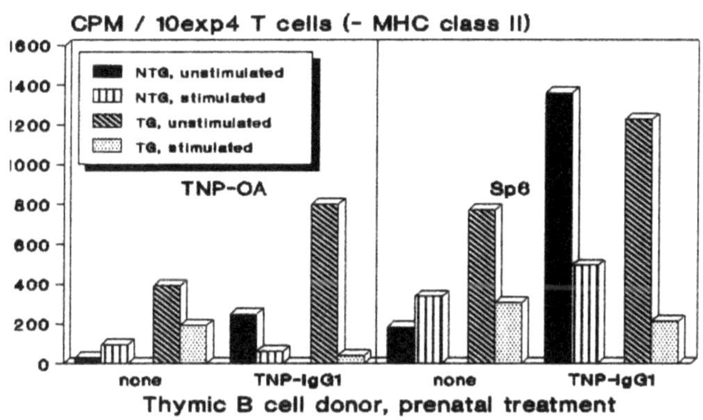

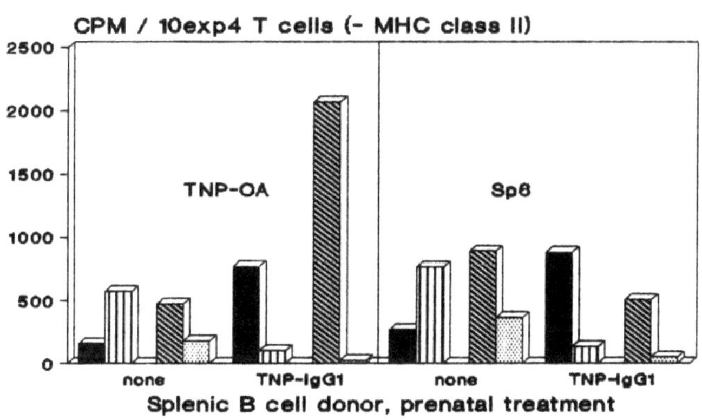

Figure 4. Antigen and idiotype presentation by B cells from prenatally untreated and TNP-IgG1-treated NTG and TG mice: Thymic and splenic B cells from 1 week old mice were cultured in the presence of either TNP-OA or Sp6 with $1x10^4$ splenic T cells from adult BALB/c mice, which were depleted of adherent cells, B cells and remaining MHC class II-positive cells. ^{3}H-thymidine incorporation was determined after 3 days of culture. CPM are shown for cultures containing $0.6x10^3$ B cells. CPM of the stimulator control were subtracted.

It is concluded that thymus- as well as spleen-derived B cells are involved in selection of antigen- and idiotype-specific T cells and that their capacity for antigen and idiotype presentation depends to a considerable degree on their stage of activation.

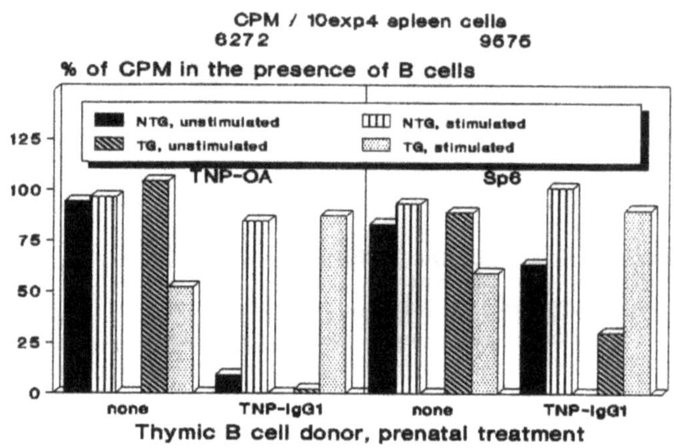

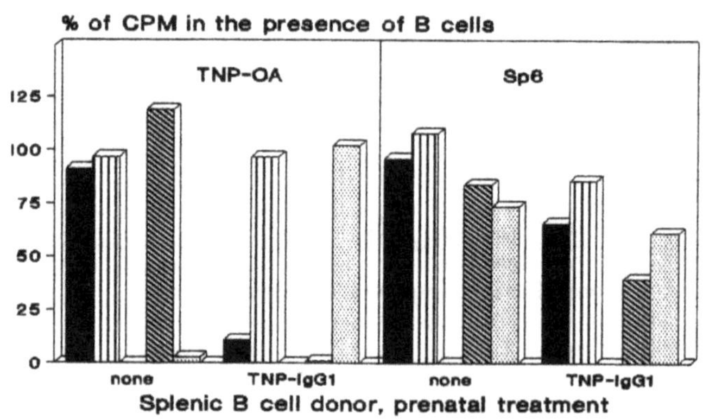

Figure 5. Down-regulation of T cell proliferation by B cells from prenatally TNP-IgG1-treated, non-transgenic and transgenic mice: Splenic T cells from adult BALB/c mice were cultured in the presence of TNP-OVA or SP6 with titrated numbers of B cells from prenatally untreated or TNP-IgG1-treated NTG and TG one week old mice. Where indicated, the B cell donor received an intraperitoneal injection of TNP-SRBC at the age of 4 days. Proliferation of T cells was determined as described in figure 4. The percentage of CPM in comparison to cultures, which did not contain B cells is shown.

Finally, it was analyzed whether hyperreactivity of B cells, i.e. secretion of antibodies, may not even be involved in "negative" selection of antigen-specific T cells. The latter phenomenon could be mediated by B cells, but by macrophages or T cells mediating antibody-dependent cellular cytotoxicity as well. So far we have not analyzed the effector mechanism responsible for B-cell-mediated suppression. Yet, addition of B cells from the thymus or the spleen from 1 week old prenatally antigen-treated NTG and TG mice severly impaired proliferation of T cells in response to TNP and Sp6. While B cells of these mice lost the capacity to "suppress" T cell proliferation after antigenic stimulation, B cells from prenatally untreated TG mice only were inhibitory after antigenic stimulation, and B cells from NTG mice did not interfere with T cell proliferation, irrespective of postnatal antigen-treatment (Figure 5).

DISCUSSION

It is well documented that T cells learn to tolerate self during their passage through the thymus by consecutive events of positive and negative selection (McDuffie et al.,1986, vBoehmer and Schubiger,1986, Lo et al.,1988). T cells, which bind to self-MHC structures will be selected in the first step, those with high avidity for selfMHC will be deleted, thereafter. How T cells learn to tolerate non-MHC self antigens is less well defined (McDuffie et al.,1986, HayGlass et al.,1986, Smith et al.,1989, Lorenz and Allen,1989). There is evidence that tolerance towards non-MHC self antigens can be established in the periphery. Alternatively, non-MHC self antigens, too, may be presented within the thymic environment. In the latter case, one candidate for presenting non-MHC self antigens would be B cells, which are present in the thymus already early during ontogeny (Isaacson et al., 1987, Kimoto et al., 1989). In addition, idiopeptides could serve as a source of a highly diverse repertoire of internal antigens (Marcos et al.,1989).

Recently, evidence was presented that intrathymic B cells are involved in shaping the T cell repertoire by presenting nominal antigen as well as idiopeptides (Zöller, 1990a, 1990b). Now the question was addressed, which function intrathymic B cells may play in selection and/or deletion of autoreactive T cells. As a model we choose immunoglobulin-transgenic mice and introduced the nominal antigen for the immunoglobulin-transgene during development. By this mean, the transgenic immunoglobulin became an autoantibody.

It could be demonstrated that the presence of a soluble autoantibody, TNP-IgG1, during B and T cell development led to expansion and, in particular, activation of autoantigen-specific B cells. This accounted for transgenic as well as non-transgenic B cells. The period of hyperreactivity was followed by an anergic stage, i.e. after renewed contact with the nominal antigen, autoreactive B cells did not expand and did not mature to PFC. This could be either due to exhaustion of the system, as described for establishment of tolerance at the T cell level (Bandeira et al.,1990), or it could be due to a high degree of idiotypic connectivity of naturally activated autoreactive B cells (Holmberg and Coutinho, 1985).

Irrespective of the mechanism of partial tolerance at the level of autoreactive, NTG and TG B cells, the data strongly support that autoreactive B cells are of utmost importance in selection and regulation of autoreactive T cells. In prenatally antigen-treated mice, where autoreactive B cells were expanded and activated early after birth, autoreactive as well as idiotype-specific T cells were expanded, too. Furthermore, after renewed contact with the nominal self-antigen, when B cells became anergic, a decrease in the frequency of "autoreactive" T cells was noted. Finally, activated autoreactive B cells were capable of presenting the autoantigen, while "anergic" B cells did not at all or insufficiently present antigen. Thus, selection of autoreactive and idiotype-specific T cells followed precisely the frequencies and activation stage of autoreactive B cells. In addition, it should be noted that selection of autoreactive T cells accounted for intrathymic as well as splenic T cells. Thus, selection of the repertoire of T cells with specificity for non-MHC antigens may take place in the thymus as well as in the periphery.

Finally, one observation, in particular, requires comment and further experimental exploration. Although activated, autoreactive B cells presented the autoantigen and apparently initiated maturation / expansion of autoreactive T cells, in cocultures with spleen cells, which were depleted of B cells, purified "autoreactive" B cells severely inhibited proliferation of T cells. Since inhibition of T cell proliferation was only observed with B cells in an activated stage, but not with anergic B cells from mice challenged with the autoantigen after birth, it is likely that secreted antibodies may play a role in downregulation of T cell proliferation. Furthermore, when purified B cells were added to purified T cells, they initiated rather than inhibited proliferation. From there it can be concluded that a direct suppressive effect of B cells is unlikely. We favor the hypothesis that antibody-mediated apoptosis (Smith et al., 1989) or a T cell or macrophage mediated antibody-dependent cellular cytotoxicity mechanism may be involved. Irrespective of the precise mechanism of suppression of T cell proliferation by activated B cells, the dichotomic function of autoreactive B cells during postnatal maturation of the immune system - i.e. they present the autoantigen and downmodulate T cell activation, probably by secreted antibodies - could guarantee selection of T cells with specificity for non-MHC self antigens as well as establishment of a stage of tolerance. The latter would be based - different from tolerance to self-MHC antigens - on a non-deletional mechanism.

REFERENCES

Bandeira A, Coutinho A, Carnaud C, Jacquemart F and Forni L (1989) Transplantation tolerance correlates with high levels of T- and B-lymphocyte activity. Proc.Natl.Acad.Sci. USA 86:272.

Bevan M (1977) In a radiation chimera host H-2 antigens determine the thymus responsiveness of donor cytotoxic cells. Nature 269:417.

Blackman MA, Kappler JW and Marrack P (1988) T cell specificity and repertoire. Immunol.Rev. 101:5.

vBoehmer H and Schubiger K (1986) Thymocytes appear to ignore class I major histocompatibbility complex antigens expressed on thymus epithelial cells. Eur.J.Immunol. 14:1048.

Dighiero G, Lymberi P, Holmberg D, Lundkvist I, Coutinho A and Avrameas S (1985) High frequencies of natural auto-antibodies in normal newborn mice. J.Immunol. 134:765.
HayGlass KT, Naides SH, Scott CF, Benacerraf B and Sy MS (1986) T cell development in a B cell-deficient mice: IV. The role of B cells as antigen-presenting cells in vivo. J.Immunol. 136:823.
Holmberg D, Forsgren S, Ivars F and Coutinho A (1984) Reactions amongst IgM antibodies derived from normal neonatal mice. Eur.J.Immunol. 14:435.
Holmberg D and Coutinho A (1985) Natural antibodies and autoimmunity. Immunol. today 6:356.
Isaacson PG, Norton AJ and Addis BJ (1987) The human thymus contains a novel population of B lymphocytes. Lancet, Dec.26:1488.
Jenkinson EJ, Jhittay P, Kingston R and Owen JJT (1985) Studies on the role of the thymic environment in the induction of tolerance to MHC antigens. Transplant. 39:331.
Jerne NK and Nordin AA (1963) The hemolytic plaque assay. Science 140:405.
Kappler JW, Roehm N and Marrack P (1987) T cell tolerance by clonal elimination in the thymus. Cell 49:273.
Kimoto H, Shirasawa T, Taniguchi M and Takemori T (1989) B cell precursors are present in the thymus during early development. Eur.J.Immunol. 19:97.
Kisielow P, Blüthmann H, Staerz U, Steinmetz M and vBoehmer H (1988) Tolerance in T cell receptor transgenic mice involves deletion of nonmature $CD4^{+}8^{+}$ thymocytes. Nature 332:742.
Köhler G and Shulman MJ (1980) Immunoglobulin M mutants. Eur.J.Immunol. 10:467.
Lefkovits I and Waldmann H (1979) Limiting dilution analysis of cells in the immune system. Cambridge University press, Cambridge.
Lo D, Burkly LC, Widera G, Cowing C, Flavell RA, Palmiter RD and Brinster RL (1988) Diabetes and tolerance in transgenic mice expressing class II MHC molecules in pancreatic beta cells. Cell 53:159.
Lorenz RG and Allen PM (1989) Thymic cortical epithelial cells can present self-antigens in vivo. Nature 337:560.
Marcos MAR, DeLaHera A, Pereira P, Marquez C, Toribio M, Coutinho A and Martinez-A.C (1988) B cell participation in the recursive selection of T cell repertoires. Eur.J.Immunol. 18:1015.
Marrack P, Kushnir E, Born W, McDuffie M and Kappler J (1988a) The development of helper T cell precursors in mouse thymus. J.Immunol. 140:2508.
Marrack P, Lo D, Brinster R, Palmiter R, Burkly L, Flavell RH and Kappler K (1988b) The effect of thymus environment on T cell development and tolerance. Cell 53:627.
Matzinger P and Grueder S (1989) Does T cell tolerance require a dedicated antigen-presenting cell? Nature 338:74.
McDonald HR, Schneider R, Lees RK, Howe RC, Acha-Orbea H, Festenstein H, Zinkernagel RM and Hen-gartner H (1988) T cell receptor Vß use predicts reac-tivity and tolerance to Mls^{a}-encoded antigens. Nature 332:40.
McDonald HR (1988) Intrathymic elimination of Mls^{a}-reactive (vß6) cells during neonatal tolerance induction of Mls^{a}-encoded antigens. J.Exp.Med. 167:2005.
McDuffie M, Born W, Marrack P and Kappler JW (1986) The role of the T cell receptor in thymocyte maturation. Effects in vivo of anti-receptor antibody. Proc.Natl.Acad.Sci.USA 83:8728.

Rusconi S and Köhler G (1985) Transmission and expression of a specific pair of rearranged immunoglobulin μ and k genes in a transgenic mouse line. Nature 314:330.

Smith H, Chen IM, Kubo R and Tung KSK (1989) Neonatal thymectomy results in a repertoire enriched in T cells deleted in adult thymus. Science 245:849.

Smith CA, Williams GT, Hingston R, Jenkinson EJ and Owen JJT (1989) Antibodies to CD3/T cell receptor complex induce death by apoptosis in immature T cells in thymic cultures. Nature 337:181.

Sprent J, Lo D, Gao EK and Ron Y (1988) T cell selection in the thymus. Immunol.Rev. 101:173.

Talor E and Rose NR (1988) The induction of antigen-specific thymic regulatory cells in the mouse. Cell.Immunol. 116:24.

Wysocki LJ and Sato VL (1978) Panning for lymphocytes: a method for cell selection. Proc.Natl.Acad.Sci.USA 75:2844.

Zinkernagel RN, Callahan GN, Althage A, Cooper S, Steilein JW and Klein J (1978) The thymoreticular system in triggering virus plus self-specific cytotoxic T cells: evidence for T help. J.Exp.Med. 147:897.

Zöller M and Andrighetto G (1984) Comparison between in vitro activities of in vivo induced hapten-specific suppressor cells and supernatants of a hapten-specific suppressor hybridoma. Cell.Immunol. 89:310.

Zöller M (1991) 2,4,6-trinitrophenyl (TNP) responsiveness of anti-TNP (Sp6) transgenic mice. Eur.J.Immunol. 21:1601.

Zöller M (1990a) Intrathymic T cell repertoire after prenatal trinitrobenzene-sulfonic-acid-treatment. Cell.Immunol. 126:31.

Zöller M (1990b) Intrathymic presentation of nominal antigen by B cells. Int.Immunol. 2:427.

EXPERIMENTAL AND THEORETICAL INVESTIGATIONS ON IDIOTYPIC MIMICRY

by

Alf Grandien°*, Antonio Coutinho* and John Stewart*

°Department of Immunology, University of Uppsala,
*Unité d'Immunobiologie, CNRS URA 359, Institut Pasteur, Paris

Introduction

This will be a progress report on some intriguing data concerning idiotypic mimicry in immunoglobulin transgenic mice, and on the genesis of a model of the immune system where similar phenomena can be simulated. We think that the observations on idiotypic mimicry are something more than a curiosity of "idiotypic cascades" in transgenic mice; rather, we believe that such simplified model systems can help us understand the basic rules on which an extremely complex immune system is built and learns self-tolerance through V-region repertoire selection.

We also regard mathematical modelling of the immune system as much more than an entertaining exercise or an activity restricted to theoreticians. Thus, it might well be necessary to work with dynamic explanations of phenomena such as tolerance, immune responses, and memory - areas which cover the majority of immunologists' concerns.

NATO ASI Series, Vol. H 66
Theoretical and Experimental Insights into Immunology
Edited by A. S. Perelson and G. Weisbuch

Experimental observations

Most manipulations of the immune system will result in some kind of alteration of its dynamics and/or structure. The introduction of a transgene coding for a complete immunoglobulin heavy chain into the germline of a mouse should not be an exception. In 1986, an article was published in which such an alteration was reported (Weaver et al, 1986). The finding was that the repertoire of B cells expressing endogenously encoded immunoglobulins was altered, such that many B cells expressed an idiotype that was similar to that of the transgene. We decided to analyse another immunoblobulin-transgenic mouse, carrying a different genetic construct. This transgenic mouse line was "made" in the laboratory of Michael Neuberger and is described in two articles (Neuberger et al, 1989; Pettersson et al, 1989). It contains a transgene encoding both heavy and light chains of an antibody, the V-regions of which were taken from a hybridoma (B1.8) that had been isolated in a C57BL/6 mouse immunized with the hapten NP (Reth et al, 1978). The constant region of the heavy chain (a μ-chain in the original antibody), has been replaced by the constant region of a human IgA that had its transmembrane part deleted. This allows the molecule to be secreted but neither to be expressed on the membrane of B cells, nor to inhibit rearrangement and expression of endogenous antibody genes. The predominant light chain used in a primary anti-NP response is lambda1, which is expressed by the hybridoma B1.8, and is also used in this transgenic construct. The light chain can associate either with the transgene or with endogenous immunoglobulins, regardless of subclass.

A schematic representation of the transgene is shown in Figure 1a. One more advantage of this model is that there existed already several anti-idiotypic antibodies directed against B1.8, e.g., Ac38 and Ac146, both of which require expression of the lambda1 light chain (Reth et al, 1979).

Our question was whether, in this transgenic mouse, the frequency of B cells expressing IgM which mimic B1.8 would be altered. By *mimicry,* we mean idiotypes which also bind to the Ac38 and Ac146 anti-idiotypes; see Figure 1b. Since V-regions associated with a μ-chain should in principle be the product of endogenous gene expression, this simple strategy could allow for a rapid detection of repertoire alterations due to the expression of the transgene. Our analyses can be summarized as follows (see Table 1). In the serum of transgenic mice, IgM which mimic B1.8 can be detected, but the titers are not dramatically elevated as compared to normal mice. In the compartment of immunocompetent resting B cells, however, mitogen driven limiting dilution analysis (Andersson et al, 1977) revealed surprisigly high frequencies (up to 45%) of LPS responsive, clonable B cells expressing IgM molecules which bind to the anti-idiotypic antibodies and to the hapten NIP (which can be used in place of NP). Moreover, the frequency of lambda1-producing B cells is even higher, which probably reflects the fact that the transgenic lambda1 can associate with endogenous μ-chains, but alone fails to "reproduce" the idiotypic profile of the original V-region.

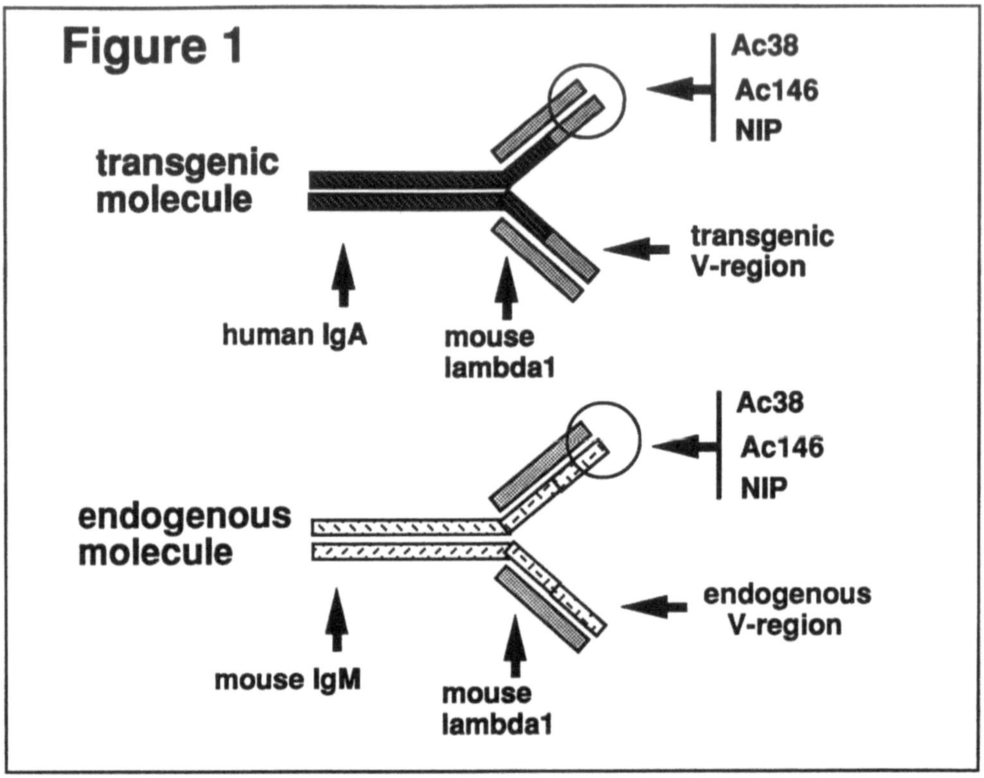
Figure 1
Ac38
Ac146
NIP
transgenic
molecule
transgenic
V-region
human IgA
mouse
lambda1
Ac38
Ac146
NIP
endogenous
molecule
endogenous
V-region
mouse IgM
mouse
lambda1

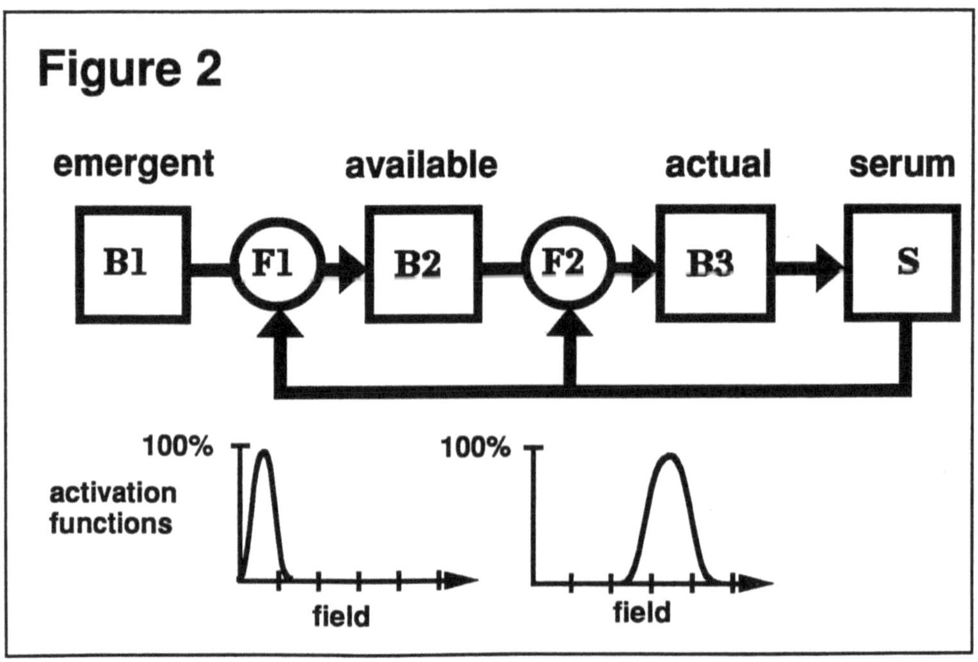
Figure 2
emergent
available
actual
serum
B1
F1
B2
F2
B3
S
100%
activation
functions
field
100%
field

Table 1

	Repertoire of resting immunocompetent B cells Frequency of cells producing IgM binding to :			
	Ac38	Ac146	B1.8	anti-lambda 1
non transgenic	±	-	±	+
transgenic	+++	+++	±	++++
transgenic nude	+++	+++	±	++++

	Repertoire of plasma cells Frequency of cells producing IgM binding to :			
	Ac38	Ac146	B1.8	anti-lambda 1
non transgenic	±	-	±	+
transgenic	++	++	±	++
transgenic nude	++	++	±	++

Although the phenomenon of idiotypic mimicry is both difficult to analyze and quite controversial for the time being, we were strongly impressed by this very high frequency of B cells expressing endogenous Ig mimicking the idiotype of the transgene in this transgenic mouse line. A trivial explanation for the high frequencies observed would be that our reagents cannot distiguish between IgM and human IgA. Another possibility would be the existance of mixed molecules, that is immunoglobulins consisting of both an endogenous μ-chain and a transgenic heavy chain with a human $C\alpha$. In order to check these points, we made B cell hybridomas from the spleen of a transgenic mouse. We then analysed IgM expressing hybrids that failed to produce molecules bearing human $C\alpha$, and observed, as in the limiting dilution analysis, very high frequencies of clones binding Ac38 and Ac146 anti-idiotypes.

Furthermore, even considering the possibility that some clones produce both types of constant regions, the multiparameter analysis of one hundred hybridomas showed that there is no correlation between the levels of expression of Ac38-reactive molecules, and the fact of bearing human Cα versus mouse Cμ. We conclude, therefore, that our assays can indeed distinguish between human IgA and mouse IgM, and that the contribution of mixed molecules to the above results can be excluded.

We are, therefore, in the presence of a phenomenon that looks like idiotypic mimicry, but by what mechanism does it occur? Is this a "real" idiotypic mimicry, mediated by intercellular selection, or is it a "quasi-mimicry", dependent on intracellular mechanisms? For example, could it be that the expression of a lambda1 light chain, in association with any heavy chain whatsoever, would give an immunoglobulin that binds to the anti-idiotypic antibodies Ac38 and Ac146? We have analyzed two antibodies expressing lambda1 (one designated J558 and specific for dextran, the other produced by J558L cells that had been transfected with a μ-chain from a NP-binding hybridoma) and found both of them negative for binding to Ac38 and Ac146 anti-idiotypes. It could also be that the presence, in most B cells, of a transgene-encoded lambda1 light chain that associates preferentially with certain heavy chains, would select for survival of those cells producing a heavy chain carrying an idiotype similar to B1.8. We did not consider this to be a very likely possibility but, to be certain, we have analysed a mouse transgenic for the lambda1 chain alone, and could not reproduce the observations of high frequencies of idiotype-related B cells.

Yet another possibility for a "pseudo-mimicry" would be either some kind of gene conversion or trans-splicing, where the transgenic V-region would function as donor sequence, and the endogenous IgM-associated V-regions as recipients. Basically this would mean that all or part of the nucleotide sequences of Cμ-associated V-regions would be identical to the transgene. So far, there is one report describing isotype switching of an immunoglobulin transgene as a mechanism for idiotypic mimicry (Durdik et al, 1989). In order to investigate this possibility, we have started to sequence the heavy chain V-regions of idiotype positive and IgM positive hybridomas. Both the V_H-genes used, as well as the VDJ joinings of the first two such clones sequenced, are different from the transgenic B1.8 V-region. This argues against, but does not exclude gene conversion or trans-splicing as a general explanation for the observed idiotypic mimicry. More sequences have to be done, and the present results are to be regarded as preliminary.

If, as it seems, the idiotypic mimicry described here results from intercellular selection, the mechanisms implicated in this process must be, in one way or another, specific for the transgenic V-region. The nature of such specificity, however, would be quite different depending on whether the phenomenon is T cell-dependent (e.g., idiopeptides), or whether it is due to anti-idiotypic antibodies (e.g., conformational determinants) available in the transgenic mouse. In order to decide between these major alternatives, we crossed the transgenic line into nude mice, animals lacking a thymus and having hardly any TCRαβ T cells at all. The frequencies of B1.8-like IgM, however, was found to be similar in both athymic and euthymic transgenic mice, indicating that the mimicry is not T cell-dependent. Furthermore, the frequency of B cells which bind specifically to B1.8 i.e. the transgene idiotype, is low, regardless of transgene or thymus.

Our data, although preliminary and incomplete, certainly favour the possibility that the observed idiotypic mimicry is mediated by intercellular selection, but seemingly independent of T cells. In addition it does not seem to depend on increased levels of anti-idiotypic antibodies. Let us, for the time being accept this possibility and try to understand by what mechanism it could work.

Theoretical investigations

There are some observations concerning the immune system that we have to take into consideration, and which gave inspiration for the model which will be presented below.

The immune system is characterized both by an enormous diversity of V-regions and by a very high turnover of its molecular and cellular components. Out of all the B cells produced in the bone marrow every day, only 25-50% reach functional maturity and migrate to the periphery (Osmond, 1986), and of these, only a very small fraction will differentiate into high-rate Ig-secreting plasma cells, which produce the Ig in the serum and other fluids (Jerne, 1971). Cellular selection, therefore, might act and determine the repertoire of variable region molecules, at several steps in differentiation. Previously, we have shown that such selection actually takes place and is very powerful (Grandien et al, 1990, 1991). Our data principally concerned positive selection in the periphery, and especially the selection of which cells differentiate into high-rate Ig-secreting cells. In what concerns selection in the bone marrow for differentiation into mature resting B cells, we have several observations arguing that this is principally the result of suppression or "negative selection" (Freitas et al, 1991; Sundblad et al, 1991). Furthermore, we have shown that V-regions in the normal immune system are "connected", i.e. that they cross-react in varying degrees with each other (Holmberg et al, 1984). This connectivity is high during early life and decreases to reach a stable value in the mature animal.

By simply putting these notions together, we obtain a very simple model of the immune system which is depicted schematically in Figure 2. There are four compartments: B1, newly formed B cells in the bone marrow, which forms the "emergent repertoire"; B2, immunocompetent resting cells or "available repertoire"; B3, effector high-rate Ig-secreting plasma cells or "actual repertoire"; and S, soluble immunoglobulin circulating in the serum. Each compartment is occupied by n clones, each with a specific idiotype, represented here as a bitstring as previously introduced by others (Farmer et al, 1986). (In the simulations to be reported here n=30. The bitstrings are of length 10; bitstrings with perfect complementarity have 100% affinity; each mismatch reduces the affinity by 70%). In the simple version of the model used here, the concentrations of circulating immunoglobulins are simply proportional to the size of the corresponding clone of plasma cells, so that there are effectively only three independant compartments. In the model, cells are formed in B1, and pass progressively into B2 and then into B3. In the simulations, the unitary operation of the algorithm consists of eliminating clone i in B3 (corresponding to cell-death) and replacing it by clone i from B2 (subject to filter F2 as will be explained); clone i in B2 is then similarly replaced by clone i from B1 subject to filter F1; and clone i in B1 is similarly replaced by a new clone with a randomly generated idiotype. This operation is repeated in turn for all n clones (i=1...n), forming a "cycle".

A key feature of the model consists of the two "filters" F1 and F2, which are the locus of idiotypic interactions. We define h_i, the "field" experienced by idiotype i, as : $h_i = \Sigma\ m_{ij}S_j$, where m_{ij} is the affinity between idiotypes i and j, S_j is the concentration in the serum compartment of idiotype j, and the summation is over all the idiotypes (j=1...n) in the serum compartment. Following Varela et al (1988) and De Boer & Perelson (1991), we suppose that lymphocyte activation is a log-normal bell-shaped function of the field experienced by the idiotype of the lymphocyte in question. Thus, in the passage from one compartment to the next, the clone size in the new compartment is the clone size in the old compartment multiplied by the activation function. For filter F1, the curve is displaced to the left, so that cells in B1 with a low field are favoured in the passage to B2, all others being negatively selected (Figure 2, lower left). For filter F2, the curve is further to the right, so that the passage of resting B cells (B2) to plasma cells (B3) is subject to positive selection and is favoured by a moderately high field (Figure 2, lower right).

In Figure 3, three different situations are depicted. In the first, the filters are almost completely off, and nearly any specificity goes through. In the second situation, the filters are on, and F2 is displaced towards a higher field, as compared to F1. The third case is similar to

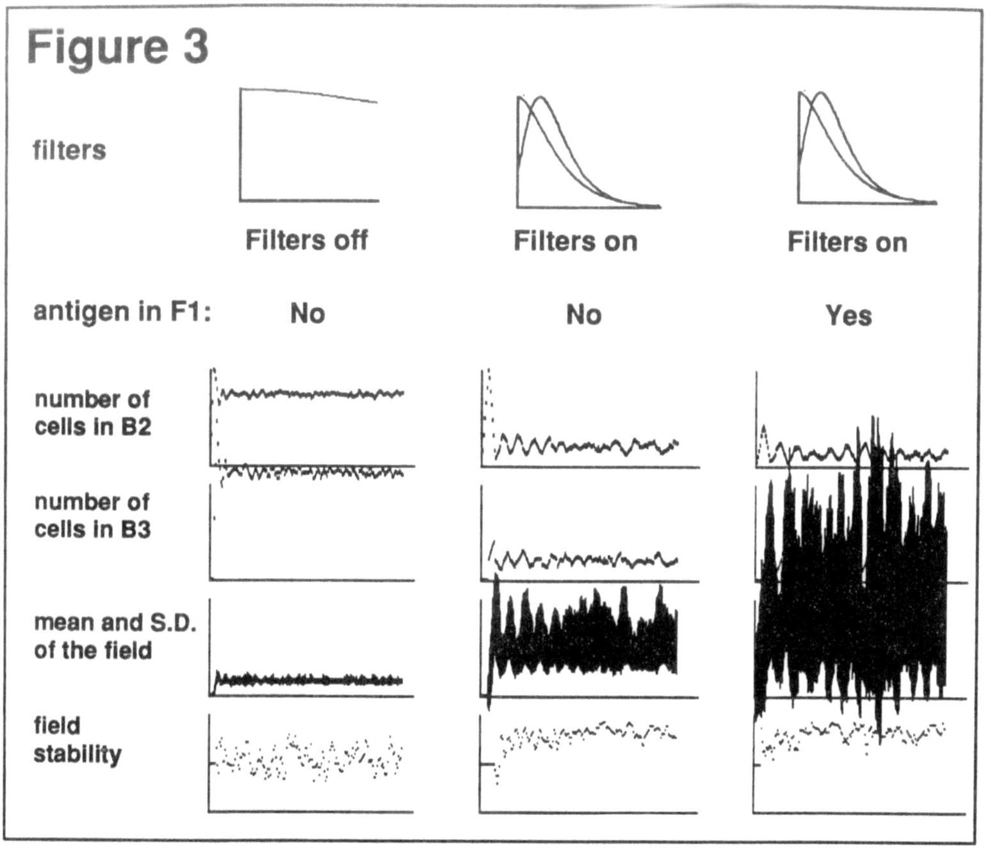

Figure 3. The results of three simulations. Filter diagrams: the horizontal axis is the field h as defined in the text (log scale); the vertical axis is the activation function (% of cells passing through the filter). Lower part of the figure: the horizontal axis is time; successive points correspond to intervals of five unit operations, so that six points correspond to a complete "cycle". In the model, one cycle corresponds to the life-time of a cell in a given compartment, which we may estimate as being of the order of 3 days. The total duration of the simulations, 200 points, thus corresponds to 33 cycles and 100 days. The vertical axes, in order are as follows. The first two graphs show the numbers of cells in compartments B2 and B3 (expressed as % of cells of a given idiotype generated in B1). (In the third simulation, the number of cells in B3, hardly visible due to the high standard deviation of the field, is similar to the second simulation). The third graph shows the value of the field h on a logscale, with a vertical bar running from m+sd to m-sd, where m is the mean and sd the standard deviation of h as calculated for the 30 idiotypes in B2. The scale runs from 0.1S2 to 3S2, where S2 is the field giving maximal activation in Filter 2. The fourth graph shows the "stability" of the field, calculated as follows. Each idiotype experiences a field when passing through filter F1, and then one cycle later experiences another field when passing through filter F2. Since each compartment is completely renewed in the space of a single cycle, a priori these two fields should be completely uncorrelated. The vertical axis here shows the values of the correlation coefficient between these two fields, the scale running from -1 to +1.

the second, except that we have now imposed in F1 the presence of a fixed structure; this corresponds to the situation where high concentrations of a monoclonal Ig (the product of the transgene) are present. This is done by adding an additional term in the calculation of the field for F1 : $h_i = \sum m_{ij}S_j + m_{ik}S_k$, where the index k refers to the monoclonal Ig, to which we attribute arbitrarily the idiotypic bitstring (1111111111).

The lower part of Figure 3 shows the results of the three simulations. When the filters are off, there is obviously little or no selection; both the values for the field (s) and the standard deviation of their mean (or s) are low, indicating that the repertoire in S (soluble, circulating immunoglobulins) is random; in this null reference situation, there is no field stability (correlation values fluctuate randomly above and below 0). With the filters on, the picture changes dramatically. Repertoire selection is now in operation, as can be seen from the fact that the numbers of cells in B2 and B3 are reduced. The standard deviation of the field is now high, which can be taken as indicating that certain idiotypes have a strong field while others have a low field, i.e. that the composition of the compartment S has a definite non-random structure. Finally, the "stability" of repertoire interactions is high: the correlation coefficients are consistently around 0.8, which indicates that this "structure" of the field is maintained even though the idiotypes making up each compartment have been completely renewed. In the third case, in the presence of a fixed string in F1, the most drastic change, as compared to the second case, is that the standard deviation of the field is even higher; its stability remains at comparable levels as above.

In Figure 4, we take a closer look at the idiotypic composition of S, in the second and the third cases. We probe the field created by compartment S with 4 bitstrings: one that is identical to that forced upon Fl (i.e. 1111111111); another that is complementary to the first; and two unrelated bitstrings that are also complementary to each other. When there is no structure deliberately imposed on Fl (the bone marrow filter), the "structure" of the field drifts rather slowly: major fluctuations occur, but take much longer than a single cycle (6 consecutive points). This is indeed to be expected, in view of the significant "field stability". It is also to be noted that the fields for complementary idiotypes vary inversely.

Finally, it is clear that the imposed structure in Fl (modelling the transgenic V-region) results in a repertoire of idiotypes in S that is selected to contain an overrepresentation of idiotypes that are similar to the string imposed in F1. These simulations, therefore, seem to reproduce the experimental observations on idiotypic mimicry. Qualitatively, this can be understood as resulting from the following two-step mechanism. Firstly, a given idiotype present in high concentrations suppresses idiotypes that are complementary to it (this is clearly illustrated in Figure 4). Secondly, since negative selection at high fields is a major feature of this model, (principally in Fl, but also in F2 at excessively high field values), this suppression in return favourises idiotypes that mimic the first idiotype. In this way, idiotypic mimicry on the one hand, and a certain self-sustaining stability of the structure of the field on the other, are conjointly explained by the model we propose here.

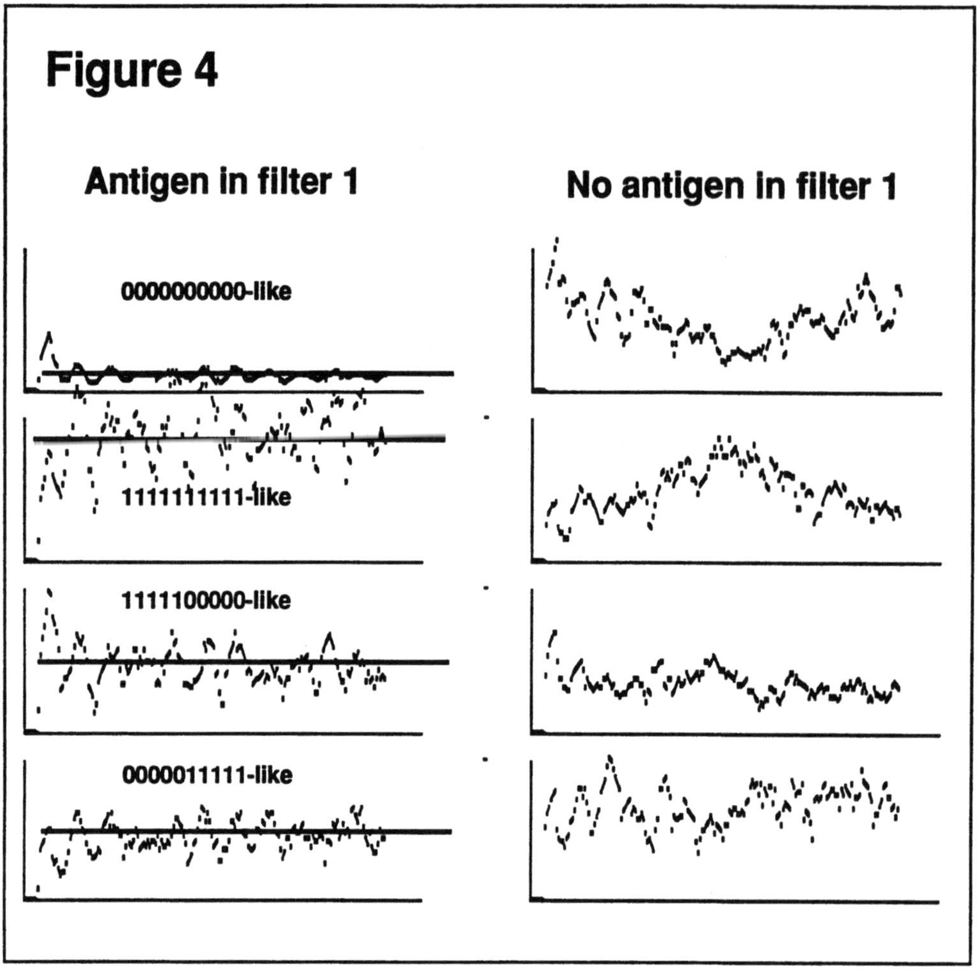

Figure 4. Vertical axis: the composition of the serum compartment S, as measured by the fields experienced by four idiotypic "probes". In order: the field for the idiotype 1111111111 (due to the presence in S of idiotypes similar to the complementary form 0000000000); probe 0000000000; probe 0000011111; probe 1111100000. Horizontal axis: time, as in Figure 3.

It may be noted that our model shares certain features with those proposed by Varela, Stewart and Coutinho (Varela et al,1988; Stewart and Varela, 1989), and by Perelson, De Boer and colleagues (Perelson, 1989; DeBoer et al, 1990), in particular the postulate that lymphocyte activation is a bell-shaped function of the "field" created by idiotypic interactions. Important differences, however, are the compartmentalization of cells in consecutive stages of differentiation; and the postulate that all clones have a pre-determined, limited life-span. Our intention in deriving and testing this model was to investigate whether we could reproduce the data on idiotypic mimicry. The variable region repertoire of our computer model is very small (10^3), when compared to that in the mouse (more than 10^{12}), and the B1.8 idiotype is also certainly very different from a bitstring. Nevertheless, it seems that this simple model is indeed able to provide an explanation for the phenomena of idiotypic mimicry. This encourages us to plan future explorations of the possibility that an extended version of this model might be capable of providing dynamic explanations for other features of the immune system, such as tolerance, immune responses, and memory.

Acknowledgements

We thank Dr. Jose Faro and Dr Jan Andersson for helpful discussions and criticism. This work was supported by grants from The Swedish Medical Research Council, EEC, AFM, CNRS, INSERM, DRET.

References

Andersson, J., A. Coutinho, W. Lernhardt and F. Melchers. (1977) Clonal growth and maturation to immunoglobulin secretion in vitro of every growth-inducible B lymphocyte. Cell. 10:27.

De Boer, R. J., and Perelson, A. S. (1990) Size and connectivity as emergent properties of a developing immune network. J. Theoret. Biol.

Durdik, J., Gerstein, R. M., Rath, S., Robbins, P. F., Nisonoff, A., and Selsing, E. (1989) Isotype swithing by a microinjected mu immunoglobulin heavy chain gene in trangenic mice. Proc. Natl. Acad. Sci. 86: 2346.

Farmer, J. D., Packard, N. H., and Perelson, A. S. (1986) The immune system, adaptation, and machine learning. Physica D. 22:187.

Freitas, A. A., A. C. Viale, A. Sundblad, C. Heusser and A. Coutinho. (1991) Normal serum immunoglobulins participate in the selection of peripheral B cell repertoires. Proc. Natl. Acad. Soc. 88:5640.

Grandien, A., A. Coutinho and J. Anderson. (1990) Selective peripheral expansion and activation of B cells expressing endogenous immunoglobulin in mu-transgenic mice. Eur. J. Immunol. 20:991.

Grandien, A., A. Coutinho, I. Andersson and A. Freitas. (1991) Endogenous VH-gene families expression in immunoglobulin-transgenic mice : evidence for selection of antibody repertoires. Intern. Immunol. 3:67.

Holmberg, D., S. Forsgren, F. Ivars and A. Coutinho. (1984) Reactions amongst IgM antibodies derived from normal, neonatal mice. Eur. J. Immunol. 14:435.

Jerne, N. K. (1971) What precedes clonal selection ? In Ciba Foundation Symposium : Ontogeny of acquired immunity. Elsevier, Amsterdam. pp. 1.

Neuberger, M. S., Caskey, H. M., Pettersson, S., Williams, G. T., and Surani, M.A. (1989) Isotype exclusion and transgene down-regulation in immunoglobulin-lambda transgenic mice. Nature 338: 350.

Osmond, D. G. (1986) Population dynamics of bone marrow B lymphocytes. Immunol. Rev. 93:103.

Perelson, A. S. (1989) Immune network theory. Immunol. Rev. 110:5.

Pettersson, S., Sharpe, M. J., Gilmore, D. R., Surani, M. A., and Neuberger, M. S. (1989) Cellular selection leads to age-dependent and reversible down-regulation of transgenisc immunoglobulin light chain genes. Int. Immunol. 1: 509.

Reth, M., Hammerling, G. J., and Rajewsky, K. (1978) Analysis of the repertoire of anti-NP antibodies in C57BL/6 mice by cell fusion. I. Eur. J. Immunol. 8:393.

Reth, M., Imanishi-Kari, T., and Rajewsky, K. (1979) Analysis of the repertoire of anti-(4-hydroxy-3-nitrophenyl) acetyl (NP) antibodies in C57BL/6 mice by cell fusion. II. Eur. J. Immunol. 9:1004.

Sundblad, A., M. A. R. Marcos, F. Huetz, A. Freitas, C. Heusser, D. Portnoï and C. A. (1991) Normal serum immunoglobulins regulates the numbers of bone marrow pre-B and B cells. Eur. J. Immunol. 21:1155.

Varela, F., A. Coutinho, B. Dupire and N. Vaz. (1988) Cognitive networks: immune, neural and otherwise. In In Theoretical Immunology. A. S. Perelson, A. S. Perelson. Addison-Wesley Publ. Co., Inc., Redwood City. 359.

Varela, F. and J. Stewart. (1989) Exploring the meaning of connectivity in the immune network. Immunol. Rev. 110:37.

Weaver, D., Reis, M. H., Albanese, C., Constantini, F., Baltimore, D., and Imanishi-Kari, T.(1986) Altered repretoire of endogenous immunoglobulin gene expression in transgenic mice containing a rearranged mu heavy chain gene. Cell. 45:247.

Index

NATO ASI Series H

Vol. 1: Biology and Molecular Biology of Plant-Pathogen Interactions. Edited by J.A. Bailey. 415 pages. 1986.

Vol. 2: Glial-Neuronal Communication in Development and Regeneration. Edited by H.H. Althaus and W. Seifert. 865 pages. 1987.

Vol. 3: Nicotinic Acetylcholine Receptor: Structure and Function. Edited by A. Maelicke. 489 pages. 1986.

Vol. 4: Recognition in Microbe-Plant Symbiotic and Pathogenic Interactions. Edited by B. Lugtenberg. 449 pages. 1986.

Vol. 5: Mesenchymal-Epithelial Interactions in Neural Development. Edited by J. R. Wolff, J. Sievers, and M. Berry. 428 pages. 1987.

Vol. 6: Molecular Mechanisms of Desensitization to Signal Molecules. Edited by T M. Konijn, P J. M. Van Haastert, H. Van der Starre, H. Van der Wel, and M.D. Houslay. 336 pages. 1987.

Vol. 7: Gangliosides and Modulation of Neuronal Functions. Edited by H. Rahmann. 647 pages. 1987.

Vol. 8: Molecular and Cellular Aspects of Erythropoietin and Erythropoiesis. Edited by I.N. Rich. 460 pages. 1987.

Vol. 9: Modification of Cell to Cell Signals During Normal and Pathological Aging. Edited by S. Govoni and F Battaini. 297 pages. 1987.

Vol. 10: Plant Hormone Receptors. Edited by D. Klämbt. 319 pages. 1987.

Vol. 11: Host-Parasite Cellular and Molecular Interactions in Protozoal Infections. Edited by K.-P. Chang and D. Snary. 425 pages. 1987.

Vol. 12: The Cell Surface in Signal Transduction. Edited by E. Wagner, H. Greppin, and B. Millet. 243 pages. 1987.

Vol. 13: Toxicology of Pesticides: Experimental, Clinical and Regulatory Perspectives. Edited by L.G. Costa, C.L. Galli, and S.D. Murphy. 320 pages. 1987.

Vol. 14: Genetics of Translation. New Approaches. Edited by M.F. Tuite, M. Picard, and M. Bolotin-Fukuhara. 524 pages. 1988.

Vol. 15: Photosensitisation. Molecular, Cellular and Medical Aspects. Edited by G. Moreno, R. H. Pottier, and T. G. Truscott. 521 pages. 1988.

Vol. 16: Membrane Biogenesis. Edited byJ.A.F Op den Kamp. 477 pages. 1988.

Vol. 17: Cell to Cell Signals in Plant, Animal and Microbial Symbiosis. Edited by S. Scannerini, D. Smith, P. Bonfante-Fasolo, and V. Gianinazzi-Pearson. 414 pages. 1988.

Vol. 18: Plant Cell Biotechnology. Edited by M.S.S. Pais, F. Mavituna, and J. M. Novais. 500 pages. 1988.

Vol. 19: Modulation of Synaptic Transmission and Plasticity in Nervous Systems. Edited by G. Hertting and H.-C. Spatz. 457 pages. 1988.

Vol. 20: Amino Acid Availability and Brain Function in Health and Disease. Edited by G. Huether. 487 pages. 1988.

Vol. 21: Cellular and Molecular Basis of Synaptic Transmission. Edited by H. Zimmermann. 547 pages. 1988.

NATO ASI Series H

Vol. 22: **Neural Development and Regeneration. Cellular and Molecular Aspects.** Edited by A. Gorio, J. R. Perez-Polo, J. de Vellis, and B. Haber. 711 pages. 1988.

Vol. 23: **The Semiotics of Cellular Communication in the Immune System.** Edited by E.E. Sercarz, F. Celada, N.A. Mitchison, and T. Tada. 326 pages. 1988.

Vol. 24: **Bacteria, Complement and the Phagocytic Cell.** Edited by F. C. Cabello und C. Pruzzo. 372 pages. 1988.

Vol. 25: **Nicotinic Acetylcholine Receptors in the Nervous System.** Edited by F. Clementi, C. Gotti, and E. Sher. 424 pages. 1988.

Vol. 26: **Cell to Cell Signals in Mammalian Development.** Edited by S.W de Laat, J.G. Bluemink, and C.L. Mummery. 322 pages. 1989.

Vol. 27: **Phytotoxins and Plant Pathogenesis.** Edited by A. Graniti, R. D. Durbin, and A. Ballio. 508 pages. 1989.

Vol. 28: **Vascular Wilt Diseases of Plants. Basic Studies and Control.** Edited by E. C. Tjamos and C. H. Beckman. 590 pages. 1989.

Vol. 29: **Receptors, Membrane Transport and Signal Transduction.** Edited by A. E. Evangelopoulos, J. P. Changeux, L. Packer, T. G. Sotiroudis, and K.W.A. Wirtz. 387 pages. 1989.

Vol. 30: **Effects of Mineral Dusts on Cells.** Edited by B.T. Mossman and R.O. Begin. 470 pages. 1989.

Vol. 31: **Neurobiology of the Inner Retina.** Edited by R. Weiler and N.N. Osborne. 529 pages. 1989.

Vol. 32: **Molecular Biology of Neuroreceptors and Ion Channels.** Edited by A. Maelicke. 675 pages. 1989.

Vol. 33: **Regulatory Mechanisms of Neuron to Vessel Communication in Brain.** Edited by F. Battaini, S. Govoni, M.S. Magnoni, and M. Trabucchi. 416 pages. 1989.

Vol. 34: **VectorsasTools for the Study of Normal and Abnormal Growth and Differentiation.** Edited by H. Lother, R. Dernick, and W. Ostertag. 477 pages. 1989.

Vol. 35: **Cell Separation in Plants: Physiology, Biochemistry and Molecular Biology.** Edited by D. J. Osborne and M. B. Jackson. 449 pages. 1989.

Vol. 36: **Signal Molecules in Plants and Plant-Microbe Interactions.** Edited by B.J.J. Lugtenberg. 425 pages. 1989.

Vol. 37: **Tin-Based Antitumour Drugs.** Edited by M. Gielen. 226 pages. 1990.

Vol. 38: **The Molecular Biology of Autoimmune Disease.** Edited by A.G. Demaine, J-P. Banga, and A.M. McGregor. 404 pages. 1990.

Vol. 39: **Chemosensory Information Processing.** Edited by D. Schild. 403 pages. 1990.

Vol. 40: **Dynamics and Biogenesis of Membranes.** Edited by J. A. F. Op den Kamp. 367 pages. 1990.

Vol. 41: **Recognition and Response in Plant-Virus Interactions.** Edited by R. S. S. Fraser. 467 pages. 1990.

NATO ASI Series H

Vol. 42: Biomechanics of Active Movement and Deformation of Cells.
Edited by N. Akkas. 524 pages. 1990.

Vol. 43: Cellularand MolecularBiologyof Myelination.
Edited by G. Jeserich, H. H. Althaus, and T. V. Waehneldt. 565 pages. 1990.

Vol. 44: Activation and Desensitization of Transducing Pathways.
Edited by T. M. Konijn, M. D. Houslay, and P. J. M. Van Haastert. 336 pages. 1990.

Vol. 45: Mechanism of Fertilization: Plants to Humans.
Edited by B. Dale. 710 pages. 1990.

Vol .46: Parallels in Cell to Cell Junctions in Plants and Animals.
Edited by A. W Robards, W. J . Lucas, J . D. Pitts, H . J . Jongsma, and D. C. Spray. 296 pages. 1990.

Vol. 47: Signal Perception andTransduction in Higher Plants.
Edited by R. Ranjeva and A. M. Boudet. 357 pages. 1990.

Vol. 48: Calcium Transport and Intracellular Calcium Homeostasis.
Edited by D. Pansu and F. Bronner. 456 pages. 1990.

Vol. 49: Post-Transcriptional Control of Gene Expression.
Edited by J. E. G. McCarthy and M. F. Tuite. 671 pages. 1990.

Vol. 50: Phytochrome Properties and Biological Action.
Edited by B. Thomas and C. B. Johnson. 337 pages. 1991.

Vol. 51: Cell to Cell Signals in Plants and Animals.
Edited by V. Neuhoff and J. Friend. 404 pages. 1991.

Vol. 52: Biological Signal Transduction.
Edited by E. M . Ross and K . W. A. Wirtz. 560 pages. 1991.

Vol. 53: Fungal Cell Wall and Immune Response.
Edited by J. P. Latge and D. Boucias. 472 pages. 1991.

Vol. 54: The Early Effects of Radiation on DNA.
Edited by E. M. Fielden and P. O'Neill. 448 pages. 1991.

Vol. 55: The Translational Apparatus of Photosynthetic Organelles.
Edited by R. Mache, E. Stutz, and A. R. Subramanian. 260 pages. 1991.

Vol. 56: Cellular Regulation by Prote in Phosphorylation.
Edited by L. M. G. Heilmeyer, Jr. 520 pages. 1991.

Vol. 57: MolecularTechniques in Taxonomy.
Edited by G . M . Hewitt, A. W. B. Johnston, and J. P. W. Young . 420 pages. 1991.

Vol. 58: Neurocytochemical Methods.
Edited by A. Calas and D. Eugene. 352 pages. 1991.

Vol. 59: Molecular Evolution of the Major Histocompatibility Complex.
Edited by J. Klein and D. Klein. 522 pages. 1991.

Vol. 60: Intracellular Regulation of Ion Channels.
Edited by M. Morad and Z. Agus. 261 pages. 1992.

Vol. 61: Prader-Willi Syndrome and Other Chromosome 15q Deletion Disorders.
Edited by S. B. Cassidy. 277 pages. 1992.

NATO ASI Series H

MIX
Papier aus verantwortungsvollen Quellen
Paper from responsible sources
FSC® C105338

If you have any concerns about our products,
you can contact us on
ProductSafety@springernature.com

In case Publisher is established outside the EU,
the EU authorized representative is:
Springer Nature Customer Service Center GmbH
Europaplatz 3, 69115 Heidelberg, Germany

Printed by Libri Plureos GmbH
in Hamburg, Germany